Chemistry

for CBSE Class X

Atul Kumar Singhal
Ph.D. (Organometallic Chemistry)
Academic Head, Aakash Institute,
Modipuram, Meerut

Universities Press

CHEMISTRY FOR CBSE CLASS X

UNIVERSITIES PRESS (INDIA) PRIVATE LIMITED

Registered Office
3-6-747/1/A & 3-6-754/1, Himayatnagar, Hyderabad 500 029, Telangana, India
info@universitiespress.com; www.universitiespress.com

Distributed by
Orient Blackswan Private Limited

Registered Office
3-6-752 Himayatnagar, Hyderabad 500 029, Telangana, India

Other Offices
Bengaluru / Chennai / Guwahati / Hyderabad / Kolkata
Mumbai / New Delhi / Noida / Patna / Visakhapatnam

First published 2024

Cover and book design:

501964

ISBN 978-93-93330-63-5

Typeset in Minion Pro 9.5 points *by*
Cameo Corporate Services Limited, Coimbatore

Printed in India by
Shiva Printech Pvt Ltd, Tronica city (U.P.)

Published by
Universities Press (India) Private Limited
3-6-747/1/A & 3-6-754/1, Himayatnagar, Hyderabad 500 029, Telangana, India

Dedicated to
My parents and teachers

Preface

Chemistry for CBSE Class X is an excellent preparatory book for students. This book provides class-tested material and practice problems that help students understand the theory and concepts and, as a consequence, develop problem-solving skills to attempt exams with full confidence.

The book is written in lucid language with a variety of solved examples, and aims to assist students in understanding concepts even without the help of an instructor, as solutions are provided for most of the questions. In addition, all the questions, exercises and activities from the NCERT textbook are solved.

Salient features

- Structured according to the latest syllabi and exam pattern of the CBSE
- Includes MCQs, Very Short, Short, Long, HOTS and Case-study/Application-based questions (including Fill in the blanks, Match the following, True or false, Para-based)
- Includes NCERT exercises and Exemplar questions along with solutions as well as the Activities with discussions
- Related advanced level concepts highlighted in boxes (***Competition Edge***)
- ***Test Your Knowledge*** includes concept-based questions with answers
- The App accompanying this book provides additional chapter-wise MCQs for competitive exam practice

The purpose of this book is two-fold: to strengthen the students' foundation in Chemistry in order to excel in school exams and to prepare students in Chemistry for competitive exams.

I welcome suggestions and constructive criticism from the readers. Students and teachers can share their feedback at singhal.atul1974@gmail.com.

Atul Kumar Singhal

Acknowledgments

The contentment that accompanies the successful completion of my work would remain essentially incomplete if I fail to mention the people whose constant support has encouraged me.

I am grateful to all my revered teachers, especially the late J K Mishra, Dr D K Rastogi, the late A K Rastogi and my honorable guide, Dr S K Agarwala. Their knowledge and wisdom assisted me in no small measure in presenting this work.

I express my immense gratitude to my colleagues for collaborating so patiently and constructively during the various stages of this project. I am extremely thankful to Ronit Singh Kushwaha for typing out the manuscript.

I am indebted to my father, B K Singhal, mother, Usha Singhal, and brothers Amit Singhal and Katar Singh, who have been my motivators at every step. Their never-ending affection has provided me with moral support and encouragement while writing this book. Last but not the least, I express my deepest gratitude to my wife, Urmila, my daughters Khushi and Shanvi, and my little but witty-beyond-years son Shashwat, who always supported me during my work.

I would like to express my heartfelt gratitude to Dr Gita S Dattatri for her brilliant editing and upgradations. I would also like to show my appreciation to Kallol Das, Thomas Mathew Rajesh and Madhu Reddy for providing all the necessary help and assistance.

Atul Kumar Singhal

Contents

CHAPTER 1

Chemical Reactions and Equations

Learning Objectives

After studying this unit, you will be able to:

- understand what a chemical reaction, reactant and product are
- define a chemical equation and write a balanced equation
- study and differentiate between different types of reactions—combination, decomposition, displacement, double displacement, and oxidation and reduction
- understand the concepts of rancidity and corrosion

1.1 Introduction

Let us consider some of the common processes which take place in our daily lives, such as, (i) melting of ice, (ii) dissolution of sugar or salt in water, (iii) evaporation of water, (iv) burning of coke in air, (v) burning of LPG (Fig. 1.1), (vi) souring of milk during summer, (vii) rusting of iron objects such as nails, and so on. Out of these examples, in the first three, no new chemical substance is formed which means they involve only *physical changes*. In the rest, new chemical substances are formed; thus, we can say they involve *chemical changes*. We can differentiate between a physical and chemical change by considering that if no new substance is formed, it is a physical change and if a new chemical substance is formed it is a chemical change. Here are few more examples in our daily lives involving chemical changes.

Fig. 1.1 Burning of LPG

Source: Ivan Radic, CC BY 2.0 <https://creativecommons.org/licenses/by/2.0>, via Wikimedia Commons

(i) Fermentation of grapes
(ii) Respiration
(iii) Digestion of food in our body
(iv) Cooking of food
(v) Browning of a freshy cut apple in the air
(vi) Burning of fuels such as kerosene, petrol, LPG
(vii) Burning of candle wax
(viii) Formation of curd from milk (Fig. 1.2)

Milk Curd

Fig. 1.2 Formation of curd

Source: Daria-Yakovleva, CC0, via Wikimedia Commons; Vivekdoshi11, CC BY-SA 4.0 <https://creativecommons.org/licenses/by-sa/4.0>, via Wikimedia Commons

1.2 Chemical Reactions

Chemical reactions are the processes by which new substances with new properties are formed. Chemical reactions involve chemical changes. During a chemical reaction, a rearrangement of atoms takes place between the reacting species to form new species having entirely different properties. Chemical reactions involve breaking of old chemical bonds which exist between the atoms of reacting substances, and then forming of new chemical bonds between the rearranged atoms of the new substances. During a chemical reaction, atoms of one element do not change into those of another element; however, only a rearrangement of atoms takes place. Let us now discuss reactants and products of a chemical reaction.

Reactant: The substances which take part in a chemical reaction are known as *reactants*.

For example, in the breaking up of calcium carbonate into calcium oxide and carbon dioxide, calcium carbonate is the reactant (Fig. 1.3). Similarly, sodium and water are the reactants when they react to give sodium hydroxide and hydrogen.

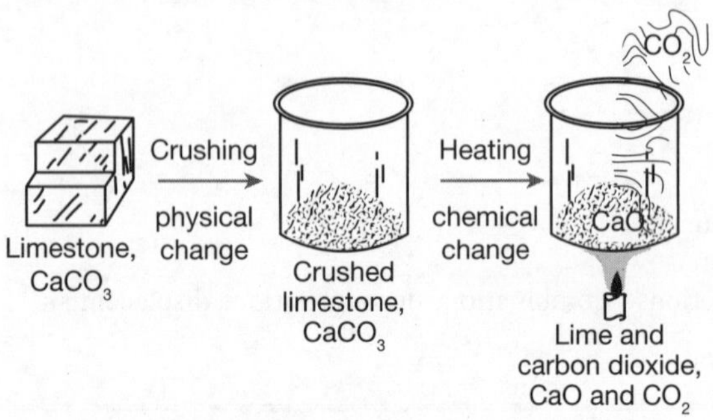

Fig. 1.3 Decomposition of $CaCO_3$

$$\underset{\text{Calcium carbonate}}{CaCO_3} \rightarrow \underset{\text{Calcium oxide}}{CaO} + \underset{\text{Carbon dioxide}}{CO_2}$$

$$\underset{\text{Sodium}}{Na} + \underset{\text{Water}}{H_2O} \rightarrow \underset{\text{Sodium hydroxide}}{NaOH} + \underset{\text{Hydrogen}}{H_2}$$

Product: The new substances produced as a result of a chemical reaction are known as products. In the breaking of calcium carbonate, calcium oxide and carbon dioxide are products and hydrogen and sodium hydroxide are the products of the reaction between sodium and water.

1.2.1 Examples of Chemical Reactions

(i) Burning of magnesium ribbon in air

When a magnesium ribbon is heated, it burns in air with a dazzling white flame to form a white powder known as magnesium oxide.

$$\underset{\text{(As ribbon)}}{\text{Magnesium}} + \underset{\text{(From air)}}{\text{Oxygen}} \xrightarrow{\text{Heat}} \underset{\text{(White powder)}}{\text{Magnesium oxide}}$$

$$2Mg + O_2 \rightarrow 2MgO$$

Process and precautions

Take a 2 cm long magnesium ribbon; clean it by rubbing its surface with a sandpaper. Now hold the ribbon with a pair of tongs at one end and heat the other end with a burner (Fig. 1.4). It burns with a dazzling white light. Hold it over a watch glass so that the magnesium oxide powder can be collected over it.

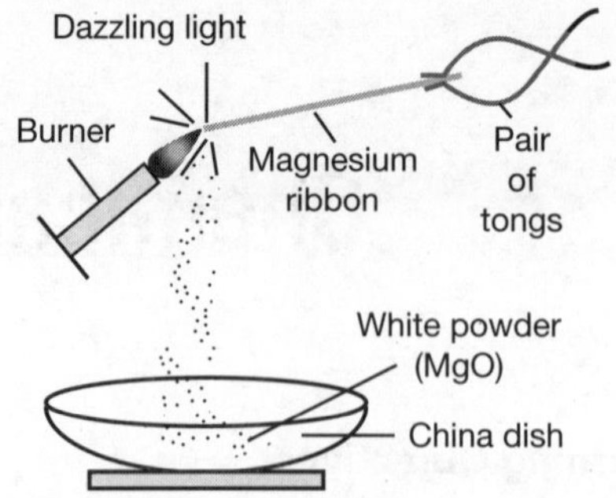

Fig. 1.4 Burning of magnesium wire

The magnesium ribbon used here usually has a coating of 'magnesium oxide' on its surface which is formed by the slow action of the oxygen of air on it. So, before burning in air, the magnesium ribbon is cleaned by rubbing with a sandpaper. This is done to remove the protective layer of magnesium oxide from the surface of the magnesium ribbon so that it may readily combine with the oxygen of air (on heating). The dazzling (very bright) white light given out during the burning of a magnesium ribbon is harmful to the eyes. So, the magnesium ribbon should be burned by keeping it as far as possible from the eyes.

(ii) Reaction between zinc granules and dilute sulfuric acid

When zinc granules are heated with dilute sulfuric acid or dilute hydrochloric acid, a colourless solution of zinc sulfate or zinc chloride is obtained along with brisk effervescence of hydrogen gas (Fig. 1.5).

$$\text{Zinc} + \text{Sulfuric acid (dilute)} \rightarrow \underset{\text{Colourless solution}}{\text{Zinc sulfate}} + \underset{\text{Gas}}{\text{hydrogen}}$$

$$\underbrace{Zn + H_2SO_4}_{\text{Reactants}} \rightarrow \underbrace{ZnSO_4 + H_2}_{\text{Products}}$$

Process and precautions

Take few pieces of granulated zinc in a small test tube and add little dilute sulfuric acid into it. On heating the tube, there was a brisk evolution of a gas from the surface of zinc. On bringing a lighted candle near the upper end of the tube, this gas burns with a popping sound which means it is hydrogen gas. Here, zinc sulfate solution can also be observed in the test tube. The test tube will be very hot, so do not touch it immediately.

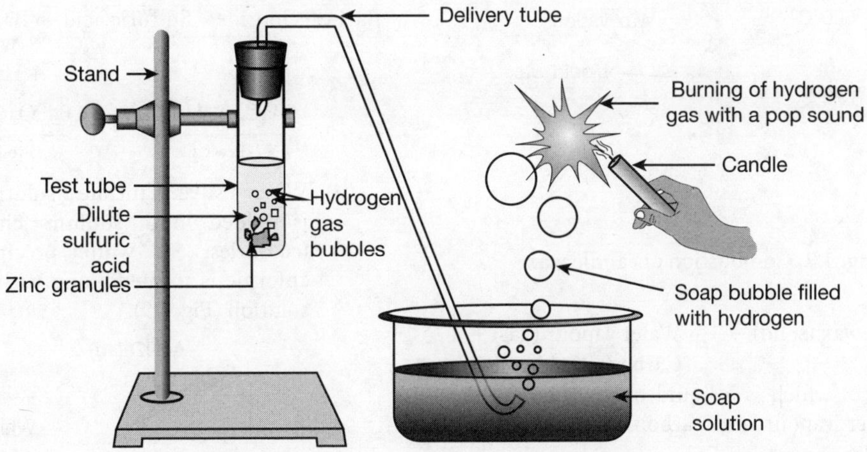

Fig. 1.5 Reaction of zinc granules with H_2SO_4

(iii) Reaction between lead nitrate and potassium iodide

When lead nitrate (colourless) is treated with potassium iodide (colourless), a yellow solid precipitate of lead iodide and colourless solution of potassium nitrate are formed (Fig. 1.6).

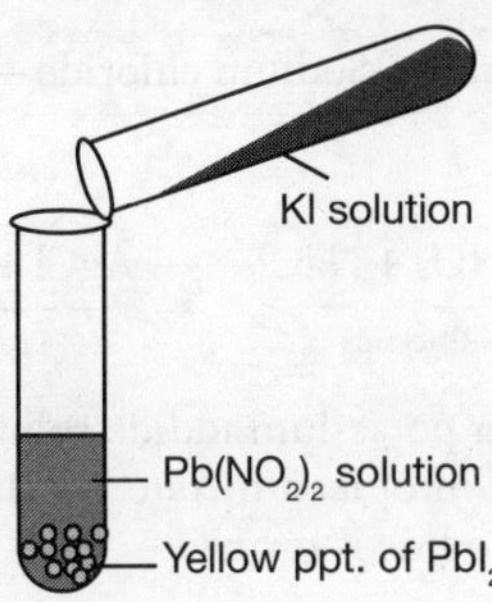

Fig. 1.6 Formation of PbI_2

$$\underset{\text{Colourless solution}}{\text{Lead nitrate}} + \underset{\text{Colourless solution}}{\text{Potassium iodide}} \rightarrow \underset{\text{Yellow solid}}{\text{Lead iodide}} + \underset{\text{Colourless solution}}{\text{Potassium nitrate}}$$

$$\underbrace{Pb(NO_3)_2 + 2KI}_{\text{Reactants}} \rightarrow \underbrace{PbI_2 + 2KNO_3}_{\text{Products}}$$

Process and precautions

First take two separate test tubes, one having lead nitrate solution and another having potassium iodide solution. Now add the potassium nitrate solution into the lead nitrate solution; soon a yellow precipitate of lead iodide can be seen along with colourless potassium nitrate solution.

1.2.2 Characteristics of Chemical Reactions

The conversion of reactants into products in a chemical reaction is often accompanied by some specific features which can be noticed easily. The easily observable features or changes which take place as result of chemical reactions are called *characteristics* or features of chemical reactions. The important characteristics of chemical reactions are:

- change in state,
- formation of a precipitate,
- change in temperature,
- change in colour, and
- evolution of a gas.

Change in state

Many chemical reactions are characterised by a change in state; the physical state changes from solid to liquid and gas. For example, (i) the combustion reaction of candle wax is characterised by a change in state from solid to liquid and gas (Fig. 1.7). Wax is a solid, its combustion leads to a liquid at room temperature and carbon dioxide produced by the combustion of wax is a gas.

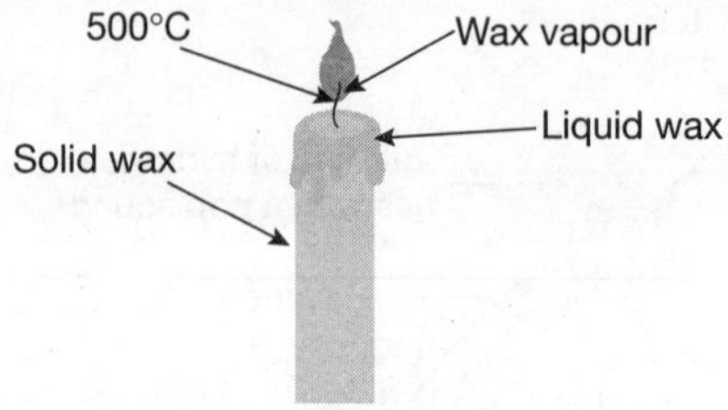

Fig. 1.7 Combustion of candle wax

$$\text{Candle wax (solid)} \xrightarrow{\Delta} \text{Water vapour (gas)} + \text{Carbon dioxide (gas)}$$

(ii) Petrol which is a liquid on combustion gives water vapour and carbon dioxide which are gases.

$$\text{Petrol (liquid)} \xrightarrow{\Delta} \text{Water vapour (gas)} + \text{Carbon dioxide (gas)}$$

Formation of a precipitate

A chemical reaction is also characterised by the formation of a precipitate. A precipitate is a solid product which separates out from the solution during a chemical reaction. A precipitate can be formed by mixing an aqueous solution of the reactants when one of the products is insoluble in water. A precipitate can also be formed by passing a gas into an aqueous solution of a substance, such as, passing carbon dioxide gas into lime water. Let us consider a few examples.

(i) When dilute sulfuric acid (colourless) is added to barium chloride solution (colourless) taken in a test tube, a white precipitate of barium sulfate is formed (Fig. 1.8). Thus, the chemical reaction between sulfuric acid and barium chloride solution is characterised by the formation of a white precipitate of barium sulfate.

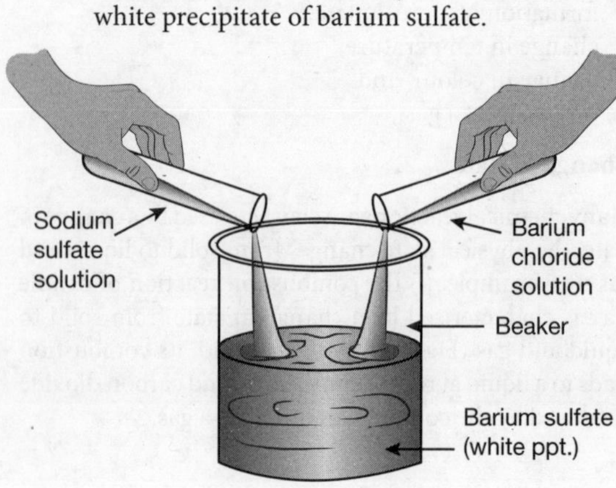

Fig. 1.8 Formation of precipitate of barium sulfate

$$\text{Barium chloride} + \text{Sulfuric acid} \rightarrow \underset{\text{(White precipitate)}}{\text{Barium sulfate}} + \text{Hydrogen chloride}$$

$$\underbrace{BaCl_2 + H_2SO_4}_{\text{Reactants}} \longrightarrow \underbrace{BaSO_4 + 2HCl}_{\text{Products}}$$

(ii) When silver nitrate solution (colourless) is added into sodium chloride solution (colourless), a white precipitate of silver chloride is formed along with sodium nitrate solution (Fig. 1.9).

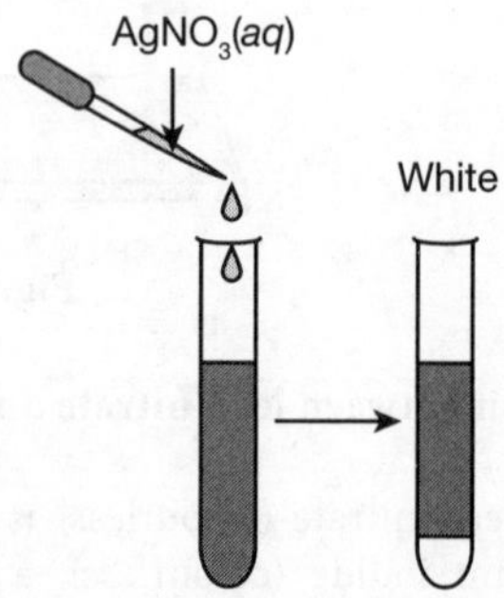

Fig. 1.9 Formation of precipitate of silver chloride

$$\text{Silver nitrate} + \text{Sodium chloride} \rightarrow \underset{\text{(White precipitate)}}{\text{Silver chloride}} + \text{Sodium nitrate}$$

$$\underbrace{AgNO_3 + NaCl}_{\text{Reactants}} \longrightarrow \underbrace{AgCl + NaNO_3}_{\text{Products}}$$

(iii) When potassium iodide solution is added to a solution of lead nitrate, a yellow precipitate of lead iodide is formed.

Change in temperature

Many chemical reactions are characterised by a change in temperature, which means, there is a rise or fall in temperature during the reaction. A reaction which is accompanied by a rise in temperature is known as an *exothermic reaction*. In an exothermic reaction, heat is given out. A reaction which is accompanied by a fall or decrease in temperature is known as an *endothermic reaction*. In an endothermic reaction, heat is absorbed.

For example, (i) when quicklime reacts with water, slaked lime is formed and a huge amount of heat energy is produced. This heat raises the temperature, due to which the reaction mixture becomes hot. This is an exothermic reaction. To perform this chemical reaction, take a little quicklime in a hard glass beaker and add water into it slowly (Fig. 1.10). On touching the beaker carefully, heat can be observed as the beaker is quite hot.

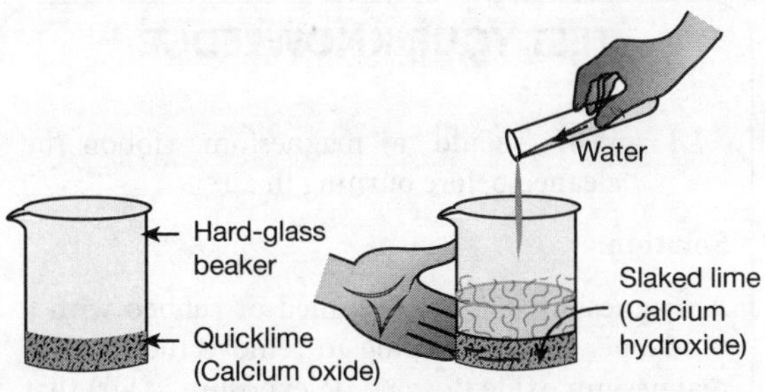

Fig. 1.10 Reaction of quicklime and water

Quicklime (Calcium oxide, solid) + Water (liquid) → Slaked lime (Calcium hydroxide) + Heat

$$\underbrace{CaO + H_2O}_{\text{Reactants}} \longrightarrow \underbrace{Ca(OH)_2 + \text{Heat}}_{\text{Products}}$$

(ii) When barium hydroxide [$Ba(OH)_2$] is added to ammonium chloride (NH_4Cl) taken in a test tube and mixed with a glass rod, then barium chloride, ammonia and water are formed. A lot of heat energy is absorbed during this reaction due to which the temperature of the reaction mixture falls and the bottom of the test tube becomes very cold. It is a case of an endothermic reaction.

Barium hydroxide + Ammonium chloride → Barium chloride + Ammonia + Water – Heat

$$\underbrace{Ba(OH)_2 + 2NH_4Cl}_{\text{Reactants}} \longrightarrow \underbrace{BaCl_2 + 2NH_3 + 2H_2O - \text{Heat}}_{\text{Products}}$$

(iii) Reaction between zinc granules and dilute sulfuric acid is also an exothermic reaction in which heat is evolved.

Change in colour

Some chemical reactions are characterised by a change in colour. Examples are given below.

(i) When citric acid (lemon juice) or oxalic acid reacts with potassium permanganate solution, the purple colour of potassium permanganate solution fades and finally becomes colourless (Fig. 1.11).

(ii) When sulfur dioxide gas is passed through acidified potassium dichromate solution, then the orange colour of potassium dichromate solution changes to green (Fig. 1.11).

Fig. 1.11 Colour change of potassium permanganate and potassium dichromate

Evolution of a gas

Many chemical reactions are characterised by the evolution of a specific gas. Examples are given below.

(i) When magnesium reacts with a dilute acid such as dilute hydrochloric acid or dilute sulfuric acid, hydrogen gas is evolved. Take a magnesium strip in a test tube and add dilute hydrochloric acid into it. Bubbles of hydrogen gas are formed (Fig. 1.12).

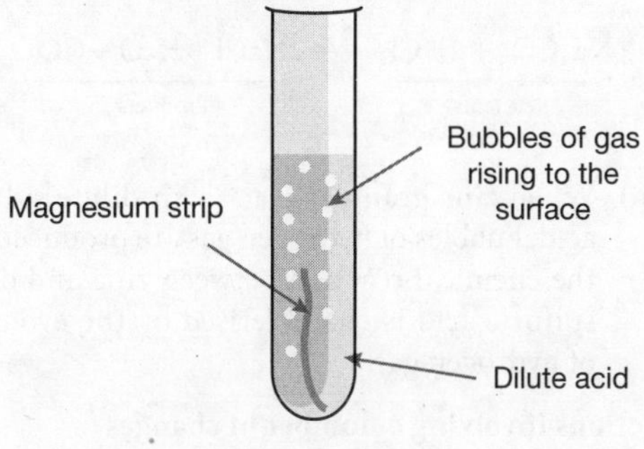

Fig. 1.12 Reaction of magnesium and dilute hydrochloric acid

Magnesium + Hydrogen chloride → Magnesium chloride + Hydrogen

$$\underbrace{Mg + 2HCl}_{\text{Reactants}} \longrightarrow \underbrace{MgCl_2 + H_2}_{\text{Products}}$$

(ii) When sodium carbonate is treated with dilute hydrochloric acid, the reaction is characterised by the evolution of carbon dioxide gas. Take little amount of sodium carbonate in a test tube and add dilute hydrochloric acid into it. There is a brisk effervescence of carbon dioxide (Fig. 1.13).

Sodium carbonate + Hydrogen chloride →
Sodium chloride + Water + Carbon dioxide

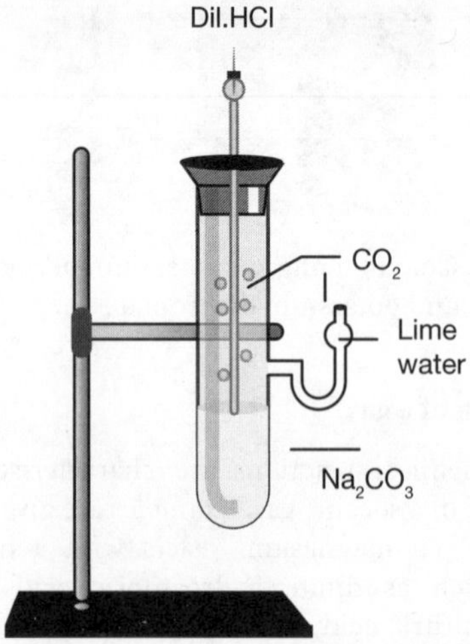

Fig. 1.13 Reaction of sodium carbonate and hydrochloric acid

$$\underbrace{Na_2CO_3 + 2HCl}_{\text{Reactants}} \longrightarrow \underbrace{2NaCl + H_2O + CO_2}_{\text{Products}}$$

(iii) When zinc granules react with dilute sulfuric acid, bubbles of hydrogen gas are produced. So, the chemical reaction between zinc and dilute sulfuric acid is characterised by the evolution of hydrogen gas.

Reactions involving a number of changes

There are many reactions in which many of the changes mentioned above are observed. Examples are given below.

(i) When zinc granules react with dilute sulfuric acid, bubbles of hydrogen gas are produced along with a rise in temperature (evolution of heat).

(ii) When lead nitrate and potassium iodide solutions are mixed, a yellow precipitate of lead iodide is formed—precipitate formation and change of colour occur.

(iii) When hydrochloric acid is added into calcium carbonate (solid), CO_2 gas is evolved along with a rise in temperature. Here, a change of state also takes place as the solid disappears to give a solution.

TEST YOUR KNOWLEDGE

1.1 Why should a magnesium ribbon be cleaned before burning in air?

Solution:

A magnesium ribbon is cleaned or rubbed with a sandpaper before burning to remove the layer of magnesium oxide (formed on exposure to air) that interferes in the further reaction of the pure metal.

1.2 Which one is a chemical change—rusting of iron or melting of iron?

Solution:

Rusting of iron is a chemical change.

1.3 Why is respiration considered as an exothermic reaction?

Solution:

Respiration results in oxidation of glucose which produces heat energy. Hence, it is an exothermic reaction.

1.3 Chemical Equations

All chemical reactions can be represented by chemical equations. The method of representing a chemical reaction with the help of symbols and formulae of the substances involved in it is called a *chemical equation.* There are two ways of representing a chemical reaction.

- **In terms of words:** Here, the names of the reactants and products are given to express the equation. For example, zinc metal reacts with dilute sulfuric acid to give zinc sulfate and hydrogen gas.

$$\text{Zinc} + \text{Sulfuric acid (dilute)} \rightarrow \underset{\text{Colourless solution}}{\text{Zinc sulfate}} + \underset{\text{Gas}}{\text{Hydrogen}}$$

This is known as the word equation.

- **In terms of symbols and formulae:** By putting the symbols and formulae of all substances in the word equation, a chemical equation is obtained. For example, the above word equation can be changed into a chemical equation as follows.

$$\underbrace{Zn + H_2SO_4}_{\text{Reactants}} \rightarrow \underbrace{ZnSO_4 + H_2}_{\text{Products}}$$

The short hand method of representing a chemical reaction in terms of symbols and formulae of the different reactants and products is known as a chemical equation.

In a chemical equation, these three points are of great concern.

(i) The substances which combine or react are called reactants. For example, zinc and sulfuric acid are the reactants here. The reactants are always written on the left hand side in an equation with a plus sign (+) between them.
(ii) The new substances produced in a reaction are called products. For example, zinc sulfate and hydrogen are the products in this case. The products are always written on the right hand side in an equation with a plus sign (+) between them.
(iii) An arrow (→) is put between the reactants and products, pointing from the reactants side to the product side.

1.3.1 Rules for Writing a Chemical Equation

The following rules have to be followed while writing a chemical equation.

(i) The reactants taking part in the reaction are written in terms of their symbols or molecular formulae on the left hand side of the equation.
(ii) A plus (+) sign is added between the formulae of the reactants.
(iii) The products of reaction are written in terms of their symbols or molecular formulae on the right hand side of the equation.
(iv) A plus (+) sign is added between the formulae of the products.
(v) In between the reactants and the products, an arrow sign (→) is inserted to show which way the reaction is occurring.

$$P + Q \rightarrow R + S$$

In this chemical equation, P and Q are the reactants, and R and S are the products. The arrow indicates that the reaction proceeds towards the formation of R and S. For example, when potassium nitrate is heated, it gives potassium nitrite and oxygen. This reaction may be represented in the form of a chemical equation as follows.

$$\underset{\text{Potassium nitrate}}{KNO_3} \rightarrow \underset{\text{Potassium nitrite}}{KNO_2} + \underset{\text{Oxygen}}{O_2}$$

Here potassium nitrate is the reactant while potassium nitrite and oxygen are products.

Competition Edge

The amount of product actually obtained in a reaction is known as the *actual yield*.

Balancing of a chemical equation refers to establishing the mathematical relationship between the quantity of reactants and products. These quantities are expressed as grams or moles.

1.3.2 Balanced and Unbalanced Chemical Equations

Balanced equation: A balanced chemical equation has the same number of atoms of different elements in the reactants and products. In other words, in a balanced chemical equation the number of atoms of every element on both the sides of the equation are equal. For example,

$$Zn + H_2SO_4 \rightarrow ZnSO_4 + H_2$$

Let us count the number of atoms of all the elements in the reactants and products separately.

	In reactants	In products
Number of Zn atoms	1	1
Number of H atoms	2	2
Number of S atoms	1	1
Number of O atoms	4	4

Unbalanced: An unbalanced chemical equation has an unequal number of atoms of one or more elements in the reactants and products. For example,

$$H_2 + O_2 \rightarrow H_2O$$

Let us count the number of hydrogen atoms and oxygen atoms in the reactants as well as products:

	In reactants	In products
Number of H atoms	2	2
Number of O atoms	2	1

1.3.3 Balancing of Chemical Equations

The process of making the number of different types of atoms equal on both the sides of an equation is known as *balancing of an equation*. The importance of a balanced chemical equation lies in the fact that it satisfies the law of conservation of mass, that is, in a chemical equation, the total mass of all the products must be equal to the total mass of the all the reactants. This is only possible when the number of atoms is equal on both the sides.

Simple equations can be balanced by hit and trial method by the following steps.

Step 1: To write the word equation: First write the equation in the word form by writing the names of the reactants and products on the left and right hand sides respectively. However, if the equation is given in terms of symbols and formulae ignore this step.

Step 2: To write the skeletal chemical equation: Write down the symbols of all the reactants and the products on the left and right hand sides respectively.

Step 3: To enclose the formulae in boxes: Enclose the formulae of every reactant and product in a box in order to remember that during balancing, the formula of any reactant or product may not be changed.

Step 4: To list the number of atoms of different elements: List the number of atoms of all the reactants and the products on the LHS and RHS respectively.

Step 5: To select the biggest formula: Start the process of balancing by selecting the compound having the maximum number of atoms ignoring the fact whether it is a reactant or a product.

Step 6: To start balancing of different elements: After balancing the compound with the maximum number of atoms, start balancing the other compounds to one by one. In order to balance the atoms of an element put a small whole number (natural number) coefficient before the formula of the compound (or symbol of the element for elementary substances).

Step 7: Now examine the number of atoms of different elements present in the balanced equation for all reactants and products.

Let us consider an example.

(i) Hydrogen burns in oxygen or air to form water. This reaction can be written in an equation as follows:

Step 1: Word equation: Hydrogen + Oxygen → Water

Step 2: Skeletal equation: Now use the formula of hydrogen (H_2), oxygen (O_2) and water (H_2O) to write the skeletal equation.

$$H_2 + O_2 \rightarrow H_2O$$

Step 3: To count number of atoms: Let us count the number of hydrogen atoms and oxygen atoms in the reactants and product.

	In reactants	In products
Number of H atoms	2	2
Number of O atoms	2	1

Step 4: To write balanced equation: Balance the number of atoms on both sides. For example, the number of hydrogen atoms is equal on both the sides (2 each), but the number of oxygen atoms is unequal. There are 2 oxygen atoms on the left side but only 1 oxygen atom on the right side. To have 2 oxygen atoms on the right side, we multiply H_2O by 2 and write $2H_2O$, so that

$$H_2 + O_2 \rightarrow 2H_2O$$

Let us count the number of various atoms on both the sides again

	In reactants	In products
Number of H atoms	2	4
Number of O atoms	2	2

Although the number of oxygen atoms has become equal, that is, 2 on both sides, the number of hydrogen atoms has now become unequal. There are 2 hydrogen atoms on the left side but 4 hydrogen atoms on the right side. To have 4 hydrogen atoms on the left side, we multiply H_2 by 2 and write $2H_2$, so that

$$2H_2 + O_2 \rightarrow 2H_2O$$

Now, let us count the number of various atoms on both the sides once again.

	In reactants	In products
Number of H atoms	4	4
Number of O atoms	2	2

This chemical equation contains an equal number of atoms of hydrogen and oxygen on both the sides, so this is a balanced equation.

Example 1.1 Write the balanced chemical equation for the following reaction:

Steam is passed over heated iron to form magnetic oxide of iron (Fe_3O_4) and hydrogen.

Solution: Step 1: Word equation.

Iron + Steam → Magnetic oxide of iron + Hydrogen

Step 2: Skeletal chemical equation.

$$Fe + H_2O \rightarrow Fe_3O_4 + H_2$$

Step 3: Listing number of atoms of different elements.

Element	No. of atoms on LHS	No. of atoms of RHS
Fe	1	3
H	2	2
O	1	4

Step 4: Selecting the biggest formula (Fe_3O_4) and balancing the element with the highest number of atoms, that is, oxygen. There are 4 O atoms on the RHS and 1 O atom on the LHS. To balance the O atoms, multiply H_2O on LHS by 4.

We get, $$Fe + 4\,H_2O \rightarrow Fe_3O_4 + H_2$$

Step 5: To balance Fe atoms on both sides. There are 3 Fe atoms on the RHS and 1 Fe atom on the LHS. To balance the Fe atoms, multiply Fe on the LHS by 3.

We get, $$3\,Fe + 4\,H_2O \rightarrow Fe_3O_4 + H_2$$

Step 6: To balance H atoms. There are 8 H atoms on the LHS and 2 H atoms on the RHS. To balance the H atoms, multiply H_2 by 4. Finally, we get the balanced equation.

$$3\,Fe + 4\,H_2O \rightarrow Fe_3O_4 + H_2$$

Step 7: To check the correctness of the balanced equation.

	LHS	RHS
Fe atoms	3	3
H atoms	8	8
O atoms	4	4

In this equation all the atoms on both sides are balanced.

$$3\,Fe + 4\,H_2O \rightarrow Fe_3O_4 + 4\,H_2 \quad \text{(Balanced equation)}$$

Example 1.2 Write the balanced equation for the reaction involving the combustion of methane in oxygen to form carbon dioxide and water.

Solution: Step 1: Word equation.

Methane + Oxygen → Carbon dioxide + Water

Step 2: Skeletal chemical equation.

$$CH_4 + O_2 \rightarrow CO_2 + H_2O$$

Step 3: Selecting the element which occurs at a minimum number of places, that is, C or H. Carbon atoms are already balanced on the two sides (one on each side). There are 4 H atoms on the LHS and 2 H atoms on the RHS. To balance the H atoms, multiply H_2O on the RHS by 2. We get,

$$CH_4 + O_2 \rightarrow CO_2 + 2\,H_2O$$

Step 4: To balance the O atoms: There are 4 O atoms on the RHS and 2 O atoms on the LHS. To balance the O atoms, multiply O_2 on the LHS by 2. We get,

$$CH_4 + 2\,O_2 \rightarrow CO_2 + 2\,H_2O$$

Step 5: To check the correctness of the balanced equation.

	LHS	RHS
C atoms	1	1
H atoms	4	4
O atoms	4	4

Hence, the equation is balanced.

Example 1.3 Nitrogen and hydrogen combine to form ammonia. Write the balanced chemical equation for the reaction involved.

Solution: Step 1: Word equation.

Nitrogen + Hydrogen → Ammonia

Step 2: Skeletal chemical equation.

$$N_2 + H_2 \rightarrow NH_3$$

Step 3: Balancing of atoms on both sides: There are 2 N atoms on the LHS and 1 N atom on the RHS. To balance the N atoms, multiply NH_3 on the RHS by 2. We get,

$$N_2 + H_2 \rightarrow 2NH_3$$

Now, there are 6 H atoms on the RHS and 2 H atoms on the LHS. To balance the H atoms, multiply H_2 on the LHS by 3. We get,

$$N_2 + 3H_2 \rightarrow 2NH_3$$

Step 4: To check the correctness of the balanced equation.

	LHS	RHS
N atoms	2	2
H atoms	6	6

Hence, the equation is balanced.

Ionic Equations

An ionic equation involves ionic compounds in solution or sometimes, only the ions of the ionic compounds that are actually taking part in the reaction. For example, the reaction of zinc with aqueous copper sulfate.

$$Zn(s) + CuSO_4(aq) \rightarrow ZnSO_4(aq) + Cu(s)$$

In the aqueous solution, $CuSO_4$ exists as Cu^{2+} ions and SO_4^{2-} ions and $ZnSO_4$ as Zn^{2+} ions and SO_4^{2-} ions, thus, the above equation can be written as,

$$Zn + Cu^{2+} + SO_4^{2-} \rightarrow Zn^{2+} + SO_4^{2-} + Cu$$

SO_4^{2-} ions exist on both the sides, this means that SO_4^{2-} ions do not take part in the reaction. Cancelling out these ions on both the sides, the equation can be written as

$$Zn + Cu^{2+} \rightarrow Zn^{2+} + Cu$$

It means an ionic equation contains not only atoms or molecules but also contains ions and the ions are actually taking part in the reaction.

Balancing of Ionic Equations

In the balancing of an ionic equation, the number of atoms of different elements must be equal on both sides, also, the total charges present on the ions must be equal on both sides. For example, in the reaction between cupric ion (Cu^{2+}) and H_2S,

Initial unbalanced equation:

$$Cu^{2+}(aq) + H_2S \rightarrow CuS(s) + H^+(aq)$$

Balanced equation:

$$Cu^{2+}(aq) + H_2S \rightarrow CuS(s) + 2H^+(aq)$$

Here we have balanced the number of atoms of hydrogen along with charges. It was +2 on the LHS and we have made it +2 on the RHS. The number of hydrogen, Cu and sulfur atoms are also balanced on both sides.

Example 1.4 Write the following equation in the ionic form and then balance the ionic equation.

$$Fe + HCl \rightarrow FeCl_3 + H_2$$

Solution: The given equation in the ionic form can be written as

$$Fe + H^+ \rightarrow Fe^{3+} + H_2$$

To balance the number of atoms, we shall multiply H^+ by 2 (as Fe atoms are already balanced).

Now, we get $Fe + 2H^+ \rightarrow Fe^{3+} + H_2$

Now, the charge on the LHS is +2 (on $2H^+$) while on the RHS, it is +3 (on Fe^{3+}). For charge balance, we multiply $2H^+$ by 3 and Fe^{3+} by 2 so that the charge on each side becomes +6.

Now, we get $Fe + 6H^+ \rightarrow 2Fe^{3+} + H_2$

As mass balance has been disturbed, we should multiply Fe on the LHS by 2 and H_2 on the RHS by 3.

Now, we get $2Fe + 6H^+ \rightarrow 2\,Fe^{3+} + 3\,H_2$

This is the required balanced equation in which both the atoms and the charges are balanced on both sides.

Applications of a Balanced Chemical Equation

One of the most important applications of a balanced chemical equation is that it provides a quantitative relationship between the various reactants and the products in terms of masses, moles, volumes and molecules. This is known as *stoichiometry* (in Greek, stoichiometry means, to measure an element). The coefficients present in the balanced chemical equation are known as *stoichiometric coefficients*.

Quantitatively, a balanced chemical equation gives the following information.

- Number of moles of the reactants and products taking part in the reaction.
- Number of molecules of each of the reactants and products taking part in the reaction.
- Mass–mass relationship, that is, relative masses of the reactants reacting and the masses of the products formed in an equation.
- Mass–volume relationship, that is, relative masses of the reactants and products and volumes of the gaseous reactants or products (don't forget that one mole of any gaseous substance has a volume of 22.4 litres at STP, that is, standard temperature and pressure conditions, namely, 0°C and 1 atmospheric pressure).
- Volume–volume relationship in the case of gaseous reactions only.
For example, (i)

	$MgCO_3$	+	2HCl	→	$MgCl_2$	+	H_2O	+	CO_2
	Magnesium carbonate		Hydrochloric acid		Magnesium chloride		Water		Carbon dioxide
Parts by mass:	(24+12+3×16)=84		2(1+35.5)=73.5		24+2×35.5=95		2×1+16=18		12+2×16=44

This equation provides the following information.

- One molecule of magnesium carbonate reacts with two molecules of hydrochloric acid to give one molecule of magnesium chloride, one molecule of water and one molecule of carbon dioxide.
- One mole of $MgCO_3$ reacts with two moles of HCl acid to give one mole of $MgCl_2$, one mole of H_2O and one mole of CO_2.
- 84 g of $MgCO_3$ react with 73 g of HCl acid to yield 95 g of $MgCl_2$, 18 g of H_2O and 44 g of CO_2.
- 84 g of $MgCO_3$ reacts with 73 g of HCl acid to produce 95 g of $MgCl_2$, 18 g of H_2O and 22.4 litres of CO_2 at STP.

(ii) Formation of ammonia from nitrogen and hydrogen involves gaseous reactants and products only.

- One molecule of N_2 reacts with 3 molecules of H_2 to give two molecules of NH_3.
- One mole of N_2 reacts with 3 moles of H_2 to produce two moles of NH_3.
- 28 g of N_2 reacts with 6 g of H_2 to yield 2 (14 + 3) = 34 g of NH_3.
- One volume of N_2 reacts with 3 volumes of H_2 to give two volumes of NH_3.

To Make Equations More Informative

Although a balanced chemical equation provides us with a lot of useful information as discussed above, it fails to tell us about the physical states of the reactants and the products, dilute or concentrated nature of the acid and heat changes occurring during the reaction. This means, it is important to do few more things for getting this information. The equation which gives more information about the chemical reaction is known as a more informative (or information giving) equation. A chemical equation can be made more informative by these three ways.

(i) By indicating the *physical states* of the reactants and products.
(ii) By indicating the *heat changes* taking place in the reaction.
(iii) By indicating the *conditions* under which the reaction takes place.

We will discuss these three points in detail one by one.

(i) To indicate the physical states of reactants and products in an equation: There can be four physical states for the reactants and products of a chemical reaction—solid, liquid, aqueous solution and gas. Solid state is indicated by the symbol (*s*), liquid state is indicated by the symbol (*l*), aqueous solution (solution made in water) is indicated by the symbol (*aq*) and gaseous state is indicated by the symbol (*g*). The physical states of the reactants and products are shown by putting the above *state symbols* just after their symbols or formulae in an equation. For example,

$$Zn(s) + H_2SO_4(aq) \rightarrow ZnSO_4(aq) + H_2(g)$$

This equation is more informative as it tells us the physical states of the various substances involved here. It tells us that zinc is in the solid state, sulfuric acid is in the form of an aqueous solution, zinc sulfate is also an aqueous solution but hydrogen is in the gaseous state. For example,

$$\underset{\substack{\text{Calcium hydroxide}\\ \text{(Lime water)}}}{Ca(OH)_2(aq)} + \underset{\text{Carbon dioxide}}{CO_2(g)} \rightarrow \underset{\substack{\text{Calcium carbonate}\\ \text{(white ppt.)}}}{CaCO_3(s)} + H_2O(l)$$

In this reaction, calcium carbonate is formed as a solid product (precipitate), so its physical state is indicated by the symbol (*s*). Water is a liquid, so its physical state has been indicated by the symbol (*l*).

(ii) To indicate heat changes in an equation: There are two types of reactions on the basis of heat changes involved—exothermic and endothermic reactions.

- Those reactions in which heat is evolved are known as exothermic reactions. For example, when carbon burns in oxygen to form carbon dioxide, a lot of heat is produced.

$$\underset{\text{Carbon}}{C(s)} + \underset{\text{Oxygen}}{O_2(g)} \rightarrow \underset{\text{(Carbon dioxide)}}{CO_2(g)} + \text{Heat}$$

Natural gas is mainly methane (CH_4). When natural gas burns in the oxygen of air, it forms carbon dioxide and water vapour. A large amount of heat energy is also produced. This can be written as,

$$\underset{\substack{\text{Methane}\\ \text{(Natural gas)}}}{CH_4(g)} + \underset{\substack{\text{Oxygen}\\ \text{(From air)}}}{2O_2(g)} \rightarrow \underset{\text{Carbon dioxide}}{CO_2(g)} + \underset{\text{Water}}{2H_2O(g)} + \text{Heat energy}$$

The burning of natural gas is an exothermic reaction because heat is produced in this reaction. Please note that all the combustion reactions are exothermic reactions.

During respiration, glucose combines with oxygen in the cells of our body to form carbon dioxide and water along with the production of energy.

$$\underset{\text{Glucose}}{C_6H_{12}O_6(aq)} + \underset{\text{Oxygen}}{6O_2(g)} \rightarrow \underset{\text{Carbon dioxide}}{6CO_2(g)} + \underset{\text{Water}}{6H_2O(l)} + \text{Energy}$$

- Those reactions in which heat is absorbed are known as endothermic reactions. For example, when nitrogen and oxygen are heated to a very high temperature (of about 3000°C) they combine to form nitrogen monoxide, and a lot of heat is absorbed in this reaction.

$$\underset{\text{Nitrogen}}{N_2(g)} + \underset{\text{Oxygen}}{O_2(g)} + \text{Heat} \rightarrow \underset{\text{Nitrogen monoxide}}{2NO(g)}$$

All the decomposition reactions are endothermic reactions. For example, when calcium carbonate is heated, it decomposes to form calcium oxide and carbon dioxide.

$$\underset{\text{Calcium carbonate}}{CaCO_3(s)} + \text{Heat} \rightarrow \underset{\text{Calcium oxide}}{CaO(s)} + \underset{\text{Carbon dioxide}}{CO_2(g)}$$

The electrolysis of water to form hydrogen and oxygen is also an endothermic reaction.

- To indicate the condition under which the reaction is possible: If heat is needed for a reaction to occur, then the heat sign, delta (Δ) is put over the arrow of the equation. If the reaction takes place in the presence of a catalyst, then the symbol or formula of the catalyst is also written above or below the arrow sign in the equation. For example,

$$\underset{\text{Potassium chlorate}}{2KClO_3(s)} \xrightarrow[MnO_2]{\Delta} \underset{\text{Potassium chloride}}{2KCl(s)} + \underset{\text{Oxygen}}{3O_2(g)}$$

$$\underset{\text{Carbon monoxide}}{CO(g)} + \underset{\text{Hydrogen}}{2H_2(g)} \xrightarrow[ZnO+CrO_3]{300\text{ atm; }300^\circ C} \underset{\substack{\text{Methanol}\\\text{(Methyl alcohol)}}}{CH_3OH(l)}$$

Green plants make food by photosynthesis. During photosynthesis, carbon dioxide combines with water in the presence of sunlight and chlorophyll (the green pigment in leaves) to make food (glucose) and oxygen gas is given out.

$$\underset{\text{Carbon dioxide}}{6CO_2(g)} + \underset{\text{Water}}{6H_2O(l)} \xrightarrow[\text{Chlophyll}]{\text{Sunlight}} \underset{\text{Glucose}}{C_6H_{12}O_6(aq)} + \underset{\text{Oxygen}}{6O_2(g)}$$

Competition Edge

Exothermic reaction	Endothermic reaction
In this reaction heat is evolved or given out.	In this reaction heat is absorbed or taken in.
Heating effect is observed here.	Colling effect is observed here.
For example, (i) Burning of methane (natural gas)	For example, (i) Formation of nitric oxide (NO)

Exothermic reaction

$$\underset{\text{Methane}}{CH_4(g)} + \underset{\text{Oxygen}}{2O_2(g)} \rightarrow \underset{\text{Carbon dioxide}}{CO_2(g)} + \underset{\text{Water}}{2H_2O(l)} + \text{Heat}$$

(ii) Rotting of vegetable matter: The decomposition of vegetable matter into compost is also an exothermic reaction.

(iii) Respiration: We get the energy for our life processes from the food that we eat. The food like wheat, rice and potatoes contains carbohydrates. During digestion, carbohydrates are broken down into their simpler forms (like glucose). The air we breathe in during respiration oxidises glucose in the cells of our body. This reaction is exothermic and provides us with the required energy.

$$\underset{\text{Glucose}}{C_6H_{12}O_6(aq)} + 6O_2(g) \rightarrow 6CO_2(g) + 6H_2O(l) + \text{Energy}$$

Endothermic reaction

$$\underset{\text{Nitrogen}}{N_2(g)} + \underset{\text{Oxygen}}{O_2(g)} + \text{Heat} \rightarrow \underset{\text{Nitric oxide}}{2NO(g)}$$

Or

$$\underset{\text{Nitrogen}}{N_2(g)} + \underset{\text{Oxygen}}{O_2(g)} \rightarrow \underset{\text{Nitric oxide}}{2NO(g)} - \text{Heat}$$

(ii) Formation of ammonia (NH_3)

$$N_2(g) + 3H_2(g) \xrightarrow[200\text{ atm, Fe}]{500^\circ C} 2NH_3(g)$$

Photochemical reaction: A reaction which occurs in presence of light is known as a photochemical reaction. For example, photosynthesis of glucose, formation of HCl.

$$H_2(g) + Cl_2(g) \xrightarrow{\text{Sunlight}} 2HCl(g)$$

$$\underset{\text{Carbon dioxide}}{6CO_2(g)} + \underset{\text{Water}}{6H_2O(l)} \xrightarrow[\text{Chlorophyll}]{\text{Sunlight}} \underset{\text{Glucose}}{C_6H_{12}O_6}$$

Catalyst: A substance which can increase the speed of a chemical reaction without taking part in it. A catalyst does not change in mass and composition before and after the reaction. In the preparation of ammonia, iron filings are used as the catalyst.

TEST YOUR KNOWLEDGE

1.4 Write the balanced equation for the reaction between magnesium and hydrochloric acid. Name the product obtained and identify the type of reaction.

Solution:

$$Mg(s) + 2HCl(dil.) \rightarrow MgCl_2(aq) + H_2(g)$$

The products obtained are magnesium chloride and hydrogen gas. The reaction is a displacement reaction.

1.5 Do the following chemical equations represent correct balanced chemical equations or not?

(i) $Cu(s) + 2H_2SO_4 \rightarrow CuSO_4 + SO_2 + 2H_2O$

(ii) $\underset{\text{Zinc nitrate}}{Zn(NO_3)_2} \rightarrow 2ZnO + 4NO_2 + O_2$

Solution:

(i) No, the equation is not correctly balanced as it is not specified whether the acid is dilute or concentrated. Moreover, the state of reactants and products is not mentioned here. Hence, the correct balanced equation will be:

$$\underset{\text{Copper}}{Cu(s)} + \underset{\text{(Conc.)}}{2H_2SO_4} \rightarrow \underset{\text{Copper sulfate}}{CuSO_4(s)} + \underset{\text{Sulfur dioxide}}{SO_2(g)} \uparrow + \underset{\text{Water}}{2H_2O(l)}$$

(ii) No, the equation is not correctly balanced as the number of atoms of different elements is not the same in the reactants and the products. Moreover, the states of the reactants and products are not specified here. So the correct balanced equation will be:

$$\underset{\text{Zinc nitrate}}{2Zn(NO_3)_2(s)} \rightarrow \underset{\text{Zinc oxide}}{2ZnO(s)} + 4NO_2(g) + O_2(g)$$

1.6 Write balanced chemical equations from the following word equations. Also include the physical state of each element or compound involved in the reaction.

(i) Sodium metal reacts with water to give hydrogen gas and aqueous hydroxide solution.

(ii) Potassium chlorate on heating gives potassium chloride and oxygen gas.

Solution:

(i) $2Na(s) + 2H_2O(l) \rightarrow H_2(g) + 2NaOH(aq)$

(ii) $2KClO_3(s) \rightarrow 2KCl(s) + 3O_2(g)$

1.7 What is meant by a skeletal type chemical equation? What does it represent? Using the equation for electrolytic decomposition of water, differentiate between a skeletal chemical equation and a balanced chemical equation.

Solution:

An equation in which the substances involved in the reaction are written in the atomic form instead of the molecular form and the equation is not balanced, is called a skeletal type chemical equation.

$$H_2O \xrightarrow{\text{Electrolytic}} H + O \text{ (Skeletal equation)}$$

Hydrogen and oxygen are written in the atomic form and the equation is not balanced.

$$2H_2O \rightarrow 2H_2 + O_2 \text{ (Balanced chemical equation)}$$

$$H_2O \rightarrow H_2 + O_2$$

This is also a skeletal equation.

Quick Review

- **Chemical reactions** are the processes by which new substances with new properties are formed.
- The substances which take part in a chemical reaction are known as **reactants**.
- The new substances produced as a result of a chemical reaction are known as **products**.
- A chemical reaction is possible by following these conditions—mixing, heat, light, electricity, pressure and presence of catalyst.
- The important characteristics of chemical reactions are change in state, formation of precipitate, change in temperature, change in colour and evolution of a gas.
- The method of representing a chemical reaction with the help of symbols and formulate of the substances involved in it is called a **chemical equation**.
- A **balanced chemical equation** has an equal number of atoms of different elements in the reactants and products.

- An **unbalanced chemical equation** has an unequal number of atoms of one or more elements in the reactants and products.
- A reaction in which heat or energy is released during the course of a reaction is known as an **exothermic** reaction.
- A reaction in which heat or energy is absorbed during the course of a reaction is known as an **endothermic** reaction.
- A **catalyst** is a substance which is usually used to increase the rate of a reaction. However, it does not take part in the reaction and remains unchanged in mass and properties.

Exercise 1

Section A: Multiple Choice Questions (1 Mark)

1. Which of the following is not a physical change?
(a) boiling of water to give water vapour
(b) melting of ice to give water
(c) dissolution of salt in water
(d) combustion of liquified petroleum gas (LPG)

2. A chemical reaction does not involve
(a) changing of the atoms of an element into those of another element to form new products
(b) formation of new substances having entirely different properties than those of the reactants
(c) rearrangement of the atoms of the reactants to form new products
(d) breaking of old chemical bonds and formation of new chemical bonds

3. One of the following processes does not involve a chemical reaction. That is,
(a) digestion of food in our stomach
(b) burning of candle wax when heated
(c) ripening of banana
(d) melting of candle wax when heated

4. Which one of the following processes involve chemical reactions?
(a) heating copper wire in the presence of air at high temperature
(b) keeping petrol in a china dish in the open
(c) liquefaction of air
(d) storing of oxygen gas under pressure in a gas cylinder

5. In which of the following chemical equations, do the abbreviations represent the correct states of the reactants and products involved, at the reaction temperature?
(a) $2H_2(l) + O_2(l) \rightarrow 2H_2O(g)$
(b) $2H_2(g) + O_2(l) \rightarrow 2H_2O(l)$
(c) $2H_2(g) + O_2(g) \rightarrow 2H_2O(l)$
(d) $2H_2(g) + O_2(g) \rightarrow 2H_2O(g)$

6. A researcher adds barium hydroxide to hydrochloric acid to form white-coloured barium chloride. Which option gives the balanced chemical equation of the reaction?
(a) $HCl + Ba(OH)_2 \rightarrow BaCl_2 + 2HOH$
(b) $2HCl + Ba(OH)_2 \rightarrow BaCl_2 + 2HOH$
(c) $2HCl + Ba(OH)_2 \rightarrow BaH_2 + 2HCl + O_2$
(d) $HCl + 2Ba(OH) \rightarrow 2BaCl_2 + 2HOH + O_2$

7. When carbon dioxide is passed through lime water,
(a) calcium hydroxide is formed
(b) a white precipitate of CaO is formed
(c) lime water turns milky
(d) the colour of lime water disappears

8. In which of the following, will heat energy be evolved?
(a) electrolysis of water
(b) dissolution of NH_4Cl in water
(c) burning of LPG
(d) decomposition of AgBr in the presence of sunlight

9. Silver articles turn black when kept in the open for a few days due to the formation of
(a) H_2S (b) AgS
(c) $AgSO_4$ (d) Ag_2S

10. When crystals of lead nitrate are heated strongly in a dry test tube
(a) the crystals immediately melt
(b) a brown residue is left
(c) white fumes appear in the tube
(d) a yellow residue is left

11. Dilute hydrochloric acid is added to granulated zinc taken in a test tube. The following observations are recorded. Point out the correct observation.
(a) the surface of the metal becomes shiny
(b) the reaction mixture turns milky
(c) the pungent odour of a gas is recorded
(d) a colourless and odourless gas is evolved

12. On immersing an iron nail in $CuSO_4$ solution for few minutes, you will observe that
(a) no reaction takes place
(b) the colour of the solution fades away
(c) the surface of the iron nails acquires a black coating
(d) the colour of the solution changes to green

13. The electrolytic decomposition of water gives H_2 and O_2 in the ratio of
(a) 1 : 2 by volume (b) 2 : 1 by volume
(c) 8 : 1 by mass (d) 1 : 2 by mass

14. In the decomposition of lead(II) nitrate to give lead(II) oxide, nitrogen dioxide and oxygen gas, the coefficient of nitrogen dioxide (in the balanced equation) is
(a) 1 (b) 2
(c) 3 (d) 4

15. Chemically, rust is
(a) hydrated ferrous oxide
(b) only ferric oxide
(c) hydrated ferric oxide
(d) none of these

16. When Ag is exposed to air, it gets a black coating of
(a) $AgNO_3$ (b) Ag_2S
(c) Ag_2O (d) Ag_2CO_3

17. Which of the following reactions is an endothermic reaction?
(a) burning of coal
(b) decomposition of vegetable matter into compost
(c) process of respiration
(d) decomposition of calcium carbonate to form quicklime and carbon dioxide
[Board SQP, 2020]

18. When SO_2 gas is passed through a saturated solution of H_2S, which of the following reactions occurs?
(a) $SO_2 + 2H_2S \rightarrow 2H_2O + 3S$
(b) $SO_2 + 2H_2S \rightarrow H_2O + 3S$
(c) $SO_2 + H_2S \rightarrow H_2O + S$
(d) $SO_2 + H_2O \rightarrow SO_3 + H_2$

19. Name the products formed when iron filings are heated with dilute hydrochloric acid
(a) Fe(III) chloride and water
(b) Fe(II) chloride and water
(c) Fe(II) chloride and hydrogen gas
(d) Fe(III) chloride and hydrogen gas

20. Which among the following are physical changes?
(i) sublimation of solid ammonium chloride
(ii) burning of liquified petroleum gas (LPG)
(iii) heating of an iron rod to red heat
(iv) evaporation of petrol
(a) (i), (iii) (b) (i), (iii), (iv)
(c) (ii), (iii), (iv) (d) (i), (ii), (iii)

21. Which of the following are exothermic processes?
(i) reaction of water with quick lime
(ii) dilution of an acid
(iii) evaporation of water (iv) sublimation of camphor (crystals)
(a) (i) and (ii) (b) (ii) and (iii)
(c) (i) and (iv) (d) (ii) and (iv)

22. A substance X is used in white washing and is obtained by heating limestone in the absence of air. Identify X.
(a) $CaOCl_2$ (b) $Ca(OH)_2$
(c) CaO (d) $CaCO_3$

23. What happens when lead nitrate reacts with potassium iodide?
(a) they will not react
(b) a large amount of hydrogen is released
(c) a yellow precipitate of lead iodide and potassium nitrate will be produced
(d) evolution of gas will occur

24. A student took sodium sulfate solution in a test tube and added barium chloride solution to it. He observed that an insoluble substance has formed. The colour and molecular formula of the insoluble substance is **(CBSE_2021)**
(a) grey, Ba_2SO_4 (b) yellow, $Ba(SO_4)_2$
(c) white, $BaSO_4$ (d) pink, $BaSO_4$

25. Sodium reacts with water to form sodium hydroxide and hydrogen gas. The balanced equation which represents the above reaction is **(CBSE_2021)**

(a) $Na(s) + 2H_2O(l) \rightarrow 2NaOH(aq) + 2H_2(g)$
(b) $2Na(s) + 2H_2O(l) \rightarrow 2NaOH(aq) + H_2(g)$
(c) $2Na(s) + 2H_2O(l) \rightarrow NaOH(aq) + 2H_2(g)$
(d) $2Na(s) + H_2O(l) \rightarrow 2NaOH(aq) + 2H_2(g)$

26. A student while burning a magnesium ribbon in air, collected the products in a wet watch glass. The new product obtained was: **[CBSE 2021]**

(a) magnesium oxide
(b) magnesium carbonate
(c) magnesium hydroxide
(d) magnesium chloride

27. Consider the following chemical equation: **[CBSE 2021]**

$$2NaOH + H_2SO_4 \rightarrow Na_2SO_4 + 2H_2O$$

The information conveyed by the chemical equation is:

I. NaOH reacts with H_2SO_4 to produce Na_2SO_4 and H_2O
II. for every one molecule of H_2SO_4, two molecules of NaOH are required
III. acids and bases are non-ionic in nature
IV. this is not a redox reaction

The correct statements are:

(a) I and II (b) II and III
(c) III and IV (d) I and IV

28. When lead nitrate powder is heated in a boiling tube, we observe **[CBSE 2021]**

(a) brown fumes of nitrogen dioxide
(b) brown fumes of lead oxide
(c) yellow fumes of nitrogen dioxide
(d) brown fumes of nitric oxide

29. It is important to balance the chemical equations to satisfy the law of conservation of mass. Which of the following statements of the law is incorrect? **[CBSE 2021]**

(a) The total mass of the elements present in the reactants is equal to the total mass of the elements present in the products.
(b) The number of atoms of each element remains the same before and after a chemical reaction.
(c) The chemical composition of the reactants is the same before and after the reaction.
(d) Mass can neither be created nor be destroyed in a chemical reaction.

30. Reena took 5 mL of lead nitrate solution in a beaker and added approximately 4 mL of potassium iodide solution to it. What would she observe? **[CBSE 2021]**

(a) the solution turned red
(b) a yellow precipitate was formed
(c) a white precipitate was formed
(d) the reaction mixture became hot

31. Three beakers labelled as A, B and C each containing 25 mL of water were taken. A small amount of NaOH, anhydrous $CuSO_4$ and NaCl were added to the beakers A, B and C respectively. It was observed that there was an increase in the temperature of the solution contained in beakers A and B, whereas in the case of beaker C, the temperature of the solution falls. Which one of the following statement(s) is(are) correct?

(i) In beakers A and B, exothermic processes have occurred.
(ii) In beakers A and B, endothermic processes have occurred.
(iii) In beaker C, an exothermic process has occurred.
(iv) In beaker C, an endothermic process has occurred.

(a) (i) only (b) (ii) only
(c) (i) and (iv) (d) (iv), (ii) and (iii)

32. Match the following.

Column I	Column II
1. Burning of magnesium ribbon in air	(i) physical change
2. Addition of sulfuric acid to zinc granules	(ii) chemical change
3. Souring of milk in summer	(iii) white dazzling light
4. Dissolution of sugar in water	(iv) hydrogen gas is evolved

(a) (1)—(ii, iv), (2)—(ii, iii), (3)—(i), (4)—(ii)
(b) (1)—(iii), (2)—(iv), (3)—(ii), (4)—(i)
(c) (1)—(ii, iii), (2)—(ii, iv), (3)—(ii), (4)—(i)
(d) (1)—(i, iii), (2)—(i, iv), (3)—(ii, i), (4)—(iv)

Assertion Reason Questions

Directions: In the following questions, two statements, an Assertion (A) and a Reason (R) are given. Mark

(a) if A and R are both correct and R is the correct explanation of A.
(b) if A and R are both correct but R is not the correct explanation of A.
(c) A is true but R is false.
(d) A is false but R is true.

Assertion	Reason
1. A chemical reaction becomes faster at higher temperatures.	1. At higher temperatures, molecular motion becomes more rapid.
2. Burning of a candle is a physical change.	2. In a physical change, no new substance is formed.
3. Sodium metal is stored under kerosene.	3. Metallic sodium melts when exposed to air.
4. To dilute sulfuric acid, the acid is added to water and not water to acid.	4. Specific heat of water is quite large.
5. Burning of natural gas is an endothermic reaction. **[CBSE 2021]**	5. Methane gas combines with oxygen to produce carbon dioxide and water.
6. Silver salts are used in black and white photography. **[CBSE 2021]**	6. Silver salts do not decompose in the presence of sunlight.
7. After white washing the walls a shiny white finish on walls is obtained after 2 to 3 days. **[CBSE 2021]**	7. Calcium oxide reacts with carbon dioxide to form calcium hydrogen carbonate which gives a shiny white finish.

Section B: Very Short Answer Questions (2 Marks)

1. What is meant by a chemical reaction?
2. Name and state the law which is kept in mind while we balance a chemical equation.
3. State one basic difference between a physical change and a chemical change.
4. Which one is a chemical change—rusting of iron or melting of iron?
5. Identify the chemical change—melting of ice or conversion of milk into curd.
6. Is burning of a candle, a physical change or a chemical change?
7. Write the chemical equation involved when magnesium metal reacts with aqueous hydrochloric acid to produce a solution of magnesium chloride and hydrogen gas.
8. What happens when quicklime is added to water?
9. What happens when sodium reacts with water?
10. What happens when $ZnCO_3$ is heated in the absence of air? Give the relevant equation.
11. Write the reaction when potassium chlorate ($KClO_3$) is heated.
12. Write a balanced chemical equation for the following reaction. Ethanol is warmed with ethanoic acid to form ethyl acetate in the presence of concentrated H_2SO_4.
13. Why should a magnesium ribbon may be cleaned before burning it in air?
14. Write the chemical equation for reactions that take place when lead nitrate and potassium iodide solutions are mixed.
15. Complete and balance the following equations.
    (i) $NaOH + \_\_\_\_\_\_ \rightarrow Na_2SO_4 + H_2O$
    (ii) $Ca(OH)_2 + \_\_\_\_\_\_ \rightarrow CaCO_3 + H_2O$
16. Write a balanced chemical equation:
    $FeSO_4(s) \xrightarrow{\text{Heat}} Fe_2O_3(s) + SO_2(g) + SO_3(g)$
17. List any two observations when ferrous sulfate is heated in a dry test tube. **[CBSE sample paper, 2021]**
18. Identify the products formed when 1 mL of dil. hydrochloric acid is added to 1 g of sodium metal. **[CBSE sample paper, 2021]**

Case Study or Passage Based Questions

When a chemical equation is written, a few important things are kept in mind. The symbols and formulae of the reactants are taken on the left hand side while those of the product are taken on the right hand side and an arrow is placed in between both the sides. In a balanced chemical equation, the number of atoms of every element must be equal on both the sides. A balanced chemical equation satisfies law of conservation of mass also.

19. What do you mean by a balanced chemical equation?

20. In the reaction $2KClO_3 \xrightarrow[MnO_2]{\Delta} 2KCl + 3O_2$ what do the symbols Δ and MnO_2 written on the arrow signify?

21. Write the missing product in the given equation.

$$2\ FeSO_4 \rightarrow Fe_2O_3 + SO_2 + \_\_\_\_\_\_$$

22. Write the missing product in the given equation.

$$Na_2SO_4 + BaCl_2 \rightarrow \_\_\_\_\_\_ + 2NaCl$$

23. Complete the missing components/variables given as x and y in the following reactions:
   (i) $Pb(NO_3)_2(aq) + 2KI(aq) \rightarrow PbI_2(x) + 2KNO_3(y)$
   (ii) $Cu(s) + 2AgNO_3(aq) \rightarrow Cu(NO_3)_2(aq) + x(s)$

Section C: Short Answer Questions (3 Marks)

1. Define a chemical reaction. Which observation helps you to determine whether a chemical reaction has taken place?

2. List four observations that help us to determine whether a chemical reaction has taken place.

3. (i) Define a balanced chemical equation. Why should an equation be balanced?
   (ii) Write the balanced chemical equation for the following reactions.
   (a) Phosphorus burns in the presence of chlorine to form phosphorus pentachloride.
   (b) Burning of natural gas.
   (c) The process of respiration.
   [CBSE 2015]

4. (i) State the law which is followed in balancing a chemical equation.
   (ii) Balance the following chemical equation.

$$Fe + H_2O \rightarrow Fe_3O_4 + H_2$$

5. Give one example of each.
   (i) Chemical reaction showing the evolution of a gas.
   (ii) Change in a substance's colour during a chemical reaction.

6. Write one equation each for decomposition reactions where energy is supplied in the form of heat, light and electricity.

7. Translate the following statements into chemical equations and then balance them.
   (i) Hydrogen gas combines with nitrogen to form ammonia.
   (ii) Hydrogen sulfide gas burns in air to give water and sulfur dioxide.
   (iii) Barium chloride reacts with aluminium sulfate to give aluminium chloride and a precipitate of barium sulfate.
   (iv) Potassium metal reacts with water to give potassium hydroxide and hydrogen gas.

8. What is observed when carbon dioxide gas is passed through lime water
   (i) for a short duration?
   (ii) for a long duration? Also write the chemical equations for the reaction involved.

9. *A copper plate was dipped into a solution of silver nitrate. After sometime, a black layer was observed on the surface of the copper plate. State the reason for it and write a chemical equation of the reaction involved.

10. *When an iron rod is kept dipped in copper sulfate solution for some time, a brown coating is formed on the iron rod. What change will be observed in the colour of the solution? Also write the chemical equation for the reaction involved.

11. Balance the following chemical equations.
   (i) $HNO_3 + Ca(OH)_2 \rightarrow Ca(NO_3)_2 + H_2O$
   (ii) $NaOH + H_2SO_4 \rightarrow Na_2SO_4 + H_2O$
   (iii) $NaCl + AgNO_3 \rightarrow AgCl + NaNO_3$
   (iv) $BaCl_2 + H_2SO_4 \rightarrow BaSO_4 + HCl$

12. When hydrogen gas is passed over heated copper(II) oxide, copper and steam are formed. Write the balanced chemical equation with physical states for this reaction. State what kind of chemical reaction is this?

13. Write the skeletal equation for the following reactions.

(i) Hydrogen sulfide reacts with sulfur dioxide to form sulfur and water.
(ii) Methane on burning combines with oxygen to produce carbon dioxide and water.
What is the need of balanced equations?

14. Translate the following statement into a chemical equation and then balance it.
A metal in the form of a ribbon burns with a dazzling white flame and changes into a white powder.

15. *State what happens when zinc granules are heated with sodium hydroxide solution. Write the balanced chemical equation for the reaction. Name the main product formed in this reaction.

16. Give the reactions of calcium and magnesium with dilute nitric acid.

17. Write balanced chemical equations for the following reactions.
(i) Silver bromide on exposure to sunlight decomposes into silver and bromine.
(ii) Sodium metal reacts with water to form sodium hydroxide and hydrogen gas.

18. *In a test tube, hydrochloric acid is poured over a few zinc granules. List two observations that suggest that a chemical reaction has occurred.

$$2HCl + Zn \rightarrow ZnCl_2 + H_2$$

19. Convert the following statements into balanced chemical equations.:
(i) Zinc reacts with sulfuric acid to form zinc sulfate and hydrogen gas.
(ii) Magnesium burns in oxygen to form magnesium oxide.

20. Write the balanced chemical equations for the following reactions.
(i) Calcium hydroxide + Carbon dioxide → Calcium carbonate + Water
(ii) Zinc + Silver nitrate → Zinc nitrate + Silver
(iii) Aluminium + Copper chloride → Aluminium chloride + Copper
(iv) Barium chloride + Potassium sulfate → Barium sulfate + Potassium chloride

21. *Metal compound X reacts with dilute hydrochloric acid to produce effervescence. The gas evolved extinguishes a burning candle. Write a balanced chemical equation for the reaction if one of the compounds formed is calcium chloride.

Section D: Long Answer Questions (5 Marks)

1. (i) Mention four types of information given by an equation.
(ii) State the law of conservation of mass as applicable in a chemical reaction.

2. Write the steps for balancing the chemical equation for the formation of ammonia by the combination of nitrogen and hydrogen.

3. (i) List any three observations which determine that a chemical reaction has taken place. Also list three types of information that cannot be obtained about a chemical reaction, merely from its chemical equation.
(ii) Balance the following chemical equations.
(a) $Fe + H_2O \rightarrow Fe_3O_4 + H_2$
(b) $CO_2 + H_2O \rightarrow C_6H_{12}O_6 + O_2$

4. Discuss the importance of decomposition reactions in the metal industry with three points. **[CBSE 2015]**

5. (i) Write one example for each decomposition reaction carried out with the help of (a) electricity (b) heat (c) light

Or

Decomposition reactions require energy either in the form of heat, light or electricity for breaking down the reactants. Write one equation each for decomposition reactions where energy is supplied in the form of heat, light and electricity. **[CBSE 2018]**
(ii) Which of the following statements is correct?
(a) Copper can displace silver from silver nitrate.
(b) Silver can displace copper from copper sulfate solution.

6. (i) What happens when:
(a) dilute hydrochloric acid is added to solid sodium carbonate?
(b) quicklime is treated with water?

(c) sodium chloride solution is added to lead nitrate solution?

Also write the chemical equation in each case.

(ii) Complete and balance the following chemical equations.

(a) $CaCO_3 + HCl$

(b) $Al + HCl$

(c) $MnO_2 + HCl$

7. Write the chemical equation of the reaction with an example each in which the following change has taken place.
   (i) Change in colour
   (ii) Change in temperature
   (iii) Formation of precipitate

8. (i) Write balanced chemical equations for the following reactions.
   (a) Dilute sulfuric acid reacts with aluminium powder.
   (b) Dilute hydrochloric acid reacts with sodium carbonate.
   (c) Carbon dioxide is passed through lime water.

   (ii) Balance the following chemical equations and state whether the reactions are exothermic or endothermic.
   (a) $Na + H_2O \rightarrow NaOH + H_2$
   (b) $FeSO_4 \rightarrow Fe_2O_3 + SO_2 + SO_3$

9. Define a chemical reaction. State four observations which help us to determine that a chemical reaction has taken place. Write one example of each observation with a balanced chemical equation.

10. What is a balanced chemical equation? Why should a chemical equation be balanced?

Section E: Case Study or Passage Based Questions

I. Some processes in which no new chemical substances are formed are known as physical changes. For example, melting of ice, evaporation of water, and so on. However, if during a process the original substance loses its identity and nature and forms a new substance, it is called a chemical change and involves a chemical reaction. In a reaction, reactants change into products. In a chemical reaction, law of conservation of mass is always followed. In order to represent a chemical change, a balanced equation is given in which the number of atoms of all the elements remains same on both the sides.

1. The reaction in which a substance or substances undergo changes to produce new substances with new properties is called
   (a) a biochemical reaction
   (b) a nuclear reaction
   (c) a physical reaction
   (d) a chemical reaction

2. Which of the following conditions is necessary for a chemical reaction?
   (a) it must follow the law of conservation of mass
   (b) at least one of the reactants must be in a fixed quantity
   (c) it must be accompanied with a change in temperature and pressure
   (d) all of these

3. There is an equation E, which contains an equal number of atoms of each element on both the sides. What is E?
   (a) an unbalanced equation
   (b) a balanced equation
   (c) a chemical equation (d) all of these

4. In the given equation, what does *n* stand for?

   $$(2)Al + (n)H_2SO_4 \rightarrow Al_2(SO_4)_3 + (3)H_2$$

   (a) 2 (b) 3
   (c) 4 (d) 5

5. What happens when dilute hydrochloric acid is added to iron filings? Choose the correct answer.
   (a) hydrogen gas and iron chloride are produced
   (b) chlorine gas and iron hydroxide are produced
   (c) no reaction takes place
   (d) iron salt and water are produced

II. Marble became popular in ancient Rome and Greece, where white and off-white marble were used to construct a variety of structures, from hand-made sculptures to massive pillars and buildings. In India the world-famous Taj Mahal is also made of white marble. The white colour of the Taj Mahal is damaged due to many factors such as acid rain.

Source: Sam Hawley, CC BY 2.0 <https://creativecommons.org/licenses/by/2.0>, via Wikimedia Commons

6. The substance not likely to contain $CaCO_3$ is
(a) sea shells (b) a marble statue
(c) calcined gypsum (d) dolomite

7. A student added 10 g of calcium carbonate into a rigid container, secured it tightly and started to heat it. After some time, an increase in pressure was observed, the pressure reading was then noted at intervals of 5 minutes and plotted against time, in a graph as shown below. During which time interval did maximum decomposition take place?

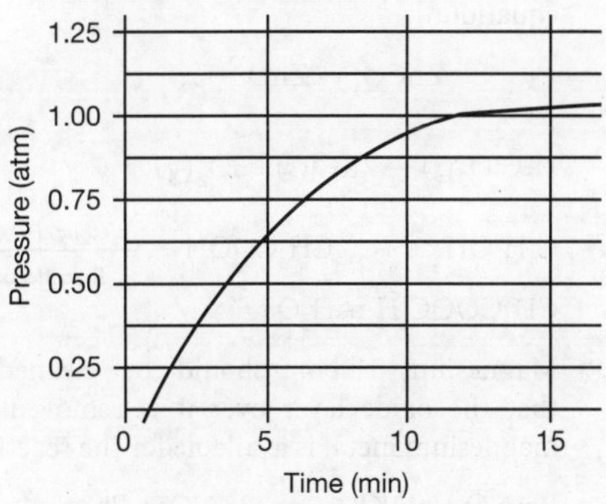

(a) 0–5 min (b) 5–10 min
(c) 10–15 min (d) 15–20 min

8. Gas X, obtained above is a reactant for a very important biochemical process which occurs in the presence of sunlight. Which of the following processes is this?
(a) photolysis (b) photosynthesis
(c) transpiration (d) respiration

9. Marble statues are corroded or stained by rain water. Identify the main reason.
(a) polluted water is acidic in nature hence it reacts with calcium carbonate
(b) polluted water is basic in nature hence it reacts with calcium carbonate
(c) decomposition of calcium carbonate to calcium oxide
(d) calcium carbonate dissolves in water to give calcium hydroxide

10. Calcium oxide can be reduced to calcium, by heating with sodium metal. Which compound would act as an oxidising agent in the above process?
(a) sodium (b) calcium
(c) sodium oxide (d) calcium oxide

High Order Thinking Skills (HOTS) Questions

1. *When a copper wire was left in silver nitrate solution for some time, it was observed that the solution turned bluish green.
(i) Explain the observation.
(ii) Write the balanced chemical equation to represent the change taking place.
[Board Term-I Set-1ZHNPNO, 2016]

2. *2 g ferrous sulfate crystals are heated in a dry boiling tube.
(i) List any two observations.
(ii) Name the type of chemical reaction taking place.
(iii) Write the chemical equation of the reaction.
[Board Term-I Set-NS9SX1D, 2016]

3. *(i) Write chemical equations.
(a) When carbon dioxide gas is passed through lime water.
(b) When excess of carbon dioxide gas is passed through lime water.
(ii) List two natural forms of calcium carbonate.

Or

(i) Solution of a substance X is used for testing carbon dioxide. Write the equation of the reaction of X with carbon dioxide.
(ii) How is X obtained? Write the chemical equation.

4. Write the chemical equations involved in the following chemical reactions.
   (i) White washing
   (ii) Black and white photography
   [Board Term-I Set-OQKPLGV, 2016]

5. Write balanced chemical equations for the following statements.
   (i) Bleaching powder is kept open in air.
   (ii) Blue crystals of copper sulfate are heated.
   (iii) Chlorine gas is passed through dry slaked lime.
   (iv) Carbon dioxide gas is passed through lime water.
   (v) NaOH solution is heated with zinc granules.

6. *What happens when zinc granules are treated with dilute solutions of H_2SO_4, HCl, HNO_3, NaCl and NaOH? Also write the chemical equation.

Answers

Section A: Multiple Choice Questions

1. (d)	**2.** (a)	**3.** (d)	**4.** (a)
5. (d)	**6.** (b)	**7.** (c)	**8.** (c)
9. (d)	**10.** (b)	**11.** (d)	**12.** (d)
13. (b)	**14.** (d)	**15.** (c)	**16.** (b)
17. (d)	**18.** (a)	**19.** (d)	**20.** (b)
21. (a)	**22.** (c)	**23.** (c)	**24.** (c)
25. (b)	**26.** (c)	**27.** (a)	**28.** (a)
29. (c)	**30.** (b)	**31.** (c)	**32.** (c)

Assertion Reason Questions

1. (a)	**2.** (d)	**3.** (c)	**4.** (a)
5. (d)	**6.** (c)	**7.** (c)	

Section B: Very Short Answer Questions

1. A process in which two or more substances react to form some other new substances with new set of properties is called a chemical reaction.

2. Law of conservation of mass. Mass can neither be created nor be destroyed during a chemical reaction.

3. In a physical change, no new substance is formed. In a chemical change, one or more new substances are formed.

4. Rusting of iron.

5. Conversion of milk into curd.

6. Both, chemical change and physical change.

7. $Mg(s) + 2HCl(aq) \rightarrow MgCl_2(aq) + H_2(g).$

8. Quicklime reacts with water vigorously to produce slaked lime and a large amount of heat.

$$\underset{\text{(Quicklime)}}{CaO(s)} + H_2O \rightarrow \underset{\text{(Slaked lime)}}{Ca(OH)_2} + \text{Heat}$$

9. Sodium being more reactive than hydrogen, displaces hydrogen from water.

$$2Na(s) + H_2O(l) \rightarrow 2NaOH(aq) + H_2(g)$$

10. ZnO(*s*) and $CO_2(g)$ are formed. Chemical equation:

$$ZnCO_3 \rightarrow ZnO + CO_2$$

11. $2KClO_3(s) \rightarrow 2KCl(s) + 3O_2(g)$

12. $C_2H_5OH + CH_3COOH \xrightarrow[\text{Esterification}]{\text{Conc. } H_2SO_4} CH_3COOC_2H_5 + H_2O$

13. Magnesium ribbon should be cleaned so that the oxide layer over it is removed and magnesium metal is available for the reaction.

14. $\underset{\text{(Lead nitrate)}}{Pb(NO_3)_2} + \underset{\text{(Potassium iodide)}}{2KI} \rightarrow \underset{\text{(Potassium nitrate)}}{2KNO_3} + \underset{\text{(Lead iodide)}}{PbI_2}$

15. (i) $2NaOH + H_2SO_4 \rightarrow Na_2SO_4 + 2H_2O$
    (ii) $Ca(OH)_2 + CO_2 \rightarrow CaCO_3 + H_2O$

16. Balanced equation:

$$2FeSO_4(s) \xrightarrow{\text{Heat}} Fe_2O_3(s) + SO_2(g) + SO_3(g)$$

17. (i) Light green coloured ferrous sulfate crystals decompose to produce a reddish-brown residue of ferric oxide.
(ii) Colourless gas with choking smell (SO_2) is produced.

18. Sodium chloride and hydrogen gas.

Case Study or Passage Based Type Question

19. See text, Section 1.2.2.

20. Here symbol Δ signifies that *heat* is needed to make this reaction possible and MnO_2 represents the catalyst which speeds up the reaction.

21. SO_3

22. $BaSO_4$

23. (i) x → (*s*); y → (*aq*), (ii) x → 2Ag

Section C: Short Answer Questions

1. See text, Section 1.2.2.

2. (i) Evolution of a gas
(ii) Change in temperature
(iii) Change in state
(iv) Change in colour

3. (i) A balanced chemical equation consists of an equal number of atoms of different elements in the reactants and products. According to the law of conservation of mass, matter can neither be created nor be destroyed in a chemical reaction. To follow this law, an equation should be balanced.
(ii) (a) $P_4(s) + 10Cl_2(g) \rightarrow 4PCl_5(s)$
(b) $CH_4(g) + 2O_2(g) \rightarrow CO_2(g) + 2H_2O(l) + \text{Heat energy}$
(c) $C_6H_{12}O_6(aq) + 6O_2(g) \rightarrow 6CO_2(aq) + 6H_2O(l) + \text{Heat energy}$

4. (i) Mass can neither be created nor destroyed in a chemical reaction: Law of conservation of mass.
(ii) $3Fe + 4H_2O \rightarrow Fe_3O_4 + 4H_2$

5. (i) Chemical reaction showing evolution of gas:

$$Zn + H_2SO_4 \rightarrow ZnSO_4 + H_2\uparrow$$

(ii) Change in colour of a substance during a chemical reaction:

$$\underset{\text{Ferrous sulfate (geen)}}{2FeSO_4(s)} \rightarrow \underset{\text{Ferric oxide (red-brown)}}{Fe_2O_3(s)} + SO_2(g) + SO_3(g)$$

6. (i) $\underset{\text{(Calcium carbonate)}}{CaCO_3(s)} \xrightarrow{\text{Heat}} \underset{\text{(Calcium oxide)}}{CaO(s)} + \underset{\text{(Carbon carbonate)}}{CO_2(g)}$
(ii) $\underset{\text{(Silver chloride)}}{2AgCl(s)} \xrightarrow{\text{Sunlight}} \underset{\text{(Silver)}}{2Ag(s)} + \underset{\text{(Chlorine)}}{Cl_2(g)}$
(iii) $\underset{\text{(Water)}}{2H_2O(l)} \xrightarrow{\text{Electric current}} \underset{\text{(Hydrogen)}}{2H_2(g)} + \underset{\text{(Oxygen)}}{O_2(g)}$

7. (i) $2N_2 + 3H_2 \rightarrow 2NH_3$
(ii) $2H_2S + 3O_2 \rightarrow 2SO_2\uparrow + 2H_2O$
(iii) $3BaCl_2 + Al_2(SO_4)_3 \rightarrow 2AlCl_3 + 3BaSO_4$
(iv) $2K + 2H_2O \rightarrow 2KOH + H_2\uparrow$

8. (i) **For short duration:** Lime water turns milky due to formation of $CaCO_3$, which is insoluble in water.

$$\underset{\text{Lime water}}{Ca(OH)_2(aq)} + CO_2(g) \rightarrow CaCO_3(s) + H_2O(l)$$

(ii) **For long duration:** A clear solution is obtained due to the formation of calcium bicarbonate, $Ca(HCO_3)_2$, which is soluble in water.

$$CaCO_3(s) + H_2O(l) + CO_2(g) \rightarrow Ca(HCO_3)_2(aq)$$

9. A black layer is deposited due to a coating of silver, because copper being more reactive than silver, displaces silver from silver nitrate solution.

$$2AgNO_3(aq) + Cu \rightarrow Cu(NO_3)_2(aq) + 2Ag(s)$$

10. The blue colour of the solution changes to light green. A reddish-brown deposit is formed on the iron nail.

$$Fe(s) + CuSO_4(aq) \rightarrow FeSO_4(aq) + Cu(s)$$

11. (i) $2HNO_3 + Ca(OH)_2 \rightarrow Ca(NO_3)_2 + 2H_2O$
(ii) $2NaOH + H_2SO_4 \rightarrow Na_2SO_4 + 2H_2O$
(iii) $NaCl + AgNO_3 \rightarrow AgCl + NaNO_3$
(iv) $BaCl_2 + H_2SO_4 \rightarrow BaSO_4 + 2HCl$

12. (i) $CuO(s) + H_2(g) \xrightarrow{Heat} Cu(s) + H_2O(g)$
(ii) Redox reaction.

13. (i) $4H_2S + 2SO_2 \rightarrow 6S + 4H_2O$
(ii) $CH_4 + 2O_2 \rightarrow CO_2 + 2H_2O$
To obey the law of conservation of mass, so that the number of atoms of each element before and after the reaction remain the same.

14. Magnesium combines with atmospheric oxygen to form magnesium oxide.

$$2Mg + O_2 \rightarrow 2MgO$$

15. When zinc granules are heated with NaOH solution, sodium zincate is formed with the evolution of hydrogen gas.

$$2NaOH(aq) + Zn(s) \xrightarrow{Heat} Na_2ZnO_2(aq) + H_2(g)$$

The main product formed in this reaction is H_2 gas.

16. $Ca + 2HNO_3 \rightarrow Ca(NO_3)_2 + H_2$
$Mg + 2HNO_3 \rightarrow Mg(NO_3)_2 + H_2$

17. (i) $2AgBr(s) \xrightarrow{Sunlight} 2Ag(s) + Br_2(g)$
(ii) $2Na(s) + 2H_2O(l) \rightarrow 2NaOH(aq) + H_2(g)$

18. (i) A colourless gas is evolved.
(ii) The test tube becomes hot.

19. (i) $Zn + H_2SO_4 \rightarrow ZnSO_4 + H_2$
(ii) $2Mg + O_2 \rightarrow 2MgO$

20. (i) $Ca(OH)_2 + CO_2 \rightarrow CaCO_3 + H_2O$
(ii) $Zn + 2AgNO_3 \rightarrow Zn\ (NO_3)_2 + 2Ag$
(iii) $2Al + 3CuCl_2 \rightarrow 2AlCl_3 + 3Cu$
(iv) $BaCl_2 + K_2SO_4 \rightarrow BaSO_4 + 2KCl$

21. The compound is calcium carbonate and the gas evolved is carbon dioxide which extinguishes a burning candle.

$$\underset{\text{Calcium carbonate}}{CaCO_3(s)} + 2HCl(dil) \rightarrow CaCl_2(aq) + H_2O(l) + CO_2(g)$$

Section D: Long Answer Questions

1. (i) (a) Physical state of reactants and products.
(b) Conditions such as temperature, pressure, heat and so on.
(c) Catalyst involved.
(d) Change in state.
(ii) Total mass of the elements present in the products in a chemical reaction has to be equal to the total mass of elements present in the reactants. Or, mass can neither be created nor destroyed in a chemical reaction.

2. $N_2 + H_2 \rightarrow NH_3$ (unbalanced equation)
(i) Examine the number of atoms of different elements present in the unbalanced equation.
(ii) In the above reaction, N_2 and H_2 of both the sides are unbalanced.
(iii) To balance hydrogen, H_2 is multiplied by 3 on the left side. It makes 6 H atoms on the left hand side.
(iv) Now to balance hydrogen atoms on the right hand side, NH_3 should be multiplied by 2. It makes 6 H atoms on this side.
(v) Now to balance nitrogen atoms, they are counted separately for both the sides and we will find that, nitrogen atoms are 2 on both the sides.
(vi) The balanced chemical equation will be $N_2 + 3H_2 \rightarrow 2NH_3$

3. (i) Three observations which determine that a chemical reaction has taken place are:
(a) Change in state and colour.
(b) Evolution of gas.
(c) Change in temperature.
Three types of information that cannot be obtained about a chemical reaction, merely from its chemical equation are:
(a) The state of matter of the reactants and products.
(b) Amount of heat evolved or absorbed in a reaction.
(c) The change of colour after the chemical reaction.
(ii) Balanced chemical equations:
(a) $3Fe + 4H_2O \rightarrow Fe_3O_4 + 4H_2$
(b) $6CO_2 + 6H_2O \rightarrow C_6H_{12}O_6 + 6O_2$

4. (i) See text, Section 1.2.2.
(ii) (a) Molten NaCl is electrolytically decomposed to form sodium metal.
(b) Aluminium metal is obtained by electric decomposition of bauxite ore mixed with cryolite.
(c) Carbonate ores are thermally decomposed to give the metal oxide which on reduction gives the metal.

5. (i) (a) Electricity (electrochemical decomposition)

$$2H_2O \xrightarrow{Electricity} 2H_2 + O_2$$

(b) Heat (thermal decomposition)

$$2FeSO_4 \xrightarrow{\text{Heat}} Fe_2O_3 + SO_2 + SO_3$$

(c) Light (photochemical decomposition)

$$2AgBr \xrightarrow{\text{Sunlight}} 2Ag + Br_2$$

(ii) The correct statement is

(a) Copper can displace silver from silver nitrate as copper is more reactive than silver.

$$Cu + 2AgNO_3(aq) \rightarrow Cu(NO_3)_2(aq) + 2Ag(s)$$

6. (i) (a) $Na_2CO_3(s) + 2HCl(aq) \rightarrow 2NaCl(aq) + H_2O(l) + CO_2(g)$

(b) $CaO(s) + H_2O(l) \rightarrow Ca(OH)_2(aq) +$ Heat

(c) $Pb(NO_3)_2(aq) + NaCl(aq) \rightarrow PbCl_2(s) + 2NaNO_3(aq)$

(ii) (a) $CaCO_3(s) + 2HCl(aq) \rightarrow CaCl_2(s) + H_2O(l) + CO_2(g)$

(b) $2Al(s) + 6HCl(aq) \rightarrow 2AlCl_3(aq) + 3H_2(g)$

(c) $MnO_2 + 4HCl(aq) \rightarrow MnCl_2 + 2H_2O(l) + Cl_2(g)$

7. (i) **Change in colour:** Reaction between lead nitrate solution and potassium iodide solution.

$$Pb(NO_3)_2(aq) + 2KI \rightarrow PbI_2(s) + 2KNO_3(aq)$$

In this reaction, the colour changes from colourless to yellow.

(ii) **Change in temperature:** Action of dilute sulfuric acid on zinc.

$$Zn(s) + H_2SO_4(aq) \rightarrow ZnSO_4(aq) + H_2$$

In this reaction, heat is evolved.

(iii) **Formation of precipitate:** Action of barium chloride on sodium sulfate.

$$BaCl_2(aq) + Na_2SO_4(aq) \rightarrow BaSO_4(s) + 2NaCl(aq) \text{ ppt}$$

8. (i) (a) $2Al + 3H_2SO_4 \rightarrow Al_2(SO_4)_3 + 3H_2$

(b) $Na_2CO_3 + 2HCl \rightarrow 2NaCl + H_2O + CO_2$

(c) $Ca(OH)_2 + CO_2 \rightarrow CaCO_3 + H_2O$

(ii) (a) $2Na + 2H_2O \rightarrow 2NaOH + H_2$

It is an exothermic reaction.

(b) $2FeSO_4(s) \rightarrow Fe_2O_3 + SO_2(g) + SO_3(g)$

It is an endothermic reaction.

9. A process in which new substances with new properties are formed by the rearrangement of atoms is known as a chemical reaction.

(i) **Evolution of gas:** The chemical reaction between zinc and dilute H_2SO_4.

$$Zn(s) + H_2SO_4(aq) \rightarrow ZnSO_4(sq) + H_2(g)\uparrow$$

(ii) **Change in colour:** The chemical reaction between potassium iodide solution and lead nitrate solution.

$$Pb(NO_3)_2(aq) + 2KI \rightarrow 2KNO_3(aq) + PbI_2(s)$$

Colourless to yellow

(iii) **Formation of precipitate:** The chemical reaction between sulfuric acid and barium chloride solution.

$$BaCl_2(aq) + H_2SO_4(aq) \rightarrow 2HCl(aq) + BaSO_4(s)$$

White precipitate

(iv) **Change in temperature:** The chemical reaction between quicklime and water.

$$CaO(s) + H_2O(l) \rightarrow Ca(OH)_2(aq) + \text{Heat}$$

10. The equation in which atoms of various elements on both sides of a chemical equation are equal in accordance with the law of conservation of mass is known as a balanced chemical equation. The chemical equation is to be balanced in accordance with the law of conservation of mass. It means the total mass of the elements present in the products of a chemical reaction has to be equal to the total mass of the elements present in the reactants. In other words, the number of atoms of each element remains the same, before and after a chemical reaction. Hence, we need to balance a chemical equation.

Section E: Case Study or Passage Based Questions

I.

1. (d) **2.** (a) **3.** (b) **4.** (b)
5. (a)

II.

6. (c) **7.** (a) **8.** (b) **9.** (b)
10. (d)

High Order Thinking Skills (HOTS) Questions

1. (i) Copper is more reactive than silver. Hence, when copper wire is dipped in silver nitrate solution, it displaces silver from $AgNO_3$ solution forming copper nitrate which is bluish green in colour.

(ii) $Cu + 2AgNO_3 \rightarrow \underset{\text{Copper (II) nitrate (bluish-green)}}{Cu(NO_3)_2} + \underset{\text{Silver}}{2Ag}$

2. (i) Before heating, it is pale green. After heating, it is brown or reddish brown.

Two observations are:

(a) Change in state and colour.

(b) Evolution of gas.

(ii) Decomposition reaction.

(iii) $2FeSO_4(s) \rightarrow Fe_2O_3(s) + SO_2(g) + SO_3(g)$

3. (i) (a) $Ca(OH)_2 + CO_2 \rightarrow CaCO_3 + H_2O$

(b) $Ca(OH)_2(aq) + CO_2(g) \rightarrow CaCO_3(s) + H2O(l)$

$CaCO_3(s) + H_2O(l) + CO_2(g) \rightarrow Ca(HCO_3)_2(aq)$

(ii) Aragonite and calcite.

4. (i) In white washing, quicklime reacts with water to form slaked lime.

$\underset{\text{Quick lime}}{CaO} + H_2O \rightarrow \underset{\text{Slaked lime}}{Ca(OH)_2} + Heat$

(ii) Silver bromide, when exposed to light decomposes to silver and bromine.

$\underset{\text{(Silver bromide)}}{2\,AgBr(s)} \xrightarrow{\text{Sunlight}} \underset{\text{(Silver)}}{2\,Ag(s)} + \underset{\text{(Bromine)}}{Br_2(g)}$

5. (i) $\underset{\text{Bleaching powder}}{CaOCl_2(s)} + \underset{\text{Carbon dioxide}}{CO_2(g)} \rightarrow \underset{\text{Calcium carbonate}}{CaCO_3(s)} + \underset{\text{Chlorine}}{Cl_2(g)}$

(ii) $CuSO_4.5H_2O \rightarrow \underset{\text{Blue}}{CuSO_4} + \underset{\text{White}}{5H_2O}$

(iii) $\underset{\text{Dry slaked lime}}{Ca(OH)_2} + \underset{\text{Chlorine}}{Cl_2} \rightarrow \underset{\text{Bleaching powder}}{CaOCl_2} + H_2O$

(iv) $\underset{\text{Lime water}}{Ca(OH)_2} + CO_2 \rightarrow \underset{\text{Calcium carbonate (Milky)}}{CaCO_3} + H_2O$

(v) $2NaOH + Zn \rightarrow \underset{\text{Sodium zincate}}{Na_2ZnO_2} + \underset{\text{Hydrogen}}{H_2}$

6. (i) **With dilute H_2SO_4:** H_2 gas is evolved.

$Zn + H_2SO_4 \rightarrow ZnSO_4 + H_2\uparrow$

(ii) **With dilute HCl:** H_2 gas is evolved.

$Zn + HCl \rightarrow ZnCl_2 + H_2\uparrow$

(iii) **With dilute HNO_3:** H_2 gas is evolved.

$Zn + 2HNO_3 \rightarrow Zn(NO_3)_2 + 5H_2O + H_2$

(iv) **With dilute NaCl:** No chemical reaction takes place.

(v) **With dilute NaOH:** Sodium zincate is formed and H_2 gas is evolved.

$Zn + 2NaOH \rightarrow Na_2ZnO_2 + H_2\uparrow$

Hints

Section A: Multiple Choice Questions

1. During combustion of liquified petroleum gas (LPG), it forms CO_2 and H_2O.

4. Chemical changes involve the formation of new compounds from one or more substances. On heating copper wire in the presence of air at high temperature, copper (II) oxide is formed.

21. In both the cases, heat energy is evolved.

22.

$\underset{\text{Lime stone}}{CaCO_3} \xrightarrow{\text{Heat}} \underset{\text{Quick lime}}{CaO} + CO_2$

$\underset{\text{Quick lime}}{CaO} + H_2O \xrightarrow{\text{Heat}} \underset{\text{Slaked lime}}{Ca(OH)_2}$

$\underset{\text{Slaked lime}}{Ca(OH)_2} + Cl_2 \xrightarrow{\text{Heat}} \underset{\text{Quick lime}}{CaOCl_2} + H_2O$

Section E: Case Study or Passage Based Questions

I.

1. A reaction in which a substance is changed to one or more new substances is called a chemical reaction.

2. A chemical reaction must follow the law of conservation of mass.

3. A balanced equation is the one in which the number of atoms of every element will be the same on both the sides of the equation.

4. Balanced equation is $(2)Al + (3)H_2SO_4 \rightarrow Al_2(SO_4)_3 + (3)H_2$

Hence, $n = 3$ here.

5. $Fe(s) + 2HCl(aq) \rightarrow FeCl_2(aq) + H_2$

II.

6. The composition of gypsum is $CaSO_4{\cdot}2H_2O$, so it does not have $CaCO_3$.

7. The maximum decomposition occurs when the pressure is maximum and we can see in the graph that from 0 to 5 minutes, the pressure increases from 0 to 0.625 atm.

8. When $CaCO_3$ is heated, the gas evolved is carbon dioxide, which is utilised in the process of photosynthesis.

$$CaCO_3 \rightarrow CaO + CO_2$$

9. The atmosphere contains many oxides which dissolve in water forming acids such as sulfuric acid (H_2SO_4) and nitric acid (HNO_3) which are common due to air pollution. Even carbon dioxide forms carbonic acid (H_2CO_3) which also damages marble objects. These acids will react with the marble and result in the formation of calcium salt, carbon dioxide and water. So, under extended periods, the wear of marble statues is expected.

10. A substance that undergoes reduction is an oxidising agent. Here, CaO is losing oxygen and undergoing reduction. So, CaO is the oxidising agent.

Oxidising agent

$$CaO \longrightarrow Ca + O_2$$

1.4 Types of Chemical Reactions

Depending upon the nature of the reaction or the type of chemical change occurring, chemical reactions are mainly of these types.

- Combination reactions
- Decomposition reactions
- Displacement reactions
- Double displacement reactions
- Oxidation and reduction reactions

1.4.1 Combination Reactions

The reactions in which two or more substances combine to form a single compound are known as *combination reactions*. In a combination reaction, two or more elements can combine to form a compound or two or more compounds can combine to form a new compound or an element and a compound can combine to form a new compound.

For example, burning of a magnesium ribbon. Magnesium and oxygen combine when heated, to form magnesium oxide.

$$\underset{\text{Magnesium}}{2Mg(s)} + \underset{\text{Oxygen}}{O_2(g)} \xrightarrow{\text{Combination}} \underset{\text{Magnesium oxide}}{2MgO(s)}$$

Combination reactions can be further studied as follows:

- **Combination reactions between two elements:** Here are a few examples.
  (i) Hydrogen burns in oxygen to form water.

$$\underset{\text{Hydrogen}}{2H_2(g)} + \underset{\text{Oxygen}}{O_2(g)} \xrightarrow{\text{Combination}} \underset{\text{Water}}{2H_2O(l)}$$

  (ii) Carbon (coal) burns in air to form carbon dioxide.

$$\underset{\text{Carbon (coal)}}{C(s)} + \underset{\text{Oxygen (From air)}}{O_2(g)} \xrightarrow{\text{Combination}} \underset{\text{Carbon dioxide}}{CO_2(g)}$$

  (iii) Hydrogen combines with chlorine to give hydrogen chloride.

$$\underset{\text{Hydrogen}}{H_2(g)} + \underset{\text{Chloride}}{Cl_2(g)} \xrightarrow{\text{Combination}} \underset{\text{Hydrogen chloride}}{2HCl(g)}$$

  (iv) Sodium metal burns in chloride to give sodium chloride.

$$\underset{\text{Sodium}}{2Na(s)} + \underset{\text{Chlorine}}{Cl_2(g)} \xrightarrow{\text{Combination}} \underset{\text{Sodium chloride}}{2NaCl(s)}$$

  (v) When iron powder is heated with sulfur, iron sulfide is formed.

$$\underset{\text{Iron}}{Fe(s)} + \underset{\text{Sulfur}}{S(s)} \xrightarrow{\text{Combination}} \underset{\text{Iron sulfide}}{FeS(s)}$$

- **Combination reactions between two molecules:** Here are a few examples.
  (i) Calcium oxide (lime or quicklime) reacts vigorously with water to form calcium hydroxide (slaked lime).

$$\underset{\text{Calcium oxide (Lime or quicklime)}}{CaO(s)} + \underset{\text{Water}}{H_2O(l)} \xrightarrow{\text{Combination}} \underset{\text{Calcium hydroxide (Slaked lime)}}{Ca(OH)_2(s)}$$

(ii) Ammonia reacts with hydrogen chloride to form ammonium chloride.

$$\underset{\text{Ammonia}}{NH_3(g)} + \underset{\text{Hydrogen chloride}}{HCl(g)} \xrightarrow{\text{Combination}} \underset{\text{Ammonium chloride}}{NH_4Cl(s)}$$

(iii) Ammonia reacts with sulfuric acid to give ammonium sulfate.

$$\underset{\text{Ammonia}}{2NH_3(g)} + \underset{\text{Sulfuric acid}}{H_2SO_4(aq)} \xrightarrow{\text{Combination}} \underset{\text{Ammonium sulfate}}{(NH_4)_2SO_4(aq)}$$

(iv) Calcium hydroxide reacts with carbon dioxide to give calcium carbonate.

$$\underset{\text{Calcium hydroxide}}{Ca(OH)_2(aq)} + \underset{\substack{\text{Carbon dioxide}\\\text{(From air)}}}{CO_2(g)} \longrightarrow \underset{\text{Calcium carbonate}}{CaCO_3(s)} + \underset{\text{Water}}{H_2O(l)}$$

- **Combination reactions between an element and a compound:** Here are a few examples.

  (i) Carbon dioxide reacts with carbon to give carbon monoxide.

$$\underset{\text{Carbon dioxide}}{CO_2(g)} + \underset{\text{Carbon}}{C(s)} \xrightarrow{\Delta} \underset{\text{Carbon monoxide}}{2CO}$$

  (ii) Nitric oxide reacts with oxygen to form nitrogen dioxide.

$$\underset{\text{Nitric oxide}}{2NO(g)} + \underset{\text{Oxygen}}{O_2(g)} \rightarrow \underset{\text{Nitrogen dioxide}}{2NO_2(g)}$$

  (iii) Sulfur dioxide reacts with oxygen to give sulfur trioxide.

$$\underset{\text{Sulphur dioxide}}{2SO_2(g)} + \underset{\text{Oxygen}}{O_2(g)} \rightarrow \underset{\text{Sulphur trioxide}}{2SO_3(g)}$$

  (iv) Carbon monoxide reacts with oxygen to give carbon dioxide.

$$\underset{\text{Carbon monoxide}}{2CO(g)} + \underset{\text{Oxygen}}{O_2(g)} \rightarrow \underset{\text{Carbon dioxide}}{2CO_2(g)}$$

TEST YOUR KNOWLEDGE

1.8 Write balanced chemical equations for the following reactions.

(i) Sodium metal reacts with water to form sodium hydroxide and hydrogen gas.

(ii) Magnesium reacts with dil. HCl.

Solution:

The reactions for these equations are

(i) $\underset{\text{Sodium}}{2Na(s)} + \underset{\text{Water}}{2H_2O(l)} \rightarrow \underset{\text{Sodium hydroxide}}{2NaOH(aq)} + \underset{\text{Hydrogen}}{H_2(g)}$

(ii) $Mg + 2HCl \rightarrow MgCl_2 + H_2$

1.9 Write chemical equations for the reactions taking place when:

(i) iron reacts with steam.

(ii) copper is heated in air.

Solution: (i) $3Fe(s) + 4H_2O(g) \rightarrow Fe_3O_4(s) + 4H_2(g)$

(ii) $2Cu + O_2 \xrightarrow{\text{Heat}} 2CuO(s)$

1.10 Write a balanced chemical equation for the reaction that take place during respiration. Identify the type of combination reaction that takes place during this process and justify the name. Give one more example of this type of reaction.

Solution:

The reaction taking place during respiration is:

$$C_6H_{12}O_6 + 6O_2 \rightarrow 6CO_2 + 6H_2O + \text{Heat}$$

Exothermic combination reaction takes place during this process due to the evolution of heat.

Combustion of methane is another example of such a reaction.

$$CH_4(g) + 2O_2(g) \rightarrow CO_2(g) + 2H_2O$$

1.4.2 Decomposition Reactions

The reactions in which a compound splits up into two or more simpler substances is called a decomposition reaction. Decomposition reactions are carried out by applying heat, light, electricity, and so on. Heat, light and electricity provide energy which breaks a compound into two or more simpler compounds. A decomposition reaction is just the opposite of a combination reaction. A decomposition reaction is always endothermic in nature as it need energy. Decomposition reactions can be further classified depending on the source of energy.

(i) Thermal Decomposition Reactions

When a decomposition reaction is carried out by heating, it is known as thermal decomposition (thermal means, relating to heat). Here are few examples of thermal decomposition reactions.

- When calcium carbonate is heated, it decomposes to give calcium oxide and carbon dioxide.

$$\underset{\substack{\text{Calcium carbonate}\\\text{(Limestone)}}}{CaCO_3(s)} \xrightarrow[\text{(Decomposition)}]{\text{Heat}} \underset{\substack{\text{Calcium oxide}\\\text{(Lime)}}}{CaO(s)} + \underset{\text{Carbon dioxide}}{CO_2(g)}$$

- When lead nitrate is heated strongly, it breaks down to form simpler substances like lead monoxide, nitrogen dioxide and oxygen. Here, lead nitrate is taken in a dry test tube and on heating, brown fumes of nitrogen dioxide are evolved and a yellow residue of lead oxide is left behind in the test tube.

$$\underset{\substack{\text{Lead nitrate}\\\text{(Colourless)}}}{2Pb(NO_3)_2(s)} \xrightarrow[\text{(Decomposition)}]{\text{Heat}} \underset{\substack{\text{Lead monoxide}\\\text{(Yellow)}}}{2PbO(s)} + \underset{\substack{\text{Nitrogen dioxide}\\\text{(Brown fumes)}}}{4NO_2(g)} + \underset{\text{Oxygen}}{O_2(g)}$$

- When hydrated ferrous sulfate (green vitriol) is heated, it first changes into anhydrous ferrous sulfate (white) with the removal of water. Anhydrous ferrous sulfate on further strong heating gives a red-brown precipitate of ferric oxide along with sulfur dioxide and sulfur trioxide.

$$\underset{\text{Ferrous sulfate crystals}}{FeSO_4.7H_2O(s)} \xrightarrow{\text{Heat}} \underset{\text{Anhydrous ferrous sulfate}}{FeSO_4(s)} + 7H_2O(l)$$

$$\underset{\substack{\text{Ferrous sulfate}\\\text{(Green)}}}{2FeSO_4(s)} \xrightarrow{\text{Heat}} \underset{\substack{\text{Ferric oxide}\\\text{(Reddish brown)}}}{Fe_2O_3(s)} + \underset{\substack{\text{Sulfur dioxide}\\\text{(Having smell of}\\\text{burning sulfur)}}}{SO_2(g)\uparrow} + \underset{\text{Sulfur trioxide}}{SO_3(g)\uparrow}$$

- When potassium chlorate is heated in the presence of manganese dioxide catalyst, it decomposes to give potassium chloride and oxygen.

$$\underset{\text{Potassium chlorate}}{2KClO_3(s)} \xrightarrow[\text{(Decomposition)}]{\text{Heat}} \underset{\text{Potassium chloride}}{2KCl(s)} + \underset{\text{Oxygen}}{3O_2(g)}$$

Uses of thermal decomposition reactions: Such reactions form one of the steps in the extraction of metals. For example, zinc carbonate (ore of zinc) is first decomposed to give zinc oxide and carbon dioxide. Zinc oxide on further reduction gives zinc metal.

(ii) Electrolytic Decomposition Reaction or Electrolysis

Electric current or electric energy is used for the decomposition reaction. This is also known as electrolysis. Here are few examples of electrolytic decomposition reactions.

- **Electrolysis of water:** When electric current is passed through acidified water, it decomposes to give hydrogen gas and oxygen gas. This reaction can be represented as,

$$\underset{\text{Water}}{2H_2O(l)} \xrightarrow[\text{(Decomposition)}]{\text{Electricity}} \underset{\text{Hydrogen}}{2H_2(g)} + \underset{\text{Oxygen}}{O_2(g)}$$

Process: In order to electrolyse water (Fig. 1.14), we have to use a plastic mug containing water and few drops of sulfuric acid. In it, two graduated test tubes having H_2 and O_2 respectively are placed. A 6 volt battery is used to connect the cathode and anode. This complete arrangement is known as a voltmeter which is used for electrolysis. On switching the battery on, it is observed that bubbles of the gases (H_2, O_2) start rising from the cathode and anode respectively and the level of water starts decreasing in both the test tubes. Keep the experiment on till sufficient amount of gases can be collected in both the test tubes. In the test tube labelled the cathode, the amount of gas collected is found to be double the amount of gas present in the test tube labelled as the anode. On testing the gases, the gas present in the cathodic tube is found to be hydrogen as it gives a popping sound when a match stick is burnt near it. In the anodic tube the gas is oxygen as it supports combustion.

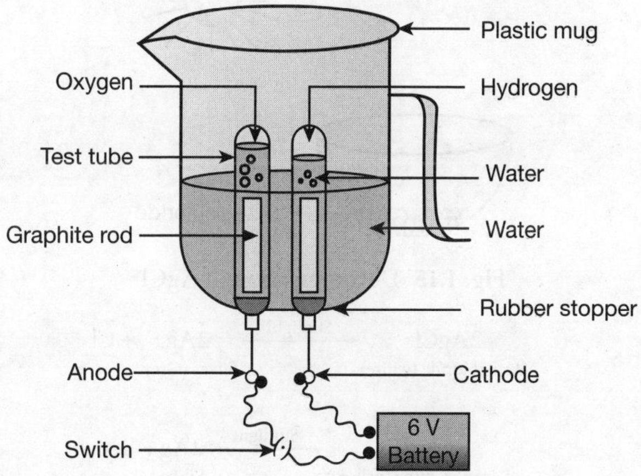

Fig. 1.14 Electrolysis of water

- When electric current is passed through molten sodium chloride, it decomposes into sodium metal and chlorine gas.

$$\underset{\text{Sodium chloride (Molten)}}{2NaCl(l)} \xrightarrow[\text{(Decomposition)}]{\text{Electricity}} \underset{\text{Sodium metal}}{2Na(s)} + \underset{\text{Chlorine gas}}{Cl_2(g)}$$

- When electric current is passed through molten aluminium oxide, it decomposes into aluminium metal and oxygen gas.

$$\underset{\text{Aluminium oxide (Molten)}}{2Al_2O_3(l)} \xrightarrow[\text{(Decomposition)}]{\text{Electricity}} \underset{\text{Aluminium metal}}{4Al(l)} + 3O_2(g)$$

Uses of electrolysis: Electrolysis is used in the extraction of metals like sodium from molten sodium chloride, aluminium from molten alumina, and so on.

(iii) Photo-decomposition Reaction

Light (light energy) is used for the decomposition reaction. This is also known as *photolysis*. Here are a few examples of photo-decomposition reactions.

- **Decomposition of silver chloride and silver bromide**: When a pinch of silver chloride or silver bromide is taken on a watch glass and kept in the presence of light for some time, photo-decomposition of white silver chloride or yellow silver bromide occurs to give a grey precipitate of silver along with the removal of chlorine or bromine gas respectively (Fig. 1.15).

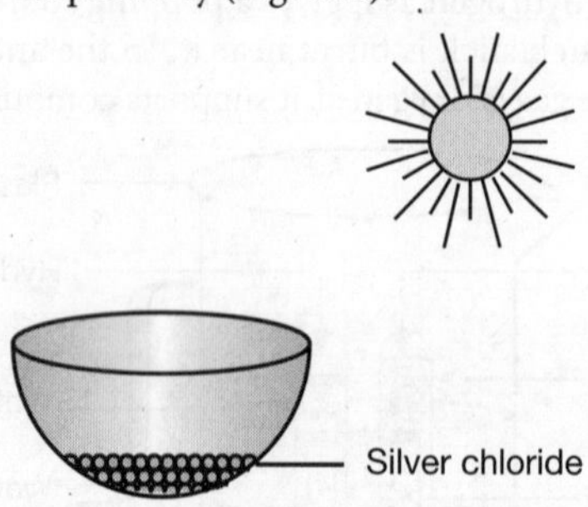

Fig. 1.15 Decomposition of AgCl

$$\underset{\text{Silver chloride (white)}}{2AgCl} \xrightarrow{\text{Sunlight}} \underset{\text{Silver (Grey)}}{2Ag} + Cl_2$$

$$\underset{\text{Silver bromide (yellow)}}{2AgBr} \xrightarrow{\text{Sunlight}} \underset{\text{Silver}}{2Ag} + Br_2$$

- **Photolytic decomposition of hydrogen peroxide**: In the presence of light, hydrogen peroxide decomposes into water and oxygen.

$$\underset{\text{Hydrogen peroxide}}{2H_2O_2} \xrightarrow{\text{Light}} \underset{\text{Water}}{2H_2O} + \underset{\text{Oxygen}}{O_2} \uparrow$$

Therefore, hydrogen peroxide is kept in coloured bottles in order to cut off light and to avoid its decomposition.

- **Photolysis of hydrogen iodide**: Hydrogen iodide decomposes in the presence ultraviolet light into hydrogen and iodine.

$$\underset{\text{Hydrogen iodide}}{2HI} \xrightarrow{\text{UV light}} \underset{\text{Hydrogen}}{H_2} + \underset{\text{Iodine}}{I_2}$$

Uses of photo-decomposition: Photo decomposition reaction is used in black and white photography. The photographic plate is made of silver bromide or coated with silver bromide.

Competition Edge

Decomposition reaction in our body: The digestion of food in our body can take place only by decomposition reactions. The food that we ingest mainly contains starch (wheat, rice, potatoes), proteins (pulses, paneer) and fat. In the presence of biological catalysts (enzymes) in our body, starch decomposes into the simple carbohydrate glucose and proteins decompose into simple amino acid molecules.

TEST YOUR KNOWLEDGE

1.11 Why are decomposition reactions called the opposite of combination reactions? Write equations for these reactions.

Solution:

A decomposition reaction is one in which a compound is broken down into simpler elements or compounds. For example,

$$\underset{\text{Calcium carbonate}}{CaCO_3(s)} \xrightarrow{\Delta} CaO(s) + CO_2(g)$$

It is just the opposite of a combination reaction in which two or more reactants combine to form a single product.

1.12 Why do we store silver chloride in dark-coloured bottles?

Solution:

We store silver chloride in dark-coloured bottles to avoid its decomposition in the sunlight.

1.13 Write balanced chemical equations for the following reactions.

(i) Silver bromide on exposure to sunlight decomposes into silver and bromine.

(ii) What is the colour of ferrous sulfate crystals? How does this colour change after heating?

Solution:

The reactions for these equations are

(i) $2AgBr(s) \xrightarrow{\text{Sunlight}} 2Ag(s) + Br_2(g)$

Silver bromide → Silver + Bromine

(ii) The colour of ferrous sulfate crystals is pale green which changes to reddish-brown on heating due to the formation of iron(III) oxide.

$$2FeSO_4(s) \xrightarrow{\text{Heat}} Fe_2O_3(s) + SO_2(g) + SO_3(g)$$

Iron sulfate (Pale green) → Iron (III) oxide (Reddish-brown)

1.4.3 Displacement Reactions

A reaction in which one element takes the place of another element in a compound is called a *displacement reaction*. In general, a more reactive element displaces a less reactive element from its compound. It is decided on the basis of the reactivity series of metals and reactivity order of non-metals. Displacement reactions are exothermic by nature.

Reactivity or Activity Series

	Element	Symbol	
Metals more reactive than hydrogen	Potassium	K	← Most reactive
	Sodium	Na	Reactivity decreases downwards
	Barium	Ba	
	Calcium	Ca	
	Magnesium	Mg	
	Aluminium	Al	
	Zinc	Zn	
	Iron	Fe	
	Nickel	Ni	
	Tin	Sn	
	Lead	Pb	
	Hydrogen	H	
Metals less reactive than hydrogen	Copper	Cu	
	Mercury	Hg	
	Silver	Ag	
	Gold	Au	
	Platinum	Pt	← Least reactive

It is the arrangement of metals in a vertical column in the decreasing order of their activity or reactivity. It is based upon the study of the reactions of the metal with different reagents. From the series it is clear that potassium is the most reactive metal while platinum is the least. It is important to note that a higher placed element in the series can easily displace a lower placed element from its compound while the reverse is not true. For example, zinc can displace copper from copper sulfate due to more reactivity but silver cannot displace copper due to its less reactivity than copper. All elements above hydrogen can displace hydrogen from hydrogen compounds like HCl, while elements placed below hydrogen cannot do that. Here are a few examples of displacement reactions.

- When a strip of zinc metal is placed in copper sulfate solution, then zinc sulfate solution and copper are obtained (Fig. 1.16).

$$CuSO_4(aq) + Zn(s) \rightarrow ZnSO_4(aq) + Cu(s)$$

Copper sulfate (Blue solution) + Zinc (Silvery-white) → Zinc sulfate (Colourless solution) + Copper (Red-brown)

Fig. 1.16 Zinc strip in copper sulfate solution

The displacement reaction takes place because zinc is more reactive than copper. The blue colour of the copper sulfate solution disappears or fades due to the formation of white zinc sulfate solution. The reddish-brown copper particles get deposited on the surface of the zinc or also fall at the bottom of the reaction vessel (Fig. 1.16).

- When a piece of magnesium metal is placed in copper sulfate solution, magnesium sulfate solution and copper metal are formed.

$$CuSO_4(aq) + Mg(s) \rightarrow MgSO_4(aq) + Cu(s)$$

Copper sulfate (Blue solution) + Magnesium (Silvery white) → Magnesium sulfate (Colourless solution) + Copper (Red-brown)

Displacement reaction occurs because magnesium is more reactive than copper. The blue colour of copper sulfate solution disappears or fades due to the formation of colourless magnesium sulfate solution (Fig. 1.17).

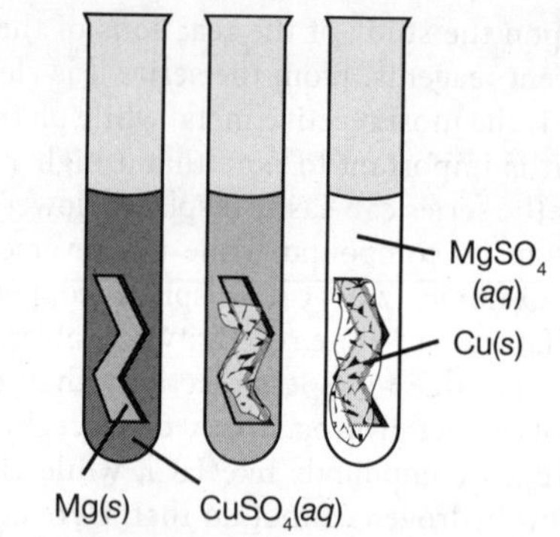

Fig. 1.17 Magnesium strip in copper sulfate solution

- When a piece of iron metal like an iron nail is placed in copper sulfate, iron sulfate solution and copper metal are formed (Fig. 1.18).

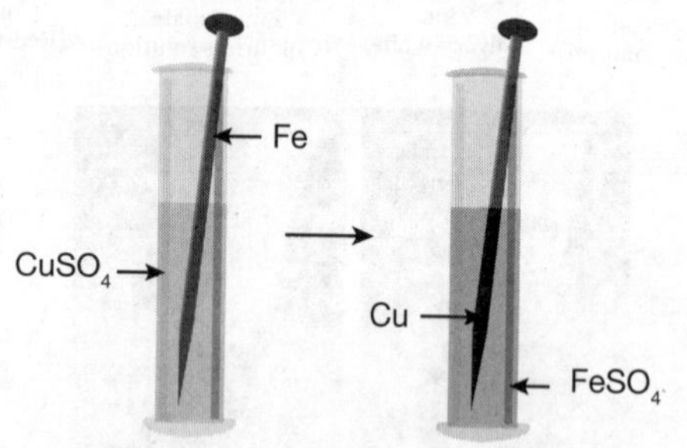

Fig. 1.18 Iron strip in copper sulfate solution

$$\underset{\text{Copper sulfate (Blue solution)}}{CuSO_4(aq)} + \underset{\text{Iron (Grey)}}{Fe(s)} \rightarrow \underset{\text{Iron sulfate (Greenish solution)}}{FeSO_4(aq)} + \underset{\text{Copper (Red-brown)}}{Cu(s)}$$

Displacement reaction takes place because iron is more reactive than copper. The blue colour of copper sulfate solution disappears or fades due to the formation of light green iron sulfate solution. A red-brown layer of copper metal is deposited on the surface of iron metal (Fig. 1.18).

- When a strip of lead metal is placed in a solution of copper chloride, lead chloride solution and copper metal are formed.

$$\underset{\text{Copper chloride (Green solution)}}{CuCl_2(aq)} + \underset{\text{Lead (Bluish grey)}}{Pb(s)} \rightarrow \underset{\text{Lead chloride (Coloureless solution)}}{PbCl_2(aq)} + \underset{\text{Copper (Red-brown)}}{Cu(s)}$$

Displacement reaction takes place because lead is more reactive than copper. The blue colour of copper sulfate solution disappears or fades due to the formation of white or colourless lead chloride sulfate solution.

- When a copper strip is placed in a solution of silver nitrate, copper nitrate solution and silver metal are formed.

$$\underset{\text{Silver nitrate (Colourless solution)}}{2AgNO_3(aq)} + \underset{\text{Copper (Red-brown)}}{Cu(s)} \rightarrow \underset{\text{Copper nitrate (Blue solution)}}{Cu(NO_3)_2(aq)} + \underset{\text{Silver (Greyish white)}}{2Ag(s)}$$

This displacement reaction occurs because copper is more reactive than silver. A shining greyish-white deposit of silver is formed on the copper strip and the solution becomes blue due to the formation of copper nitrate.

*It is important to note that if the silver plate or strip is dipped in copper nitrate solution, no reaction is possible as the reactivity of silver is less than that of copper. So, silver cannot displace copper.

- When iron (III) oxide is heated strongly with aluminium powder, then aluminium oxide and iron metal are formed.

$$\underset{\text{Iron (III) oxide (Ferric oxide)}}{Fe_2O_3(s)} + \underset{\text{Aluminium}}{2Al(s)} \rightarrow \underset{\text{Aluminium oxide}}{Al_2O_3(s)} + \underset{\text{Iron (molten)}}{2Fe(l)}$$

Displacement reaction occurs because aluminium is more reactive than iron. Here, the temperature is so high that the obtained iron is in the molten state. This reaction is used in thermite welding or aluminothermy in which broken parts of iron objects are welded together.

Reactivity order of non-metals: A more reactive non-metal can displace a less reactive non-metal from its compounds. In the case of halogens the reactivity order in decreasing order is $F_2 > Cl_2 > Br_2 > I_2$. Here are a few examples.

- Chlorine gas reacts with potassium iodide solution to give potassium chloride and iodine.

$$\underset{\text{Chlorine}}{Cl_2(g)} + \underset{\text{Potassium iodide}}{2KI(aq)} \rightarrow \underset{\text{Potassium chloride}}{2KCl(aq)} + \underset{\text{Iodine}}{I_2(g)}$$

This reaction occurs as chlorine is more reactive than iodine. Iodine dissolves in solution giving a violet colour.

- Bromine reacts with potassium iodide solution to give potassium bromide and iodine.

$$\underset{\text{Bromine}}{Br_2(g)} + \underset{\text{Potassium iodide}}{2KI(aq)} \rightarrow \underset{\text{Potassium bromide}}{2KBr(aq)} + \underset{\text{Iodine}}{I_2(g)}$$

This reaction occurs as bromine is more reactive than iodine. Iodine dissolves in solution giving a violet colour.

*It is important to note that bromine or iodine cannot displace chlorine from potassium chloride as their reactivity is less than that of chlorine.

*All examples given here represent single displacement reactions as only one element is displaced by another element.

TEST YOUR KNOWLEDGE

1.14 Why does the colour of copper sulfate solution change when an iron nail is dipped in it? Write two observations.

Solution:

The blue colour of copper sulfate solution changes due to the occurrence of a displacement reaction taking place. Iron displaces copper from copper sulfate solution to give following observations.

(i) Pale green coloured solution of $FeSO_4$ is formed.

(ii) Reddish brown copper metal gets deposited.

$$Fe(s) + CuSO_4(aq) \rightarrow FeSO_4(aq) + Cu(s)$$

1.4.4 Double Displacement Reactions

A reaction in which two compounds react by an exchange of ions to form two new compounds is known as a *double displacement reaction.* Such a reaction usually occurs in solution and one of the products, being insoluble, precipitates out or separates as a solid. Some examples of double displacement reactions are given below.

- When silver nitrate solution is added to sodium chloride, a white precipitate of silver chloride is formed along with sodium nitrate solution (Fig. 1.19).

Fig. 1.19 Formation of AgCl precipitate

$$\underset{\text{Silver nitrate}}{AgNO_3(aq)} + \underset{\text{Sodium chloride}}{NaCl(aq)} \rightarrow \underset{\substack{\text{Silver chloride}\\\text{(White ppt.)}}}{AgCl(s)} + \underset{\text{Sodium nitrate}}{NaNO_3(aq)}$$

In this reaction, an exchange of ions occurs as the Ag^+ ions of silver nitrate react with the Cl^- (chloride) ions of sodium chloride and the Na^+ ions react with the NO_3^- ions.

- When barium chloride solution is added to sodium sulfate solution, a white precipitate of barium sulfate is formed along with sodium chloride solution (Fig. 1.20).

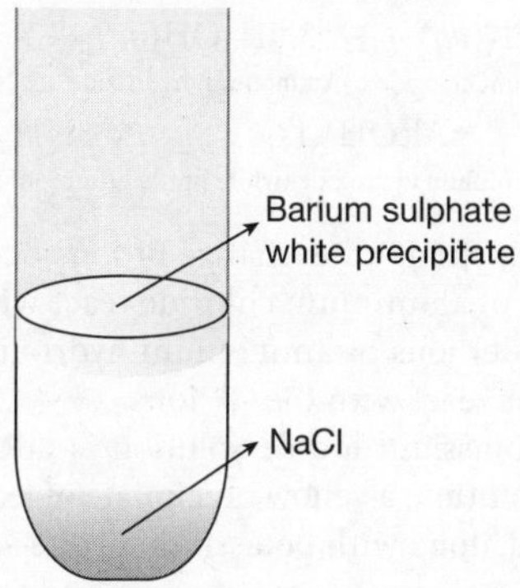

Fig. 1.20 Formation of $BaSO_4$ precipitate

$$\underset{\text{Barium chloride}}{BaCl_2(aq)} + \underset{\text{Sodium sulfate}}{Na_2SO_4(aq)} \rightarrow \underset{\text{Barium sulfate (white ppt.)}}{BaSO_4(s)} + \underset{\text{Sodium chloride}}{2NaCl(aq)}$$

In this reaction, an exchange of ions occurs as the Ba^{2+} ions of barium chloride react with the SO_4^{2-} (sulfate) ions of sodium sulfate and the Na^+ ions react with the Cl^- ions.

- If barium chloride solution is added to copper sulfate solution, a white precipitate of barium sulfate is produced along with copper chloride solution.

$$\underset{\text{Barium chloride}}{BaCl_2(aq)} + \underset{\text{Copper sulfate}}{CuSO_4(aq)} \rightarrow \underset{\text{Barium sulfate (White ppt.)}}{BaSO_4(s)} + \underset{\text{Copper chloride}}{CuCl_2(aq)}$$

In this reaction, an exchange of ions occurs as the Ba^{2+} ions of barium chloride react with the SO_4^{2-} (sulfate) ions of copper sulfate and the Cu^{2+} ions react with the Cl^- ions.

- When hydrogen sulfide gas is passed through copper sulfate solution, a black precipitate of copper sulfide is formed along with sulfuric acid solution.

$$\underset{\text{Copper sulfate}}{CuSO_4(aq)} + \underset{\text{Hydrogen sulfide}}{H_2S(g)} \rightarrow \underset{\text{Copper sulfide (Black ppt.)}}{CuS(s)} + \underset{\text{Sulfuric acid}}{H_2SO_4(aq)}$$

In this reaction, an exchange of ions occurs as the Cu^{2+} ions of copper sulfate react with the S^{2-} (sulfide) ions of hydrogen sulfide and the H^+ ions react with the SO_4^{2-} ions.

- When ammonium hydroxide solution is added to aluminium chloride solution, a white precipitate of aluminium hydroxide is formed along with ammonium chloride solution.

$$\underset{\text{Aluminium chloride}}{AlCl_3(aq)} + \underset{\text{Ammonium hydroxide}}{3NH_4OH(aq)} \rightarrow \underset{\text{Aluminium hydroxide (white ppt.)}}{Al(OH)_3(s)} + \underset{\text{Ammonium chloride}}{3NH_4Cl(aq)}$$

In this reaction, an exchange of ions occurs as the Al^{3+} ions of aluminium chloride react with the OH^- (hydroxide) ions of ammonium hydroxide and the NH_4^+ ions react with the Cl^- ions.

- When potassium iodide solution is added to lead nitrate solution, a yellow precipitate of lead iodide is produced along with potassium nitrate solution.

$$\underset{\text{Lead nitrate}}{Pb(NO_3)_2(aq)} + \underset{\text{Potassium iodide}}{2KI(aq)} \rightarrow \underset{\text{Lead iodide (Yellow ppt.)}}{PbI_2(s)} + \underset{\text{Potassium nitrate}}{2KNO_3(aq)}$$

In this reaction, an exchange of ions occurs as the Pb^{2+} ions of lead nitrate react with the I^- (iodide) ions of potassium iodide and the K^+ ions react with the NO_3^- ions.

1.4.5 Types of Double Displacement Reactions

Double displacement reactions can be further classified as precipitation, gas formation and acid–base neutralisation reactions.

(i) **Precipitation reaction:** A precipitation reaction is possible when two solutions are mixed together and a solid separates out from the solution. The solid part that forms and separates from the solutions is called the precipitate. Some examples of precipitation reactions are given.

- Lead nitrate solution on reaction with potassium iodide solution gives a yellow precipitate of lead iodide along with potassium nitrate solution (Fig. 1.21).

$$Pb(NO_3)_2(aq) + 2KI(aq) \rightarrow PbI_2(s)\downarrow + 2KNO_3(aq)$$

- Sodium carbonate on reaction with hydrochloric acid gives an aqueous solution of sodium chloride along with water and carbon dioxide.

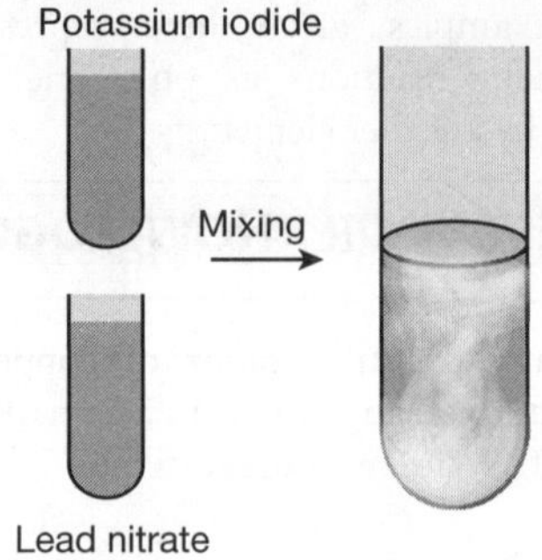

Fig. 1.21 Formation of PbI_2 precipitate

$$\underset{\text{Sodium carbonate}}{Na_2CO_3(s)} + \underset{\text{Hydrochloric acid}}{2HCl(aq)} \rightarrow \underset{\text{Sodium chloride}}{2NaCl(aq)} + \underset{\text{Water}}{H_2O(l)} + \underset{\text{Carbon dioxide}}{CO_2(g)}$$

(ii) **Neutralisation reaction:** Neutralisation is a special kind of double displacement reaction. In an acid–base reaction, an acid reacts with an equal quantity of a base. The acid–base reaction results in the formation of salt and water (Fig. 1.22). It is always an exothermic reaction, that is, heat is evolved during neutralisation.

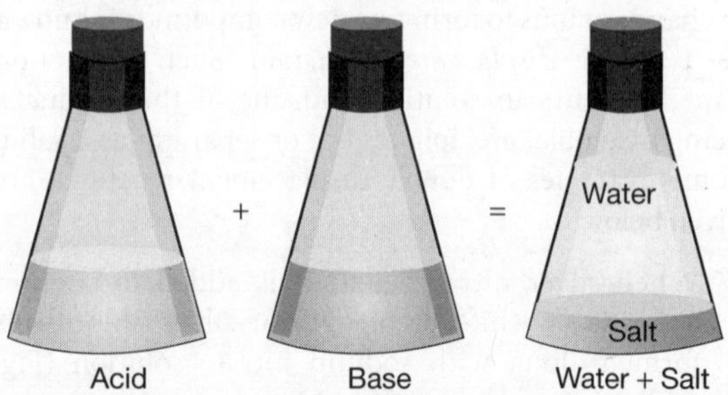

Fig. 1.22 Neutralisation reaction

Acid + Base → Salt + water + heat

Some examples of neutralisation reactions are given below.

- $$\underset{\text{Hydrochloric acid}}{HCl(aq)} + \underset{\text{Sodium hydroxide}}{NaOH(aq)} \rightarrow \underset{\text{Sodium chloride}}{NaCl(aq)} + \underset{\text{Water}}{H_2O(l)}$$

- $$\underset{\text{Sulfuric acid}}{H_2SO_4} + \underset{\text{Sodium hydroxide}}{2NaOH} \rightarrow \underset{\text{Sodium sulfate}}{Na_2SO_4} + \underset{\text{Water}}{2H_2O}$$

- $$\underset{\text{Acetic acid (Acid)}}{CH_3COOH} + \underset{\text{Sodium hydroxide (base)}}{NaOH} \rightarrow \underset{\text{Sodium acetate}}{CH_3COONa} + \underset{\text{Water}}{H_2O}$$

- $\underset{\text{oxalic acid}}{(COOH)_2} + \underset{\text{Sodium hydroxide}}{2NaOH} \rightarrow \underset{\text{Sodium oxalate}}{(COONa)_2} + \underset{\text{Water}}{2H_2O}$

- $\underset{\text{Hydrochloric acid (acid)}}{HCl} + \underset{\text{Ammonium hydroxide (base)}}{NH_4OH} \rightarrow \underset{\text{Ammonium chloride}}{NH_4Cl} + \underset{\text{Water}}{H_2O}$

(iii) **Gas formation reaction:** A double displacement reaction is possible when an insoluble gas is formed. All gases are soluble in water to some extent, but only a few gases [$HCl(g)$ and $NH_3(g)$] are highly soluble. For example, many sulfide salts react with acids to form gaseous hydrogen sulfide.

$$ZnS(s) + 2HCl(aq) \rightarrow ZnCl_2(aq) + H_2S(g)$$

TEST YOUR KNOWLEDGE

1.15 Translate the following statement into a chemical equation and then balance it. Barium chloride reacts with aluminium sulfate to give aluminium chloride and a precipitate of barium sulfate. State the two types in which this reaction can be classified.

Solution:

Chemical equation:

$$\underset{\text{Barium chloride}}{3BaCl_2(aq)} + \underset{\text{Aluminium sulfate}}{Al_2(SO_4)_3(aq)} \rightarrow \underset{\text{Barium sulfate}}{3BaSO_4(s)} + \underset{\text{Aluminium chloride}}{2AlCl_3(aq)}$$

This reaction can be classified as a double displacement reaction and a precipitation reaction.

1.16 What is observed when a solution of potassium iodide solution is added to a solution of lead nitrate? Name the type of reaction. Write a balanced chemical equation to represent the above chemical reaction.

Solution:

A yellow precipitate of lead iodide is formed. It is a double displacement reaction and is also a precipitation reaction.

$$Pb(NO_3)_2(aq) + 2KI(aq) \rightarrow PbI_2(s) + 2KNO_3(aq)$$

1.17 When a solution of potassium chloride is mixed with silver nitrate solution, an insoluble white substance is formed. Write the chemical reaction involved and also mention the type of the chemical reaction.

Solution:

$$KCl(aq) + AgNO_3(aq) \rightarrow \underset{\text{Silver chloride (white ppt.)}}{AgCl(s)} + \underset{\text{Potassium nitrate}}{KNO_3(aq)}$$

The reaction is a double displacement reaction. It is also called a precipitation reaction due to the formation of a white precipitate of AgCl.

1.18 Write balanced chemical equations from the following word equations. Also include the physical state of each element or compound involved in the reaction.

(i) An aqueous sodium chloride solution reacts with an aqueous silver nitrate solution to give a silver chloride precipitate (solid) and a sodium nitrate solution.

(ii) An aqueous phosphoric acid solution reacts with an aqueous calcium hydroxide solution to give water and solid calcium phosphate.

Solution:

(i) $NaCl(aq) + AgNO_3(aq) \rightarrow AgCl(s) + NaNO_3(aq)$

(ii) $3H_3PO_4(aq) + 3Ca(OH)_2(aq) \rightarrow Ca_3(PO_4)_2(s) + 6H_2O(l)$

1.4.6 Oxidation and Reduction Reactions

The earlier concept of oxidation and reduction is based on the addition or removal of oxygen or hydrogen. So, in terms of oxygen or hydrogen, oxidation and reduction reactions can be defined as follows.

Oxidation

Combination of oxygen: The addition of oxygen to a substance is called oxidation. For example,

$$2Mg + O_2 \rightarrow 2MgO$$

Removal of hydrogen: The removal of hydrogen from a substance is also called oxidation. For example,

$$4HCl + MnO_2 \rightarrow MnCl_2 + Cl_2 + 2H_2O$$

Oxidation also involves the following processes.

- Addition of a electronegative element or an increase in its ratio. For example,

$$2FeCl_2 + Cl_2 \rightarrow 2FeCl_3$$

- Removal of electropositive element or decrease in its ratio. For example,

$$Hg_2Cl_2 \rightarrow HgCl_2 + Hg$$

- Loss of electrons by an atom or ion or molecule, that is, de-electronation. For example,

$$Na - 1e^- \rightarrow Na^+$$

$$Fe^{2+} - 1e^- \rightarrow Fe^{3+}$$

*Rusting of iron and combustion of fuels are examples of oxidation.

Reduction

Removal of oxygen: The removal of oxygen from a substance is called reduction. For example,

$$ZnO + C \rightarrow Zn + CO$$

Addition of hydrogen: The addition of hydrogen to a substance is called reduction. For example,

$$H_2 + Cl_2 \rightarrow 2HCl$$

Reduction also involves the following processes.

- Removal of an electronegative element or decrease in its ratio. For example,

$$2FeCl_3 \rightarrow 2FeCl_2 + Cl_2$$

- Addition of an electropositive element or increase in its ratio. For example,

$$2HgCl_2 \rightarrow Hg_2Cl_2 + Cl_2$$

- Gain of electrons by an atom or ion, that is, electronation. For example,

$$Cu^{2+} + e^- \rightarrow Cu^+$$

Oxidising agent: It oxidises others but itself gets reduced. An oxidant can be the substance which gives oxygen for oxidation or the substance which removes hydrogen. For example,

$$PbS + \underset{\text{Oxidising agent}}{4O_3} \rightarrow PbSO_4 + 4O_2$$

Some common oxidants

- Molecules of most electronegative elements such as O_3, O_2, F_2, Cl_2 and so on.
- Some compounds (oxidants) such as $KMnO_4$, $K_2Cr_2O_7$, H_2SO_4, HNO_3, SO_3, CO_2, $FeCl_3$ and so on.

Reducing agent: It reduces others but itself gets oxidised. A reductant can be the substance which gives hydrogen for reduction or the substance which removes oxygen.

Some common reductants

- Metals of group 1 (Li, Na), group 2 (Mg, Ca).
- Non-metals such as C, S, P, H and so on.
- HX such as HI, HBr, HCl.
- Metal hydrides (MH) such as LiH and CaH_2.
- Molecules or compounds as reductants such as Hg_2Cl_2, H_2S, Cu_2O, $SnCl_2$, $FeCl_2$, and some organic acids like HCOOH (formic acid).

Redox

The oxidation and reduction reactions are also called *redox reactions* (in the name redox, the term 'red' stands for reduction and 'ox' stands for oxidation). Here, oxidation and reduction take place simultaneously, that is, one species gets oxidised while another one gets reduced at the same time. In general, a redox reaction is given as $A_1 + A_2^{n+} \rightarrow A_1^{2+} + A_2$.

Here A_1 is oxidised into A_1^{n+} while A_2^{n+} is reduced into A_2. Thus, A_1 is the reducing agent as it gets oxidised while A_2 is the oxidising agent as gets reduced.

Let us consider a few examples of redox reactions in which we will decide which substance undergoes oxidation and which undergoes reduction and we shall also find which one is the oxidant and which one is the reductant.

- When ferric chloride ($FeCl_3$) is treated with stannous chloride ($SnCl_2$), ferrous chloride ($FeCl_2$) and stannic chloride ($SnCl_4$) are formed.

$$2FeCl_3 + SnCl_2 \rightarrow 2FeCl_2 + SnCl_4$$

$SnCl_2$ is oxidised into $SnCl_4$ while $FeCl_3$ is reduced into $FeCl_2$. Here, $FeCl_3$ is the oxidant as it undergoes reduction while $SnCl_2$ is the reductant as it undergoes oxidation.

- When copper oxide is heated with hydrogen, copper metal and water are formed.

$$\underset{\text{Copper oxide}}{CuO} + \underset{\text{Hydrogen}}{H_2} \xrightarrow{\text{Heat}} \underset{\text{Copper}}{Cu} + \underset{\text{Water}}{H_2O}$$

Copper oxide is reduced into copper while hydrogen is oxidised into water. Here the oxidising agent is copper oxide as gets reduced and the reducing agent is hydrogen as it gets oxidised.

- When hydrogen sulfide reacts with chlorine, sulfur and hydrogen chloride are formed.

$$\underset{\text{Hydrogen sulphide}}{H_2S} + \underset{\text{Chlorine}}{Cl_2} \rightarrow \underset{\text{Sulfur}}{S} + \underset{\text{Hydrogen chloride}}{2HCl}$$

Hydrogen sulfide gets oxidised to sulfur while chlorine gets reduced to hydrogen chloride. The oxidation–reduction reaction between hydrogen sulfide and chlorine can be written more clearly as follows:

Removal of hydrogen: Oxidation

$$H_2S + Cl_2 \rightarrow S + 2HCl$$

Addition of hydrogen: Reduction

Here, chlorine is the oxidising agent while hydrogen sulfide is the reducing agent.

- When manganese dioxide reacts with hydrochloric acid, manganese dichloride, chlorine and water are formed.

$$\underset{\text{Manganese dioxide}}{MnO_2} + \underset{\text{Hydrochloric acid}}{4HCl} \rightarrow \underset{\text{Manganese dichloride}}{MnCl_2} + \underset{\text{Chlorine}}{Cl_2} + 2H_2O$$

MnO_2 loses oxygen to form $MnCl_2$, so manganese dioxide (MnO_2) gets reduced to manganese dichloride ($MnCl_2$). On the other hand, HCl loses hydrogen to form Cl_2, so hydrochloric acid (HCl) is being oxidised to chlorine (Cl_2). In this reaction, manganese dioxide (MnO_2) is the oxidising agent whereas hydrochloric acid (HCl) is the reducing agent.

TEST YOUR KNOWLEDGE

1.19 What is a redox reaction? Identify the substance oxidised and the substance reduced in the following reactions.

(i) $2PbO + C \rightarrow 2Pb + CO_2$

(ii) $2MnO_2 + 4HCl \rightarrow MnCl_2 + 2H_2O + Cl_2$

Solution:

The reaction in which oxidation and reduction take place simultaneously is called a redox reaction.

(i) Substance oxidised: C; Substance reduced: PbO

(ii) Substance oxidised: HCl; Substance reduced: MnO_2

Competition Edge

Oxidation number: It is the residual charge (±) present on an atom in a given molecule or ion when the shared electrons are counted towards more electronegative atoms.

Rules to find oxidation number

(i) The oxidation number of all neutral atoms, molecules, compounds, complexes, polymeric forms is zero. For example, Na, O_2, P_4, O_3, $KMnO_4$.

(ii) The oxidation number of any ion = charge on ion. For example, $CN^{-1} = -1$, $NH_4^+ = +1$, $SO_4^{-2} = -2$ and $PO_4^{3-} = -3$.

(iii) H has a +1 oxidation state with non-metals but with metals in metal hydrides (MH) the oxidation state is –1. For example, H_2O = +1, LiH = –1 and CaH_2 = +1.

(iv) The oxidation number of the alkali metals (group 1) is always +1 and the oxidation number of the alkaline earth metals (group 2) is always +2 in their compounds. For example,

$$\underset{+1}{\mathbf{Na}}Cl,\ \underset{+1}{\mathbf{Na}_2}O,\ \underset{+2}{\mathbf{Mg}}O,\ \underset{+2}{\mathbf{Ca}}(OH)_2$$

(v) The oxidation number of oxygen is –2 with metals and non-metals. For example,

$$C\underset{-2}{\mathbf{O}},\ Mg\underset{-2}{\mathbf{O}},$$

(vi) Oxidation number of halogens: The common oxidation number of halogens is –1. However, chlorine, bromine and iodine can show +1 to +7 oxidation states as well.

Fluorine always exhibits the –1 oxidation state in its compounds. For example, HF, OF_2.

Chlorine has +1 to +7 oxidation states with fluorine or oxygen but most commonly it is –1 with metals and non-metals. For example, HCl, NaCl, $MgCl_2$ (–1 state),

$$\underset{+2}{ClO} \rightarrow \underset{+7}{Cl_2O_7}$$

1.5 Effects of Oxidation Reactions in Everyday Life

Oxidation has a damaging effect on metals as well as on food. There are two common effects of oxidation reactions which we observe in daily life. The damaging effect of oxidation on metals is *corrosion* and that on food is *rancidity*.

1.5.1 Corrosion

Corrosion is the phenomenon in which metals are eaten up gradually by the action of air, moisture or a chemical like an acid on their surface. Corrosion is caused mainly by the oxidation of metals by the oxygen of air. Actually, corrosion is just the opposite to extraction of metals. In corrosion, a metal is changed into its compound. For example, rusting of iron is the most common form of corrosion. When an iron object is left in damp air for a considerable time, it gets covered with a red-brown flaky substance known as 'rust' (Fig. 1.23) and this phenomenon is known as rusting of iron; it is a redox reaction.

Fig. 1.23 Rusting of iron

Source: Anna Frodesiak, CC0, via Wikimedia Commons

During the corrosion of iron, iron metal is oxidised by the oxygen of air in the presence of water or moisture to form hydrated iron(III) oxide (rust).

$$\underset{\text{Iron}}{4Fe} + \underset{\text{Oxygen}}{3O_2} + \underset{\text{Water}}{2xH_2O} \rightarrow \underset{\text{Hydrated iron (III) oxide (Rust)}}{2Fe_2O_3xH_2O}$$

Drawback of rusting: Rust is a soft and a porous material which gradually falls of from the surface of an iron object. This occurs continuously as a result of which the whole object is damaged or destroyed. Corrosion weakens the iron and steel objects and structures such as railings, car bodies, bridges and ships, and cuts short their lives.

Prevention: Rusting can be prevented by (i) painting iron objects with antirust solutions or paints, (ii) by greasing and oiling, (iii) by galvanisation or coating of zinc on iron objects. As zinc is more reactive than iron, zinc undergoes reaction with moisture and prevents iron from rusting. This process is called galvanisation.

Like iron, copper objects also undergo corrosion and lose their lustre or shine after some time. A green coating of basic copper carbonate develops on the surface of the copper object.

$$\underset{\text{Iron}}{2Fe(s)} + \underbrace{\underset{\text{Oxygen}}{\frac{3}{2}O_2(g)} + \underset{\text{Moisture}}{xH_2O(l)}}_{\text{From moist air}} \rightarrow \underset{\text{Hydrated ferric oxide (rust)}}{Fe_2O_3.xH_2O(s)}$$

The surface of silver metal also loses its lustre and becomes dull (tarnished) in air due to corrosion by the hydrogen sulfide present in air due to which a black coating of silver sulfide is formed.

$$\underset{\text{Copper}}{2Cu(s)} + \underbrace{CO_2(g) + O_2(g) + H_2O(l)}_{\text{From moist air}} \rightarrow \underset{\text{Basic copper carbonate (green)}}{CuCO_3.Cu(OH)_2}$$

1.5.2 Rancidity

Oxidation also has a damaging effect on foods having fats and oils. You would have noticed that when such food material is kept for a long time in the open, it starts giving an unpleasant smell or odour and bad taste as well. That is, it becomes sour or stale (common parlance) or rancid (technical term).

When the fats and oils present in food material get oxidised by the oxygen in air, their oxidation products have an unpleasant odour and taste. This condition produced by aerial oxidation of fats and oils in foods marked by unpleasant smell and taste is known as *rancidity.* Rancidity spoils such food materials and makes them unsuitable for eating. The characteristics of a rancid food are unpleasant odour and unpleasant taste.

Prevention: The development of rancidity of food can be prevented or retarded as follows.

- **By adding anti-oxidants:** An antioxidant is a substance which prevents oxidation and it is actually a reducing agent. When such antioxidants are added to foods, the fats and oils present in the food do not get oxidised easily and hence do not turn rancid. It means the foods remain good to eat for a much longer time. The two common most antioxidants used in foods to prevent the development of rancidity are BHA (butylated hydroxy anisole) and BHT (butylated hydroxy toluene).
- **Packing with unreactive gas:** When the packed food is surrounded by an unreactive gas like nitrogen, there is no oxygen or air to cause its oxidation and make it rancid. The manufacturers of potato chips and other similar food products fill the plastic bags having chips with nitrogen gas to prevent the chips from being oxidised and turning rancid (Fig. 1.24).

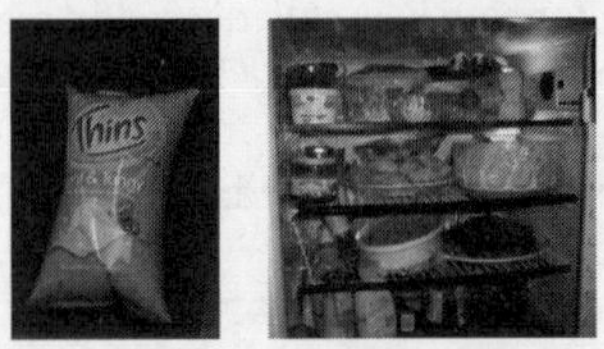

Fig. 1.24 Food item kept in packets with nitrogen and food items kept in a refrigerator

Source: Edoddridge, CC BY-SA 3.0 <https://creativecommons.org/licenses/by-sa/3.0>, via Wikimedia Commons; Seasonsgreetings at English Wikipedia, Public domain, via Wikimedia Commons

- **By keeping food in a refrigerator:** As the refrigerator has a low temperature inside, when the food is kept in it, the oxidation of fats and oils in it is slowed down. The development of rancidity due to oxidation is retarded.

- **By storing food in air-tight containers**: When food is stored in air-tight containers, there is little exposure to oxygen or air. That is, due to reduced exposure to oxygen, the oxidation of fats and oils present in food is slowed down and hence the development of rancidity is retarded.
- **By storing foods away from light:** In the absence of light, the oxidation of fats and oils present in food is slowed down and hence the development of rancidity is retarded.

TEST YOUR KNOWLEDGE

1.20 What type of reaction is involved in rusting? Write the balanced chemical equation for rusting of iron.

Solution:

The rusting of iron is a redox reaction and it can be given as follows

$$\underset{\text{Iron}}{4Fe} + \underset{\text{Oxygen}}{3O_2} + \underset{\text{Water}}{2xH_2O} \rightarrow \underset{\text{Hydrated iron (III) oxide (Rust)}}{2Fe_2O_3xH_2O}$$

1.21 Name two antioxidants that are usually added to fat and oil-containing foods to prevent rancidity.

Solution:

BHA (butylated hydroxy anisole) and BHT (butylated hydroxy toluene) are two antioxidants that are usually added to fat and oil-containing foods to prevent rancidity.

Quick Review

- The reactions in which two or more substances combine to form a single compound, are known as **combination reactions.**
- The reactions in which a compound splits up into two or more simpler substances are called **decomposition reactions.**
- On the basis of their reaction with water, metals can be arranged in the decreasing order of reactivity. This arrangement of metals in the decreasing order of reactivity in the form of a series is known as the **activity** or **reactivity series** of metals.

$$\xrightarrow{\text{K, Na, Ca, Mg, Al, Zn, Fe, Pb [H], Cu, Hg, Ag, Au}}$$

- A reaction in which one element takes the place of another element in a compound is called a **displacement reaction.**
- A reaction in which two compounds react by an exchange of ions to form two new compounds is known as a **double displacement reaction.**
- A precipitation reaction is possible when two solutions are mixed together and a solid separates out from the solution. The solid part that forms and separates from the solutions is called the **precipitate.**
- A **neutralisation** is an acid–base reaction in which an acid reacts with a quantity of a base. The acid–base reaction results in the formation of salt and water. It is an exothermic reaction in which heat is always evolved.
- **Oxidation** involves addition of oxygen or removal of hydrogen or loss of electrons.
- **Reduction** involves removal of oxygen or addition of hydrogen or addition of electrons.
- An **oxidising agent** oxidises others but itself gets reduced. For example, O_3, $KMnO_4$, HNO_3.
- A **reducing agent** reduces others but itself gets oxidised. For example, metals of group 1, H, C.
- **Redox reaction:** Oxidation and reduction take place simultaneously, that is, one species gets oxidised while another one gets reduced.
- **Oxidation number** is the residual charge (±) present on an atom in a given molecule or ion when the shared electrons are counted towards more electronegative atoms.
- **Corrosion** is the process in which metals are eaten up gradually by the action of air, moisture or a chemical on their surface. Corrosion is caused mainly by the oxidation of metals by the oxygen of air.
- The condition produced by aerial oxidation of fats and oils in foods marked by unpleasant smell and taste is called **rancidity.**

Exercise 2

Section A: Multiple Choice Questions (1 Mark)

1. The reaction of H_2 gas with oxygen gas to form water is an example of a/an
   (a) combination reaction
   (b) redox reaction
   (c) exothermic reaction
   (d) all of these reactions

2. The reaction in which two compounds exchange their ions to form two new compounds is called a
   (a) displacement reaction
   (b) combination reaction
   (c) double displacement reaction
   (d) redox reaction

3. Oxidation is a process which involves the
   (a) addition of oxygen
   (b) addition of hydrogen
   (c) removal of oxygen
   (d) removal of hydrogen

4. $Pb + CuCl_2 \rightarrow PbCl_2 + Cu$
   The above reaction is an example of
   (a) combination
   (b) double displacement
   (c) decomposition
   (d) displacement

5. What type of chemical reactions take place when electricity is passed through water?
   (a) displacement
   (b) combination
   (c) decomposition
   (d) double displacement

6. Which of the following statements is **Not True** with respect to the neutralisation reaction?
   (a) The reaction occurs between an acid and a base.
   (b) The more reactive element displaces the less reactive element.
   (c) The reactants are in the gaseous state.
   (d) A salt is formed.

7. A student writes a chemical equation of the reaction between carbon monoxide and hydrogen.

   $$CO_2 + 2H_2 \rightarrow CH_3OH$$

   How can the reaction be classified?
   (a) The reaction is an example of a combination reaction as one compound separates into two compounds.
   (b) The reaction is an example of a decomposition reaction as a compound dissociates into two compounds.
   (c) The reaction is an example of a combination reaction as two compounds react to form a single compound.
   (d) The reaction is an example of a decomposition reaction as two compounds react to form a single compound.

8. A student adds lead and silver to two different test tubes containing an equal amount of copper sulfate solution. The student observes that the colour of the solution in the test tube with lead changes. What explains the change in the colour of the solution?
   (a) A displacement reaction takes place as lead replaces copper from the solution.
   (b) A combination reaction takes place as lead combines with the sulfate in the solution.
   (c) A decomposition reaction takes place as copper dissociates from the sulfate in the solution.
   (d) A double displacement reaction takes place as copper dissociates from the sulfate and lead combines with the sulfate in the solution.

9. Which of the reactions is used in black and white photography?
   (a) combination reaction
   (b) decomposition reaction
   (c) displacement reaction
   (d) oxidation reaction

10. The process of reduction involves
    (a) addition of oxygen
    (b) addition of hydrogen
    (c) removal of oxygen
    (d) removal of hydrogen

11. Select the oxidising agent for the following reaction: $H_2S + I_2 \rightarrow 2HI + S$
    (a) I_2 (b) H_2S
    (c) HI (d) S

12. Which of the following does show an oxidation reaction?

(a) gain of oxygen (b) loss of oxygen
(c) gain of hydrogen (d) none of these

13. $MnO_2 + 4HCl \rightarrow MnCl_2 + 2H_2O + Cl_2$
Identify the substance oxidised in the above equation.
(a) $MnCl_2$ (b) HCl
(c) H_2O (d) MnO_2

14. A dilute ferrous sulfate solution was gradually added to a beaker containing acidified permanganate solution. The light purple colour of the solution fades and finally disappears. Which of the following statements is the correct explanation for the observation?
(a) $KMnO_4$ is an oxidising agent, it oxidises $FeSO_4$.
(b) $FeSO_4$ acts as an oxidising agent and oxidises $KMnO_4$.
(c) The colour disappears due to dilution; no reaction is involved.
(d) $KMnO_4$ is an unstable compound and decomposes in the presence of $FeSO_4$ to a colourless compound.

15. The image shows a reaction between zinc and hydrogen.

$$Zn + 2H^+ \rightarrow Zn^{2+} + H_2$$

Which option shows oxidation?
(a) $Zn \rightarrow Zn^{+2}$ (b) $2H^+ \rightarrow H_2$
(c) $Zn^{+2} \rightarrow Zn$ (d) $H_2 \rightarrow 2H^+$

16. An element X on exposure to moist air turns reddish-brown and a new compound Y is formed. The substances X and Y are
(a) X = Fe, Y = Fe_2O_3
(b) X = Ag, Y = Ag_2S
(c) X = Cu, Y = CuO
(d) X = Al, Y = Al_2O_3

17. Which of the following gases can be used for storing a fresh sample of an oil for a long time?
(a) carbon dioxide or oxygen
(b) nitrogen or helium
(c) helium or oxygen
(d) nitrogen or oxygen

18. Fatty foods become rancid due to the process of
(a) oxidation (b) corrosion
(c) reduction (d) hydrogenation

19. Which of the following gases can be used for storage?
(a) carbon dioxide or oxygen
(b) nitrogen or oxygen
(c) carbon dioxide or helium
(d) helium or nitrogen

20. We store silver chloride in a dark-coloured bottle because it is
(a) a white solid
(b) undergoes redox reaction
(c) to avoid action by sunlight
(d) none of the above

21. The condition produced by aerial oxidation of fats and oils in foods marked by unpleasant smell and taste is called,
(a) antioxidation (b) reduction
(c) rancidity (d) corrosion

22. Rancidity can be prevented by
(a) adding antioxidants
(b) storing food away from light
(c) keeping food in a refrigerator
(d) all of these

23. A substance added to food containing fats and oils is called, a/an
(a) oxidant (b) rancid
(c) coolant (d) antioxidant

24. The displacement reaction between iron(III) oxide and a metal E is used for welding rail tracks. Here E is,
(a) magnesium ribbon (b) aluminum dust
(c) copper granules (d) sodium pellets

25. The chemical reaction between hydrogen sulfide and iodine to give hydrogen iodide and sulfur is given below.

$$H_2S + I_2 \rightarrow 2HI + S$$

The oxidising and reducing agents involved in this redox reaction are,
(a) sulfur and iodine respectively
(b) iodine and hydrogen sulfide respectively
(c) iodine and sulfur respectively
(d) hydrogen sulfide and sulfur

26. In the double displacement reaction between aqueous potassium iodide and aqueous lead nitrate, a yellow precipitate of lead iodide is formed. While performing the activity if lead nitrate is not available, which of the following can be used in place of lead nitrate?
(a) lead sulfate (insoluble)
(b) lead acetate
(c) ammonium nitrate

(d) potassium sulfate

27. Which among the following is(are) double displacement reaction(s)?
(i) $Pb + CuCl_2 \rightarrow PbCl_2 + Cu$
(ii) $Na_2SO_4 + BaCl_2 \rightarrow BaSO_4 + 2NaCl$
(iii) $C + O_2 \rightarrow CO_2$
(iv) $CH_4 + 2O_2 \rightarrow CO_2 + 2H_2O$
(a) (i) and (iv) (b) (ii) only
(c) (i) and (ii) (d) (iii) and (iv)

28. The following reaction is an example of a ________.

$$4NH_3(g) + 5O_2(g) \rightarrow 4NO(g) + 6H_2O(g)$$

(i) displacement reaction
(ii) combination reaction
(iii) redox reaction
(iv) neutralisation reaction
(a) (i) and (iii) (b) (iii) and (iv)
(c) (i) and (iv) (d) (ii) and (iii)

29. Calcium oxide reacts vigorously with water to produce slaked lime

$$CaO(s) + H_2O(l) \rightarrow Ca(OH)_2(aq)$$

This reaction can be classified as
(i) combination reaction
(ii) exothermic reaction
(iii) endothermic reaction
(iv) oxidation reaction
[CBSE sample paper, 2020]
(a) (i), (iii) (b) (iii), (iv)
(c) (i), (iii), (iv) (d) (i), (ii)

30. Which of the following statements about the given reaction are correct?

$$3Fe(s) + 4H_2O(g) \rightarrow Fe_3O_4(s) + 4\,H_2(g)$$

(i) iron metal is getting oxidised
(ii) water is getting reduced
(iii) water is acting as a reducing agent
(iv) water is acting as an oxidising agent
(a) (i), (ii) and (iii) (b) (iii) and (iv)
(c) (i), (ii) and (iv) (d) (ii) and (iv)

31. Ammonia reacts with hydrogen chloride to form a compound. Which of the following statements is correct?
(i) NH_4Cl is formed.
(ii) It is a displacement reaction.
(iii) It is a combination reaction.
(iv) NH_4Cl has dense white fumes.
(a) (i), (iii), (iv) (b) (i), (ii), (iii), (iv)
(c) (i), (ii), (iv) (d) (i), (ii), (iii)

32. Match the following.

Column I	Column II
1. $MgO + H_2O \rightarrow Mg(OH)_2$	(i) Redox reaction
2. $3BaCl_2 + Al_2(SO_4)_3 \rightarrow 2AlCl_3 + 3BaSO_4$	(ii) Decomposition reaction
3. $2FeSO_4 \rightarrow Fe_2O_3 + SO_2 + SO_3$	(iii) Double displacement reaction
4. $CuSO_4 + Zn \rightarrow Cu + ZnSO_4$	(iv) Combination reaction

(a) (1)—(iv), (2)—(iii), (3)—(i), (4)—(ii)
(b) (1)—(iv), (2)—(iii), (3)—(ii), (4)—(i)
(c) (1)—(iii), (2)—(iv), (3)—(ii), (4)—(i)
(d) (1)—(i), (2)—(ii), (3)—(iii), (4)—(iv)

Assertion Reason Questions

Directions: In the following questions, two statements (Assertion) A and Reason (R) are given. Mark
(a) if A and R are both correct and R is the correct explanation of A;
(b) if A and R are both correct but R is not the correct explanation of A;
(c) A is true but R is false;
(d) A is false but R is true.

Assertion	Reason
1. In the reaction $MnO_2 + 4HCl \rightarrow MnCl_2 + 2H_2O + Cl_2$ HCl is getting oxidised while MnO_2 is getting reduced.	**1.** The process in which oxygen is added to a substance is called oxidation, whereas the process in which oxygen is removed from a substance is called reduction.
2. Chips manufacturers usually flush bags of chips with gases such as nitrogen.	**2.** Nitrogen gas prevents the oil and fats of the chips from being oxidised.
3. The oxidation of oils or fat in a food resulting in a bad smell and bad taste is known as rancidity.	**3.** Rancidity can be slowed down by using antioxidants or vacuum packing.

Assertion	Reason
4. When calcium carbonate reacts with CO_2 and H_2O, soluble calcium bicarbonate is formed.	4. Milkyness is observed.
5. When chlorine reacts with potassium bromide, reddish-brown bromine gas is liberated.	5. It is a case of displacement reaction in which chlorine being more reactive displaces bromine from KBr.
6. Copper, silver and gold can easily displace H_2 from acids.	6. In the reactivity series copper, silver and gold are below hydrogen.
7. When electric current is passed into molten alumina (Al_2O_3), oxygen is released.	7. This reaction is used for the extraction of aluminium metal.
8. When a silver plate is dipped in copper nitrate, silver displaces copper.	8. Silver is present below copper in the reactivity series.

Section B: Very Short Answer Questions (2 Marks)

1. What is a combination reaction? State one example giving a balanced chemical equation for the reaction.
2. $N_2 + 3H_2 \rightarrow 2NH_3$, name the type of this reaction.
3. What is a precipitation reaction? Give an example.
4. Write a chemical equation for a double displacement reaction.
5. Why is respiration considered as an exothermic reaction?
6. The addition of hydrogen to a substance is known as ________ whereas removal of hydrogen is known as ________.
7. What type of chemical reactions take place when
   (i) limestone is heated?
   (ii) a magnesium wire is burnt in air?
8. What type of chemical reactions take place when
   (i) electricity is passed through water?
   (ii) ammonia and hydrogen chloride are mixed?
9. Why are decomposition reactions called the opposite of combination reactions?
10. Why is photosynthesis considered an endothermic reaction?
11. State the type of chemical reaction used for the extraction of metals from their naturally occurring chlorides or oxides.
12. A shiny brown element X on heating in air becomes black in colour. Name the element X and the black coloured compound formed.
13. Food items having oil and fat can be flushed by using nitrogen. Why?
14. Give reasons why chips manufacturers usually flush bags of chips with a gas such as nitrogen?
15. Is photosynthesis an endothermic or an exothermic reaction?
16. Identify the substance that is oxidised and substance that is reduced in the reaction.

    $$CuO(s) + H_2(g) \rightarrow Cu(s) + H_2O(l)$$
17. What is the oxidation number of Mn in $KMnO_4$?
18. Write the full form of BHT (a well-known antioxidant).
19. What are two conditions which promote corrosion?
20. Name any two methods used to prevent rusting of iron.

Case Study or Passage Based Questions

Chemical reactions can be further divided into the following types—combination, decomposition, displacement, precipitation, neutralisation and redox reactions. Every type of reaction is unique in its own way by which it can be distinguished from the rest. For example, in decomposition reactions, a single compound on heating decomposes into two or more simple substances while in a displacement reaction a more active metal displaces a less reactive metal from its salt. The reactions such as corrosion and rancidity are of great importance in our everyday life.

21. Precipitation reactions produce insoluble ________.

22. Reactions in which energy is given out are known as ________.

23. A reaction in which an element displaces another element from its compound is called a ________.

24. The addition of oxygen to a substance is known as ________ whereas removal of oxygen is known as ________ respectively.

25. Two antioxidants which are usually added to fat and oil-containing foods to prevent rancidity, are ________ and ________.

26. ________ is the process in which metals are eaten up gradually by the action of air, moisture or a chemical on their surface.

Section C: Short Answer Questions (3 Marks)

1. What does one mean by exothermic and endothermic reactions? Give examples.

2. What is meant by a precipitation reaction? Explain by giving an example. Also give a balanced chemical equation for the reaction stating the states of the reactants and the products formed.

3. What is a redox reaction? Write down a chemical reaction representing it.

4. Classify the following reactions as combination, decomposition, displacement and double displacement.
   (i) $BaCl_2 + H_2SO_4 \rightarrow BaSO_4 + 2HCl$
   (ii) $C + O_2 \rightarrow CO_2$
   (iii) $3CuSO_4 + 2Al \rightarrow Al_2(SO_4)_3 + 3Cu$
   (iv) $ZnCO_3 \rightarrow ZnO + CO_2$

5. Why is the amount of gas collected in one of the test tubes in the electrolysis of water, double of the amount collected in the other? Name this gas.

6. Define the term decomposition reaction. Give one example each of thermal decomposition and electrolytic decomposition.

7. (i) Write a balanced chemical equation for the process of photosynthesis.
   (ii) When do desert plants take up carbon dioxide and perform photosynthesis? **[CBSE 2015]**

8. (i) $2PbO + C \rightarrow 2Pb + CO_2$
   (ii) $MnO_2 + 4HCl \rightarrow MnCl_2 + 2H_2O + Cl_2$
   What is a redox reaction? Identify the substance oxidised and the substance reduced in the above reactions.

9. Differentiate between a combination reaction and a decomposition reaction. Write one chemical equation each for these reactions.

10. Give an example each for thermal decomposition and photochemical decomposition reactions. Write balanced equations for the same.

11. Name the reducing agent in the following reaction.

$$3MnO_2 + 4Al \rightarrow 3Mn + 2Al_2O_3$$

State which is more reactive, Mn or Al, and why? **[CBSE 2015]**

12. Name the salts that are used in black and white photography. Give reactions for the reactions occurring when they are exposed to light. Define the type of chemical reaction taking place.

13. State the type of chemical reactions with chemical equations that take place when ammonia and hydrogen chloride gases are mixed. **[CBSE 2015]**

14. What happens when water is added to quicklime? Write the chemical equation. **[CBSE 2015]**

15. Write the chemical equation of the reaction in which the following changes have taken place with an example of formation of precipitate.
    (i) Change in colour
    (ii) Change in temperature
    (iii) Formation of precipitate **[CBSE 2015]**

16. Name the type of chemical reaction represented by the following equations.
    (i) $CaO + H_2O \rightarrow Ca(OH)_2$
    (ii) $3BaCl_2 + Al_2(SO_4)_3 \rightarrow 2AlCl_3 + 3BaSO_4$
    (iii) $2FeSO_4 \rightarrow Fe_2O_3 + SO_2 + SO_3$
    [CBSE 2015]

17. Define the term decomposition reaction. Give one example each of thermal decomposition and electrolytic decomposition. **[CBSE 2016]**

18. Silver chloride turns grey when exposed to sunlight. Give a chemical equation to explain it. **[CBSE 2017]**

19. Identify the type of each of the following reactions. Also write a balanced chemical equation for each.
(i) The reaction mixture becomes warm.
(ii) An insoluble substance is formed.
[Board Term-I, Set-A85V2IL, 2015]

20. 2 g of ferrous sulfate crystals are heated in a boiling tube.
(i) State the colour of the ferrous sulfate crystals both before heating and after heating.
(ii) Name the gases produced during heating.
(iii) Write the chemical equation for the reaction.

21. What is rancidity? Mention any two ways by which rancidity can be prevented.

22. Name the term used to indicate the development of unpleasant smell and taste in fat and oil-containing food due to oxidation. What are antioxidants? Why are they added to fat and oil-containing food?
[Board Term-I Set-1ZHNPNO, 2016]

23. State reasons for the following.
(i) Potato chips manufacturers fill the packet of chips with nitrogen gas.
(ii) Iron articles are shining when new, but get coated with a reddish-brown powder, when left for some time.

24. (i) List any two changes which take place when oily food gets oxidised.
(ii) Mention a measure which prevents or slows down its oxidation.

25. Barium chloride reacts with aluminium sulfate to give aluminium chloride and barium sulfate.
(i) State the two ways in which the above reaction can be classified.
(ii) Translate the above statement into a chemical equation.

26. A solution of copper sulfate was kept in an iron pot. After few days, the iron pot was found to have a number of holes in it. Explain the reaction with the help of a chemical equation.
[Board Term-I 2015]

27. State the kind of chemical reactions in the following examples.
(i) Digestion of food in the stomach.
(ii) Combustion of coal in air.
(iii) Heating of limestone.

28. Name two metals which do not corrode easily. Give an example in each of the following cases to support that.
(i) Corrosion of some metals is an advantage.
(ii) Corrosion of a metal is a serious problem.

29. (i) Give an example for a combination reaction which is exothermic.
(ii) Identify the oxidising agent and reducing agent in the following reaction.

$$H_2S + Cl_2 \rightarrow 2HCl + S$$

(iii) Name the phenomenon due to which the taste and smell of oily food changes when kept for a long time in the open. Suggest one method to prevent it.

30. Write a balanced chemical equation for the reactions that take place during respiration. Identify the type of combination reaction that takes place during this process and justify the name. Give one more example of this type of reaction.

Section D: Long Answer Questions (5 Marks)

1. What is observed when a solution of sodium sulfate is added to a solution of barium chloride taken in a test tube? Write the equation for the chemical reaction involved and name the type of reaction in this case. **[CBSE 2018]**

2. 2 g of ferrous sulfate crystals are heated in a dry boiling tube.
(i) List any two observations.
(ii) Name the type of chemical reaction taking place.
(iii) Write the balanced chemical equation for the reaction and name the products formed.

OR

You might have noted that when copper powder is heated in a china dish, the reddish-brown surface of copper powder becomes coated with a black substance.
(i) Why has this black substance formed?
(ii) What is this black substance?

(iii) Write the chemical equation of the reaction that takes place.

(iv) How can the black coating on the surface be turned reddish-brown? **[CBSE 2019]**

3. Identify the type of chemical reaction in the following statements and define each of them.
   (i) Digestion of food in our body.
   (ii) Rusting of iron.
   (iii) Heating of manganese dioxide with aluminium powder.
   (iv) The blue colour of copper sulfate solution disappears when iron filings are added to it.
   (v) Dilute hydrochloric acid is added to sodium hydroxide solution to form sodium chloride and water.

4. (i) In the following reactions, name the reactants which undergo oxidation and reduction.
   (a) $CuO(s) + H_2(g) \rightarrow Cu(s) + H_2O(g)$
   (b) $CuO(s) + Zn(s) \rightarrow ZnO(s) + Cu(s)$
   (ii) State one industrial application of reduction.

5. In an industrial process used for the manufacture of sodium hydroxide, a gas A is formed as a byproduct. The gas A reacts with limewater to give a compound B which is used as a bleaching agent in the chemical industry. Identify A and B. Also give the chemical equations of the reactions involved. **[CBSE 2019]**

6. What would you observe on adding zinc granules to freshly prepared ferrous sulfate solution? Give reasons for your answer. **[CBSE 2019]**

7. Lead nitrate solution is added to a test tube containing potassium iodide solution.
   (i) Write the name and colour of the compound precipitated.
   (ii) Write the balanced chemical equation for the reaction involved.
   (iii) Name the type of this reaction justifying your answer. **[CBSE 2020]**

OR

What happens when food materials containing fats and oils are left for a long time? List two observable changes and suggest three ways by which this phenomenon can be prevented.

8. In the electrolysis of water,
   (i) Name the gas collected at the cathode and anode respectively.
   (ii) Why is the volume of one gas collected at one electrode double that at the other? Name this gas.
   (iii) How will you test the evolved gases?

9. In the following chemical reaction 'zinc oxide reacts with carbon to produce zinc metal and carbon',

$$ZnO + C \rightarrow Zn + CO$$

   (i) Identify the substance getting oxidised and the one getting reduced.
   (ii) State the reason for choosing the substances in (i).
   (iii) Name the type of reaction and give another example of similar type of reaction.

10. (i) Define corrosion.
   (ii) What is corrosion of iron called?
   (iii) How will you recognise the corrosion of silver?
   (iv) Why is the corrosion of iron a serious problem?
   (v) How can we prevent corrosion of iron? **[Board Term-I Set-NS9SX1D, 2016]**

OR

   (i) Define corrosion; what name is given to the corrosion of iron?
   (ii) Name the colour of the coating formed on silver and copper articles when exposed to air.
   (iii) List two damages caused by corrosion and suggest how corrosion can be prevented. **[Board Term-I 2016, Set-L7ZSVLH]**

11. (i) Most of the metals acquire a dull surface when exposed to air. Name the chemical phenomenon responsible for this process.
   (ii) State the conditions under which iron articles get rusted. Design an activity to investigate the conditions necessary for rusting. Suggest any two methods to prevent rusting of iron.

12. *Spoilage is the process in which food deteriorates to the point when it is not edible.
   (i) Why does the food get spoilt?
   (ii) How can the spoilage of food be prevented?

(iii) How are chips prevented from getting spoiled in packets for so many days?

13. *Observe the following figures A and B.

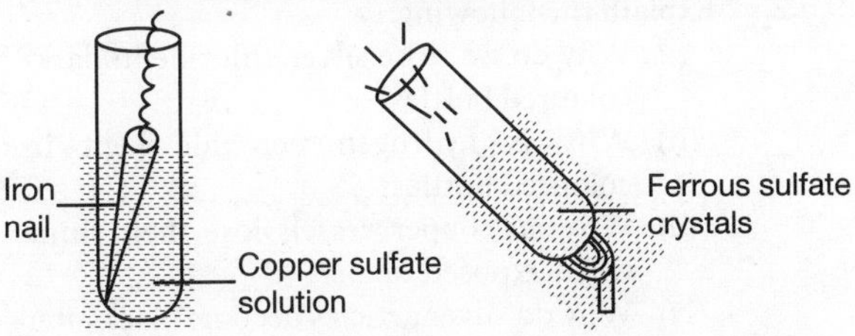

State the change in substances taken in the test tubes A and B. Mention the type of reaction in each case. **[Board Term-I Set-1ZHNPNO, 2016]**

Section E: Case Study or Passage Based Questions (4 Marks)

I. Some of the important types of chemical reactions are decomposition (single compound decomposes into two or more compounds), combination (two or more compounds combine together), displacement (one species displaces another), neutralisation (acid and base react), redox (one species gets oxidised while another one gets reduced). Such reactions are very common and have a great role in our daily lives. Processes like respiration and photosynthesis have a lot of biological applications.

1. When hydrogen sulfide gas is passed through a blue solution of copper sulfate, a black precipitate of copper sulfide is obtained and the sulfuric acid so formed remains in the solution. The reaction is an example of **[CBSE sample paper, 2020]**

(a) combination reaction
(b) displacement reaction
(c) decomposition reaction
(d) double displacement reaction

2. In a double displacement reaction such as the reaction between sodium sulfate solution and barium chloride solution, **[CBSE sample paper, 2020]**

(i) exchange of atoms takes place
(ii) exchange of ions takes place
(iii) a precipitate is produced
(iv) an insoluble salt is produced

(a) (ii), (iv) (b) (i), (iii)
(c) only (ii) (d) (ii) (iii), (iv)

3. Which among the following statement(s) is(are) true? Silver chloride on exposure to sunlight for a long duration turns grey due to

(i) the formation of silver by the decomposition of silver chloride
(ii) the sublimation of silver chloride
(iii) the decomposition of chlorine gas from silver chloride
(iv) the oxidation of silver chloride

(a) (i) only (b) (i) and (iii)
(c) (ii) and (iii) (d) (iv) only

4. $C_6H_{12}O_6(aq) + 6O_2(aq) \rightarrow 6CO_2(aq) + 6H_2O(l)$ **[CBSE 2021]**

The above reaction is a/an

(a) displacement reaction
(b) endothermic reaction
(c) exothermic reaction
(d) neutralisation reaction

5. Which of the following statements about the reaction given below are correct? **[CBSE 2021]**

$$MnO_2 + 4HCl \rightarrow MnCl_2 + 2H_2O + Cl_2$$

(i) HCl is oxidised to Cl_2
(ii) MnO_2 is reduced to $MnCl_2$
(iii) $MnCl_2$ acts as an oxidising agent
(iv) HCl acts as an oxidising agent

(a) (ii), (iii) and (iv)
(b) (i), (ii) and (iii)
(c) (i) and (ii) only
(d) (iii) and (iv) only

6. Which one of the following reactions is categorised as a thermal decomposition reaction? **[CBSE 2021]**

(a) $2H_2O(l) \rightarrow 2H_2(g) + O_2(g)$
(b) $2AgBr(s) \rightarrow 2Ag(s) + Br_2(g)$
(c) $2AgCl(s) \rightarrow 2Ag(s) + Cl_2(g)$
(d) $CaCO_3(s) \rightarrow CaO(s) + CO_2(g)$

II. For an internal combustion engine to move a vehicle down the road, it must convert the energy stored in the fuel into mechanical energy to drive the wheels. In your car, the distributor and battery provide this starting energy by creating an electrical spark, which helps in combustion of fuels like gasoline. Below is the reaction depicting complete combustion of gasoline in full supply of air:

$2C_8H_{18}(l) + 25O_2(g) \rightarrow 16X + Y$ **[CBSE QB 2021]**

7. 'Although nitrogen is the most abundant gas in the atmosphere, it does not undergo combustion'. Identify the correct reason for this statement.
   (a) nitrogen is an explosive gas
   (b) nitrogen is an inert gas
   (c) nitrogen is a reactive gas
   (d) only hydrocarbons can take part in combustion

8. Which of the following are the products obtained from the reaction mentioned in the above case? Product X, Product Y.
   (a) CO_2, H_2O_2 (b) H_2O, CO
   (c) CO_2, H_2O (d) CH_3OH, H_2O

9. Identify the types of chemical reaction occurring during the combustion of fuels.
   (a) oxidation and endothermic reaction
   (b) combination and endothermic reaction
   (c) oxidation and exothermic reaction
   (d) decomposition and exothermic reaction

10. On the basis of evolution/absorption of energy, which of the following processes are similar to the combustion of a fuel?
    (i) decomposition of ferric sulfate
    (ii) respiration in the human body
    (iii) decomposition of vegetable matter
    (iv) photosynthesis in plants
    (a) (ii), (iii) (b) (i), (ii)
    (c) (iii), (iv) (d) (ii), (i)

11. A student while walking on the road observed that a cloud of black smoke belched out from the exhaust stack of moving trucks on the road. Choose the correct reason for the production of black smoke.
    (a) limited supply of air leads to incomplete combustion of fuel
    (b) rich supply of air leads to a combination reaction
    (c) rich supply of air leads to complete combustion of fuel
    (d) limited supply of air leads to complete combustion of fuel

High Order Thinking Skills (HOTS) Questions

1. Consider the following chemical reaction:

$$X + \text{Barium chloride} \rightarrow \underset{\text{White ppt.}}{Y} + \text{Sodium chloride}$$

   (i) Identify X and Y.
   (ii) Identify te type of reaction.

[Hots; CBSE 2015]

2. Explain the following:
   (i) Why do we store silver chloride in dark-coloured bottles?
   (ii) Why is hydrogen peroxide kept in coloured bottles?
   (iii) Why do copper vessels lose their shine when exposed to air?
   (iv) Why do silver articles become black after some time, when exposed to air?

3. *(i) What happens when milk is left open at room temperature during summers?
   (ii) What changes do you observe in the iron nails and colour of copper sulfate solution, if iron nails are dipped in $CuSO_4$ solution for 15 minutes?

4. *(i) A student prepares aqueous solutions of the following salts: copper sulfate, ferrous sulfate, sodium sulfate, barium chloride. Write the colour of each solution thus formed.
   (ii) Mention the colour of $FeSO_4.7H_2O$ crystals. How does this colour change upon heating?

   Give a balanced chemical equation for the change.

5. Some articles made of silver, copper and iron get a coloured coating over them when they are exposed to air. Identify the colour and chemical name of the substance in each case.

[Board Term-I Set-2ZGOVVV, 2015]

6. *The following diagram displays a chemical reaction. Observe carefully and answer the following questions.

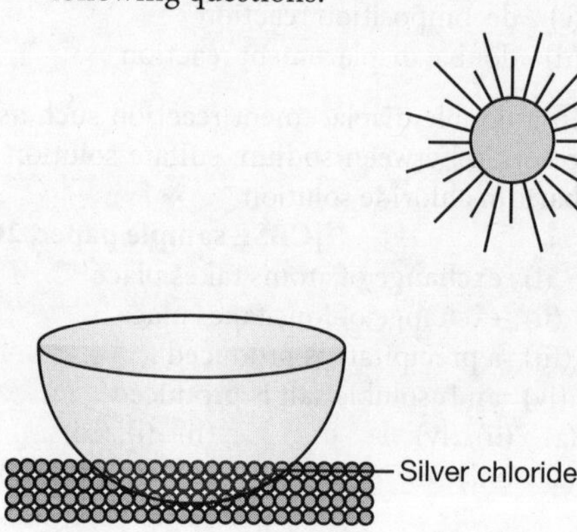

(i) Identify the type of chemical reaction that will take place and define it. How will the colour of the salt change?
(ii) Write the chemical equation for the reaction that takes place.
(iii) Mention one commercial use of this salt.

7. *A small amount of calcium oxide is taken in a beaker and water is added slowly to it.
(i) Will there be any change in the temperature of the contents? Explain.
(ii) Name and define the type of reaction taking place.
(iii) Write the chemical equation for the above reaction.

8. *2 g of lead nitrate powder is taken in a boiling tube. The boiling tube is heated over a flame. Now answer the following.
(i) State the colour of the fumes evolved and the residue left.
(ii) Name the type of chemical reaction that has taken place, providing its balanced equation.

9. Account for the following.
(i) White silver chloride turns grey in sunlight.
(ii) Brown coloured copper powder on heating in air turns into a black coloured substance.

10. *A compound X is a constituent of baking powder. It is used as an antacid. When X is heated, it gives out a gas Y which when passed through lime water turns it milky and a salt Z is formed which is the main constituent of washing power. Identify X, Y and Z. Write balanced chemical equations for the reactions involved.

11. *Shashwat is a 3-year-old boy. He had purchased a packet of tomato flavoured potato chips more than three months ago. Shashwat opened the packet at that time but could eat only few of the potato chips. He then kept the open packet having the remaining potato chips on his table. He wanted to eat these remaining potato chips today. Just when Shashwat was about to put these potato chips into his mouth, his elder sister Shanvi, who is a student of class X, entered the room. She observed that these potato chips had an unpleasant smell. When she put one potato chip in her mouth, it also had an unpleasant taste. Shanvi took away the packet from Shashwat and did not allow him to eat these potato chips. She threw away the potato chips into a dustbin.
(i) What name is given to the condition which the potato chips turn into when kept open for a considerable time and give out an unpleasant smell and taste?
(ii) Which chemical reaction is responsible for the spoilage of potato chips kept exposed by Shashwat for a considerable time?
(iii) What produces the unpleasant smell as well as unpleasant taste in the potato chips? Explain.
(iv) With which gas are the packs containing potato chips filled and sealed by the manufacturers? How does it help?
(v) What values are displayed by Shanvi in this episode?

Answers

Section A: Multiple Choice Questions

1. (a)	**2.** (c)	**3.** (a)	**4.** (d)
5. (c)	**6.** (b)	**7.** (c)	**8.** (a)
9. (b)	**10.** (b)	**11.** (a)	**12.** (a)
13. (b)	**14.** (a)	**15.** (a)	**16.** (a)
17. (b)	**18.** (a)	**19.** (d)	**20.** (c)
21. (c)	**22.** (d)	**23.** (d)	**24.** (b)
25. (b)	**26.** (b)	**27.** (b)	**28.** (d)
29. (d)	**30.** (c)	**31.** (a)	**32.** (b)

Assertion Reason Questions

1. (a)	**2.** (a)	**3.** (b)	**4.** (c)
5. (a)	**6.** (d)	**7.** (b)	**8.** (d)

Section B: Very Short Answer Questions

1. A reaction in which two or more simple substances combine to form a single product.
   Example: $C + O_2 \rightarrow CO_2$
2. It is a combination reaction.
3. The reaction in which an insoluble substance or precipitate is formed is called a precipitation reaction.
   Example: $Na_2SO_4 + BaCl_2 \rightarrow 2NaCl + BaSO_4\downarrow$
4. Double displacement reaction: $Na_2SO_4(aq) + BaCl_2(aq) \rightarrow BaSO_4(s) + 2NaCl(aq)$
5. Respiration is considered as an exothermic reaction because heat is released along with the formation of products.
6. Reduction, oxidation.
7. (i) Decomposition, (ii) combination.
8. (i) Decomposition, (ii) combination.
9. In combination reactions, two substances combine to form one compound and in decomposition reactions, a compound breaks down into two or more substances, so they are opposite to each other.
10. Heat is absorbed in this process.
11. Electrolytic reduction.
12. Element X is copper and the black coloured compound is cupric oxide, Cu_2O.
13. Food stuff that contain oil and fat can be flushed by using nitrogen because they can be preserved an atmosphere of nitrogen; it keeps them from being oxidised.
14. To prevent the oil and fats of the chips from being oxidised thus becoming rancid.
15. Photosynthesis is an endothermic reaction as light energy is used for this process.
16.

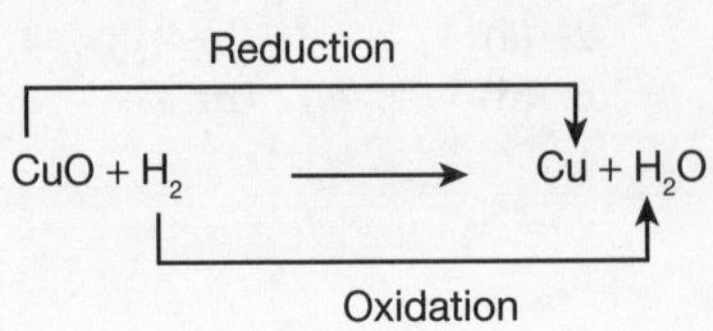

Substance reduced = CuO
Substance oxidised = H_2

17. $KMnO_4$
    $1 + x + 4(-2) = 0, \quad x = +7$
18. BHT is butylated hydroxy toluene.
19. Two conditions which promote corrosion are: (i) **the** substance is exposed to air, water, (ii) the metal gets oxidised to metal oxide.
20. Rusting of iron can be prevented by (i) zinc coating on the iron object, (ii) by applying paints and varnish on the iron objects.

Case Study or Passage Based Questions

21. Salts
22. Exothermic reactions
23. Displacement reaction
24. Oxidation, reduction
25. BHA and BHT
26. Corrosion

Section C: Short Answer Questions

1. Exothermic reactions: Reaction in which heat is released along with the formation of products.
   Example: $CH_4(g) + O_2(g) \rightarrow CO_2(g) + 2H_2O(g)$ + Heat
   Endothermic reaction: A reaction in which the energy is required in the form of heat, light or electricity to break reactants are called an endothermic reaction.
   $$2\,AgCl(s) \rightarrow 2\,Ag(s) + Cl_2(g)$$
2. On mixing the clear solutions of two ionic compounds, a substance which is insoluble in water is formed. This insoluble substance is known as a precipitate and the reaction in which precipitate is formed is called a precipitation reaction.
   Example: When sodium sulfate solution is mixed with barium chloride solution, a white precipitate of $BaSO_4$ is formed.
   $Na_2SO_4(aq) + BaCl_2(aq) \rightarrow BaSO_4(\downarrow) + 2NaCl(aq)$
3. The addition of oxygen to a substance and the removal of hydrogen from a substance is called an oxidation reaction. The addition of hydrogen to a substance or the removal of oxygen from a substance is called a reduction

reaction. So, when oxidation and reduction take place together it is called a redox reaction.
Example:

Oxidation

$$CuO + H_2 \xrightarrow{\text{Heat}} Cu + H_2O$$

Reduction

4. (i) Double displacement reaction
(ii) Combination reaction
(iii) Displacement reaction
(iv) Decomposition reaction

5. Water is formed when hydrogen and oxygen combine in a ratio of 2 : 1 by volume. The gas with double volume is hydrogen.

6. A reaction in which a single reactant breaks down to give simpler products.
Thermal decomposition:
$CaCO_3 \rightarrow CaO + CO_2$ (Or any other)
$$\underset{\text{(Water)}}{2H_2O(l)} \xrightarrow{\text{Electric current}} \underset{\text{(Hydrogen gas)}}{2H_2(g)} + \underset{\text{(Oxygen gas)}}{O_2(g)}$$

7. (i) $6CO_2(g) + 6H_2O(l) \xrightarrow{\text{Sunlight}} \underset{\text{Glucose}}{C_6H_{12}O_6(s)} + 6O_2(g)$
(ii) In desert plants, the stomata open during the night and take in carbon dioxide which they store in the vacuoles. This stored CO_2 is utilised in the presence of sunlight during the daytime to perform photosynthesis.

8. Those reactions in which both oxidation and reduction occur simultaneously.
(i) PbO is reduced, C is oxidised.
(ii) HCl is oxidised, MnO_2 is reduced.

9. The chemical reaction in which a single product is formed from two or more reactants and energy is evolved, is known as a combination reaction.
Example : $2Mg + O_2 \rightarrow 2MgO$
When a single reactant decomposes on gaining energy to give two or more simpler products, such a reaction is called a decomposition reaction.
Example: $CaCO_3 \xrightarrow{\Delta} CaO + CO_2$

10. Thermal decomposition: Heating of limestone
$$CaCO_3 \rightarrow CaO + CO_2$$
Photochemical decomposition: Action of light on silver bromide.
$$2\,AgBr \rightarrow 2\,Ag + Br_2$$

11. Al is a reducing agent. Al is more reactive than Mn because Al displaces Mn from its oxide.

12. Silver chloride (AgCl) and silver bromide (AgBr)
$2AgCl \xrightarrow{\text{Sunlight}} 2Ag + Cl_2$
$2AgBr \xrightarrow{\text{Sunlight}} 2Ag + Br_2$
Those reactions in which energy is absorbed (in the form of heat, light or electricity) to break down the reactants is called a decomposition reaction.

13. Ammonia and hydrogen chloride gases are mixed to form ammonium chloride.
$$NH_3 + HCl \rightarrow NH_4Cl$$
The reaction is a combination reaction.

14. Slaked lime is formed with a hissing sound and a large amount of heat is evolved.
$$CaO + H_2O \rightarrow Ca(OH)_2 + \text{Heat}$$

15.
$$\underset{\text{Lead nitrate}}{Pb(NO_3)_2(aq)} + 2KI(aq) \rightarrow \underset{\substack{\text{Lead iodide}\\\text{(Yellow ppt.)}}}{PbI_2(s)} + 2KNO_3(aq)$$
A yellow precipitate of PbI_2 is formed.

16. The types of chemical reactions are:
(i) combination reaction
(ii) double displacement reaction
(iii) decomposition reaction

17. Decomposition reaction: The reaction in which a compound is broken down into simpler elements or simple compounds is known as a decomposition reaction.
Thermal decomposition:
$$CaCO_3(s) \xrightarrow{\Delta} CaO(s) + CO_2(g)$$
Electrolytic decomposition:
$$2H_2O(l) \xrightarrow{\text{Electricity}} 2H_2(g) + O_2(g)$$

18. The energy from sunlight causes silver chloride (AgCl) to decompose into silver (Ag^+) and chloride (Cl^-) ions. The metallic silver appears as small flakes which are essentially black.

These tiny black spots evenly distributed in the mass of white AgCl and makes it look grey.

$$2AgCl(s) \rightarrow 2Ag(s) + Cl_2(g)$$

19. (i) Exothermic reaction.
$CaO + H_2O \rightarrow Ca(OH)_2$
(ii) Double displacement reaction.
$BaCl_2 + Na_2SO_4 \rightarrow BaSO_4 + 2NaCl$

20. (i) Before heating: Pale green.
After heating: Brown or reddish-brown.
(ii) SO_2 and SO_3.
(iii) $2FeSO_4(s) \xrightarrow{\Delta} Fe_2O_3(s) + SO_2(g) + SO_3(g)$

21. The oxidation of oils or fats in food resulting in bad smell and bad taste is called rancidity.
It can be prevented by (i) adding antioxidants, and (ii) flushing with nitrogen gas.

22. Rancidity.
Antioxidants are substances that inhibit oxidation, especially ones used to counteract the deterioration of stored food products.
Antioxidants are added to food materials containing fats and oils to prevent their oxidation. Oxidation of food materials containing fats and oils is known as rancidity because of which they become unfit for consumption and develop bad odour and bad taste.

23. (i) To prevent rancidity.
(ii) Rusting and corrosion.

24. (i) (a) Food becomes rancid.
(b) Its taste and smell changes.
(ii) Keeping food in air tight containers/ flushing the container with an inert gas like nitrogen (N_2).

25. (i) Double displacement reaction and precipitation reaction.
(ii) $3BaCl_2 + Al_2(SO_4)_3 \rightarrow 2AlCl_3 + 3BaSO_4(\downarrow)$

26. Iron is more reactive than copper. Iron displaces copper from copper sulfate solution and forms iron sulfate, hence holes appear on the pot.
Equation: $Fe + CuSO_4(aq) \rightarrow FeSO_4(aq) + Cu$

27. (i) Decomposition reaction and exothermic reaction.
$$C_6H_{12}O_6 + 6O_2 \rightarrow 6CO_2 + 6H_2O + \text{Energy}$$
(ii) Combination reaction.
$$C + O_2 \rightarrow CO_2$$
(iii) Decomposition reaction.
$$CaCO_3(s) \xrightarrow{\Delta} CaO(s) + CO_2(g)$$

28. Gold and platinum.
(i) Corrosion of aluminium is useful. A protective layer of aluminium oxide is formed on the surface of the metal which renders the metal passive and prevents is further corrosion.
(ii) Corrosion of iron is a serious problem. Every year large amount of money is spent to replace damaged iron and steel structures. Here, corrosion is a serious problem.

29. (i) Any suitable exothermic reaction like
$$CaO + H_2O \rightarrow Ca(OH)_2 + \text{heat}$$
(ii) Oxidising agent —Cl_2
Reducing agent —H_2S
(iii) Rancidity, keep food in airtight containers.

30. $C_6H_{12}O_6 + 6O_2 \rightarrow 6CO_2 + 6H_2O + \text{Energy}$
It is an exothermic reaction because a large amount of heat is released.
Example: Decomposition of vegetable matter into compost.

Section D: Long Answer Questions

1. When a solution of sodium sulfate is added to a solution of barium chloride taken in a test tube, a white precipitate of barium sulfate is formed.
Chemical reaction Involved:
$$BaCl_2(aq) + Na_2SO_4(aq) \rightarrow \underset{\text{White ppt.}}{BaSO_4 \downarrow} + 2NaCl(aq)$$
Type of reaction: This reaction is a double displacement reaction.

2. (i) Observations: The colour changes from green to white.
Formation of reddish-brown ferric oxide (Fe_2O_3)/evolution of SO_2/SO_3 gas.
(ii) Decomposition reaction.
(iii) $2FeSO_4 \xrightarrow{\text{Heat}} \underset{\text{Ferric oxide}}{Fe_2O_3} + \underset{\text{Sulfuric oxide}}{SO_2} + \underset{\text{Sulfuric trioxide}}{SO_3}$

Or

(i) When copper is heated in air, oxidation takes place.
(ii) CuO/copper oxide
(iii) $2Cu + O_2 \rightarrow 2CuO$
(iv) On passing hydrogen gas over the heated material.

3. (i) Decomposition reaction: Carbohydrates are broken down to form glucose.
(ii) Oxidation reaction: When an iron object is left in moist air for a considerable time, it gets covered with a red-brown flaky substance called rust.
(iii) Displacement reaction: A more reactive metal displaces a less reactive metal from its salt solution.
(iv) Displacement reaction: A more reactive metal displaces a less reactive metal from its salt solution.
(v) Double displacement reaction: A reaction in which two compounds react by an exchange of ions to form two new compounds.

4. (i) (a) The reactants are copper oxide and hydrogen. Copper oxide undergoes reduction and hydrogen (H_2) undergoes oxidation.
(b) The reactants are copper oxide and zinc. Zinc undergoes oxidation and copper oxide undergoes reduction.
(ii) Calcium carbonate is reduced in industries to get CaO and CO_2.

5. A = Cl_2 (Chlorine gas) B = $CaOCl_2$ (Calcium oxychloride)
$2NaCl + 2H_2O \rightarrow 2NaOH + Cl_2 + H_2$
$Cl_2 + Ca(OH)_2 \rightarrow CaOCl_2 + H_2O$

6. The solution turns
(i) green to colourless
(ii) black coating is formed on zinc.
Reason: Zinc is more reactive than iron so it displaces the iron from its salt solution.

7. (i) Name of compound precipitated: Lead iodide; Colour of the compound precipitated: Yellow.
(ii) $2KI(s)+Pb(NO_3)_2(aq) \rightarrow 2KNO_3(aq) + PbI_2(s)$
(iii) This is a double displacement reaction because two compounds exchange their ions to form two new compounds. Here, both lead nitrate and potassium iodide are exchanging their ions.

Or

When food materials containing fats and oils are left for a long time they undergo oxidation and their taste and smell changes.
This can be prevented by, (i) using airtight and light protective packing, (ii) using antioxidants like vitamin A and C and (iii) filling nitrogen gas in the chips container.

8. (i) At cathode: hydrogen gas (H_2)
At anode : oxygen gas (O_2)
(ii) Since $2H_2O \xrightarrow{\text{Electric Current}} 2H_2 + O_2$
2 molecules of H_2 combine with 1 mol of O_2 to form H_2O, so the volume of H_2 liberated is double that of O_2.
(iii) When a burning splinter is brought near the mouth of the liberated gases, the burning splinter gets extinguished near the H_2 gas while the burning splinter keeps burning more near the O_2 gas.

9. (i) C is getting oxidised to CO, ZnO is getting reduced to Zn.
(ii) Carbon is gaining oxygen and ZnO is losing oxygen.
(iii) It is a redox reaction or oxidation and reduction reaction.

$$CuO + H_2 \rightarrow Cu + H_2O$$

10. (i) Corrosion is a process in which metals are deteriorated by the action of air, moisture, chemicals and so on.
(ii) Corrosion of iron is called rusting.
(iii) Silver turns black as it reacts with the H_2S present in air and forms a layer of Ag_2S.
(iv) Corrosion of iron is a serious problem because it leads to wastage of tonnes of iron every year and a lot of money is spent to repair or replace it.
(v) The iron articles should be painted.

11. (i) Corrosion (ii) Conditions are: air and moisture.
Activity: Take three test tubes. Place clean iron nails in each test tube.
(i) Pour some water in test tube 1, cork it.
(ii) Pour water (boiled/distilled) in test tube 2, add some oil and cork it.

(iii) Put some anhydrous calcium chloride in test tube 3 and cork it.

(iv) After 2–3 days, we observe that the nails in test tube 1 get rusted because they are exposed to both air and water, while the nails in test tubes 2 and 3 do not get rusted. This shows that rusting of iron takes place in the presence of air and moisture, both.

Methods to prevent rusting: Alloying, galvanisation, painting, lubrication (any two).

12. (i) When fats and oils are oxidised, they become rancid and their smell and taste changes.

(ii) To prevent spoilage of food, special types of substances called antioxidants are added to fatty foods.

(iii) Chips are packed in packets flushed with nitrogen to prevent spoilage by oxidation.

13. In A, the green colour of the solution fades away and a brown-red deposition on the iron nail occurs. It is a displacement reaction. In B, a reddish-brown substance is left in the test tube. It is a decomposition reaction.

Section E: Case Study or Passage Based Questions

I.

1. (d)	**2.** (d)	**3.** (a)	**4.** (c)
5. (c)	**6.** (d)		

II.

7. (b)	**8.** (c)	**9.** (c)	**10.** (a)
11. (a)			

High Order Thinking Skills (HOTS) Questions

1. (i) X is Na_2SO_4 and Y is $BaSO_4$.

(ii) $\underset{\text{Sodium sulfate}}{Na_2SO_4(aq)} + BaCl_2(aq) \rightarrow \underset{\text{White ppt.}}{BaSO_4 \downarrow} + 2NaCl(aq)$

The reaction is a double displacement reaction.

It is also called a precipitation reaction.

2. (i) Dark coloured bottles interrupt the path of light such that light cannot reach the silver chloride in the bottles and its decomposition is prevented.

(ii) Hydrogen peroxide decomposes into H_2O and O_2 in the presence of sunlight. To prevent decomposition, they are kept in coloured bottles.

$$2H_2O_2 \xrightarrow{\text{Sunlight}} 2H_2O + O_2$$

(iii) When a copper vessel is exposed to air, it loses it shine due to corrosion because a green coating is deposited on copper.

(iv) They get tarnished by reacting with atmospheric air to form silver sulfide.

3. (i) When milk is left open at room temperature during summers, it gets oxidised and its smell and taste changes.

(ii) The iron nail becomes brownish in colour and the blue colour of the copper sulfate solution fades, because iron has displaced copper from copper sulfate solution.

4. (i) Copper sulfate: Blue; Ferrous sulfate: Pale green; Sodium sulfate: Milky white; Barium chloride: White.

(ii) Light green is the colour of crystalline $FeSO_4.7H_2O$. Light green changes to reddish-brown or brown upon heating.

(a) $FeSO_4.7H_2O \rightarrow FeSO_4 + 7H_2O$

(b) $2FeSO_4 \xrightarrow{\text{Heat}} Fe_2O_3 + SO_2 + SO_3$

5. (i) Silver: Colour: Black; Chemical name: Silver sulfide

(ii) Copper: Colour: Green; Chemical name: Copper oxide

(iii) Iron: Colour: Reddish Brown; Chemical name: Ferric oxide

6. (i) Photochemical decomposition: A single reactant breaks down to give simpler products.
White silver chloride changes to grey, as it decomposes to silver and chlorine in the presence of sunlight.

(ii) $2AgCl(s) \xrightarrow{\text{Sunlight}} 2Ag(s) + Cl_2(g)$

(iii) Black and white photography.

7. (i) Yes, the temperature rises as the beaker gets hot. This is because a large amount of heat is released, that is, the reaction is exothermic.

(ii) Combination reaction. A reaction in which a single product is obtained from two or more reactants.

(iii) $\underset{\text{(Quicklime)}}{CaO} + H_2O \rightarrow \underset{\text{(Slaked lime)}}{Ca(OH)_2} + \text{heat}$

8. (i) Brown fumes, white residue.
(ii) Decomposition reaction
$2Pb(NO_3)_2 \rightarrow 2PbO + 4NO_2 + O_2$

9. (i) Due to the decomposition of silver chloride into silver and chlorine by sunlight.
$2AgCl \xrightarrow{\text{Sunlight}} 2Ag + Cl_2$
(ii) Due to the oxidation of copper powder to copper oxide, brown colour turns into black.
$2Cu(s) + O_2(g) \rightarrow 2CuO(s)$

10. Baking powder ($NaHCO_3$) salt X is commonly used in bakery products. On heating, it forms sodium carbonate (Na_2CO_3), Y and CO_2 gas, Z is evolved. When CO_2 gas, Z is passed through lime water, it forms calcium carbonate ($CaCO_3$), which is slightly soluble in water making it milky.
X: $NaHCO_3$; Y: Na_2CO_3; Z: CO_2
$2\ NaHCO_3 \rightarrow Na_2CO_3 + H_2O + CO_2$
$Na_2CO_3 \cdot 10H_2O \rightarrow Na_2CO_3 + 10H_2O$

11. (i) Rancidity.
(ii) Oxidation reaction.
(iii) Potato chips have oil and this gets oxidised by the oxygen in air and its oxidation products have an unpleasant smell and taste. The potato chips are then said to have turned rancid. These chips became unsuitable for eating.
(iv) The packs containing potato chips are filled with nitrogen gas and then sealed. When the potato chips are surrounded by an unreactive gas like nitrogen, there is no oxygen or air in this pack to cause oxidation of oil present in potato chips and to make them rancid.
(v) Here the values displayed by Shanvi are (i) awareness or knowledge that oil and fat present in foods get spoiled on keeping exposed in the open or in air for a long time and become unsuitable to eat, and (ii) great concern for the health of her younger brother as she tasted the potato chips herself first and finding them rancid, she did not allow her brother to eat them.

Hints

Section A: Multiple Choice Questions

5. $2H_2O(l) \xrightarrow{\text{Current}} 2H_2(g) + O_2(g)$
Electrolysis of water is a decomposition reaction.

13. In this reaction, HCl is oxidised to Cl_2, whereas MnO_2 is reduced to $MnCl_2$.

31. $NH_3 + HCl \rightarrow NH_4Cl$ (dense white fumes)
It is a combination reaction in which dense white fumes of NH_4Cl are seen.

Assertion Reason Questions

2. Chips manufacturers usually flush bags of chips with a gas such as nitrogen to prevent the oil and fats of the chips from being oxidised or becoming rancid.

Section E: Case Study or Passage Based Questions

I.

3. In the presence of sunlight, the large amount of energy of light decomposes AgCl to silver (Ag+) and chloride (Cl–) ions. The silver flakes are truly black which when fully spread over white silver chloride looks grey.

II.

7. The triple bond in nitrogen is too strong to be broken and hence it is an inert gas which does not take part in combustion.

8. The complete combustion of gasoline in full supply of air results in the production of carbon dioxide and water. The chemical reaction is:
$2C_8H_{18}(l) + 25O_2(g) \rightarrow 16CO_2(g) + 18H_2O(g)$

9. The addition of oxygen to a substance or removal of hydrogen from a substance is called oxidation. The reaction in which heat energy is produced is called an exothermic reaction.

10. The process of respiration in the human body and decomposition of vegetable matter involves evolution of energy.

11. The limited supply of air leads to incomplete combustion of fuel, which in turn leads to the production of black smoke.

NCERT Activities with Discussion and Conclusion

Activity 1.1: To study the reaction between magnesium and oxygen to form magnesium oxide.

Caution: Perform this activity in the presence of your chemistry teacher. It would be better to wear eye protection like that used by welders.

- Clean a magnesium ribbon of about 2 cm long by rubbing it with sand paper.
- Hold it with a pair of tongs. Burn it using a sprit lamp or burner and collect the ash so formed in a watch glass (Fig. 1.4). Burn the magnesium ribbon keeping it as far as possible from your eyes.

Now answer:
- What do you observe?

Discussion: It is observed that the magnesium ribbon burns with a dazzling white light and changes into a white powder and this white powder is magnesium oxide.

Conclusion: Magnesium burns in air to combine with oxygen to give a white powder of magnesium oxide.

$$\underset{\text{Magnesium}}{2\,Mg(s)} + \underset{\substack{\text{Oxygen}\\\text{(from air)}}}{O_2(g)} \rightarrow \underset{\substack{\text{Magnesium oxide}\\\text{(White powder)}}}{2MgO(s)}$$

Activity 1.2: To study the reaction between lead nitrate solution and potassium iodide solution.
- Take lead nitrate solution in a test tube.
- Add potassium iodide solution to the above solution (Fig. 1.6)

Now answer:
- What do you observe?

Discussion: It is observed that a yellow solid precipitate is formed.

Conclusion: Lead nitrate solution reacts with potassium iodide solution to give a yellow precipitate of lead iodide.

$$\underset{\text{Lead nitrate}}{Pb(NO_3)_2(aq)} + \underset{\substack{\text{Potassium}\\\text{iodide}}}{2KI(aq)} \rightarrow \underset{\substack{\text{Lead(II) iodide}\\\text{(yellow ppt.)}}}{PbI_2(s)} + \underset{\substack{\text{Potassium}\\\text{nitrate}}}{2KNO_3}$$

Activity 1.3: To study the reaction between zinc and sulfuric acid or hydrochloric acid.
- Take a few zinc granules in a conical flask or a test tube.
- Add dilute hydrochloric acid or sulfuric acid to this (Fig. 1.5).

Caution: Handle the acid with care.

Now answer:
- Do you observe anything happening around the zinc granules?
- Touch the conical flask or test tube. It there any change in the temperature?

Discussion: Bubbles of hydrogen gas are found to rise briskly from the surface of the zinc granules. The gas evolved can be tested by bringing a lighted candle close. It is found to burn with a popping sound. Further, the flask is found to be hot.

Conclusion: Zinc reacts with dilute sulfuric acid or hydrochloric acid with the evolution of hydrogen gas and evolution of heat (that is, the reaction is exothermic).

$$\underset{\text{Zinc}}{Zn(s)} + \underset{\text{Sulfuric acid}}{\text{Dil. } H_2SO_4\,(aq)} \rightarrow \underset{\substack{\text{Zinc sulfate}\\\text{(solution)}}}{ZnSO_4(aq)} + \underset{\text{Hydrogen gas}}{H_2(g)} + \text{Heat}$$

Activity 1.4: To study the combination reaction between calcium oxide, that is, quicklime and water.
- Take a small amount of calcium oxide or quicklime in a beaker.
- Slowly add water to this.
- Touch the beaker as shown in Fig. 1.11.

Now answer:
- Do you feel any change in temperature?

Discussion: A vigorous reaction is found to occur here and the beaker is found to become very hot.

Conclusion: Calcium oxide combines with water to give calcium hydroxide or slaked lime and this reaction is highly exothermic.

$$\underset{\substack{\text{Calcium oxide}\\\text{(Quick lime)}}}{CaO(s)} + \underset{\text{Water}}{H_2O(l)} \rightarrow \underset{\substack{\text{Calcium hydroxide}\\\text{(Slaked lime)}}}{Ca(OH)_2} + \text{Heat}$$

Activity 1.5: To study the decomposition of ferrous sulfate on heating.
- Take about 2 g of ferrous sulfate crystals in a dry boiling tube.
- Note the colour of the ferrous sulfate crystals.
- Heat the boiling tube over the flame of a burner or spirit lamp.
- Observe the colour of the crystals after heating.

Discussion: It is observed that the green coloured ferrous sulfate crystals ($FeSO_4.7H_2O$) on heating, first change colour by losing water to give anhydrous ferrous sulfate ($FeSO_4$) which on further heating decomposes to leave behind a reddish-brown residue along with the evolution of sulfur dioxide and sulfur trioxide gases.

Conclusion: The reddish-brown residue is ferric oxide. Here the following decomposition reaction takes place:

$$\underset{\text{Ferrous sulfate crystals}}{FeSO_4.7H_2O(s)} \xrightarrow{\text{Heat}} \underset{\text{Anhydrous ferrous sulfate}}{FeSO_4(s)} + 7H_2O(l)$$

$$\underset{\text{Anhyd. ferrous sulfate}}{2FeSO_4(s)} \xrightarrow{\text{Heat}} \underset{\text{Ferric oxide}}{Fe_2O_3(s)} + \underset{\text{Sulfur dioxide}}{SO_2(g)}\uparrow + \underset{\text{Sulfur trioxide}}{SO_3(g)}$$

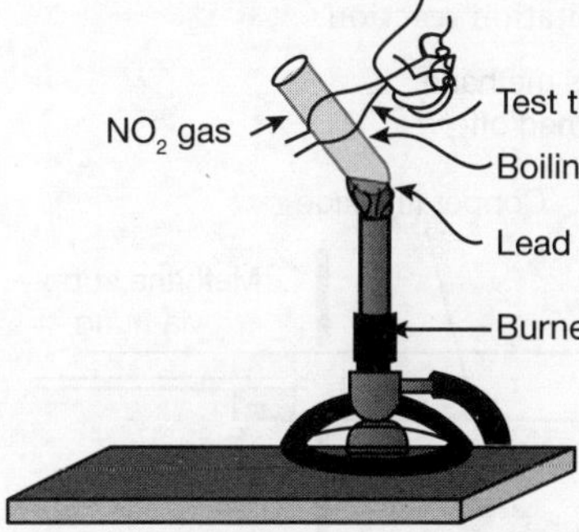

Heating of lead nitrate

Activity 1.6: To study the thermal decomposition of lead nitrate.

- Take about 2 g lead nitrate powder in a dry boiling tube.
- Hold the boiling tube with a pair of tongs and heat it over the flame as shown.

Now answer:

- What do you observe? Note down the change if any.

Discussion: Brown fumes of nitrogen dioxide (NO_2) are evolved and a yellow residue is left in the test tube.

Conclusion: Solid lead nitrate decomposes on heating to give out brown fumes of NO_2 and a yellow residue of lead(II) oxide.

$$\underset{\text{Lead nitrate (White)}}{2Pb(NO_3)_2(s)} \xrightarrow{\text{Heat}} \underset{\text{Lead (II) oxide (Yellow)}}{2PbO(s)} + \underset{\text{Nitrogen dioxide (Brown fumes)}}{4NO_2(g)} + \underset{\text{Oxygen}}{O_2(g)}$$

Activity 1.7: To study electrolytic decomposition, that is, electrolysis of water.

- Take a plastic mug. Drill two holes at its base and fit rubber stoppers in these holes. Insert carbon electrodes in these rubber stoppers as shown in Fig. 1.15.
- Connect these electrodes to a 6 volt battery.
- Fill the mug with water such that the electrodes are immersed. Add a few drops of dilute sulfuric acid to the water.
- Take two test tubes filled with water and invert them over the two carbon electrodes.
- Switch on the current and leave the apparatus undisturbed for some time.
- It will be observed that formation of bubbles takes place at both the electrodes. These bubbles displace water in the test tubes.
- Once the test tubes are partially filled with the respective gases, remove them carefully.
- Test these gases one by one by bringing a burning candle close to the mouth of the test tubes.

Now answer:

- Is the volume of the gas collected the same in both the test tubes?
- What happens in each test tube on bringing the burning candle close?
- Which gas is present in each test tube?

Discussion: It is observed that the volume of the gas collected over the cathode is double that of the gas collected over the anode. The gas with double volume burns with a popping sound whereas the other gas supports burning or combustion. So, the gas with the double volume is hydrogen whereas the gas in the other tube is oxygen.

Conclusion: Acidified water undergoes electrolysis producing H_2 and O_2 gases in a ratio of 2 : 1 by volume.

$$\underset{\text{Water}}{2H_2O(l)} \xrightarrow{\text{Electrolysis}} \underset{\substack{\text{Hydrogen} \\ \text{(burns with a popping sound)}}}{2H_2(g)} + \underset{\substack{\text{Oxygen} \\ \text{(supporter of combustion)}}}{O_2(g)}$$

Activity 1.8: To study the photodecomposition of silver chloride.

- Take about 2 g silver chloride in a china dish.
- Note its colour.
- Place this china dish in sunlight for some time (Fig. 1.16).

- Observe the colour of the silver chloride after some time.

Discussion: It is observed that the white crystals of silver chloride turn grey in the sunlight.

Conclusion: Silver chloride decomposes into grey silver and chlorine in the presence of light.

$$\underset{\text{Silver chloride (White)}}{AgCl(s)} \xrightarrow{\text{Sunlight}} \underset{\text{Silver (Grey)}}{2Ag(s)} + \underset{\text{Chlorine}}{Cl_2(g)}$$

Activity 1.9: To study the displacement of copper from copper sulfate solution by iron.

- Take two iron nails and clean them by rubbing with sandpaper.
- Take two test tubes marked as (A) and (B). In each test tube, take about 10 mL of copper sulfate solution.
- Tie one iron nail with a thread and immerse it carefully in the copper sulfate solution in test tube B for about 20 minutes (Fig. 1.19). Keep one iron nail aside for comparison.
- After 20 minutes, take out the iron nail from the copper sulfate solution.
- Compare the intensity of the blue colour of copper sulfate solutions in test tubes (A) and (B).
- Also compare the colour of the iron nails dipped sulfate solutions with the one kept aside.

Discussion: It is observed that iron nail becomes brownish in colour and the blue colour of the copper sulfate solution fades and changes to light green.

Conclusion: Iron displaces copper from copper sulfate solution forming iron(II) sulfate in the solution which has a light green colour. Thus, the blue colour of copper sulfate solution fades. The displaced copper is deposited on the iron nail, giving it a brownish colour.

$$\underset{\text{Iron}}{Fe(s)} + \underset{\text{Copper sulphate (Blue solution)}}{CuSO_4} \rightarrow \underset{\text{Ferrous sulphate (Light green solution)}}{FeSO_4(aq)} + \underset{\text{Copper (Brown)}}{Cu}$$

Activity 1.10: To study the double decomposition reaction between barium chloride solution and sodium sulfate solution.

- Take about 3 mL of sodium sulfate solution in a test tube.
- In another test tube, take about 3 mL of barium chloride solution.
- Mix the two solutions (Fig. 1.21).

Now answer:

- What do you observe?

Discussion: It is observed that a white solid, insoluble in water, is formed. This white insoluble solid is known as a precipitate.

Conclusion: Barium chloride solution reacts with sodium sulfate solution to give a white precipitate of barium sulfate along with sodium chloride in solution.

$$\underset{\text{Barium chloride}}{BaCl_2(aq)} + \underset{\text{Sodium sulfate}}{Na_2SO_4(aq)} \rightarrow \underset{\text{Barium sulfate (White ppt.)}}{BaSO_4(s)} + \underset{\text{Sodium chloride (in solution)}}{2NaCl(aq)}$$

The above reaction is a double displacement reaction as well as a precipitation reaction.

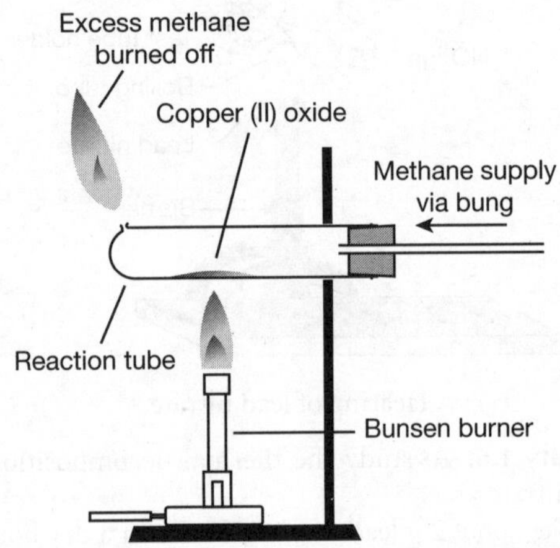

Activity 1.11: To study the oxidation of copper to copper(II) oxide.

- Heat a china dish containing about 1 g copper powder (Fig.).

Now answer:

- What do you observe?

Discussion: It is observed that the brown copper powder gets coated with black copper(II) oxide.

Conclusion: Copper (brown in colour) on heating combines with oxygen to give black copper(II) oxide.

$$\underset{\text{Copper (Brown)}}{2Cu(s)} + \underset{\text{Oxygen}}{O_2(g)} \rightarrow \underset{\text{Copper (II) oxide (Black)}}{2CuO(s)}$$

For reduction of copper(II) oxide back to copper, heat with hydrogen.

Group Activity: To determine whether a given chemical reaction or dissolution of the given salt is exothermic or endothermic.

- Take four beakers and label them A, B, C and D.
- Put 25 mL of water in A, B and C beakers and copper sulfate solution in beaker D.

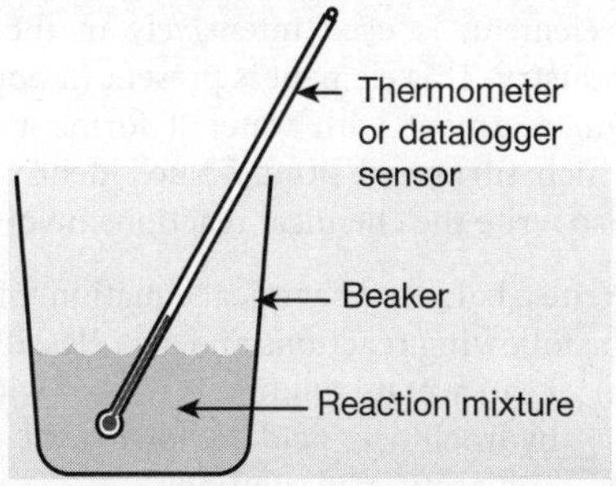

- Measure and record the temperature of each liquid contained in the beakers above.
- Add two spatulas of potassium sulfate, ammonium nitrate, anhydrous copper sulfate and fine iron filings to beakers A, B, C and D respectively and stir.
- Finally measure and record the temperature of each of the mixtures above.

Now answer:

- Which reactions are exothermic and which ones are endothermic in nature?

Discussion: It is observed that the temperature rises in beakers C and D whereas the temperature falls in beakers A and B. This shows that the dissolution of potassium sulfate and ammonium nitrate in water is endothermic while the dissolution of anhydrous copper sulfate in water is exothermic and the displacement reaction (which is a redox reaction) taking place in beaker D is also exothermic.

Conclusion: Dissolution of salts may be exothermic or endothermic but redox reactions are exothermic.

Exemplar Problems

Short Answer Questions

1. Write the balanced chemical equations for the following reactions and identify the type of reaction in each case.
(a) Nitrogen gas is treated with hydrogen gas in the presence of a catalyst at 773 K to form ammonia gas.
(b) Sodium hydroxide solution is treated with acetic acid to form sodium acetate and water.
(c) Ethanol is warmed with ethanoic acid to form ethyl acetate in the presence of concentrated H_2SO_4.
(d) Ethene is burnt in the presence of oxygen to form carbon dioxide, water and releases heat and light.

2. Write the balanced chemical equations for the following reactions and identify the type of reaction in each case.
(a) Thermite reaction, iron(III) oxide reacts with aluminium and gives molten iron and aluminium oxide.
(b) Magnesium ribbon is burnt in an atmosphere of nitrogen gas to form solid magnesium nitride.
(c) Chlorine gas is passed in an aqueous potassium iodide solution to form potassium chloride solution and solid iodine.
(d) Ethanol is burnt in air to form carbon dioxide, water and releases heat.

3. Complete the missing components/variables given as *x* and *y* in the following reactions
(a) $Pb(NO_3)_2(aq) + 2KI(aq) \rightarrow PbI_2(x) + 2KNO_3(y)$
(b) $Cu(s) + 2Ag\,NO_3(aq) \rightarrow Cu(NO_3)_2(aq) + x(s)$
(c) $Zn(s) + H_2SO_4(aq) \rightarrow ZnSO_4(x) + H_2(y)$
(d) $CaCO_3(s) \xrightarrow{x} CaO(s) + CO_2(g)$

4. Which among the following changes are exothermic or endothermic in nature?
(a) Decomposition of ferrous sulfate
(b) Dilution of sulfuric acid
(c) Dissolution of sodium hydroxide in water
(d) Dissolution of ammonium chloride in water

5. Identify the reducing agent in the following reactions
(a) $4NH_3 + 5O_2 \rightarrow 4NO + 6H_2O$
(b) $H_2O + F_2 \rightarrow HF + HOF$
(c) $Fe_2O_3 + 3CO \rightarrow 2Fe + 3CO_2$
(d) $2H_2 + O_2 \rightarrow 2H_2O$

6. Identify the oxidising agent (oxidant) in the following reactions.
(a) $Pb_3O_4 + 8HCl \rightarrow 3PbCl_2 + Cl_2 + 4H_2O$
(b) $2Mg + O_2 \rightarrow 2MgO$
(c) $CuSO_4 + Zn \rightarrow Cu + ZnSO_4$

(d) $V_2O_5 + 5Ca \rightarrow 2V + 5CaO$
(e) $3Fe + 4H_2O \rightarrow Fe_3O_4 + 4H_2$
(f) $CuO + H_2 \rightarrow Cu + H_2O$

7. Write the balanced chemical equations for the following reactions
   (a) Sodium carbonate on reaction with hydrochloric acid in equal molar concentrations gives sodium chloride and sodium hydrogencarbonate.
   (b) Sodium hydrogencarbonate on reaction with hydrochloric acid gives sodium chloride, water and liberates carbon dioxide.
   (c) Copper sulfate on treatment with potassium iodide precipitates cuprous iodide (Cu_2I_2), liberates iodine gas and also forms potassium sulfate.

8. A solution of potassium chloride when mixed with silver nitrate solution, forms an insoluble white substance. Write the chemical reaction involved and also mention the type of the chemical reaction.

9. Ferrous sulfate decomposes with the evolution of a gas having a characteristic odour of burning sulfur. Write the chemical reaction involved and identify the type of reaction.

10. Why do fire flies glow at night?

11. Grapes hanging on the plant do not ferment but after being plucked from the plant can be fermented. Under what conditions do these grapes ferment? Is it a chemical or a physical change?

12. Which among the following are physical or chemical changes?
    (a) Evaporation of petrol
    (b) Burning of Liquefied Petroleum Gas (LPG)
    (c) Heating of an iron rod to red heat
    (d) Curdling of milk
    (e) Sublimation of solid ammonium chloride

13. During the reaction of some metals with dilute hydrochloric acid, the following observations were made.
    (a) Silver metal does not show any change.
    (b) The temperature of the reaction mixture rises when aluminium (Al) is added.
    (c) The reaction of sodium metal is found to be highly explosive.
    (d) Some bubbles of a gas are seen when lead (Pb) is reacted with the acid.

    Explain these observations giving suitable reasons.

14. A substance X, which is an oxide of a group 2 element, is used intensively in the cement industry. This element is present in bones also. On treatment with water it forms a solution which turns red litmus blue. Identify X and also write the chemical reactions involved.

15. Write a balanced chemical equation for each of the following reactions and also classify them.
    (a) Lead acetate solution is treated with dilute hydrochloric acid to form lead chloride and acetic acid solution.
    (b) A piece of sodium metal is added to absolute ethanol to form sodium ethoxide and hydrogen gas.
    (c) Iron(III) oxide on heating with carbon monoxide gas reacts to form solid iron and liberates carbon dioxide gas.
    (d) Hydrogen sulfide gas reacts with oxygen gas to form solid sulfur and liquid water.

16. Why do we store silver chloride in dark coloured bottles?

17. Balance the following chemical equations and identify the type of chemical reaction.
    (a) $Mg(s) + Cl_2(g) \rightarrow MgCl_2(s)$
    (b) $HgO(s) \xrightarrow{\text{Heat}} Hg(l) + O_2(g)$
    (c) $Na(s) + S(s) \xrightarrow{\text{Fuse}} Na_2S(s)$
    (d) $TiCl_4(l) + Mg(s) \rightarrow Ti(s) + MgCl_2(s)$
    (e) $CaO(s) + SiO_2(s) \rightarrow CaSiO_3(s)$
    (f) $H_2O_2(l) \xrightarrow{\text{UV}} H_2O(l) + O_2(g)$

18. A magnesium ribbon is burnt in oxygen to give a white compound X accompanied by emission of light. If the burning ribbon is now placed in an atmosphere of nitrogen, it continues to burn and forms a compound Y.
    (a) Write the chemical formulae of X and Y.
    (b) Write a balanced chemical equation, when X is dissolved in water.

19. Zinc liberates hydrogen gas when reacted with dilute hydrochloric acid, whereas copper does not. Explain why.

20. A silver article generally turns black when kept in the open for a few days. The article when rubbed with toothpaste again starts shining.
    (a) Why do silver articles turn black when kept in the open for a few days? Name the phenomenon involved.

(b) Name the black substance formed and give its chemical formula.

Long Answer Questions

1. On heating the blue coloured powder of copper(II) nitrate in a boiling tube, copper oxide (black), oxygen gas and a brown gas X is formed.
   (a) Write a balanced chemical equation of the reaction.
   (b) Identify the brown gas X evolved.
   (c) Identify the type of reaction.
   (d) What could be the pH range of the aqueous solution of the gas X?

2. Give the characteristic tests for the following gases
   (a) CO_2 (b) SO_2
   (c) O_2 (d) H_2

3. What happens when a piece of
   (a) zinc metal is added to copper sulfate solution?
   (b) aluminium metal is added to dilute hydrochloric acid?
   (c) silver metal is added to copper sulfate solution?

   Also, write the balanced chemical equation if the reaction occurs.

4. What happens when zinc granules are treated with dilute solution of H_2SO_4, HCl, HNO_3, NaCl and NaOH? Also write the chemical equations if the reaction occurs.

5. On adding a drop of barium chloride solution to an aqueous solution of sodium sulfite, a white precipitate is obtained.
   (a) Write a balanced chemical equation of the reaction involved.
   (b) What other name can be given to this precipitation reaction?
   (c) On adding dilute hydrochloric acid to the reaction mixture, the white precipitate disappears. Why?

6. You are provided with two containers made up of copper and aluminium. You are also provided with solutions of dilute HCl, dilute HNO_3, $ZnCl_2$ and H_2O. In which of the above containers can these solutions be kept?

Answers and Solutions

Short Answer Questions

1. (a) $N_2(g) + 3H_2(g) \xrightarrow{\text{Catalyst 773 K}} 2NH_3(g)$
   Combination reaction
   (b) $NaOH(aq) + CH_3COOH(aq) \rightarrow CH_3COONa(aq) + H_2O(l)$
   Double displacement reaction/neutralisation reaction
   (c) $C_2H_5OH(l) + CH_3COOH(l) \xrightarrow{H^+} CH_3COOC_2H_5(l) + H_2O(l)$
   Double displacement reaction/esterificaton reaction
   (d) $C_2H_4(g) + 3O_2(g) \rightarrow 2CO_2(g) + 2H_2O(g) +$ Heat + Light
   Redox reaction/combustion reaction

2. (a) $Fe_2O_3(s) + 2Al(s) \rightarrow Al_2O_3(s) + 2Fe(l) +$ Heat
   Displacement reaction/redox reaction
   (b) $3Mg(s) + N_2(g) \rightarrow Mg_3N_2(s)$
   Combination reaction
   (c) $2KI(aq) + Cl_2(g) \rightarrow 2KCl(aq) + I_2(s)$
   Displacement reaction
   (d) $C_2H_5OH(l) + 3O_2(g) \rightarrow 2CO_2(g) + 3H_2O(l)$ + Heat
   Redox reaction/combustion reaction

3. (a) $x \rightarrow (s)$ $y \rightarrow (aq)$
   (b) $x \rightarrow 2$ Ag
   (c) $x \rightarrow (aq)$ $y \rightarrow (g)$
   (d) $x \rightarrow$ Heat

4. (b) and (c) are exothermic as heat is released in these changes.
   (a) and (d) are endothermic as heat is absorbed in these changes.

5. (a) Ammonia (NH_3).
   (b) Water (H_2O) as F_2 is getting reduced to HF.
   (c) Carbon monoxide (CO).
   (d) Hydrogen.

   Hint: Reducing agents are those substances which have the ability of adding hydrogen or removing oxygen from the other substances.

6. (a) Pb_3O_4
   (b) O_2
   (c) $CuSO_4$

(d) V_2O_5
(e) H_2O
(f) CuO

7. (a) $Na_2CO_3 + HCl \rightarrow NaCl + NaHCO_3$
(b) $NaHCO_3 + HCl \rightarrow NaCl + H_2O + CO_2$
(c) $2CuSO_4 + 4KI \rightarrow Cu_2I_2 + 2K_2SO_4 + I_2$

8. $KCl(aq) + AgNO_3(aq) \rightarrow AgCl(s) + KNO_3(aq)$
It is a double displacement and precipitation reaction.

9. $2FeSO_4(s) \xrightarrow{Heat} Fe_2O_3(s) + SO_2(g) + SO_3(g)$
It is a thermal decomposition reaction.

10. Fire flies have a protein which in the presence of an enzyme undergoes aerial oxidation. This is a chemical reaction which involves the emission of visible light. Therefore, fire flies glow at night.

11. Grapes when attached to the plants are living and therefore their own immune system prevents fermentation. The microbes can grow in the plucked grapes and under anaerobic conditions these can be fermented. This is a chemical change.

12. (a), (c) and (e) are physical changes.
(b) and (d) are chemical changes.

13. ***Hint***: (a) Silver metal does not react with dilute HCl.
(b) The temperature of the reaction mixture rises when aluminium is added because it is an exothermic reaction.
(c) Reaction of sodium metal is found to be highly explosive because it is an exothermic reaction.
(d) When lead is treated with hydrochloric acid, bubbles of hydrogen gas are evolved.
$Pb + 2HCl \rightarrow PbCl_2 + H_2$

14. Calcium oxide
$CaO(s) + H_2O(l) \rightarrow Ca(OH)_2(aq)$

15. (a) $Pb(CH_3COO)_2 + 2HCl \rightarrow PbCl_2 + CH_3COOH$; Double displacement reaction
(b) $2Na + 2C_2H_5OH \rightarrow 2C_2H_5ONa + H_2$; Displacement reaction
(c) $Fe_2O_3 + 3CO \rightarrow 2Fe + 3CO_2$; Redox reaction
(d) $2H_2S + O_2 \rightarrow 2S + 2H_2O$; Redox reaction

16. Silver chloride on exposure to sunlight may decompose as per the following reaction.
$2AgCl \rightarrow 2Ag + Cl_2$
Therefore, it is stored in dark coloured bottles.

17. (a) Balanced; Combination reaction.
(b) $2HgO(s) \xrightarrow{Heat} 2Hg(l) + O_2(g)$; Decomposition reaction.
(c) $2Na(s) + S(s) \xrightarrow{Fuse} Na_2S(s)$; Combination reaction.
(d) $TiCl_4(l) + 2Mg(s) \rightarrow Ti(s) + 2MgCl_2(s)$; Displacement reaction.
(e) Balanced; Combination reaction.
(f) $2H_2O_2(l) \xrightarrow{uv} 2H_2O(l) + O_2(g)$; Decomposition reaction.

18. $2Mg + O_2 \rightarrow 2MgO$
$3Mg + N_2 \rightarrow Mg_3N_2$
(a) X is MgO; Y is Mg_3N_2
(b) $MgO + H_2O \rightarrow Mg(OH)_2$

19. Zinc is above hydrogen whereas copper is below hydrogen in the activity series of metals. That is why zinc displaces hydrogen from dilute hydrochloric acid, while copper does not.
$Zn + HCl \rightarrow ZnCl_2 + H_2$
$Cu + HCl \rightarrow$ No reaction

20. (a) Metals such as silver when attacked by substances around it such as moisture, acids and gases are said to corrode and this phenomenon is called corrosion.
(b) The black substance is formed because silver (Ag) reacts with the H_2S present in air. It forms a thin black coating of silver sulfide (Ag_2S).

Long Answer Questions

1. (a) Balanced chemical equation
$2Cu(NO_3)_2(s) \xrightarrow{Heat} 2CuO(s) + O_2(g) + 4NO_2(g)$
(b) The brown gas X evolved is nitrogen dioxide (NO_2).
(c) This is a decomposition reaction.
(d) Nitrogen dioxide dissolves in water to form an acidic solution because it is an oxide of a non-metal. Therefore, the pH of this solution is less than 7.

2. The characteristic test for
(a) Carbon dioxide (CO_2) gas turns lime water milky when passed through it due to the formation of insoluble calcium carbonate.
$\underset{\text{Lime water}}{Ca(OH)_2} + \underset{\text{Carbon dioxide}}{CO_2} \rightarrow \underset{\text{Calcium carbonate}}{CaCO_3} + H_2O$

(b) Sulfur dioxide (SO_2) gas when passed through acidic potassium permanganate solution (purple in colour) turns it colourless because SO_2 is a strong reducing agent.

$2KMnO_4 + 2H_2O + 5SO_2 \rightarrow K_2SO_4$
Potasssium permanganate (Purple); Sulfur dioxide; Potassium sulfate (Colourless)

$+ 2MnSO_4 + 2H_2SO_4$
Manganese sulfate (Colourless)

Or

Sulfur dioxide gas when passed through acidic dichromate solution (orange in colour) turns it green because sulfur dioxide is a strong reducing agent.

(c) The evolution of oxygen (O_2) gas during a reaction can be confirmed by bringing a burning candle near the mouth of the test tube containing the reaction mixture. The intensity of the flame increases because oxygen supports burning.

(d) Hydrogen (H_2) gas burns with a pop sound when a burning candle is brought near it.

3. (a) Zinc being more reactive than copper displaces copper from its solution and a solution of zinc sulfate is obtained.

$Zn(s) + CuSO_4(aq) \rightarrow ZnSO_4(aq) + Cu(s)$
($CuSO_4$: Blue; $ZnSO_4$: Colourless)

This is an example of a displacement reaction.

(b) Aluminium being more reactive displaces hydrogen from dilute hydrochloric acid solution and hydrogen gas is evolved.

$2Al(s) + 6HCl(aq) \rightarrow 2AlCl_3(aq) + 3H_2(g)$
($AlCl_3$: Aluminium chloride)

(c) Silver metal being less reactive than copper cannot displace copper from its salt solution. Therefore, no reaction occurs

$Ag(s) + CuSO_4(aq) \rightarrow$ No reaction

4. The reaction of Zn granules with

(a) Dilute H_2SO_4

$Zn(s) + H_2SO_4(aq) \rightarrow ZnSO_4(aq) + H_2(g)$

(b) Dilute HCl

$Zn(s) + 2HCl(aq) \rightarrow ZnCl_2(aq) + H_2(g)$

(c) Dilute HNO_3

Reaction with dilute HNO_3 is different when compared to reactions with other acids because nitric acid is an oxidising agent and it oxidises H_2 gas evolved to H_2O.

$4\,Zn(s) + 10HNO_3(aq) \rightarrow 4\,Zn(NO_3)_2(aq) + 5\,H_2O(l) + N_2O(g)$

(d) NaCl solution

$Zn(s) + NaCl(aq) \rightarrow$ No reaction

(e) NaOH solution

$Zn(s) + 2\,NaOH(aq) \rightarrow Na_2ZnO_2(aq) + H_2(g)$
(Na_2ZnO_2: Sodium zincate)

5. (a) Balanced chemical equation

$Na_2SO_3(aq) + BaCl_2(aq) \rightarrow BaSO_3(s) + 2\,NaCl(aq)$
Sodium sulfite; Barium chloride; Barium sulfite; Sodium chloride

(b) This reaction is also known as a double displacement reaction.

(c) $BaSO_3$ is a salt of a weak acid (H_2SO_3), therefore a dilute acid such as HCl decomposes barium sulfite to produce sulfur dioxide gas which has the smell of burning sulfur.

$BaSO_3(s) + 2HCl(aq) \rightarrow BaCl_2 + H_2O + SO_2(g)$
White ppt.

$BaCl_2$ is soluble in water, hence the white precipitate disappears.

6. (A) When solutions are kept in a copper container,

(a) Dilute HCl: Copper does not react with dilute HCl. Therefore, it can be kept.

(b) Dilute HNO_3: Nitric acid acts as a strong oxidising agent and reacts with the copper vessel, therefore it cannot be kept.

(c) $ZnCl_2$: Zinc is more reactive than copper (Cu) therefore, no displacement reaction occurs and hence it can be kept.

(d) H_2O: Copper does not react with water. Therefore, it can be kept.

(B) When solutions are kept in aluminium containers,

(a) Dilute HCl: Aluminium reacts with dilute HCl to form its salt and hydrogen is evolved. Therefore, it cannot be kept.

$2\,Al + 6HCl \rightarrow 2\,AlCl_3 + 3\,H_2$

(b) Dilute HNO_3: Aluminium gets oxidised by dilute HNO_3 to form a layer of Al_2O_3 and therefore, it can be kept.

(c) $ZnCl_2$: Aluminium being more reactive than zinc can displace zinc

ion from the solution. Therefore, the solution cannot be kept.
$2\ Al + 3\ ZnCl_2 \rightarrow 2\ AlCl_3 + 3Zn$

(d) H_2O: Aluminium does not react with cold or hot water. Therefore, water can be kept.

Aluminium is attacked by steam to form aluminium oxide and hydrogen.
$2Al(s) + 3H_2O(g) \rightarrow Al_2O_3(s) + 3H_2(g)$

NCERT CORNER

In-Text Questions

1. Why should a magnesium ribbon be cleaned before burning in air?

2. Write the balanced equation for the following chemical reactions.
(I) Hydrogen + Chlorine → Hydrogen chloride
(II) Barium chloride + Aluminium sulfate → Barium sulfate + Aluminium chloride

3. Write the balanced chemical equation with state symbols for the following reactions.
(I) Solutions of barium chloride and sodium sulfate in water react to give insoluble barium sulfate and the solution of sodium chloride.
(II) Sodium hydroxide solution (in water) reacts with hydrochloric acid solution (in water) to produce sodium chloride solution and water.

4. A solution of a substance X is used for whitewashing.
(I) Name the substance X and write its formula.
(II) Write the reaction of the substance X named in (I) above with water.

5. Why is the amount of a gas collected in one of the test tubes in 'Electrolysis of water' activity double of the amount collected in the other? Name this gas.

6. When you mix the solution of lead(II) nitrate and potassium iodide,
   (i) What is the colour of the precipitate formed? Name the compound precipitated.
   (ii) Write the balanced chemical equation for this reaction.
   (iii) Is this also a double displacement reaction?

7. A magnesium ribbon burns with a dazzling flame in air (oxygen) and changes into a white substance, magnesium oxide. Is magnesium being oxidised or reduced in this reaction?

8. Why does the colour of the copper sulfate solution change when an iron nail is dipped in it?

9. Give an example of a double displacement reaction.

10. Identify the substances that are oxidised and the substances that are reduced in the following reactions.
   (i) $4Na(s) + O_2(g) \rightarrow 2Na_2O(s)$
   (ii) $CuO(s) + H_2(g) \rightarrow Cu(s) + H2O(l)$

Textbook Exercises

1. Which of the statements about the reaction below are incorrect?
$2PbO(s) + C(s) \rightarrow 2Pb(s) + CO_2(g)$
   (i) Lead is getting reduced.
   (ii) Carbon dioxide is getting oxidised.
   (iii) Carbon is getting oxidised.
   (iv) Lead oxide is getting reduced.
   (a) (i) and (ii)
   (b) (i) and (iii)
   (c) (i), (ii) and (iii)
   (d) all statements are incorrect

2. $Fe_2O_3 + 2Al \rightarrow Al_2O_3 + 2Fe$
The above reaction is an example of a:
   (a) combination reaction
   (b) double displacement reaction
   (c) decomposition reaction
   (d) displacement reaction

3. What happens when dilute hydrochloric acid is added to iron filings? Tick the correct answer.
   (a) hydrogen gas and iron chloride are produced
   (b) chlorine gas and iron hydroxide are produced
   (c) no reaction takes place
   (d) iron salt and water are produced

4. What is a balanced chemical equation? Why should chemical equations be balanced?

5. Translate the following statements into chemical equations and then balance them.
   (a) Hydrogen gas combines with nitrogen to form ammonia.
   (b) Hydrogen sulfide gas burns in air to give water and sulfur dioxide.
   (c) Barium chloride reacts with aluminium sulfate to give aluminium chloride and a precipitate of barium sulfate.
   (d) Potassium metal reacts with water to give potassium hydroxide and hydrogen gas.

6. Balance the following chemical equations.
   (a) $HNO_3 + Ca(OH)_2 \rightarrow Ca(NO_3)_2 + H_2O$
   (b) $NaOH + H_2SO_4 \rightarrow Na_2SO_4 + H_2O$
   (c) $NaCl + AgNO_3 \rightarrow AgCl + NaNO_3$
   (d) $BaCl_2 + H_2SO_4 \rightarrow BaSO_4 + HCl$

7. Write the balanced chemical equations for the following reactions.
   (a) Calcium hydroxide + Carbon dioxide → Calcium carbonate + Water
   (b) Zinc + Silver nitrate → Zinc nitrate + Silver
   (c) Aluminium + Copper chloride → Aluminium chloride + Copper
   (d) Barium chloride + Potassium sulfate → Barium sulfate + Potassium chloride

8. Write the balanced chemical equations for the following and identify the type of reaction in each case.
   (a) Potassium bromide(*aq*)+ Barium iodide(*aq*) → Potassium iodide(*aq*)+ Barium bromide(*s*)
   (b) Zinc carbonate(*s*) → Zinc oxide(*s*) + Carbon dioxide(*g*)
   (c) Hydrogen(*g*) + Chlorine(*g*) → Hydrogen chloride(*g*)
   (d) Magnesium(*s*) + Hydrochloric acid(*aq*) → Magnesium chloride(*aq*) + Hydrogen(*g*)

9. What does one mean by exothermic and endothermic reactions? Give examples.

10. Why is respiration considered an exothermic reaction? Explain.

11. Why are decomposition reactions called the opposite of combination reactions? Write equations for these reactions.

12. Write one equation each for decomposition reactions where energy is supplied in the form of heat, light or electricity.

13. What is the difference between displacement and double displacement reactions? Write equations for these reactions.

14. In the refining of silver, the recovery of silver from silver nitrate solution involves displacement by copper metal. Write down the reaction involved.

15. What do you mean by a precipitation reaction? Explain by giving examples.

16. Explain the following in terms of gain or loss of oxygen with two examples each.
    (a) Oxidation (b) Reduction

17. A shiny brown coloured element X on heating in air becomes black in colour. Name the element X and the black coloured compound formed.

18. Why do we apply paint on iron articles?

19. Oil and fat containing food items are flushed with nitrogen. Why?

20. Explain the following terms with one example each.
    (a) Corrosion (b) Rancidity

Answers and Solutions

In-Text Questions

1. A magnesium ribbon should be cleaned before burning in air to remove the layer of magnesium oxide formed due to the reaction of magnesium with oxygen present in the air. The layer of magnesium oxide is usually removed by rubbing the ribbon with sandpaper.

2.. The balanced equations for the chemical reactions are:
   (I) $H_2+Cl_2 \rightarrow 2HCl$
   (II) $3BaCl_2+Al_2(SO_4)_3 \rightarrow 3BaSO_4+2AlCl_3$

3. (I) $BaCl_2(aq)+Na_2SO_4(aq) \rightarrow BaSO_4(s)+2NaCl(aq)$
   (II) $NaOH(aq)+HCl(aq) \rightarrow NaCl(aq)+H_2O(l)$

4. (I) The substance X is quicklime (calcium oxide) which is used for white washing. Its chemical formula is CaO.
   (II) Calcium oxide on reaction with water gives calcium hydroxide.
   $CaO(s)+H_2O(l) \rightarrow Ca(OH)_2(aq)$

5. Water is formed when hydrogen and oxygen combine in the ratio of 2:1 by volume. So, the volume of hydrogen gas is double of that of oxygen gas. Thus, on decomposition of water during electrolysis, the volume of hydrogen formed is twice that of oxygen gas.

6. (i) The colour of the precipitate formed is yellow. The compound formed is lead iodide (PbI_2).
   (ii) $\underset{\text{Lead nitrate}}{Pb(NO_3)_2(aq)} + \underset{\text{Potassium iodide}}{2KI(aq)} \rightarrow \underset{\text{Lead iodide}}{PbI_2(s)} + \underset{\text{Potassium nitrate}}{2KNO_3(aq)}$
   (iii) Yes, it is a double displacement reaction.

7. Magnesium is oxidised to magnesium oxide (because it gains oxygen).

8. Copper sulfate solution is initially blue in colour. When an iron nail is dipped in it, the blue colour changes to green. This happens due to a displacement reaction where iron being more reactive than Cu, displaces Cu to form a green coloured iron sulfate solution while copper metal gets precipitated. Hence the colour changes from blue to green.
   $Fe(s) + CuSO_4(aq)$(blue) $\rightarrow FeSO_4(aq)$(green) + $Cu(s)$

9. Example of a double displacement reaction is the reaction of lead nitrate and hydrochloric acid. When aqueous solutions of both compounds are mixed, lead chloride precipitates out along with the formation of nitric acid. The chemical equation can be given as:
   $Pb(NO_3)_2(aq)$ (Lead nitrate) + $2HCl(aq)$ (Hydrochloric acid) $\rightarrow PbCl_2(s)$ (Lead chloride) + $2HNO_3(aq)$ (Nitric acid)

10. Oxidation is the addition of oxygen and reduction is removal of oxygen. Therefore, in reaction (i), sodium metal is oxidised to Na_2O as oxygen is added to it. Oxygen is reduced to Na_2O as the number of oxygen atoms per molecule becomes less (from 2 to 1). Hence, sodium gets oxidised while oxygen gets reduced.

    In reaction (ii), copper oxide is reduced to copper (loss of oxygen) and hydrogen is oxidised to H_2O (gain of oxygen).

Textbook Exercises

1. (a)
   In the above reaction, lead oxide gets converted to lead and carbon gets converted to carbon dioxide. Lead oxide loses oxygen whereas carbon gains oxygen. Loss of oxygen is known as reduction while the gain of oxygen is called oxidation. Hence, lead oxide gets reduced and carbon gets oxidised. Thus statements (iii) and (iv) are correct whereas statements (i) and (ii) are incorrect.

2. (d)
   In the given reaction, ferric oxide gets converted to iron while aluminium gets oxidised to aluminium oxide. Thus, iron is going from the combined state to the free state whereas aluminium is going from the free state to the combined state. Such a reaction is called a displacement reaction as in displacement reactions, the more reactive metal (Al in this case) displaces the less reactive metal (Fe in this case) from its compound.

3. (a)
   When metals react with acids, H_2 gas is evolved along with the formation of the respective salt of the metal.
   $Fe(s) + 2HCl(aq) \rightarrow FeCl_2(s) + H_2(g)\uparrow$

4. A chemical reaction in which the number of atoms of each element is equal on both, the reactants' side and products' side, is called a balanced chemical equation. An example of a balanced chemical reaction is given below:
   $N_2(g) + 3H_2(g) \rightarrow 2NH_3(g)$
   As we can see, there are 2 atoms of N and 6 atoms of H on both the sides and hence, this reaction is called a balanced chemical equation. A reaction should always be written in the balanced form as an unbalanced reaction violates the fundamental principle of law of conservation of mass. According to the law of conservation of mass, the total mass of reactants should be equal to the total mass of the products.

   The law of charge conservation: The total sum of charge on both sides, that is, reactant or product side must be zero. The law of conservation of mass states that mass can neither be created nor destroyed.

Hence, in a chemical reaction, the total mass of reactants should be equal to the mass of the products. It means that the total number of atoms of each element should be equal on both sides of a chemical equation. It is for this reason that chemical equations should be balanced.

5. (a) $3H_2(g) + N_2(g) \rightarrow 2NH_3(g)$
(b) $2H_2S(g) + 3O_2(g) \rightarrow 2H_2O(l) + 2SO_2(g)$
(c) $3BaCl_2(aq) + Al_2(aq) \rightarrow 2AlCl_3(aq) + 3BaSO_4(ppt)$
(d) $2K(s) + 2H_2O(l) \rightarrow 2KOH(aq) + H_2(g)$

6. (a) $2HNO_3 + Ca(OH)_2 \rightarrow Ca(NO_3)_2 + 2H_2O$
(b) $2NaOH + H_2SO_4 \rightarrow Na_2SO_4 + 2H_2O$
(c) $NaCl + AgNO_3 \rightarrow AgCl + NaNO_3$
(d) $BaCl_2 + H_2SO_4 \rightarrow BaSO_4 + 2HCl$

7. (A) $Ca(OH)_2 + CO_2 \rightarrow CaCO_3 + H_2O$
(B) $Zn + 2AgNO_3 \rightarrow Zn(NO_3)_2 + 2Ag$
(C) $2Al + 3CuCl_2 \rightarrow AlCl_3 + 3Cu$
(D) $BaCl_2 + K_2SO_4 \rightarrow BaSO_4 + 2KCl$

8. (a) $2KBr(aq) + BaI_2(aq) \rightarrow 2KI(aq) + BaBr_2(s)$; In this reaction, there is an exchange of ions between the reactants, so it is a double displacement reaction.
(b) $ZnCO_3(s) \rightarrow ZnO(s) + CO_2(g)$; In this reaction, the single reactant breaks down to give simpler products, so it is a decomposition reaction.
(c) $H_2(g) + Cl_2(g) \rightarrow 2HCl(g)$; In this reaction, a single product is formed from two reactants, so it is a combination reaction.
(d) $Mg(s) + 2HCl(aq) \rightarrow MgCl_2(aq) + H_2(g)$; In this reaction, magnesium has displaced the Cl^- ion, so it is a displacement reaction.

9. **Exothermic reaction**: Chemical reactions that release energy in the form of heat or sound are called exothermic reactions.
Example 1: Mixture of sodium and chlorine to yield table salt.
$Na(s) + 12Cl_2(g) \rightarrow NaCl(s) + 411$ kJ of energy
Example 2: Combustion reaction or burning of natural gas.
$CH_4(g) + 2O_2(g) \rightarrow CO_2(g) + 2H_2O(l)$
Example 3: Respiration reaction.
$C_6H_{12}O_6(aq)$ (Glucose) $+ 6O_2(aq) \rightarrow 6CO_2(aq) + 6H_2O(l) +$ energy
Endothermic reaction: Reactions that absorb energy or require energy in order to proceed are called endothermic reactions.
Example: White silver chloride turns grey in the sunlight. This is due to the decomposition of silver chloride in the presence of sunlight to give silver and chlorine. Energy is required (sunlight) for the reaction, so it is an endothermic reaction.
$2AgCl(s) +$ sunlight $\rightarrow 2Ag(s) + Cl_2(g)$

10. The carbohydrates are broken down to form glucose. This glucose combines with oxygen in the cells of our body and provides energy; so, it is an exothermic reaction.
$C_6H_{12}O_6(aq)$ (Glucose) $+ 6O_2(g) \rightarrow 6CO_2(g) + 6H_2O(l) +$ energy

11. Decomposition reactions are those in which a compound breaks down to form two or more substances. While, in a combination reaction, a single product is formed from two or more reactants. Thus, a decomposition reaction is the exact opposite of a combination reaction.
Example: $2H_2O(l) \rightarrow 2H_2(g) + O_2(g)$
$2H_2(g) + O_2(g) \rightarrow 2H_2O(l) +$ energy

12. (a) Thermal decomposition of ferrous sulfate:
$2FeSO_4(s)$ Ferrous sulfate $\rightarrow \Delta Fe_2O_3(s)$ Ferric oxide $+ SO_2(g)$ Sulfur dioxide $+ SO_3(g)$ Sulfur trioxide
(b) Decomposition of AgCl in the presence of light:
$2AgCl(s)$ Silver chloride + Light $\rightarrow 2Ag(s)$ Silver $+ Cl_2(g)$ Chlorine
(c) Decomposition of AgBr in the presence of light:
$2AgBr(s)$ Silver bromide + Light $\rightarrow 2Ag(s)$ Silver $+ Br_2(l)$ Bromine
(d) Decomposition by electricity:
$2Al_2O_3(aq)$ Aluminium oxide + Electricity $\rightarrow 4Al(s)$ Aluminium $+ 3O_2(g)$ Oxygen

13. **Displacement reaction:** A more reactive element replaces a less reactive element from a compound.
Example of a displacement reaction: $CuSO_4(aq) + Zn(s) \rightarrow ZnSO_4(aq) + Cu(s)$
Zinc is a more reactive metal as compared to copper, so zinc displaces the copper ion in copper.
Double displacement reaction: In a double displacement reaction, two compounds react and the positive ion (cation) and the negative ion (anion) of the two reactants switch places forming two new compounds or products.
Example: $Na_2SO_4(aq) + BaCl_2(aq) \rightarrow BaSO_4(s) + 2NaCl(aq)$

14. $2AgNO_3(aq)$ Silver nitrate + $Cu(s)$ Copper → $Cu(NO_3)_2(aq)$ Copper nitrate + $2Ag(s)$ Silver

15. A reaction in which an insoluble solid (called precipitate) is formed is called a precipitation reaction.

Example: $Na_2CO_3(aq)$ Sodium carbonate + $CaCl_2(aq)$ Calcium chloride
→$CaCO_3(s)$ Calcium carbonate + $2NaCl(aq)$ Sodium chloride

In this reaction, calcium carbonate is obtained as a precipitate. Hence, it is a precipitation reaction.

$Na_2SO_4(aq)$ Sodium sulfate + $BaCl_2(aq)$ Barium chloride → $BaSO_4(s)$ Barium sulfate + $2NaCl(aq)$ Sodium sulfate

In this reaction, barium sulfate is obtained as a precipitate.

16. (a) Oxidation involves the gain of oxygen.

Example: (i) $CO_2 + H_2 \rightarrow CO + H_2O$
Addition of oxygen: oxidation
(ii) $2Cu + O_2 \rightarrow 2CuO$
Addition of oxygen: oxidation

In equation (i), H_2 is oxidised to H_2O and in equation (ii), Cu is oxidised to CuO.

(b) Reduction is the loss of oxygen.

Example: (i) $CO_2 + H_2 \rightarrow CO + H_2O$
Removal of oxygen: reduction
(ii) $CuO + H_2\,\Delta \rightarrow Cu + H_2O$
Loss of oxygen: reduction

In equation (i), CO_2 is reduced to CO and in equation (ii), CuO is reduced to Cu.

17. X is copper (Cu) and the black-coloured compound formed is copper oxide (CuO). The equation of the reaction involved on heating copper is given below.

$2Cu + O_2$ (Shiny brown in colour) → $2CuO$ (Black in colour)

18. Iron, in the presence of moisture, reacts with oxygen to form hydrated iron oxide.

$4Fe + 3O_2 + nH_2O \rightarrow 2Fe_2O_3.nH_2O$ Hydrated iron oxide

This hydrated iron oxide is rust. Iron articles are painted because it prevents them from rusting. When painted, the contact of iron articles from moisture and air is cut off. Hence, rusting is prevented. Their presence is essential for rusting to take place.

19. Nitrogen is an inert gas and does not react easily with these substances. On the other hand, oxygen reacts with food substances and makes them rancid. Thus, bags used in packing food items are flushed with nitrogen gas to remove the oxygen inside the pack. When oxygen is not present inside the pack, rancidity of oil and fat containing food items is avoided.

20. (a) Corrosion: Corrosion is defined as a process where materials, usually metals, deteriorate as a result of a chemical reaction with air, moisture and chemicals. For example, iron, in the presence of moisture, reacts with oxygen to form hydrated iron oxide.

$4Fe + 3O_2 + nH_2O \rightarrow 2Fe_2O_3.nH_2O$
Hydrated iron oxide

This hydrated iron oxide is rust.

(b) Rancidity: The process of oxidation of fats and oils that can be easily noticed by the change in taste and smell is known as rancidity. For example, the taste and smell of butter changes when kept for long.

Rancidity can be avoided by:

- storing food in air tight containers,
- storing food in refrigerators,
- adding antioxidants, and
- storing food in an environment of nitrogen.

CHAPTER 2

Acids, Bases and Salts

Learning Objectives

After studying this unit, you will be able to understand

- the concept of indicators
- the characteristics, properties and how to distinguish between acids and bases
- salts and their types
- the principles of pH value and salt hydrolysis
- the laboratory preparation and uses of some important salts

2.1 Introduction

Most chemical substances are obtained mainly from two sources—plants and animals, and minerals and rocks. The substances that are obtained from plants and animals (living sources) are known as organic compounds. For example, acetic acid or vinegar is obtained from apples or grapes. The compounds that are obtained from minerals and rocks are known as inorganic compounds. For example, washing soda and lime.

In one of the earlier methods of classification, the taste of a compound was also used as a method. On the basis of taste, compounds can be further divided into three types—acids, bases and salts. In this chapter, we shall study all about acids, bases and salts. It is very important to know that many salts may have not only an unpleasant taste, but a few may be poisonous also. So, it is advisable not to taste any chemical substance present around us to avoid mishaps.

2.2 Acids and Bases

It is sometimes useful to classify compound into acids, bases and salts.

Acid: The word *acid* is derived from the Latin word *acere*, meaning sour. Acids are sour; for example, lemons and other citrus fruits contain citric acid and are quite sour. Milk turns sour due to lactic acid. We know that many non-metal oxides such as CO_2, NO_2, N_2O_5, SO_2 and SO_3 are acidic oxides and form acids on dissolving in water. Sulfuric, hydrochloric and nitric acids are generally known as mineral acids.

Base: The word *base* is not related to taste. Bases taste bitter. Most metal oxides and hydroxides, for example, Na_2O, MgO, CaO, KOH and $Mg(OH)_2$, are bases. Ammonia is also a base. The odour of fish is due to trimethylamine, an organic base similar to ammonia. Bases which are soluble in water are called alkalis, for example, NaOH and KOH.

Salt: Acids combine with bases to form salts. For example, hydrochloric acid reacts with sodium hydroxide to form sodium chloride (along with water)—the most well-known salt (common salt or table salt). Salts having a taste similar to that of common salt are known as salty salts.

2.2.1 Indicators for Testing Acids and Bases

In order to know more about a substance, weather it is an acid or a base, it is good to learn about indicators first. We can easily distinguish between an acid and a base using an indicator. Now let us understand what is an indicator. How many types of indicators are there? How do we use them?

An indicator is a dye which changes its colour when it is put into an acid or base. The characteristic of an indicator is that it exhibits different colours in acidic and basic mediums. That is, an indicator tells us whether the substance we are testing is an acid or a base by the change in its colour. The three most common indicators used to test for acids and bases are—litmus, phenolphthalein (HPh) and methyl orange (MeOH).

(i) Litmus

The most common indicator used for testing acids and bases in the laboratory is litmus. *Litmus is a natural indicator.* Litmus solution is a purple dye which is extracted from plants called lichens. Lichens are plants belonging to the division Thallophyta (Fig. 2.1). When a litmus solution is neither acidic nor basic (it is neutral), its colour is purple. It turns red in an acidic solution and blue in a basic solution. It can be used in the form of litmus solution or in the form of litmus paper. Litmus is of two types—blue and red.

Fig. 2.1 Lichen, plant source of litmus

Source: Alex Proimos from Sydney, Australia, CC BY 2.0 <https://creativecommons.org/licenses/by/2.0>, via Wikimedia Commons

It is important to note that an acid turns blue litmus to red while a base (or alkali) turns red litmus to blue. When the addition of one or two drops of the given solution changes blue litmus into red, it indicates that the given solution is acidic in nature. For example, orange juice, vinegar, lemon juice. On the other hand, if the addition of one or two drops of the given solution turns red litmus into blue, it indicates that the given solution is basic in nature. For example, sodium hydroxide or caustic soda solution, washing soda solution, soap solution, cucumber extract.

(ii) Methyl orange

The neutral colour of methyl orange is orange. The colour changes which take place are:

- methyl orange is red in colour in an acid solution, and
- methyl orange is yellow in colour in a basic solution.

(iii) Phenolphthalein

Phenolphthalein is colourless in a neutral solution. The colour changes which take place are:

- phenolphthalein is colourless in an acid solution, and
- phenolphthalein is pink in colour in a basic solution.

Some Other Natural Indicators

Turmeric is a natural indicator which is yellow in colour. It turns red brown in basic solutions. For example, while eating food, if curry falls on light coloured clothing, a yellow stain develops on the clothes. On applying soap solution (basic) on the cloth, the yellow stain turns reddish brown as soap solution is basic in nature. When excess of water is further used for washing, it turns yellow again as the soap solution is washed away.

Red cabbage extract obtained from red cabbage leaves is also a natural indicator (red in colour). This extract remains red in an acidic solution while it turns green on adding to a basic solution. The hydrangea flowers are bluish in colour but it turns pink in the presence of a basic solution. This is also a natural indicator. The colours of some natural indicators are summed up in Table 2.1.

Table 2.1 Colours of some natural indicators

Indicator	Colour in neutral solution	Colour in acidic solution	Colour in basic solution
1. Litmus	Purple	Red	Blue
2. Red cabbage leaf extract	Red	Red	Green
3. Flowers of hydrangea plant	Blue	Blue	Pink
4. Phenolphthalein	Colourless	Colourless	Pink
5. Methyl orange	Orange	Red	Yellow

Olfactory Indicators

There are few substances which give different types of odours in acidic and basic mediums that can be used as indicators. A substance whose smell or odour changes in an acidic or basic solution is known as an olfactory indicator. Olfactory means, relating to the sense of smell. It usually works on the fact that when an acid or base is added to it, its characteristic smell cannot be detected. For example, onion and vanilla extract.

(i) Onions have a characteristic smell. When a basic solution such as that of sodium hydroxide is added to a cloth strip treated with onions or onion extract, the onion smell cannot be detected. However, an acidic solution such as hydrochloric acid does not destroy the smell of onions.

(ii) Vanilla extract has a characteristic pleasant smell. If a basic solution such as that of sodium hydroxide is added to the vanilla extract, one cannot detect the characteristic smell of the vanilla extract. An acidic solution such as hydrochloric acid does not destroy the smell of vanilla.

(iii) Check the odour of clove oil. Take 1–2 mL of dilute HCl solution in one test tube and 1–2 mL of dilute NaOH solution in another test tube. Add a few drops of clove oil in each of them. We shall again observe that different odours are obtained from the two test tubes.

TEST YOUR KNOWLEDGE

2.1 Fill in the blanks.
(i) Turmeric is a ______ indicator.
(ii) In the basic medium, the colour of phenolphthalein is ______.
(iii) Onion and vanilla extract are examples of ______ indicators.

Solution: (i) Natural, (ii) pink, (iii) olfactory.

2.2 Define an acid–base indicator and provide two examples of synthetic acid–base indicators.

Solution: See text (Section 2.2.1).

2.3 Define an olfactory indicator and write the name of any three substances which can be used as olfactory indicators.

Solution: See text (Section 2.2.1).

2.4 A student prepared solutions of (i) an acid and (ii) a base in two separate beakers but forgot to label the solutions and litmus paper was not available in the laboratory. Since both the solutions are colourless, how will he distinguish between the two using (a) phenolphthalein and (b) methyl orange? **[CBSE 2019]**

Solution: (a) Acid + Phenolphthalein $\rightarrow$ Colourless
Base + Phenolphthalein $\rightarrow$ Pink colour
(b) Acid + Methyl orange $\rightarrow$ Red
Base + Methyl orange $\rightarrow$ Yellow

2.3 ACIDS

The word *acid* comes from the Latin word acidus, meaning sour. Acids are the chemical substances which have a sour taste and they turn blue litmus into red. For example, oranges (citric acid), lemon (citric acid), vinegar (acetic acid), curd (lactic acid).

An acid can be defined as a compound possessing hydrogen atoms which can be wholly or partially replaced by metal atoms or positive radicals. For example, (i) HCl is an acid having one H atom that can be replaced by Na to give NaCl or by the NH_4 radical to give NH_4Cl. (ii) H_2SO_4 is an acid having two H atoms which can be wholly or partially replaced to give Na_2SO_4 or $NaHSO_4$ respectively.

2.3.1 Classification of Acids

Acids can be classified in different ways. Some naturally occurring acids are given in Table 2.2.

Table 2.2 Some naturally occurring acids

Acids	Sources
Acetic acid	Vinegar
Citric acid	Citrus fruits (oranges, lemons)
Lactic acid	Sour milk
Malic acid	Apples
Tartaric acid	Grapes, tamarind, apples
Butyric acid	Rancid butter
Formic acid	Sting of bees and ants
Hydrochloric acid	Gastric juice
Oleic acid	Olive oil
Stearic acid	Fats
Uric acid	Urine
Oxalic acid	Tomato

On the basis of sources

(i) Organic acids: Most of the acids obtained from plants or other living organisms are organic acids. These acids are generally weak acids as they do not ionise completely in solution.

Hence, they contain ions as well as molecules. Examples are acetic acid (CH_3COOH) and oxalic acid $(COOH)_2$.

(ii) **Inorganic or mineral acids:** Such acids are obtained mainly from minerals and usually do not have carbon atoms except H_2CO_3. Being strong they are completely ionised in solution. Examples are nitric acid (HNO_3), sulfuric acid (H_2SO_4) and hydrochloric acid (HCl) (Fig. 2.2).

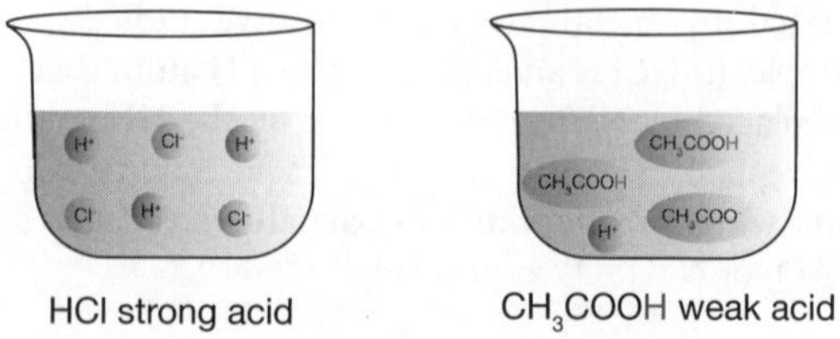

Fig. 2.2 Strong and weak acid

On the basis of acidic strength

When an acid is taken in water it undergoes dissociation or ionisation to give a cation (H^+) and an anion (A^-); this tendency of an acid is measured as acidic strength. Acidic strength indicates how easily an acid can lose H^+ ions in an aqueous solution. A strong acid loses a hydrogen ion faster than a weak acid. An acid gets ionised in an aqueous solution as follows.

$$\underset{\text{Acid}}{HA} \rightarrow H^+ + A^- \quad \underset{\text{Acid}}{HCl} \rightarrow H^+ + Cl^- \quad \underset{\text{Acid}}{HNO_3} \rightarrow H^+ + NO_3$$

(i) **Strong acids:** A strong acid is easily and more ionised in aqueous solution. Most of the mineral acids are strong acids. For example, nitric acid (HNO_3), sulfuric acid (H_2SO_4) or hydrochloric acid (HCl). Strong acids cannot be used in our daily lives even in the dilute form (Fig. 2.3).

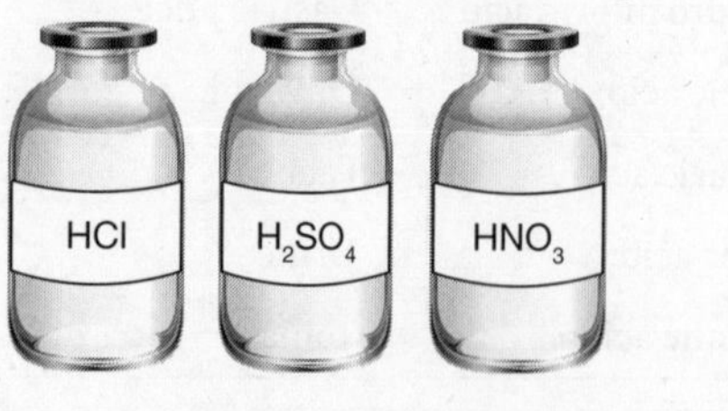

Fig. 2.3 Mineral acids

(ii) **Weak acid:** A weak acid is less ionised and ionisation occurs slowly. Most of the organic acids are weak acids. For example, H_2CO_3 (carbonic acid), CH_3COOH (acetic acid or vinegar), citric acid (lemon juice) (Fig. 2.4).

A few inorganic acids are also weak, such as, hydrogen sulfide (H_2S), carbonic acid (H_2CO_3). The dilute solutions of these weak acids are drinkable and safe too. For example, carbonic acid is used in fizzy soft drinks, acetic acid as vinegar is very commonly used in making pickles, ketchups, tartaric acid is used as an ingredient of baking powder.

Fig. 2.4 Vinegar (apple cider) and carbonic acid (soft drink)

Source: JFVelasquez Floro, CC0, via Wikimedia Commons; W.J.Pilsak, CC BY-SA 3.0 <http://creativecommons.org/licenses/by-sa/3.0/>, via Wikimedia Commons

> **Competition Edge**
>
> Acidic strength is practically measured in terms of degree of ionisation, that is, the extent to which it is ionised. The more the value of degree of ionisation, the stronger is the acid. It can be found out as follows.
>
> Degree of ionisation (α)
>
> $$= \frac{\text{No. of acid molecules ionised}}{\text{Total no. of acid molecules present in aqueous solution}} \times 100$$
>
> If the degree of ionisation (α) for an acid, base or salt in aqueous solution is more than 30%, it is strong and if it is less than 30%, it is considered weak.

On the basis of their concentration

An acid having a relatively higher proportion in its aqueous solution is a concentrated acid while an acid having a relatively lower proportion in its aqueous solution is a dilute acid. If in an acid, the minimum possible amount of water is present, it is considered concentrated and if a maximum amount of water is present, it is a dilute acid. It is very important to note that the dilution of a concentrated acid is highly exothermic (lot of heat is evolved). Therefore, the dilution must always be carried out by adding concentrated acid to water gradually with stirring and not by adding water directly into concentrated acid to avoid any kind of explosion or mishap.

*It is interesting to note that a concentrated acid is not necessarily a strong one and a dilute acid is not necessarily a weak one, as acidic strength is decided on the basis of α of H^+. For example, dil. HCl is a stronger acid than concentrated acetic acid.

On the basis of molecular composition

(i) **Hydra-acids:** Such acids possess only hydrogen and a non-metallic element but not oxygen. For example, HX (HCl, HBr, HI), H_2S.

(ii) **Oxy-acids:** Such acids have hydrogen, oxygen and some other elements. For example, H_2SO_4, HNO_3, H_3PO_4, CH_3COOH, H_2CO_3.

On the basis of basicity of acids

The basicity or protocity of an acid is equal to the number hydronium ions that can be produced by the ionisation of one molecule of an acid in an aqueous solution.

Basicity = number of ionisable hydrogen atoms per molecule

(i) **Monobasic acid:** Such an acid gives only one hydronium ion in the aqueous solution. For example, $HX + H_2O \rightarrow H_3O^+ + X^-$

Some examples of monobasic acids are HX, HNO_3, HNO_2, HCOOH, CH_3COOH, HClO, $HClO_4$.

(ii) **Dibasic acid:** Such acids on ionisation in water give two hydronium ions per molecule of the acid. Usually their ionisation occurs in two steps.

$$H_2SO_4 + H_2O \rightarrow H_3O^+ + HSO_4^-$$

$$HSO_4^- + H_2O \rightarrow H_3O^+ + SO_4^{2-}$$

Some examples of dibasic acids are H_2CO_3, $H_2C_2O_4$, H_2SO_3, H_2SO_4.

(iii) **Tribasic acid:** Such acids on ionisation in water give three hydronium ions per molecule of the acid. They ionise in three steps. For example,

$$H_3PO_4 + H_2O \rightarrow H_3O^+ + H_2PO_4^-$$

$$H_2PO_4^- + H_2O \rightarrow H_3O^+ + HPO_4^{2-}$$

$$HPO_4^{2-} + H_2O \rightarrow H_3O^+ + PO_4^{3-}$$

2.3.2 General Characteristic Properties of Acids

All acids have some common characteristic properties.

(i) **Sour taste:** They are sour in taste.

(ii) **Action on litmus paper and methyl orange:** They turn blue litmus (a dye extracted from lichens) into red. When 1–2 drops of methyl orange are added in a solution of an acidic substance, methyl orange turns red.

(iii) **Corrosive nature:** Most acids are corrosive in nature, that is, they produce a burning sensation on the skin and holes in the clothes on which they fall. For example, concentrated sulfuric acid causes severe burns on the skin, creates holes in clothes, wood and even metal surfaces. They also attack the metal structure and stone work. That is why they are not stored in metal containers but are stored in glass or ceramic containers. Acids are very harmful for living things as they can destroy the cells of living systems. Concentrated acids are specifically more dangerous to the skin and delicate body organs like the eyes. Therefore, acids must be handled with a lot of care in the laboratory and that is why a 'danger sign' is used on the containers holding concentrated acids (Fig. 2.5).

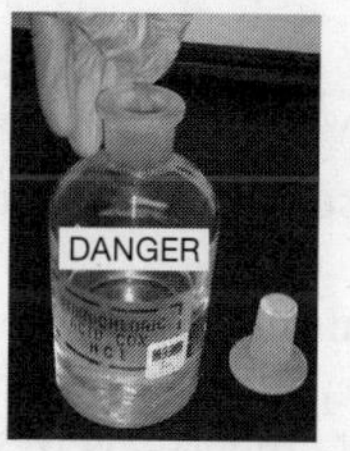

Fig. 2.5 Bottle of strong acid

Source: Walkerma at en.wikipedia, Public domain, via Wikimedia Commons

(iv) **Conduction of electricity:** Their aqueous solutions can conduct electricity due to electrolysis. When an acid is dissolved in water, the aqueous solution thus obtained, conducts electricity. A strong acid like sulfuric acid shows more electric conduction than a weak acid like acetic acid as a strong acid is more ionised than a weak acid. In the case of a strong acid, H^+ concentration is more than that in a weak acid. In the case of a strong acid, an electric bulb glows more brightly (Fig. 2.6).

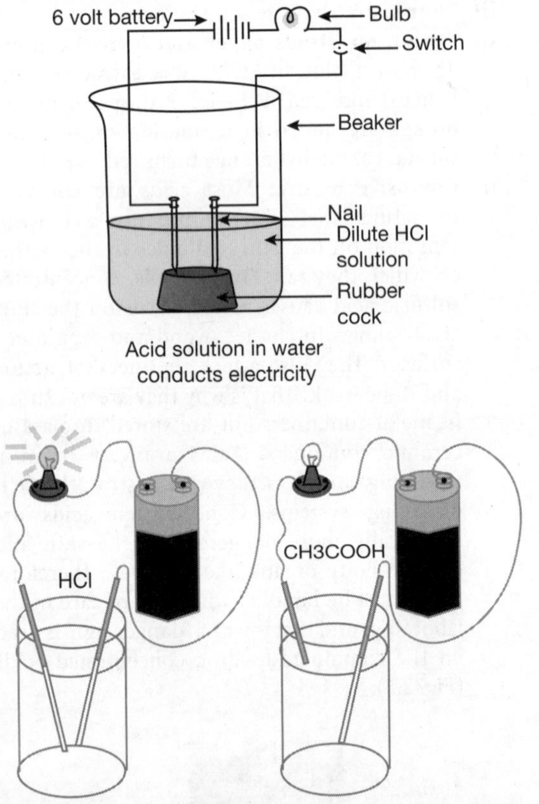

Fig. 2.6 Electric conduction in strong and weak acids

(v) **Neutralisation:** They become less acidic on combining with bases. Mixing of acids and bases in the right proportion gives a neutral solution, and the process is known as *neutralisation.* The residue obtained on the evaporation of a neutral solution is a salt. That is, neutralisation is a process in which an acid and a base react together to form a salt and water. This reaction is always an exothermic reaction, that is, heat is evolved during the reaction (Fig. 2.7).

Acid + Base $\rightarrow$ Salt + Water + Heat

$HCl + NaOH \rightarrow NaCl + H_2O$ + Heat

$HNO_3 + KOH \rightarrow KNO_3 + H_2O$ + Heat

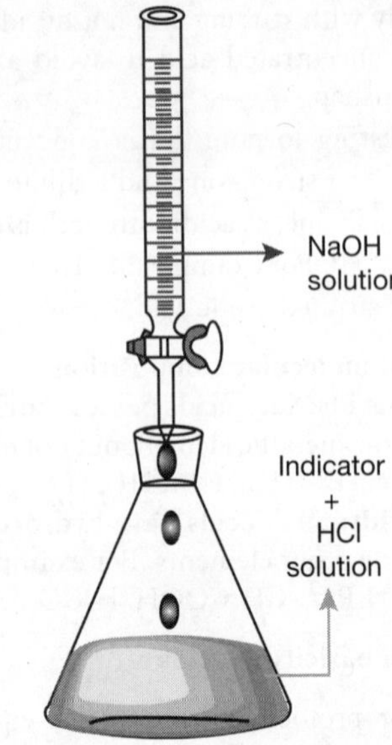

Fig. 2.7 Neutralisation (acid–base titration)

(vi) **Reaction with metals:** Acids corrode metals. Metals above hydrogen in the activity series (except Pb) can react with acids to give hydrogen gas and the corresponding metal salt. Nitric acid, HNO_3, is one acid which does not give hydrogen with any metal except magnesium and manganese. In general, metals above hydrogen react with acids to give the salt and hydrogen gas.

Metal + Acid $\rightarrow$ Salt + Hydrogen gas

For example, when zinc granules are heated with dilute sulfuric acid or dilute hydrochloric acid, a colourless solution of zinc sulfate or zinc chloride is formed along with a brisk effervescence of hydrogen gas (Fig. 2.8). When a burning candle is brought near the gas, the gas present in a soap bubble burns with a popping sound.

$$\text{Zinc + Sulfuric acid (dilute)} \rightarrow \underset{\text{Colourless solution}}{\text{Zinc sulfate}} + \underset{\text{Gas}}{\text{hydrogen}}$$

$$\underbrace{Zn + H_2SO_4}_{\text{Reactants}} \rightarrow \underbrace{ZnSO_4 + H_2}_{\text{Products}}$$

*In Chapter 1 we have discussed the reactivity of different metals towards acids.

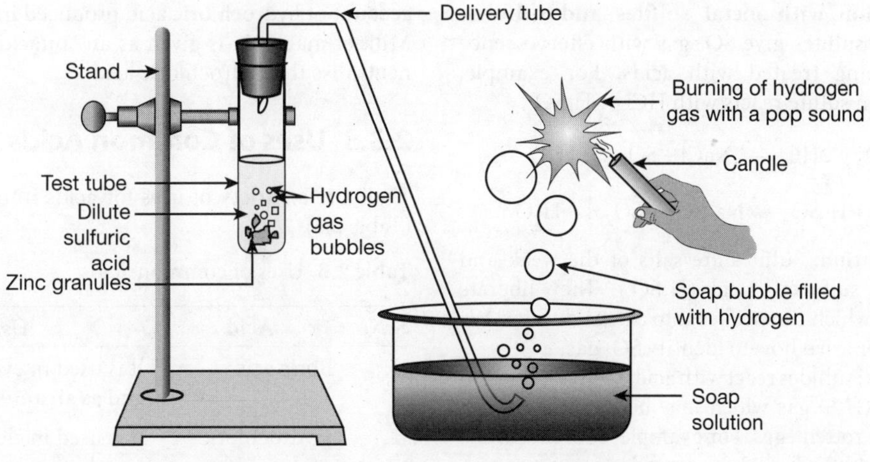

Fig. 2.8 Reaction of zinc granules with H_2SO_4

It is important to note that food stuffs with a sour taste, such as, vinegar, lemon juice and curd, should not be kept in metal vessels (copper or brass) as the acids present in the food stuff can react with the metal of the vessel to form metal compounds that can cause food poisoning.

(vii) Reaction with metal carbonates and metal bicarbonates: Acids give carbon dioxide (CO_2) gas with metal carbonates and metal hydrogen carbonates (bicarbonates). Here, CO_2 evolves with effervescence.

Metal carbonate or Bicarbonate + Acid →
Salt + Carbon dioxide + Water

For example, sodium carbonate and sodium bicarbonate react with acids as follows (Fig. 2.9).

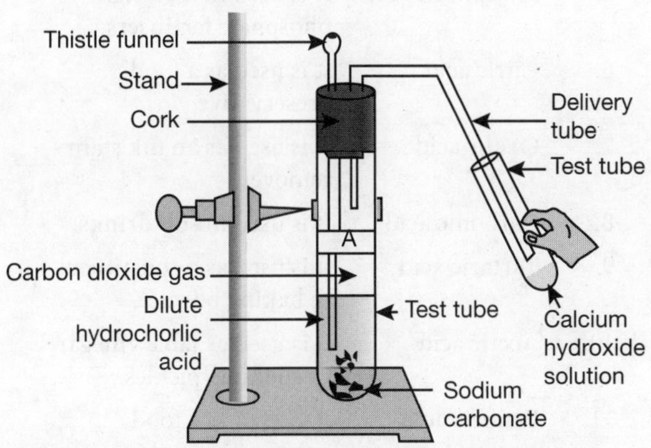

Fig. 2.9 Study of reaction of sodium carbonate with dil. hydrochloric acid

$$Na_2CO_3 + 2HCl \rightarrow 2NaCl + CO_2 \uparrow + H_2O$$

$$Na_2CO_3 + H_2SO_4 \rightarrow Na_2SO_4 + CO_2 \uparrow H_2O$$

$$NaHCO_3 + HCl \rightarrow NaCl + CO_2 \uparrow + H_2O$$

$$2NaHCO_3 + H_2SO_4 \rightarrow Na_2SO_4 + 2CO_2 \uparrow 2H_2O$$

This reaction of an acid with a metal carbonate or bicarbonate is quite useful in our daily lives. A person suffering from acidity on taking ENO or sodium bicarbonate (base) gets relief from acidity as the acid is neutralised by the base. Similarly, limestone, chalk, marble (different forms of calcium carbonate) and egg shells contain calcium carbonate as the main constituent; so, they react with the acid.

$$\underset{\substack{\text{Calcium carbonate} \\ \text{(limestone/} \\ \text{chalk/marble)}}}{CaCO_3} + \underset{\substack{\text{Hydrochloric} \\ \text{acid}}}{2HCl} \rightarrow \underset{\substack{\text{Calcium} \\ \text{chloride}}}{CaCl_2} + \underset{\text{Water}}{H_2O} + \underset{\substack{\text{Carbon} \\ \text{dioxide}}}{CO_2}$$

The carbon dioxide evolved in these reactions, on passing into lime water, forms a white precipitate or turns milky due to the formation of calcium carbonate.

$$\underset{\text{Calcium hydroxide}}{Ca(OH)_2(aq)} + \underset{\substack{\text{Carbon dioxide} \\ \text{(From air)}}}{CO_2(g)} \longrightarrow \underset{\text{Calcium carbonate}}{CaCO_3(s)} + \underset{\text{Water}}{H_2O(l)}$$

If excess of of CO_2 is passed through lime water, then the milkyness disappears and the solution becomes clear due to the formation of a soluble salt—calcium hydrogen carbonate.

$$\underset{\substack{\text{Calcium carbonate} \\ \text{(White ppt.)} \\ \text{(Insoluble in water)}}}{CaCO_3(s)} + \underset{\text{Carbon dioxide}}{CO_2(g)} + \underset{\text{Water}}{H_2O(l)} \rightarrow \underset{\substack{\text{Calcium hydrogen} \\ \text{carbonate} \\ \text{(Soluble in water)}}}{Ca(HCO_3)_2(aq)}$$

(viii) Reaction with metal sulfites and sulfides: Metal sulfites give SO_2 gas with effervescence on being treated with acids. For example, sodium sulfite reacts with HCl or H_2SO_4.

$$Na_2SO_3 + 2HCl \rightarrow 2NaCl + SO_2 \uparrow + H_2O$$

$$Na_2SO_3 + H_2SO_4 \rightarrow Na_2SO_4 + SO_2 \uparrow + H_2O$$

Explanation: Sulfites are salts of the weak and volatile sulfurous acid (H_2SO_3). They liberate H_2SO_3, which dissociates into SO_2 and water. We will later learn how to identify SO_2 gas.

Metal sulfides react with acids to give hydrogen sulfide (H_2S) gas which may be identified by its smell of rotten eggs. For example, sodium sulfide reacts with HCl or H_2SO_4 as follows.

$$Na_2S + 2HCl \rightarrow 2NaCl + H_2S \uparrow$$

$$Na_2S + H_2SO_4 \rightarrow Na_2SO_4 + H_2S \uparrow$$

(ix) Reaction with metallic oxides and hydroxides: Acids react with metal oxides to give the metal salt and water. This is a neutralisation reaction as metal oxides are basic in nature.

Metal oxides + Acid → Salt + Water + Heat

For example, quicklime (CaO) and copper oxide (CuO) react with an acid as follows (Fig. 2.10).

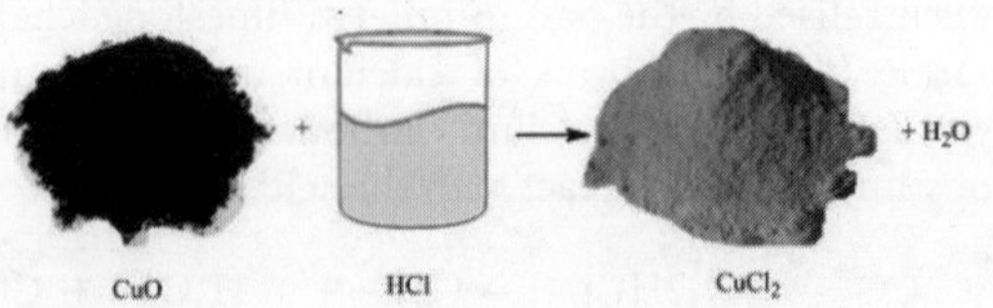

Fig. 2.10 Reaction between copper oxide and hydrochloric acid

$$\underset{\text{Calcium oxide}}{CaO(s)} + \underset{\text{Hydrochloric acid}}{2HCl(aq)} \rightarrow \underset{\text{Calcium chloride}}{CaCl_2(aq)} + \underset{\text{Water}}{H_2O}$$

$$\underset{\text{Copper(II) oxide}}{CuO(s)} + \underset{\text{Hydrochloric acid}}{2HCl(aq)} \rightarrow \underset{\text{Copper(II) chloride}}{CuCl_2} + \underset{\text{Water}}{H_2O}$$

When copper oxide reacts with dilute hydrochloric acid, the black copper oxide dissolves and a blue green cupric chloride is formed.

$$\underset{\text{Magnesium hydroxide (Milk of magnesia)}}{Mg(OH)_2} + \underset{\text{Hydrochloric acid}}{2HCl} \rightarrow \underset{\text{Magnesium chloride}}{MgCl_2} + \underset{\text{Water}}{2H_2O}$$

This reaction has an important application. The acidity produced in the stomach due to indigestion is on account of hydrochloric acid produced in the stomach. Milk of magnesia is given as an 'antacid' medicine to neutralise the hydrochloric acid.

2.3.3 Uses of Common Acids

There are a variety of uses for acids in our daily lives (Table 2.3).

Table 2.3 Uses of common acids

S.No.	Acid	Use
1.	Boric acid	It is used in eye washes and as an antiseptic.
2.	Hydrochloric acid	It is used in cleaning of metal items by removing the oxide film from their surface. It is also used in making dye stuff, leather industry and so on.
3.	Nitric acid	It is used in manufacturing explosives such as TNT (trinitro toluene), dyes and so on.
4.	Sulfuric acid (king of chemicals)	It is used in manufacturing fertilisers such as ammonium sulfate, paints, synthetic fibres, detergents, plastics, invertor batteries and so on, so it is called the king of chemicals.
5.	Phosphoric acid	It is used in making phosphate fertilisers.
6.	Citric acid	It is used as a food preservative.
7.	Oxalic acid	It is used as an ink stain remover.
8.	Carbonic acid	It is used in soft drinks.
9.	Tartaric acid	It is used as a constituent in baking powder.
10.	Acetic acid	It is used as table vinegar in cooking, pickles.
11.	Benzoic acid	It is used as a food preservative and in making of perfumes and medicines.

2.3.4 Chemical Nature of Acids

By now we have learnt a lot about acids, their types and their properties. Now, we can say that chemically, all acids have something in common. The hydrogen present in acids is such that when the acids are dissolved in water, it separates out as a cation, that is hydrogen ion (H^+) ion. This H^+ enters the solution in the form of $H^+(aq)$ ions; in the case of water, it is as H_3O^+ (hydronium) ions. Thus, an acid is a substance which dissociates or undergoes ionisation in water to give H^+ ions or H_3O^+ ions along with an anion (A^-) which is called the conjugate base.

$$HA \rightarrow H^+(aq) + A^-$$

For example,
$$\underset{\text{Hydrochloric acid}}{HCl(aq)} \rightarrow \underset{\text{Hydrogen ions}}{H^+(aq)} + \underset{\text{Chloride ions}}{Cl^-(aq)}$$

$$\underset{\text{Hydrogen ion}}{H^+} + \underset{\text{Water}}{H_2O} \rightarrow \underset{\text{Hydronium ion}}{H_3O^+}$$

$$HNO_3 + H_2O \rightarrow H_3O^+ + NO_3^-$$

$$H_2SO_4 + H_2O \rightarrow H_3O^+ + HSO_4^-$$

$$HSO_4^- + H_2O \rightarrow H_3O^+ + SO_4^{2-}$$

$$H_2CO_3 + 2H_2O \rightleftharpoons 2H_3O^+ + \underset{\text{(Carbonate ion)}}{CO_3^{2-}}$$

$$CH_3COOH + H_2O \rightleftharpoons H_3O^+ + \underset{\text{(Acetate ion)}}{CH_3COO^-}$$

Now it is clear that acids give hydrogen ions (H^+) in water which exist as H_3O^+. It means all the substances which dissolve in water to give H^+ or H_3O^+ are acids. This is the *Arrhenius concept* of an acid.

To show that all compounds having hydrogen are not acidic: There are many compounds containing hydrogen but they do not show acidic character, that is, they do not provide H^+ on dissolving in water. For example, glucose ($C_6H_{12}O_6$), sucrose ($C_{12}H_{22}O_{11}$), ethanol (C_2H_5OH). It can be practically confirmed easily as their aqueous solutions do not conduct electricity while in the case of acids like HCl, the aqueous solution can conduct electricity as ions are present due to ionisation (Fig. 2.11).

It is interesting to note that distilled water does not conduct electricity as it does not have any ionic compounds such as acid, base or salt dissolved in it. However, rain water conducts electricity as it contains dissolved gases of the air such as CO_2, SO_2 and NO_2 which dissolve in water to give their respective acids H_2CO_3, H_2SO_4 and HNO_3 which undergo ionisation to provide H^+ and anions like CO_3^{2-}, SO_4^{2-} and NO_3^-. Therefore, due to the presence of ions conduction of electricity is possible.

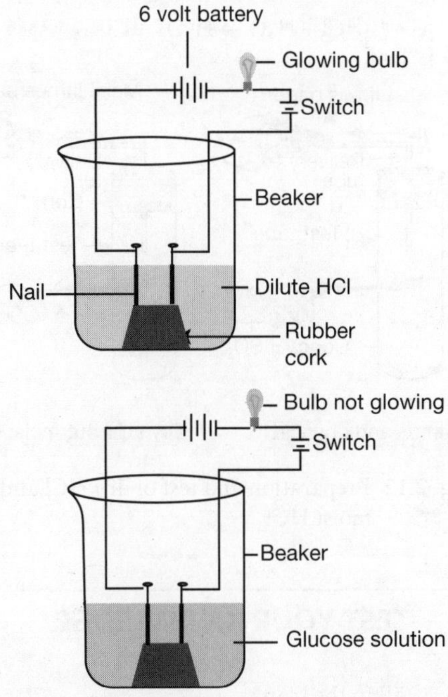

Fig. 2.11 Electrical conduction shown by HCl and not by glucose

Role of water in the dissociation of an acid: By now it is clear to us that acids dissociate or ionise to give H^+ ions in the presence of water (presence of water is mandatory). For example, dry HCl gas is not acidic but in the presence of water HCl becomes an acid. In order to demonstrate this, let us take 1–2 g of NaCl in a dry test tube and add some sulfuric acid into it. As a result, hydrogen chloride gas is formed.

$$\underset{\text{Sodium chloride}}{NaCl} + \underset{\text{Sulfuric acid}}{\text{Conc. } H_2SO_4} \rightarrow \underset{\text{Sodium hydrogen sulfate}}{NaHSO_4} + \underset{\text{Hydrogen chloride gas}}{HCl}$$

When a dry blue litmus paper is brought near the test tube having dry HCl gas, the litmus paper does not turn red, that is, it is non-acidic. However, when dry HCl gas is passed through water and then again taken in a test tube and blue litmus is brought near it, it turns red (Fig. 2.12). Now, it is clear that in the presence of water, HCl gets ionised so its acidic nature is observed. It is possible because water molecules are polar in nature and are able to pull the H^+ and Cl^- ions apart and HCl gets broken.

$$\underset{\text{Hydrochloric acid}}{HCl(aq)} \rightarrow \underset{\text{Hydrogen ions}}{H^+(aq)} + \underset{\text{Chloride ions}}{Cl^-(aq)}$$

$$HCl + H_2O \rightarrow H_3O^+ + Cl^-$$

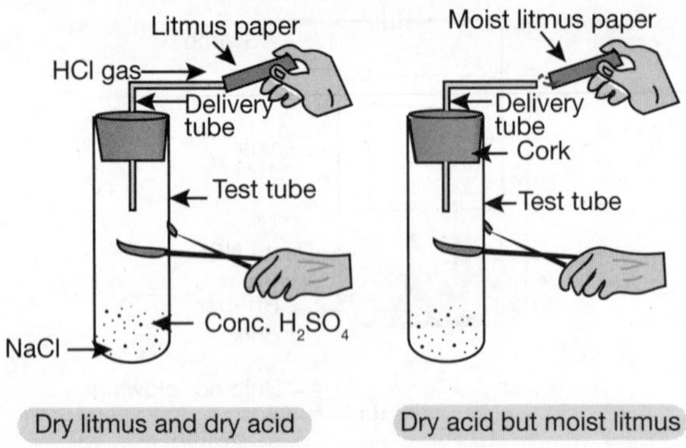

Fig. 2.12 Preparation and test of dry HCl and moist HCl

TEST YOUR KNOWLEDGE

2.5 Fill in the blanks.

(i) Acids turn __________ litmus solution __________.

(ii) The presence of __________ in acids is responsible for their acidic properties.

(iii) During indigestion, the stomach produces too much __________ and this causes pain and irritation.

(iv) Mixing an acid or a base with water results in a decrease in the concentration of _____ per unit volume. This process is called _____

(v) Among HCl, H_2SO_4 and CH_3COOH, __________ is a weak acid.

Solution: (i) blue, red, (ii) H^+, (iii) acid (HCl), (iv) OH^- ions/H_3O^+ ions, dilution, (v) CH_3COOH

2.6 Name the natural source of each of the following acids.

(i) Citric acid (ii) Oxalic acid

(iii) Lactic acid (iv) Tartaric acid

[CBSE 2014]

Solution: (i) Lemon and orange

(ii) Guava and tomatoes

(iii) Curd (sour milk) (iv) Tamarind

2.7 Answer the following.

(i) Which acid is present in milk?

(ii) SO_3 is the anhydride of which acid?

(iii) Which of the following acids is/are dibasic?
CH_3COOH, H_2SO_4, HNO_3, HCl, HOOC—COOH

Solution: (i) Lactic acid is present in milk.

(ii) SO_3 is the acidic anhydride of sulfuric acid (H_2SO_4).

(iii) H_2SO_4, HOOC—COOH are dibasic acids.

2.8 Explain why: (i) Acids do not show acidic behaviour in the absence of water.

(ii) Dry HCl gas does not change the colour of dry blue litmus paper. Give reasons.

Solution: (i) Acids do not show acidic behaviour in the absence of water because the dissociation of hydrogen ions from an acid occurs in the presence of water only.

(ii) In the dry state, HCl gas does not give any H^+ ions. It does not behave as an acid. That is why the colour of litmus paper does not change.

2.4 Bases

The solutions of substances such as caustic soda or sodium hydroxide (NaOH), lime (chuna) and washing soda are bitter in taste and soapy to the touch with a slippery nature. Such substances are known as bases. Bases are substances having a bitter taste. All bases can turn red litmus to blue, that is, the opposite of what happens with an acid. Bases are those species that react with acids to form salts, usually accompanied by water. In general, bases are oxides and hydroxides of metals (groups 1 and 2), ammonia and so on. Examples are, Na_2O (sodium oxide), NaOH (sodium hydroxide), CaO (calcium oxide or lime), $Ca(OH)_2$ (calcium hydroxide or slaked lime), NH_4OH (ammonium hydroxide) are bases as they can neutralise acids. Carbonates and bicarbonates are also bases; for example, Na_2CO_3 (sodium carbonate), $NaHCO_3$ (sodium bicarbonate).

- Na_2O is a base as it can react with acids to form salts and water.

$$Na_2O + 2HCl \rightarrow 2NaCl + H_2O$$

$$\underset{\text{Base}}{Na_2O} + \underset{\text{Acid}}{H_2SO_4} \rightarrow \underset{\text{Salt}}{Na_2SO_4} + \underset{\text{Water}}{H_2O}$$

- $Ca(OH)_2$ is a base as it can react with acids to form salt and water.

$$\underset{\text{Base}}{Ca(OH)_2} + \underset{\text{Acid}}{2HCl} \rightarrow \underset{\text{Salt}}{CaCl_2} + \underset{\text{Water}}{2H_2O}$$

- NH_3 is a base as it can react with acids to give salts.

$$\underset{\text{Base}}{2NH_3} + \underset{\text{Acid}}{H_2SO_4} \rightarrow \underset{\text{Salt}}{(NH_4)_2SO_4}$$

Bases and alkalis: Although most of the bases are insoluble in water, a few dissolve in water. Those bases which dissolve in water without any chemical reaction are known as alkalis, that is, any base soluble in water is an alkali. For example, oxides and hydroxides of groups 1 and 2—Na_2O, CaO, NaOH, $Ca(OH)_2$.

Chemical nature of bases: Something common observed in all bases is that when they are added to water, they provide hydroxide ions (OH^-). We can say that all bases are OH^- providers in water. Arrhenius defined a base as a substance which dissolves in water and gives the hydroxide ion (OH^-) in solution. For example,

$$\underset{\text{Sodium}}{NaOH(s)} + \text{Water} \rightarrow \underset{\text{Sodium ions}}{Na^+(aq)} + \underset{\text{Hydroxide ions}}{OH^-(aq)}$$

$$\underset{\text{Ammonium hydroxide}}{NH_4OH} + \text{Water} \rightarrow \underset{\text{Ammonium ions}}{NH_4^+(aq)} + \underset{\text{Hydroxide ions}}{OH^-(aq)}$$

$$\underset{\text{Magnesium hydroxide}}{Mg(OH)_2(s)} + \text{Water} \rightarrow \underset{\text{Magnesium ions}}{Mg^{2+}(aq)} + \underset{\text{Hydroxide ions}}{2OH^-(aq)}$$

Thus, a base is just opposite to an acid. An acid provides H^+ in water while a base provides OH^- in water. There are a few substances which have the –OH group but do not provide OH^- in water such as ethanol (C_2H_5OH), phenol (C_6H_5OH).

2.4.1 Classification of Bases

Bases are classified as follows.

Based on Dissociation

Bases undergo dissociation or ionisation in water to give OH^- and on this basis, bases are of two types (Fig. 2.13).

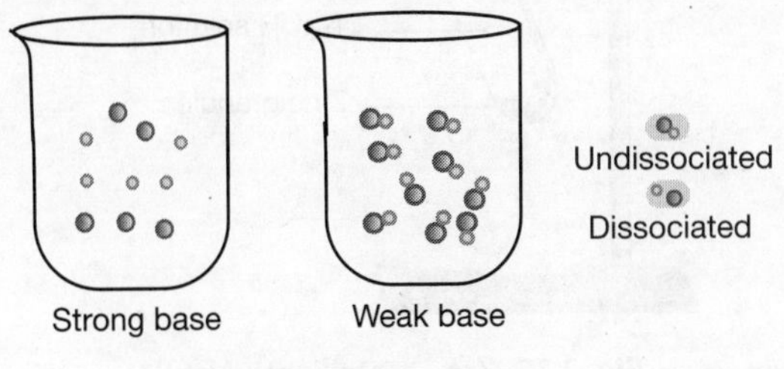

Fig. 2.13 Types of bases

(i) **Strong base:** A strong base can undergo almost complete ionisation in aqueous solution to produce a high concentration of OH^-. For example, hydroxides of groups 1 and 2 such as NaOH, KOH, $Mg(OH)_2$, $Ca(OH)_2$.

$$Na(OH)(aq) \rightarrow Na^+(aq) + OH^-(aq)$$

$$Ca(OH)_2(aq) \rightarrow Ca^{2+}(aq) + 2OH^-(aq)$$

(ii) **Weak base:** A weak base can undergo partial ionisation in an aqueous solution to produce a low concentration of OH^-. For example, NH_4OH, $Zn(OH)_2$.

$$NH_4OH(aq) \rightleftharpoons NH_4^+(aq) + OH^-(aq)$$

$$Zn(OH)_2(aq) \rightleftharpoons Zn^{2+}(aq) + 2OH^-(aq)$$

(iii) **On the basis of acidity (number of –OH^- groups):** The acidity of a base is determined in terms of the number of hydroxyl ions (OH^-) that 1 molecule of the base gives on dissociation in water (Fig. 2.14).

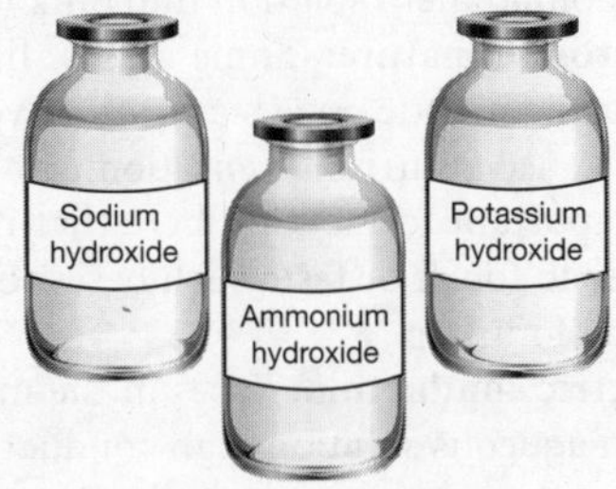

Fig. 2.14 Examples of some bases

Acidity = number of OH^- ions given by a base

- **Monoacidic base:** Such a base gives only 1 OH^- per molecule on dissociation in aqueous solution. For example, hydroxides of group 1, such as NaOH, KOH, NH_4OH.

$$NaOH \rightarrow Na^+ + OH^-$$

- (ii) **Diacidic base:** Such a base gives only 2 OH^- per molecule on dissociation in aqueous solution. For example, hydroxides of group 2, such as $Mg(OH)_2$, $Zn(OH)_2$.

$$Ca(OH)_2(aq) \rightleftharpoons Ca^{2+} + 2OH^-$$

- **Triacidic base:** Such a base gives only 3 OH^- per molecule on dissociation in aqueous solution. For example, $Fe(OH)_3$, $Al(OH)_3$, $Cr(OH)_3$.

$$Fe(OH)_3(aq) \rightleftharpoons Fe^{3+} + 3OH^-$$

2.4.2 General Characteristics of Bases

Bases have the following general properties.

(i) **Bitter taste:** Bases are bitter in taste.

(ii) **Soapy and slippery touch:** Alkali solutions are slippery with soapy touch. For example, when we rub a drop of caustic soda solution between the tips of our fingers, our fingers will soon start to feel soapy (Fig. 2.15).

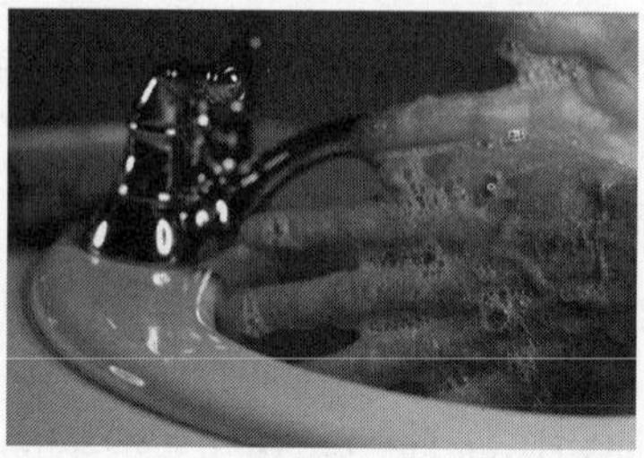

Fig. 2.15 Soapy and slippery touch

Source: MarkBuckawicki, CC0, via Wikimedia Commons

(iii) **Litmus and phenolphthalein tests:** Alkalis turn red litmus into blue and also turn phenolphthalein solution into pink (Fig. 2.20).

(iv) **Corrosive nature:** Some bases like caustic soda and caustic potash are corrosive in nature and cause a burning sensation on the skin. It is important to mention here that it advisable never to touch or taste such bases to avoid any kind of mishap.

(v) **Electric conduction:** Bases in the molten state or in aqueous solutions can conduct electricity as ions are present for conduction in solution (Fig. 2.16).

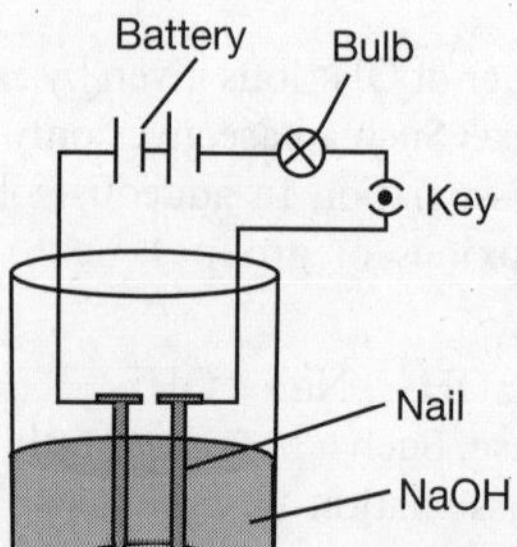

Fig. 2.16 Electric conduction

(vi) **Neutralisation:** Bases become less alkaline on combining with acids, and form salts on neutralisation. Acids and bases can neutralise each other. For example, the antacid tablets and suspensions used treat acidity contain bases like $Mg(OH)_2$ and $Al(OH)_3$. The unpleasant odour of fish (due to the base trimethylamine) can be greatly reduced by adding lemon (citric acid) or vinegar (acetic acid) owing to the neutralisation of the base by the acid.

$$\text{Acid} + \text{Base} \rightarrow \text{Salt} + \text{Water} + \text{Heat}$$

$$HCl + NaOH \rightarrow NaCl + H_2O + \text{Heat}$$

$$2HCl + Mg(OH)_2 \rightarrow MgCl_2 + 2H_2O + \text{Heat}$$

On combining an acid and a base, real neutralisation occurs when the H^+ of acid and OH^- of water combine together to give water.

$$H^+ + OH^- \rightarrow H_2O$$

(vii) **Reaction with metals:** Only few metals, like tin (Sn), aluminium (Al) and zinc (Zn) give hydrogen gas on boiling with alkalis. For example, when 2–3 pieces of granulated zinc is taken along with 2 mL of sodium hydroxide and heated (Fig. 2.17), hydrogen gas is evolved which burns with a popping sound.

$$Zn + 2NaOH \xrightarrow{\Delta} \underset{\text{Sodium zincate}}{Na_2ZnO_2} + H_2\uparrow$$

Tin and aluminium also give hydrogen on heating with sodium hydroxide.

$$Sn + 2NaOH \xrightarrow{\Delta} \underset{\text{Sodium stannate}}{Na_2SnO_2} + H_2\uparrow$$

$$2Al + 2NaOH + 2H_2O \xrightarrow{\Delta} \underset{\text{Sodium metaluminate}}{2NaAlO_2} + 3H_2\uparrow$$

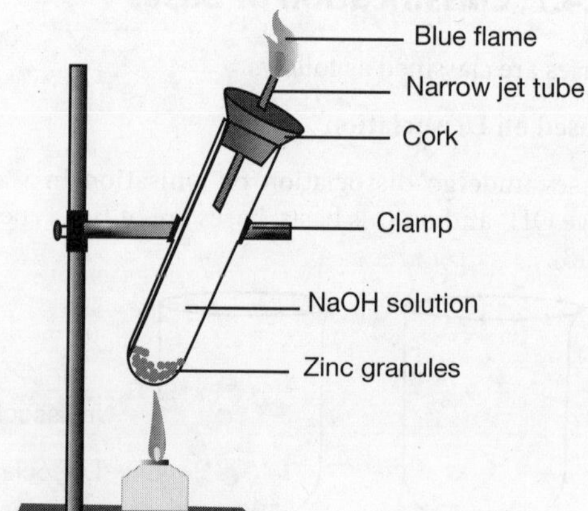

Fig. 2.17 Zinc heated with NaOH

(viii) Reaction with a non-metal oxide: Bases react with non-metal oxides such as CO_2, SO_2, SO_3, P_2O_5 to form salts and water.

Base + Non-metal oxide → Salt + Water

$$\underset{\substack{\text{Calcium hydroxide}\\\text{(Lime water)}}}{Ca(OH)_2} + \underset{\text{Carbon dioxide}}{CO_2} \rightarrow \underset{\substack{\text{Calcium carbonate}\\\text{(Salt)}}}{CaCO_3} + \underset{\text{Water}}{H_2O}$$

This reaction shows that non-metal (C, P, S) oxides are acidic in character.

(ix) Reaction with ammonium salts: Alkalis give NH_3 gas on heating with ammonium salts. A strong base removes the weak and volatile base (NH_3) from its salts.

$$NH_4Cl + NaOH \rightarrow NH_3 \uparrow + NaCl + H_2O$$

$$(NH_4)_2SO_4 + 2KOH \rightarrow 2NH_3 \uparrow + K_2SO_4 + 2H_2O$$

(x) Mixing of bases with water: Just as dilution of a concentrated acid with water is exothermic, the dissolution of bases such as sodium hydroxide (NaOH) or potassium hydroxide (KOH) in water is exothermic and the solution is found to be hotter.

2.4.3 Uses of Bases

Base	Use
1. Sodium hydroxide (NaOH)	It is used in the manufacture of soaps, artificial fibres like rayon and in the paper industry.
2. Potassium hydroxide (KOH)	It is used in the manufacture of soaps, nickel–iron storage batteries and for absorbing CO_2.
3. Magnesium hydroxide [$Mg(OH)_2$]	It is used as an antacid to neutralise acidity in the stomach.
4. Calcium hydroxide [$Ca(OH)_2$] (slaked lime)	It is used in the manufacture of bleaching powder, softening of hard water, in white washing and detecting CO_2 gas.
5. Ammonium hydroxide (NH_4OH)	It is used to remove grease stains from clothes and in cleaning window panes.
6. Aluminium hydroxide [$Al(OH)_3$]	It is used as a foaming agent in fire extinguishers.
7. Magnesia (MgO)	It is used in making refractory bricks.
8. Calcium oxide (CaO)	It is used as a drying agent, basic flux
9. Sodium carbonate (Na_2CO_3)	It is used as washing soda and for softening of hard water.
10. Sodium hydrogen carbonate ($NaHCO_3$)	It is used as baking soda in cooking food, as an antacid and in the soda–acid fire extinguisher.

Competition Edge

Effect of dilution on hydroxide (OH^-) ion; concentration of a base: As we have already discussed, when an acid is diluted with water, the H^+ ion concentration per unit volume in the solution decreases. Similarly, when a solution of a base is diluted with water, the hydroxide (OH^-) ion concentration per unit volume in the solution decreases. That is, the basic strength of the solution decreases with dilution. Further, remember that as in the case of weak acids, on dilution of the solution of a weak base, the total number of OH^- ions in the solution increases because dissociation of the weak base increases on dilution but even in this case, the concentration of the OH^- ions per unit volume decreases.

TEST YOUR KNOWLEDGE

2.9 Fill in the blanks.

(i) Bases soluble in water are known as ______.

(ii) Use of a mild base like ______ on the bee-stung area gives relief.

(iii) The acidity of calcium hydroxide is ______.

(iv) NH_4OH is known as ______.

Solution: (i) alkalis, (ii) baking soda, (iii) two, (iv) ammonium hydroxide.

2.10 Explain the following.

(i) Is strontium hydroxide $[Sr(OH)_2]$ a strong or a weak base?

(ii) On the basis of acidity of bases, what type of base is calcium hydroxide?

Solution: (i) $Sr(OH)_2$ is a strong base as it ionises almost completely in water.

(ii) We know that the acidity of a base is equal to the number of hydroxyl ions it has. So, calcium hydroxide $Ca(OH)_2$ having two hydroxyl groups, is a di-acidic base.

$$Ca(OH)_2(aq) \rightarrow Ca^{2+}(aq) + 2OH^-(aq)$$

2.11 (i) Is calcium oxide (CaO) an alkali or base or both?

(ii) An element E does not react with dilute sulfuric acid. However, it forms an oxide EO which turns red litmus into blue. Is E a metal or a non-metal?

Solution: (i) It serves as both. Alkalis are bases that dissociate in water to yield the hydroxyl ion (OH^-). Calcium oxide dissolves in water to form calcium hydroxide which further ionises to give the hydroxyl ion (OH^-).

$$CaO + 2H_2O \rightleftharpoons Ca(OH)_2 \rightleftharpoons Ca^{++} + 2OH^-$$

(ii) As the oxide EO turns red litmus into blue, it is a basic oxide. As metals form basic oxides, E is surely a metal.

2.12 A student took a small piece of solid quicklime in a china dish and poured over it a small amount of water. List two changes he is likely to observe in the china dish immediately after pouring water. **[CBSE 2015]**

Solution: When a small amount of water is poured over a piece of quicklime, the following observations are made.

(i) It reacts vigorously with water to produce a hissing sound and slaked lime is formed.

(ii) The reaction mixture becomes hot since it is an exothermic reaction.

2.5 Strength of Acid and Base in Terms of pH Scale

It is noticed that in any neutral solution, that is, a solution that is neither acidic nor basic (like water), the H^+ and OH^- ion concentrations are equal. At 25°C or 298 K, each of these concentrations is equal to 10^{-7} mol per litre, in water or any neutral solution.

Concentration of H^+ ions = Concentration of OH^- ions = 10^{-7} mol per litre

Further, it is found that for any solution (neutral, acidic or basic), the product of the H^+ ion concentration and the OH^- ion concentration (in mol per litre) is always constant and equal to 10^{-14} at 25°C. This is known as the *ionic product* of water. It is usually represented by the symbol K_w. Hence,

Ionic product of water (K_w) = Concentration of H^+ ions in mol L^{-1} × Concentration of OH^- ions in mol L^{-1}

$$K_w = [H^+]\,[OH^-] = 10^{-14} \text{ at } 25°C$$

For a solution, if the H^+ ion concentration is known, the OH^- ion concentration is automatically fixed as their product is always constant or fixed. Hence, we can express the acidic, basic or neutral character of any solution by simply expressing its H^+ ion concentration.

- As acids provide hydrogen ions in water, when an acid is added to water, the concentration of the hydrogen ions in water increases. The solution of the acid thus formed will have more of hydrogen ions and less of hydroxide ions, and it will be acidic in nature, that is, acidic solutions have an excess of hydrogen ions.
- As bases provide hydroxide ions in water, when a base is added to water, the concentration of hydroxide ions in water increases. The solution formed by dissolving a base in water will have more of hydroxide ions and less of hydrogen ions and it will be basic in nature, that is, basic solutions have an excess of hydroxide ions.

2.5.1 pH Scale

This scale was introduced by Sorenson to measure the acidity or basicity (hydrogen ion concentration) of a solution in terms of the pH value. In Danish, pH stands for *potenz de-hydrogen* (power of H^+ ions).

It can be given as $[H^+] = 10^{-pH}$

If an aqueous solution of an acid has a hydrogen ion concentration of 10^{-x} mol per litre, its pH will be x. For example, if an aqueous solution of an acid has a hydrogen ion concentration of 10^{-5} mol per litre, its pH will be 5.

In the modern way, pH can be given as: $pH = -\log_{10}[H^+]$

It means, the pH of the aqueous solution is equal to the –ve log of H^+ concentration in mol/litre or it is the reciprocal of log of the hydrogen ion concentration.

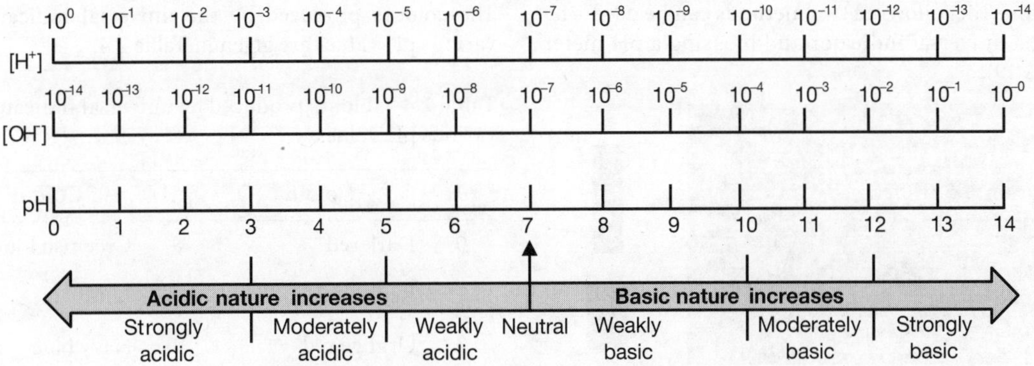

Fig. 2.18 Hydrogen ion concentration and pH range

$$\text{pH} = \log_{10}\frac{1}{[\text{H}^+]}$$

As the H^+ ion concentration is mainly in between 10^{-1} and 10^{-14}, the pH value range is in between 1 and 14 (Fig. 2.18).

2.5.2 Nature of Solution

If $[H^+] > 10^{-7}$, pH is less than 7 and the solution is acidic. If $[H^+] = 10^{-7}$, pH is 7 and the solution is neutral. If $[H^+] < 10^{-7}$, pH is more than 7 and the solution is basic.

From these points and Fig. 2.23, we can conclude the following.

- When the pH range is from 1–6 or the hydrogen ion concentration is 10^{-1} to 10^{-6} mol per litre, the solution is acidic and it turns blue litmus red. Methyl orange indicator turns red.
- When the pH is 7 or the hydrogen ion concentration is 10^{-7} mol per litre, the solution is neutral and does not change the colour of litmus.
- When the pH range is 8–14 or the hydrogen ion concentration is 10^{-8} to 10^{-14} mol per litre, the solution is basic and it turns red litmus into blue. Phenolphthalein indicator becomes pink.

pH range of some substances:

Substance	pH range
Gastric acid	1–3.0
Soft drink	2–4.0
Lemons	2.2–2.4
Vinegar	2.4–2.4
Urine	4.8–8.4
Milk	6.3–6.6
Saliva	6.5–7.5
Blood	7.3–7.5
Sea water	8.5
Tears	7.4

$$\text{Acidic nature} \propto \frac{1}{\text{pH value}} \propto \text{H}^+ \text{ concentration}$$

$$\text{Basic nature} \propto \frac{1}{\text{H}^+ \text{ concentration}} \propto \text{pH value}$$

Competition Edge

pH: $[H^+] = 10^{-pH}$, $pH = -\log_{10}[H^+]$
pOH: $[OH^-] = 10^{-pOH}$, $pOH = -\log_{10}[OH^-]$
As $[H^+] \times [OH^-] = K_w = 1 \times 10^{-14}$
$pH + pOH = pK_w = 14$
If the H^+ ion concentration of an acid is 10^{-4}, find its pH and pOH values.
As $[H^+] = 10^{-pH}$
$[H^+] = 10^{-4}$
So, pH = 4
As pH + pOH = 14
pOH = 14 – pH = 14 – 4 = 10

2.6 Measurement of Acid and Basic Strength

Solutions of all strong acids or all strong bases are not equally strong. Their strength depends upon the nature of acid or base and the extent of dilution (their concentrations). Similarly, solutions of all weak acids and all weak bases are not equally weak and their strength depends upon the nature of the acid or base and also on the extent of dilution. In order to measure the exact strength of the solution of an acid or an alkali, we have to determine the pH of the solution. The following two methods can be used—by using a universal indicator and by using a pH meter (Fig. 2.19).

Fig. 2.19 Universal indicator and pH meter

Source: Universal indicator – Toffel, Public domain, via Wikimedia Commons

(i) Universal Indicator

Indicators such as litmus, methyl orange and phenolphthalein can give an idea about the acidic or basic nature of a solution. However, they cannot tell us about the strength of the acid or base. To get an idea about acidic or basic strength, we need a specific indicator—a universal indicator.

A universal indicator is a mixture of two or more indicators which give different colours at different pH values of the entire pH scale. The common method of measuring the pH value of a solution in school laboratories is by using a universal indicator. As the pH of a solution depends upon the hydrogen ion concentration, the universal indicator shows different colours at different concentrations of hydrogen ions in the solution. When an acidic or basic solution is added to the universal indicator, the indicator gives a new colour. The colour given by the universal indicator is used to find the pH value of the acid or base by matching the colour with the colour on the pH colour chart. And knowing the pH value, it can be found out whether the given solution is a strong acid, weak acid, strong base or a weak base.

The colours produced by the universal indicator paper at various pH values are shown in Fig. 2.20.

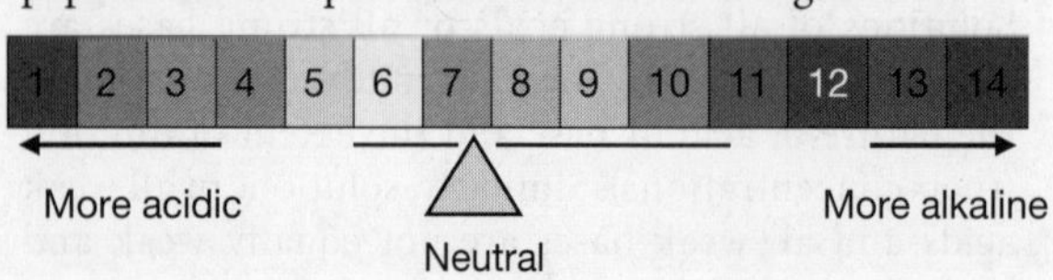

Fig. 2.20 pH colour chart showing the colours of universal indicator paper at different pH values

The colours produced by the universal indicator at various pH values are given in Table 2.4.

Table 2.4 Colours produced by universal indicator at various pH values

pH	Colour	pH	Colour
0	Dark red	8	Greenish blue
1	Red	9	Blue
3	Orange red	10	Navy blue
4	Orange	11	Purple
5	Orange yellow	12	Dark purple
6	Greenish yellow	13	Violet
7	Green	14	Violet

Process: A drop of the solution whose strength is to be found is put on a strip of the universal indicator paper. The indicator paper undergoes a change in colour where the solution has been put. This colour produced is then matched with a colour on the standard pH colour chart. By matching the colour with the corresponding pH value, we can find the strength of the solution. We can tell whether it is a strong acid or a weak acid or a strong base or a weak base. For example, if the colour is orange, the pH will be 4 and the solution will be more acidic.

(ii) pH Meter

As the pH paper can provide us only an approximate value of the pH of the solution, the pH of a solution can be measured more accurately by using an instrument known as a pH meter (Fig. 2.24).

2.7 Importance of pH in Everyday Life

The pH value plays an important role in many activities of our everyday life.

2.7.1 pH in Our Digestive System

Our stomach produces hydrochloric acid with a pH ≈ 1.4 which helps in digesting the food without harming the stomach. Sometimes an excess of acid is produced in the stomach due to various reasons such as over eating, highly spicy and oily food, and this excess acid causes indigestion which creates pain and irritation. In order to cure indigestion and get rid of the pain, one can take bases known as antacids which are a group of mild bases having no toxic effects on the body (Fig. 2.21). Being basic in nature, such antacids can react with the excess acid in the stomach and neutralise it. Some of the common antacids used for curing indigestion are magnesium hydroxide (milk of magnesia), sodium hydrogencarbonate (baking soda) and aluminium hydroxide.

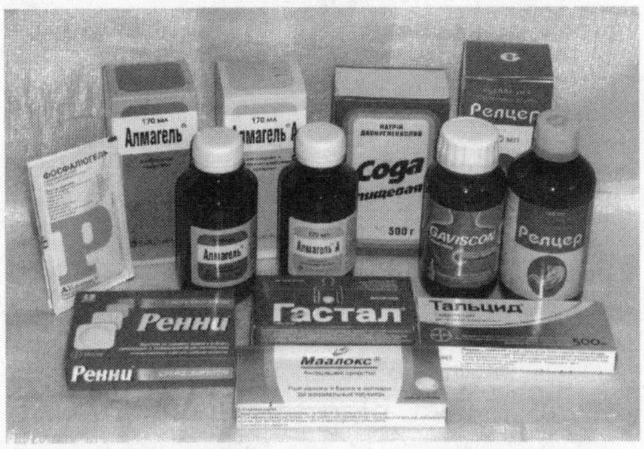

Fig. 2.21 Antacid tablets and suspensions

Source: Gastro-en, Public domain, via Wikimedia Commons

2.7.2 pH Change as the Cause of Tooth Decay

When you eat food having sugar, then the bacteria present in your mouth can break down the sugar to from acids like lactic acid (acid is formed in the mouth after sugary food has been eaten). This acid lowers the pH in the mouth by making it acidic and tooth decay starts when the pH of acid formed in the mouth falls below 5.5 (Fig. 2.22). This is because, the acid then becomes strong enough to attack the enamel (calcium phosphate) of the teeth and corrode it causing tooth decay.

The best way to prevent tooth decay is to clean the mouth thoroughly after eating food by rinsing it with lots of clean water. Most of the toothpastes contain bases to neutralise the mouth acid (the pH of toothpaste ≈ 8.0). So, using toothpaste for cleaning the teeth can neutralise the excess acid in the mouth and prevent tooth decay. Any person can lessen the chances of suffering from tooth decay by changing his/her eating habits by eating less sugary foods such as sweets, toffees, ice cream and chocolates.

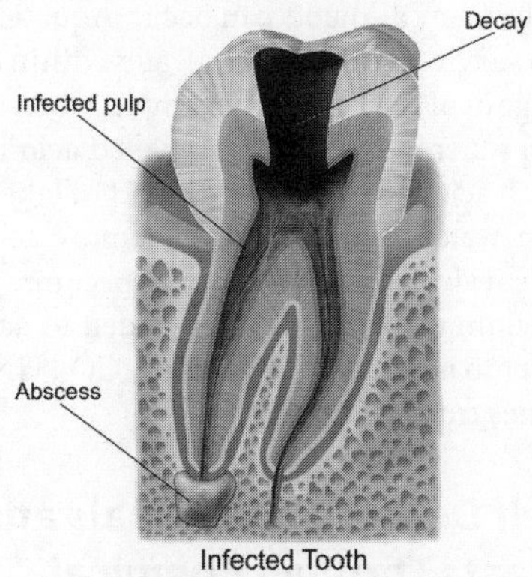

Fig. 2.22 Tooth cavity

Source: Blausen.com staff (2014). "Medical gallery of Blausen Medical 2014". WikiJournal of Medicine 1 (2). DOI:10.15347/wjm/2014.010. ISSN 2002-4436., CC BY 3.0 <https://creativecommons.org/licenses/by/3.0>, via Wikimedia Commons

2.7.3 Plants and Animals are Sensitive to pH Changes

Both plants and animals are sensitive to pH changes in the environment. In fact, the growth of plants and survival of animals depends to a large extent on the availability of proper pH conditions which suit them.

(i) **Soil pH and plant growth:** Most plants grow best when the pH of the soil is close to 7. If the soil is too acidic or too basic, the plants grow badly or cannot grow at all. The soil may be acidic or basic naturally. The pH of soil is also affected by the use of chemical fertilisers in the fields. The pH of acidic soil can be as low as 4 and that of the basic soil can be as high as 8.3. Chemicals can be added to soil to adjust its pH and to make it suitable of growing plants.

Most often, the soil in the fields is too acidic. It is then treated with materials like quicklime (calcium oxide), slaked lime (calcium hydroxide) or chalk (calcium carbonate). As these substances are bases, they react with the excess acid present in soil and reduce its acidity by neutralisation.

(ii) **pH change and survival of animals:** The pH plays an important role in the survival of animals, including human beings. Our body works well within a narrow pH range of 7.0 to 7.8. If due to some reason this pH range gets disturbed in the body, many ailments can occur. Aquatic animals can survive in river or lake water within a narrow range of pH change. For example, when the pH of rain water is about 5.6, it is called acid rain. Too much acid rain can lower the pH of lake water or river water to such an extent (more acidic) that the survival of aquatic animals becomes difficult. Calcium carbonate is often added to acidic lake water to neutralise the acids (H_2CO_3, HNO_3) that comes from acid rain.

2.7.4 Self Defence by Animals and Plants Through Chemical Warfare

Many animals and plants protect themselves from their enemies by injecting painful and irritating acids and bases into their skin. For example,

(i) when a honey bee stings, it injects an acidic liquid (formic acid) into the skin which causes immense pain and irritation. Rubbing a mild base such as baking soda solution on the stung area gives relief as the baking soda neutralises the acidic liquid injected by the sting.

(ii) When a wasp stings, it injects an alkaline liquid into the skin. So, rubbing a mild acid like vinegar on the stung area of the skin gives relief.

(iii) An ant's sting injects formic acid into the skin causing burning pain. Since an ant's sting is acidic, it can be neutralised by rubbing a mild base like baking soda on the affected area.

(iv) Few plants like the nettle can also give us painful stings (Fig. 2.23). If you touch the leaves of a nettle plant, the stinging hair of the nettle leaves inject formic acid into the skin causing burning pain which can be neutralised by rubbing baking soda on the skin.

Fig. 2.23 Stinging nettle

Source: By Kevin Thiele from Perth, Australia - Urtica incisa, CC BY 2.0, https://commons.wikimedia.org/w/index.php?curid=49645644)

TEST YOUR KNOWLEDGE

2.13 Fill in the blanks

(i) The word pH was introduced by ______.

(ii) The pH of a basic solution is always ______ than 7.

(iii) If the pH value is lower than 7, the nature of the solution is ______.

(iv) The pH value of human blood is nearly ______.

(v) If the H^+ ion concentration is 10^{-3} mol per litre, the pH value is ______.

Solution: (i) Sorenson, (ii) more/greater, (iii) acidic, (iv) 7.4, (v) 3.

2.14 Explain the following.

(i) What effect does an increase in the concentration of H^+ in a solution have on the pH of the solution?

(ii) Which of the following, 1 M HCl or 1 M HCOOH, has a higher concentration of H^+ ions?

Solution: (i) An increase in the concentration of H^+ ions in a solution decreases the pH value.

(ii) 1 M HCl has more hydrogen ion concentration than 1 M HCOOH as HCl, being a strong acid, is more ionised.

2.15 A solution has a pH value of 3. If it is diluted 10 times, what will be the value of pH after dilution?

Solution: (i) As pH = 3 and $[H^+] = 10^{-pH} = 10^{-3}$

On dilution, the new concentration of $H^+ = 10^{-3}/10 = 10^{-4}$

As pH is the negative power of $[H^+]$, the pH after dilution is 4.

2.16 Find the pH of 10^{-4} M NaOH solution.

Solution: $[H^+] \times [OH^-] = 1 \times 10^{-14}$

$$\text{So, } [H^+] = 1 \times \frac{10^{-14}}{[OH^-]} = \frac{10^{-14}}{10^{-4}} = 10^{-10}$$

As pH is negative power of $[H^+]$, pH = 10

Alternative solution

As $[OH^-] = 10^{-4}$, pOH = 4

pH + pOH = 14

So, pH = 14 – pOH = 14 – 4 = 10

2.17 A student detected the pH of four unknown solutions A, B, C and D as 11, 5, 7 and 2. Predict the nature of the solutions.

Solution: The nature of the solutions: A = basic, B = acidic, C = neutral and D = strongly acidic.

Quick Review

- An **indicator** is a dye that changes colour when it is put into an **acid** or a **base**. An indicator gives different colours in acidic and basic mediums.
- The most common indicator used for testing acids and bases in the laboratory is litmus. **Litmus is a natural indicator.** Litmus solution is a purple dye which is extracted from a type of plant called lichen.
- Universal indicator is a mixture of two or more indicators which give different colours at different pH values of the entire pH scale. The common method of measuring the pH value of a solution in school laboratories is by using a **universal indicator**.
- The word **acid** comes from the Latin word *acidus*, meaning 'sour'. An acid is a compound having hydrogen atoms which can be wholly or partially replaced by metal atoms or positive radicals.
- A **strong acid** is vigorously ionised in solution; as a result, the value of α is high (more than 30%) and the concentration of hydronium ion more. For example, nitric acid (HNO_3), sulfuric acid (H_2SO_4).
- A **weak acid** is less ionised (α is less than 30%) so H^+ concentration is low. For example, H_2CO_3, H_2S.
- **Bases** are those species that react with acids to form salts, usually accompanied by water. In general, bases are oxides and hydroxides of metals (group 1, 2). Ammonia and some related compounds such as $R–NH_2$ are also bases.
- A **strong base** can undergo almost complete ionisation in aqueous solution to produce a high concentration of OH^-. For example, hydroxides of group 1 and 2 metals such as NaOH, KOH, $Mg(OH)_2$, $Ca(OH)_2$.
- A **weak base** can undergo partial ionisation in aqueous solution to produce a low concentration of OH^-. For example, NH_4OH, $Zn(OH)_2$.
- **Arrhenius concept of acid and base:** Acids are H^+ ion donors in water and bases are OH^- ion donors in water. Number of H^+ donated = Basicity or protosity of the acid.

Exercise 1

Section A: Multiple Choice Questions (1 Mark)

1. Which one of the following is acidic?
(a) lemon juice (b) tomatoes
(c) milk (d) all of these

2. Which one of the following will turn red litmus blue?
(a) vinegar
(b) baking soda solution
(c) lemon juice
(d) soft drinks

3. Which one of the following will turn blue litmus red?
(a) vinegar
(b) lime water
(c) baking soda solution
(d) washing soda solution

4. Methyl orange is
(a) pink in acidic medium, yellow in basic medium
(b) yellow in acidic medium, pink in basic medium
(c) colourless in acidic medium, pink in basic medium
(d) pink in acidic medium, colourless in basic medium.

5. Which acid is present in tamarind?
(a) tartaric acid (b) oxalic acid
(c) lactic acid (d) citric acid

6. Lactic acid is present in
(a) orange (b) tea
(c) curd (d) vinegar

7. Alkalis are
(a) acids, which are soluble in water
(b) acids, which are insoluble in water

(c) bases, which are insoluble in water
(d) bases, which are soluble in water

8. Sodium hydroxide is a
(a) weak base (b) weak acid
(c) strong base (d) strong acid

9. Lime water is
(a) CaO (b) $Ca(OH)_2$
(c) $CaCO_3$ (d) $CaCI_2$

10. The nature of calcium phosphate that is present in tooth enamel is
(a) basic (b) amphoteric
(c) acidic (d) neutral

11. The chemical formula of caustic potash is
(a) NaOH (b) $Ca(OH)_2$
(c) NH_4OH (D) KOH

12. What is formed when zinc reacts with sodium hydroxide?
(a) zinc hydroxide and sodium
(b) sodium zincate and hydrogen gas
(c) sodium zinc oxide and hydrogen gas
(d) sodium zincate and water

13. Rain is called acid rain when its:
(a) pH falls below 7 (b) pH falls below 6
(c) pH falls below 5.6 (d) pH is above 7

14. Sodium hydroxide is used
(a) as an antacid
(b) in the manufacture of soap
(c) as a cleansing agent
(d) in alkaline batteries

15. Sodium hydroxide turns phenolphthalein solution
(a) pink (b) yellow
(c) colourless (d) orange

16. pH value less than 7 indicates that the solution is
(a) acidic
(b) basic
(c) neutral
(d) none of the above

17. Which of the following are present in a dilute aqueous solution of hydrochloric acid?
(a) $H_3O^+ + Cl^-$ (b) $H_3O^+ + OH^-$
(c) $Cl^- + OH^-$ (d) unionised HCl

18. In which pH range does our body work to survive in the atmosphere?
(a) 5.5 to 8.5 (b) 7.0 to 7.8
(c) 2.3 to 7.0 (d) 7.5 to 12.5

19. When a base reacts with a metal, it forms a salt and hydrogen gas is released. By what method can the presence of hydrogen be detected?
(a) by water
(b) by litmus paper
(c) by methyl orange
(d) by a burning candle

20. A solution gives a yellow-orange colour when a few drops of universal indicator are added to it. The solution is of
(a) sodium hydroxide
(b) milk of magnesia
(c) lemon juice
(d) sodium chloride **(CBSE 2021)**

21. Concentrated H_2SO_4 is diluted by adding dropwise
(a) water to acid followed by a base
(b) base to acid followed by cold water
(c) water to acid with constant stirring
(d) acid to water with constant stirring **(CBSE 2021)**

22. Absence of tartaric acid in baking powder makes the taste of the cake bitter due to the presence of
(a) sodium metabisulfide
(b) sodium sulfate
(c) sodium hydrogen carbonate
(d) sodium carbonate**(CBSE 2021)**

23. The pH of acid rain is approximately **(CBSE 2021)**
(a) 6.4 (b) 5.6
(c) 7.9 (d) 7.0

24. An aqueous solution of a salt shows an orange-red colour when a drop of a universal indicator is added to it. This salt is made up of **(CBSE 2021)**
(a) a strong acid a weak base
(b) a weak acid and a strong base
(c) a strong acid and a strong base
(d) a weak acid and a weak base

25. Three test tubes A, B and C contain distilled water, an acidic solution and a basic solution respectively. When red litmus solution is used for testing these solutions, the observed colour changes observed will be **(CBSE 2021)**
(a) A: becomes red, B: no change, C: becomes blue
(b) A: becomes light red, B: becomes dark red, C: becomes blue
(c) A: no change, B: becomes dark red, C: becomes blue

(d) A: becomes light red, B: becomes blue, C: becomes red

26. In the activity shown in the diagram, if the climate is humid, the role of calcium chloride taken in the guard tube is to

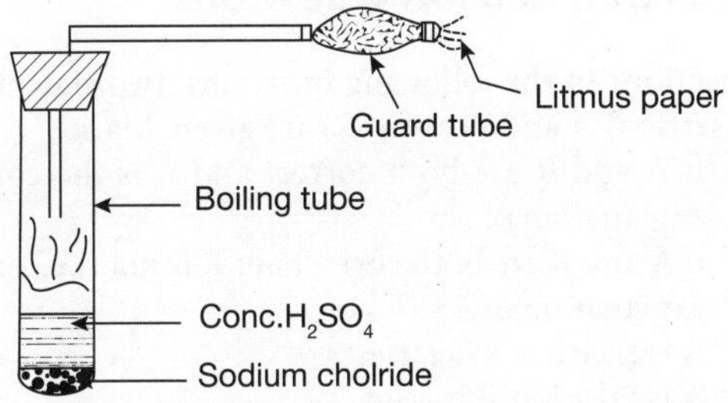

(a) dry the gas
(b) absorb the evolved gas
(c) absorb chloride ions from the evolved gas
(d) warm up the gas **(CBSE 2021)**

27. Consider the following processes.
(i) condensation of water vapour
(ii) sublimation of dry ice
(iii) dilution of sulfuric acid
(iv) dissolution of ammonium chloride in water

The endothermic process(es) is/are **(CBSE 2021)**
(a) (i) and (iii) (b) (ii) only
(c) (iii) only (d) (ii) and (iv)

28. Which of the following oxide(s) is/are soluble in water to form alkalies? **(CBSE 2021)**
(i) Na_2O (ii) SO_2 (iii) K_2O (iv) NO_2
(a) (ii) and (iv) (b) (iii) only
(c) (i) and (iii) (d) (i) only

29. Study the diagram given below and identify the gas formed in the reaction. **(CBSE 2021)**

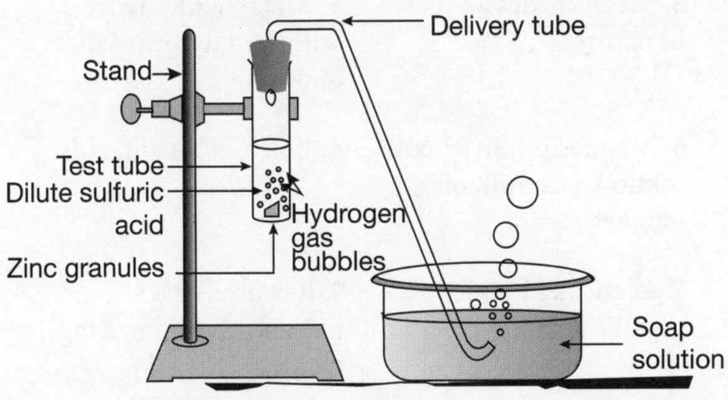

(a) Oxygen due to which the candle burns more brightly.
(b) Carbon dioxide which extinguishes the burning candle.
(c) Hydrogen which while burning produces a popping sound.
(d) Sulfur dioxide which produces a suffocating smell.

30. Which of the options in the given table are correct? **(CBSE 2021)**

Option	Natural source	Acid present
(i)	Orange	Oxalic acid
(ii)	Sour milk	Lactic acid
(iii)	Ant sting	Methanoic acid
(iv)	Tamarind	Acetic acid

(a) (i) and (iv) (b) (i) and (ii)
(c) (ii) and (iii) (d) (iii) and (iv)

31. Select from the following statements, what is true for bases. **(CBSE 2021)**
(a) Bases are sour and change red litmus to blue.
(b) Bases turn pink when a drop of phenolphthalein is added to them.
(c) Bases are bitter and turn blue litmus red.
(d) Bases have a pH value less than 7.

32. Consider the pH value of the following acidic samples: **(CBSE 2021)**

S. No	Sample	pH value
1.	Lemon juice	2.2
2.	Gastric juice	1.2
3.	Vinegar	3.76
4.	Dil. acetic acid	3.0

The decreasing order of their H^+ ion concentration is
(a) 2 > 1 > 3 > 4 (b) 3 > 4 > 1 > 2
(c) 3 > 4 > 2 > 1 (d) 2 > 1 > 4 > 3

33. 50.0 mL of tap water was taken in a beaker. Hydrochloric acid was added drop by drop to water. The temperature and pH of the solution were noted. The adjoining graph was obtained. Choose the correct statements related to the activity. **(CBSE 2021)**

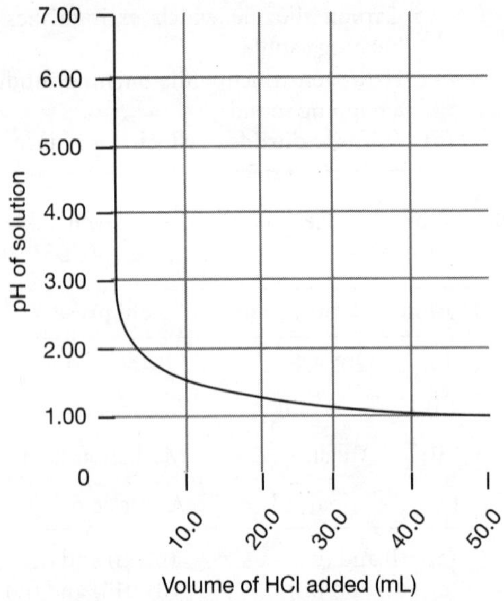

(i) The pH of tap water was around 7.
(ii) The pH of the solution decreases rapidly on addition of acid.
(iii) The pH of the solution increases rapidly on addition of acid.
(iv) The process of dissolving acid in water is highly endothermic.

(a) (i) and (ii) (b) (i) and (iii)
(c) (iii) and (iv) (d) (ii) and (iv)

34. Match the following.

Column I	Column II
1. Vinegar	(i) $Ca(OH)_2$
2. Milk of Magnesia	(ii) $Mg(OH)_2$
3. Slaked lime	(iii) CH_3COOH
4. King of chemicals	(iv) H_2SO_4

(a) (1)—(ii), (2)—(iii), (3)—(i), (4)—(iv)
(b) (1)—(iii), (2)—(iv), (3)—(ii), (4)—(i)
(c) (1)—(iii), (2))—(ii), (3)—(i), (4)—(iv)
(d) (1)—(iii), (2)—(ii), (3)—(iv), (4)—(i)

35. Match the following.

Column I (Acid)	Column II (Uses)
1. Acetic acid	(i) Cleaning the surface of the metals
2. Nitric acid	(ii) Petroleum refining
3. Sulfuric acid	(iii) In pickles
4. Hydrochloric acid	(iv) Explosive like TNT

(a) (1)—(ii), (2)—(iii), (3)—(i), (4)—(iv)
(b) (1)—(iii), (2)—(iv), (3)—(ii), (4)—(i)
(c) (1)—(iii), (2)—(ii), (3)—(i), (4)—(iv)
(d) (1)—(iii), (2)—(ii), (3)—(iv), (4)—(i)

Assertion Reason Questions

Direction: In the following questions, two statements (Assertion) A and Reason (R) are given. Mark
(a) if A and R are both correct and R is the correct explanation of A;
(b) if A and R are both correct but R is not the correct explanation of A;
(c) A is true but R is false;
(d) A is false but R is true.

Assertion	Reason
1. To dilute concentrated sulfuric acid, water is added into concentrated sulfuric acid.	**1.** Dilution of concentrated sulfuric acid is highly exothermic.
2. All the four metals Mg, Zn, Fe and Cu react with dilute HCl to produce H_2 gas though with different activity.	**2.** The activity of metals to react with dilute HCl is in the order Mg > Zn > Fe > Cu.
3. Litmus is a natural indicator.	**3.** Vinegar or lemon juice can turn the blue colour of litmus into red.
4. Zinc on reaction with dilute HCl gives chlorine gas.	**4.** Copper does not react with HCl.
5. Metal oxides are basic in nature.	**5.** Metal oxides react with acid to form salt and water.
6. Magnesium hydroxide is known as milk of magnesia.	**6.** It is used in antacids.
7. Acetic acid is an organic acid.	**7.** It is used as a preservative in making pickles.
8. On dilution, the hydrogen ion concentration decreases.	**8.** More the dilution, less is the pH value.

Section B: Very Short Answer Questions (2 Marks)

1. Define an indicator.
2. Give the names of any two indicators (other than litmus).
3. An acid turns ____ litmus into ________.
4. A base turns ______ litmus into ______.
5. Name any two natural indicators (other than litmus).
6. Define an olfactory indicator.
7. Which of the following are olfactory indicators? Onion, vanilla essence, clove oil, tomato, potato
8. Bases have ______ taste while acids have ______ taste.
9. How can you say acids are corrosive in nature?
10. How can you define the Arrhenius concept?
11. How does H^+ exist in water?
12. Name the acids present in tomatoes and oranges.
13. What is an alkali? Give an example.
14. How can you test that metal oxides are basic in nature?
15. Choose the strong acids from the following:

$$CH_3COOH, H_2SO_4, H_2CO_3, HNO_3$$

16. Write two properties of a base based on taste and reaction with metals.
17. Classify the following into acidic oxides and basic oxides.

$$Na_2O, SO_2, MgO, CO_2$$

18. Write the name given to bases that are highly soluble in water. Give an example.
19. What is the reaction called when an acid reacts with a base to produce a salt and water? Give an example.
20. Why can't dry HCl gas change the colour of dry litmus paper?
21. Is water a strong acid or a weak acid?
22. Name the acid present in ant sting.
23. How will you test a gas which is liberated when HCl acid reacts with an active metal?
24. How can you say ammonia is a base?
25. Define pH value.
26. What effect does a decrease in concentration of H^+ in a solution have on the pH of the solution?
27. Which of the following, 1 M HCl or 1 M HCOOH, have a lower pH value?
28. Define a universal indicator.
29. What is the hardest substance in our body and what is it composed of?
30. What is the pH value of a solution of an equivalent amount of HCl and NaOH?
31. How does the concentration of hydronium ions change during the dilution of an acid?
32. Name any two metals which can react with hydrochloric acid to give hydrogen.
33. Name any two metals which can react with caustic potash to give hydrogen.
34. Why is sulfuric acid called the king of chemicals?

Case Study or Passage Based Questions

I. Indicators are the substances that show different colours in the acidic and basic medium. They may be natural, such as litmus, turmeric, red cabbage and so on, or synthetic such as phenolphthalein, methyl orange and so on. Indicators giving different smells in acidic and basic medium are olfactory indicators such as clove oil.

35. To test the presence of an acid with a strip of red litmus paper, what would you do?
36. What happens when phenolphthalein is added to an acidic solution?
37. What will be the action of liquid antacid on litmus paper? Is it acidic or basic?
38. Write the colours shown by phenolphthalein and methyl orange in the basic medium.

39. Name any two natural indicators which can behave like olfactory indicators.

II. In order to determine [H^+] and [OH^-], pH value was introduced. pH value is equal to the negative power raised on 10 to express [H^+]. pH value is used to find the acidic or basic nature of a solution. If the pH value is 1–6, the solution is acidic and if it is 7–14, the solution is basic in nature. For a neutral solution, the pH is 7. In nature, every substance has a specific value of pH. For example, the pH of our blood is 7.4.

40. The pH of a solution is less than 7. What does it indicate? What is the pH of a 1 M HCl solution?

41. Which colour is produced in a pH paper when we put a drop of distilled water on it? What is the pH of distilled water?

42. Write two precautions required while finding the pH of a given samples.

43. Bottle A contains oxalic acid and bottle B contains sodium carbonate solution. When a pH paper is dipped in each of the solutions, the colour seen in A and B respectively be ________ and ________.

44. Fresh milk has a pH of 6. When it changes into curd (yogurt), will its pH value increase or decrease? Why?

Section C: Short Answer Questions (3 Marks)

1. Name one indicator which specifies the various levels of H^+ ion concentration. **[CBSE 2016]**
2. Name the acid present in the following:
   (i) Tomato (ii) Vinegar
   (iii) Tamarind **[CBSE 2015]**
3. State the meaning of a strong acid and a weak acid. Give one example of each.
4. How are bases different from alkalis? Are all bases alkalis?
5. Write one word/term for the following.
   (i) Water soluble base.
   (ii) A substance which dissociates on dissolving in water to produce hydrogen ions [$H^+(aq)$ ions].
   (iii) A reaction between an acid and a base to form salt and water.
   (iv) A substance which dissociates on dissolving in water to produce hydroxide ions (OH– ions).
6. A weak acid is added to a concentrated solution of hydrochloric acid. Does the solution become less or more acidic?
7. What will happen if lead sulfite is taken in a test tube and a few mL of aqueous nitric acid is added to it? After that litmus paper is placed at the mouth of test tube. What is observed?
8. $NH_4OH(l) \rightleftharpoons NH_4^+(l) + OH^-$

   What is the effect on the above equilibrium on addition of a base which furnishes hydroxyl (OH^-) ions?
9. What is meant by the term pH of a solution? The pH of rain water collected from two cities A and B was found to be 6 and 5 respectively. The water of which city is more acidic?
10. Explain: is a solution with pH 6 acidic, basic or neutral? Justify.
11. What is the change in the pH value of milk when it changes into curd? Explain.
12. Given below are the pH values of different liquids 7.0, 14.0, 4.0 and 2.0. Which of these could be that of (i) lemon juice, (ii) distilled water, (iii) sodium hydroxide solution, (iv) tomato juice.
13. A knife, which is used to cut a fruit, was immediately dipped into water containing drops of blue litmus solution. If the colour of the solution is changed to red, what inference can be drawn about the nature of the fruit and why?
14. Write chemical equations that shows that aluminium oxide reacts with an acid as well as a base.
15. While diluting an acid, why is it recommended that the acid should be added to water and not water to the acid?
16. When a drop of orange juice is added to pure water, how will the pH value vary? If a drop of lemon juice is also added, will there be any more change in the pH value?
17. 10 mL of water and 5 mL of sulfuric acid are to be mixed in a beaker.

(i) State the method that should be followed.
(ii) Why should this method be followed?
(iii) What is this process called?

18. Choose strong and weak acids as well as strong and weak bases from the following.
H_2CO_3, HNO_3, CH_3COOH, NaOH, NH_4OH, KOH, $Zn(OH)_2$, HCl

19. Solution A, B, C and D have pH values 2, 3, 4, and 5 respectively. Arrange them in increasing order of acidic character giving reasons.

20. Solutions P, Q and R have pH values 9, 10 and 11 respectively. Arrange them in increasing order of basic character giving reasons.

21. A liquid has a pH less than 7 which represents an acidic solution.
(i) State the nature of solution, if its pH increases from 7 to 14.
(ii) Mention the ion whose concentration increases with an increase in pH value.
(iii) Suggest a method that is generally used for measuring the pH value.

22. 15 mL of water and 10 mL of sulfuric acid are to be mixed in a beaker.
(i) State the method that should be followed giving the reason.
(ii) What is this process called? **[CBSE 2015]**

23. Explain the action of dil. HCl on the following with chemical equations.
(i) Magnesium ribbon
(ii) Sodium hydroxide
(iii) Crushed egg shells **[CBSE.2015]**

24. Explain how an antacid works. **[CBSE 2015]**

25. The soil in a field is highly acidic. List any two materials which can be added to this soil to reduce its acidity. Give the reason for your choice. **[CBSE 2016]**

26. 2 mL of sodium hydroxide solution is added to a few pieces of granulated zinc metal taken in a test tube. When the contents are warmed, a gas evolves which is bubbled through a soap solution before testing. Write the equation of the chemical reaction involved and the test to detect the gas. Name the gas which will be evolved when the same metal reacts with a dilute solution of a strong acid. **[CBSE 2018]**

27. What is the importance of pH in everyday life? **[CBSE 2017]**

28. Name the acid present in the ant sting and give its chemical formula. Also give the common method to get relief from the discomfort caused by the ant sting. **[CBSE 2019]**

29. How is the presence of an acid tested with a strip of red litmus paper? **[CBSE 2019]**

OR

A student is performing an experiment to study the properties of acetic acid. Answer the following questions.
(i) Name the substance he must add to acetic acid to produce carbon dioxide.
(ii) Give the relevant chemical equation for the reaction.
(iii) How would he test for CO_2 gas in the laboratory?

30. A cloth strip dipped in onion juice is used for testing a liquid X. The liquid X changes its odour. Which type of indicator is onion juice? The liquid X turns blue litmus red. List the observations liquid X will show on reacting with the following.
(i) Zinc granules
(ii) Solid sodium carbonate
Write the chemical equations for the reactions involved. **[CBSE 2020]**

Section D: Long Answer Questions (5 Marks)

1. Give reasons for the following.
(i) Metal oxides generally react with acids. But, zinc oxides react even with sodium hydroxide.
(ii) Heat of neutralisation value for the reaction of HCl and NH_4OH is less than 13.7 kcal/mol.
(iii) All indicators are not suitable for all titrations.

2. Explain the action of phenolphthalein indicator in the titration of acetic acid versus sodium hydroxide.

3. State the reason for the following statements.
(i) Tap water conducts electricity whereas distilled water does not.
(ii) Dry hydrogen chloride gas does not turn blue litmus red whereas dil. HCl does.
(iii) During summer, the milk man usually adds a very small amount of baking soda to fresh milk.

(iv) For dilution of an acid, the acid is added to water and not water into the acid.

(v) Ammonia is a base but does not contain a hydroxyl group. **[CBSE 2015]**

4. (i) Giving reasons for each, state which of the following will conduct electricity and which will not.
(a) A solution of glucose
(b) Dil. hydrochloric acid

(ii) If acetic acid and hydrochloric acid of the same concentration are taken, which of the two is a stronger acid and why?

(iii) How is the strength of an acid affected when some water is added to it?

5. *Sugandha prepares HCl gas in her school laboratory using certain chemicals. She puts both dry and wet blue litmus papers in contact with the gas.

(i) Name the reagents used by Sugandha to prepare HCl gas.

(ii) State the colour changes observed with the dry and wet blue litmus paper.

(iii) Show the formation of ions when HCl gas combines with water.

6. *A student takes three beakers A, B and C filled with an aqueous solution of glucose, alcohol and hydrochloric acid respectively, as shown in the following figure.

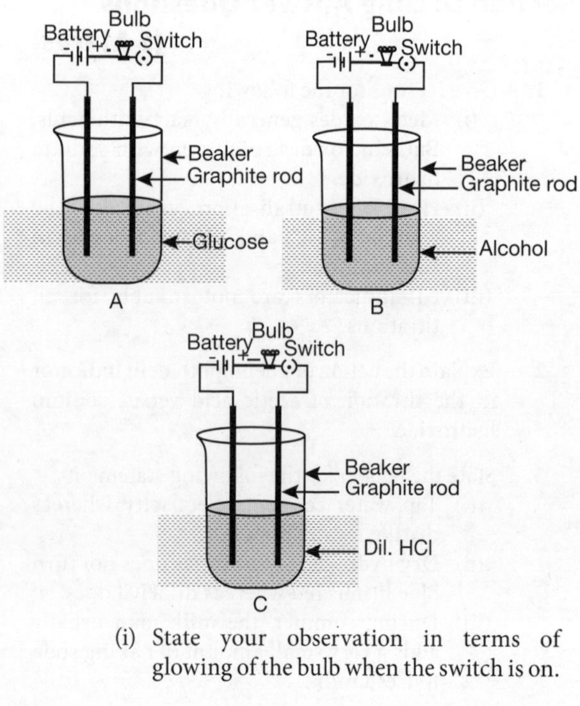

(i) State your observation in terms of glowing of the bulb when the switch is on.

(ii) Justify your observations by giving a reason in each case.

(iii) Mention the change noticed with an appropriate reason if the content of beaker B is replaced by sodium hydroxide solution.

7. Explain the action of dilute hydrochloric acid on the following with a chemical equation.
(i) magnesium ribbon
(ii) sodium hydroxide
(iii) crushed egg shells
[Board Term-I Set-OQKPLGV, 2016]

8. (i) What is the action of litmus on the following?:
(a) dry ammonia gas
(b) solution of ammonia gas in water

(ii) State the observations you would make on adding sodium hydroxide to aqueous solution of:
(a) ferrous sulfate
(b) aluminium chloride
Give balanced chemical equations.
[Board Term-I Set-1ZHNPNO, 2016]

9. In the following schematic diagram for the preparation of hydrogen gas, what would happen if the following changes are made?

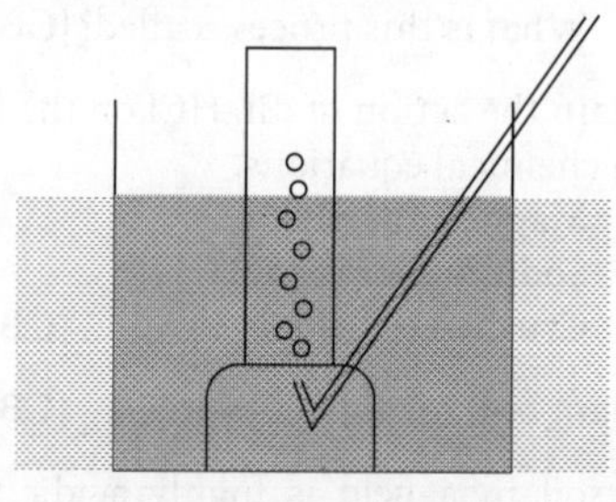

(i) Is the gas soluble or insoluble in water?
(ii) Is the gas lighter or heavier than air?
(iii) How will you test the gas?
(iv) In place of zinc granules, the same amount of zinc dust is taken in the test tube.
(v) Instead of dilute sulfuric acid, dilute hydrochloric acid is taken.
(vi) Sodium hydroxide is taken in place of dilute sulfuric acid and the test tube is heated.

10. (i) Define universal indicator. For what purpose is it used?

(ii) Two solutions A and B have pH values of 3.0 and 9.5 respectively. Which of these will turn litmus solution from blue to red and which will turn phenolphthalein from colourless to pink?

(iii) Water is a neutral substance. What colour will you get when you add a few drops of universal indicator to a test tube containing distilled water?

[Board Term-I Set-1ZHNPNO, 2016]

11. (i) A local magician was showing magic in a village street. He took an egg shell and poured a solution over it. As a result, effervescence was observed. When he took a burning matchstick over it, it went off.
    (a) Identify the solution poured and the substance present in the egg shell.
    (b) What is the reason behind the effervescence?
    (c) Write its balanced chemical equation.
    (d) Give the common name of the substance present in the egg shell.

    (ii) Draw a labelled diagram to show that acid solution in water conducts electricity.

12. (i) Five solutions A, B, C, D and E when tested with the universal indicator showed the pH as 4, 1, 11, 7 and 9 respectively. Which solution is
    (a) neutral
    (b) strongly alkaline
    (c) strongly acidic
    (d) weakly acidic
    (e) weakly alkaline
    Arrange the solutions in increasing order of their H^+ ion concentration.

    (ii) Name the acid and base from which the following salts have been formed.
    (a) Sodium acetate
    (b) Ammonium chloride

13. Ravi's father is a botanist. He extracted a purple dye from a plant in the laboratory which acts as an acid–base indicator. When he adds this purple dye to acids, it changes to red. But, when a small amount of base is added to this dye, it changes its colour to blue. Ravi was very excited to see that the colours were changing with the solution. He observed the magic and asked his father some questions.
    (i) What is an indicator?
    (ii) Name the plant from which his father extracted the dye.
    (iii) 'Some natural substances are good sources of indicators.' Suggest some examples to justify the statement.

14. *One day Lucky's mother, after a meal, felt pain and irritation in her stomach. His father was out of station. He was an intelligent boy. He remembered his teacher's statement and gave his mother some baking soda solution, which gave her relief from pain and irritation of stomach.
    (i) Which information given by the teacher helped him to select baking soda as the remedy?
    (ii) State any two values that you have learned from this passage.

Section E: Case Study or Passage Based Questions (4 Marks)

Acids are sour in nature and provide H^+ in water. Mineral acids like HCl are strong while organic acids like HCOOH are weak. There are many substances around us having acids; for example, lemon juice contains citric acid. Bases are bitter in taste and in water they provide OH^-. Bases of group 1 are strong. An acid and base react together to give a salt and water.

1. Which acid is present in tomato?
   (a) citric acid (b) oxalic acid
   (c) lactic acid (d) HCl

2. Which acid is a strong acid among the following?
   (a) HCl pH 1
   (b) CH_3COOH pH 5
   (c) lemon juice pH 2.2
   (d) pure milk pH 6

3. Which of the following is the strongest base?
   (a) NaOH (b) CsOH
   (c) NH_4OH (d) $Zn(OH)_2$

4. Which of the following pairs includes strong acids only?
   (a) HCOOH, HCl
   (b) HCOOH, H_2CO_3
   (c) HCl, H_2SO_4
   (d) CH_3COOH, HCOOH

5. Which of the following statements are correct?
   (i) Acid and base react to form salt and water.
   (ii) This reaction is called neutralisation.

(iii) It is an exothermic reaction.
(iv) It is an endothermic reaction.

(a) (i), (ii) (b) (ii), (iii)
(c) (i), (ii), (iii) (d) (i), (ii), (iii), (iv)

High Order Thinking Skills (HOTS) Questions

1. What do you observe when you add a few drops of acetic acid to a test tube containing:
   (i) phenolphthalein
   (ii) distilled water
   (iii) universal indicator
   (iv) sodium hydrogen carbonate powder [CBSE 2016]

2. P and Q are aqueous solutions of sodium chloride and sodium hydroxide respectively. Which of these will turn [CBSE 2017]
   (i) blue litmus to red
   (ii) red litmus to blue
   Justify your answer.

3. Explain the following: (i) When conc. acid is added to water, is the process exothermic or endothermic?
   (ii) Why does the flow of acid rain water into a river make the survival of aquatic life in the river difficult?
   (iii) What is the pH of gastric juices, released during digestion?
   (iv) What is the effect of an increase in concentration of H^+ ions on the addition of 1 M HCl to 1 M CH_3COOH.
   (v) Why does tooth decay start when the pH of the mouth is lower than 5.5?

4. Explain the following: (i) Why does distilled water not conduct electricity, whereas rain water does?
   (ii) Explain why sour substances are effective in cleaning copper vessels.
   (iii) Why are pickles and curd not stored in copper and brass utensils?

5. Explain the following.
   (i) Hold one moist and one dry strip of blue litmus paper over dry HCl acid gas. Which strip will turn red and why?
   (ii) HCl and HNO_3 show acidic characteristics in aqueous solution while alcohol and glucose solutions do not. Give reasons.
   (iii) While constructing a house, a builder selects marble flooring and marble table tops for the kitchen where vinegar and juice of lemon, tamarind, and so on, are often used for cooking. Will you agree to this selection and why?
   (iv) Two solutions, X and Y, are tested with universal indicator. Solution X turns orange whereas solution Y turns red. Which of the solutions is a stronger acid?

6. To the three solutions listed below, a few drops of phenolphthalein and blue litmus were added separately. Specify the colour change in each case, if any.

Name of the solution	Colour change with phenolphthalein	Colour change with blue litmus
1. Sodium carbonate		
2. Hydrochloric acid		
3. Sodium chloride		

7. (i) Explain why hydrochloric acid is a strong acid and acetic acid, a weak acid. How can it be verified?
   (ii) Explain why an aqueous solution of an acid conducts electricity.
   (iii) You have four solutions A, B, C and D. The pH of solution A is 6, B is 9, C is 12 and D is 7.
      (a) Identify the most acidic and most basic solutions.
      (b) Arrange the above four solutions in the increasing order of their H^+ ion concentrations.
      (c) State the change in colour of pH paper on dipping in solution C and D.

8. (i) In a tabular form, write the colours of the following indicators in the presence of acids and bases—litmus solution, phenolphthalein solution, methyl orange solution.
   (ii) Classify the given solutions, A and B as acidic and basic, giving reasons.
   Solution A: $[H^+]$ (<) $[OH]^-$.
   Solution B: $[H^+]$ (>) $[OH]^-$.

9. Account for the following.
   (i) State the relation between hydrogen ion concentration of an aqueous solution and its pH.
   (ii) An aqueous solution has a pH value of 7.0. Is this solution acidic, basic or neutral?
   (iii) Which has a higher pH value, 1 M HCl or 1 M NaOH solution?
   (iv) Tooth enamel is one of the hardest substances in our body. How does it undergo damage due to eating chocolates and sweets? What should we do to prevent it?
   (v) How do [H^+] ions exist in water?

10. (i) The pH of soil A is 7.5 while that of soil B is 4.5. Which of the two soils, A or B, should be treated with powdered chalk to adjust its pH and why?
    (ii) Name the chemical which is injected into the skin of a person,
    (a) during an ant's sting
    (b) during the nettle leaf sting
    (iii) Explain how the pH change in the river water can endanger the lives of aquatic animals like fish. **[CBSE 2016]**

Answers

Section A: Multiple Choice Questions

1. (d)	**2.** (b)	**3.** (a)	**4.** (a)
5. (a)	**6.** (c)	**7.** (d)	**8.** (c)
9. (b)	**10.** (a)	**11.** (d)	**12.** (b)
13. (c)	**14.** (b)	**15.** (a)	**16.** (a)
17. (a)	**18.** (b)	**19.** (d)	**20.** (c)
21. (d)	**22.** (d)	**23.** (b)	**24.** (a)
25. (b)	**26.** (a)	**27.** (d)	**28.** (c)
29. (c)	**30.** (c)	**31.** (b)	**32.** (d)
33. (a)	**34.** (c)	**35.** (b)	

Assertion Reason Questions

1. (d)	**2.** (d)	**3.** (b)	**4.** (d)
5. (a)	**6.** (b)	**7.** (b)	**8.** (c)

Section B: Very Short Answer Questions

1. See text (Section 2.2.1).
2. Phenolphthalein and methyl orange.
3. Blue and red.
4. Red and blue.
5. Turmeric and red cabbage.
6. See text (Section 2.2.1).
7. Onion, vanilla essence, clove oil
8. Bitter, sour.
9. See text (Section 2.2.1).
10. An acid is a substance which provides H^+ in water.
11. H^+ ion combines with water to give H_3O^+ (hydronium ion).
12. A tomato contains oxalic acid while an orange contains citric acid.
13. A base which is soluble in water is known as an alkali. For example, caustic soda.
14. Metal oxides are basic in nature, since, on dissolving in acids they form salts and water (neutralisation).
15. H_2SO_4 and HNO_3 are strong acids.
16. A base is bitter in taste and produces hydrogen gas when reacted with metals.
17. Acidic oxides: SO_2, CO_2
    Basic oxides: Na_2O, MgO
18. Alkali. For example, NaOH or KOH.
19. Neutralisation reaction.
    Example:
    $$\underset{\text{Sodium hydroxide}}{NaOH} + \underset{\text{Hydrochloric acid}}{HCl} \rightarrow \underset{\text{Sodium chloride}}{NaCl} + \underset{\text{Water}}{H_2O}$$
20. Dry HCl gas cannot change the colour of dry litmus paper as it remains undissociated in the absence of water. That is, it has no hydrogen ions (H^+) available.
21. Water is a weak acid as it is ionised to a small extent only and gives only a small number of H^+.

22. Formic acid (methanoic acid) HCOOH.

23. It liberates H_2 gas which burns with a pop sound.

24. Ammonia dissolves in water to give ammonium hydroxide with ionises in solution to give OH^- ions; so, ammonia is a base.

25. See text (Section 2.4.2).

26. A decrease in the concentration of H^+ ions in a solution increases the pH value.

27. 1 M HCl has a lower pH value than 1 M HCOOH. Being a strong acid, HCl is more ionised.

28. See text part (Section 2.2.4).

29. The hardest substance in our body is the enamel on our teeth and it is composed of calcium phosphate.

30. It is a neutral solution having a pH value of 7.

31. Dilution of an acid decreases the concentration of hydronium ions per unit volume.

32. Magnesium and zinc.

33. Zinc and tin.

34. Sulfuric acid is called the king of chemicals as it is used in industries a lot.

Case Study or Passage Based Questions

I.

35. Since red litmus shows no colour change with acids, it has to be first changed to blue litmus. So, first dip the strip in an alkaline solution and then use it to test the sample.

36. No colour change is observed. It gives a pink colour in a basic solution.

37. It will turn red litmus blue as it is basic in nature.

38. Pink and yellow colours respectively.

39. Onion and vanilla essence.

II.

40. It indicates that the given solution is acidic in nature. The pH of 1 M HCl solution is zero.

41. Colour: green, pH: 7, neutral.

42. (i) Wash the test tubes and other glass apparatus thoroughly with distilled water, before use.
(ii) All solutions should be freshly prepared.

43. The colour of pH paper is orange in acidic medium while it is blue in basic medium.

44. When milk changes into curd, its pH will decrease. Because curd contains acid, H^+ will increase and thus pH will decrease.

Section C: Short Answer Questions

1. Universal indicator specifies the various levels of H^+ ion concentration. Universal indicator is a mixture of different indicators. It shows different colours with different hydrogen ion concentration in a solution.

2. (i) Oxalic acid (ii) Acetic acid
(iii) Tartaric acid

3. Strong acid: An acid which dissociates completely in water and produces a large amount of hydrogen ions. For example, HCl.
Weak acid: An acid which dissociates partially in water and produces a small amount of hydrogen ions. For example, CH_3COOH.

4. Bases are substances that react with acids and neutralise them. They are usually metal oxides, metal hydroxides, metal carbonates or metal hydrogen carbonates. Many bases are insoluble, they do not dissolve in water. If a base does dissolve in water, it is known as an alkali. Alkalis as well as bases neutralise acids. Alkalis produce OH^- ions in water whereas bases such as copper oxide and ferrous oxide do not produce OH^- ions. All alkalis are bases but not all bases are not alkalis.

5. (i) Alkali (ii) Acid
(iii) Neutralisation reaction (iv) Base

6. When a weak acid is added to a concentrated solution of hydrochloric acid, the solution becomes more acidic as it increases the hydronium ion concentration of the solution.

7. Litmus paper turns red as it is exposed to the evolved gas, sulfur dioxide, which is acidic in nature. The following reaction occurs.

$$\underset{\text{Lead sulfite}}{PbSO_3(s)} + \underset{\text{Nitric acid}}{2HNO_3(aq)} \rightarrow \underset{\text{Lead nitrate}}{Pb(NO_3)_2(aq)} + \underset{\text{Water}}{H_2O(l)} + \underset{\text{Sulfur dioxide}}{SO_2(g)}$$

8. As per the principle of common ion effect, the addition of OH^- ions will shift the equilibrium towards the left side, that is, in the backward direction.

9. The pH of a solution is a measure of the H^+ concentration in the solution. Lesser the pH, more acidic is the solution. Thus, rain water of city B is more acidic.

10. A solution with pH 6 is acidic when the ionic product of water is 1×10^{-14} mol²/litre². In such a case, a solution with a pH less than 7 is acidic in nature. On increasing the temperature, the ionic product of water increases due to an increase in the ionisation of water. When the ionic product of water increases to 1×10^{-12} mol²/litre², a solution with a pH 6 will be neutral and that with a pH below 6 will be acidic in nature.

11. The pH decreases. It is due to the formation of lactic acid in curd. Since acids generally have a low pH value, when milk changes to curd its pH will decrease.

12. (i) Lemon juice: pH 2.0
(ii) Tomato juice: pH 4.0
(iii) Distilled water: pH 7.0
(iv) Sodium hydroxide solution: pH 14.0

13. The fruit may be a citrus fruit or its juice may contain an acid, that is why the traces of acid remained in the knife that turned the blue litmus red.

14. (i) $Al_2O_3 + 6HCl \rightarrow 2AlCl_3 + 3H_2O$
(ii) $Al_2O_3 + 2NaOH \rightarrow 2NaAlO_2 + H_2O$

15. Dilution of concentrated acid is an exothermic process. If water is added to a concentrated acid, the heat generated may cause the mixture to splash out and cause burns. When the acid is added to water slowly with constant stirring, the mixture will not splash out.

16. Orange and lemon are citrus fruits which contain citric acid in their juices. When these juices are added to pure water, the acid will release H^+ ions on dissociation and the pH value will decrease from 7 as the acidity increases.

17. (i) The acid must be added slowly to water.
(ii) Otherwise the mixture may splash out causing burns as a lot of heat is generated in this process.
(iii) Dilution of the acid.

18. Strong acids: HNO_3, HCl
Weak acids: H_2CO_3, CH_3COOH
Strong bases: NaOH, KOH
Weak bases: NH_4OH, $Zn(OH)_2$

19. The order of increasing acidic character will be D < C < B < A.
Reason: H^+ ion concentrations of solution A, B, C and D will be 10^{-2}, 10^{-3}, 10^{-4} and 10^{-5} M respectively. Obviously $10^{-5} < 10^{-4} < 10^{-3} < 10^{-2}$. Higher the $[H^+]$, the more acidic is the solution. Hence, the above order.

20. The increasing basic character will be P < Q < R.
Reason: H^+ ion concentrations of solutions P, Q and R will be 10^{-9}, 10^{-10} and 10^{-11} M respectively. $[H^+]\ [OH^-]$ = ionic product of water = 10^{-14}, therefore $[OH^-]$ of P, Q and R will be 10^{-5}, 10^{-4} and 10^{-3} M respectively. Obviously, $10^{-5} < 10^{-4} < 10^{-3}$. Higher the $[OH^-]$, the more basic is the solution. Hence, the above order.

21. (i) Increase in the strength of alkali so the nature of solution will be basic.
(ii) Hydroxide ion or hydroxyl ion or OH^- ion.
(iii) A paper impregnated with the universal indicator is generally used to measure the pH of a solution.

22. (i) The sulfuric acid has to be added slowly in order to prevent the mixture from getting splashed. So, it requires constant cooling as the reaction is highly exothermic.
(ii) This process is called dilution.

23. (i) Hydrogen gas is formed.

$$Mg(s) + 2HCl(dil.) \rightarrow MgCl_2(aq) + H_2(g)$$

(ii) Sodium chloride and water will be formed.

$$NaOH(aq) + HCl(aq) \rightarrow NaCl(g) + H_2O(l)$$

(iii) Crushed egg shells are made up of $CaCO_3$ that reacts with dil. HCl to give a brisk effervescence because of CO_2.

$$CaCO_3(s) + 2HCl(aq) \rightarrow CaCl_2(aq) + H_2O(l) + CO_2(g)$$

24. An antacid is basic in nature and neutralises the excess of acid formed in the stomach because of hyperacidity. This gives relief from pain in the stomach.

Note: Hyperacidity is caused by an excess secretion of HCl in the stomach.

25. The two materials that can be added are:
 (i) Lime (CaO), to neutralise the acidity.
 (ii) Chalk ($CaCO_3$) also neutralises acidity.
Since both CaO and $CaCO_3$ are basic in nature, they neutralise the acid present in soil.
Note: Chalk is a white soft earthy limestone formed from the skeletal remains of sea creatures.

26. Chemical equation involved:
$$2NaOH + Zn \rightarrow \underset{\text{Sodium zincate}}{Na_2ZnO_2} + H_2\uparrow$$
Test for detection: The hydrogen gas can be tested by holding a matchstick over the solution. The burning matchstick will extinguish with a pop sound.
Gas evolved: When Zn metal reacts with a dilute acid, zinc chloride and hydrogen gas are evolved.
$$2HCl + Zn \rightarrow ZnCl_2 + \underset{\text{Hydrogen gas}}{H_2\uparrow}$$

27. (i) Living organisms can survive only in a narrow range of pH change while the plants require a specific pH range for their healthy growth. Acidic rain water lowers the pH value of river water making aquatic life difficult.
 (ii) Tooth decay starts when the pH of the mouth is lower than 5.5.
 (iii) Our stomach and intestine work in a specific pH range. The stomach acts in a slightly acidic medium while the small intestine digests the food in a slightly alkaline medium.

28. Formic acid or methanoic acid (HCOOH) is present in ant sting. To get instant relief from it we should apply a solution of weak /mild base like baking soda.

29. No change. As acid turns blue litmus red, there is a need of blue litmus paper. To get blue litmus, dip the red litmus paper into a basic solution and then use this.

Or

 (i) Sodium hydrogen carbonate ($NaHCO_3$) or sodium carbonate (Na_2CO_3)
 (ii) $2CH_3COOH + Na_2CO_3 \rightarrow 2CH_3COONa + H_2O + CO_2$

Or

$$CH_3COOH + NaHCO_3 \rightarrow CH_3COONa + H_2O + CO_2$$
 (iii) The liberated CO_2 is passed through limewater, which turns milky.

30. Onion juice is an olfactory indicator. Since the liquid X turns blue litmus red, it is an acid.
 (i) When liquid X (an acid like HCl) reacts with zinc granules, bubbles of hydrogen will be formed.
$$2HCl(aq) + Zn(s) \rightarrow ZnCl_2(s) + H_2(g)$$
 (ii) When liquid X (an acid like HCl) reacts with sodium carbonate, carbon dioxide gas will be liberated.
$$2HCl(aq) + Na_2CO_3(aq) \rightarrow 2NaCl(aq) + H_2O(l) + CO_2(g)$$

Section D: Long Answer Questions

1. (i) Zinc oxide is amphoteric in nature so it can react both with acids as well as bases.
 (ii) As NH_4OH is a weak base, some heat is utilised to ionise the weak base.
 (iii) Only an indicator which undergoes a colour change in the pH range of the titration is suitable for that particular titration.

2. Acetic acid is a weak acid having a pH value of more than 6. Sodium hydroxide is a strong base and the addition of OH^- ions neutralises the H^+ ions formed by the dissociation of acetic acid. As H^+ ion concentration is very low, the addition of further OH^- ions increases the pH value abruptly. Hence, the titration range of this reaction lies between 6.5 and 10.0.
 As the pH range of phenolphthalein is 8.2–10.0, it undergoes a colour change in this range. Hence it acts as a suitable indicator for this reaction.

3. (i) Tap water conducts electricity as it contains ions. Distilled water does not contain ions, so it does not conduct electricity.
 (ii) Dry HCl does not turn blue litmus red as it does not form ions but dil. HCl does, since it forms H^+ and Cl^- ions.
 (iii) During summer, the milk man adds a very small amount of baking soda to fresh milk because it does not allow milk to change to lactic acid that makes milk sour.
 (iv) For dilution of an acid, the acid is added to water and not water into acid as the

addition of water into acid is highly exothermic.

(v) Ammonia does not contain a hydroxyl group but when it dissolves in water, it forms OH^-, it is basic in nature.

$$NH_3 + H_2O \rightarrow NH_4^+ + OH^-$$

4. (i) (a) A solution of glucose will not conduct electricity because it does not have ions.

(b) Dil. HCl will conduct electricity because it produces H^+ ions in water.

(ii) HCl is stronger because it gives rise to more H^+ ions than acetic acid.

(iii) The strength of the acid decreases.

5. (i) Solid sodium chloride and conc. sulfuric acid.

(ii) Dry litmus — no change
Wet blue litmus — changes to red

(iii) $HCl + H_2O \rightarrow \underset{\text{(Hydronium ion)}}{H_3O^+} + \underset{\text{(Chloride ion)}}{Cl^-}$

6. (i) Bulbs A and B do not glow but bulb C glows.

(ii) Glucose and alcohol solutions do not conduct electricity as they do not have ions.
Dil. HCl contains ions so the flow of ions is responsible for the flow of current.

(iii) After replacement, the bulb B glows as NaOH solution contains ions (Na^+ and OH^- ions).

7. (i) Dilute hydrochloric acid reacts with magnesium to form magnesium chloride and H_2 gas is liberated.

$$Mg(s) + 2HCl(aq) \rightarrow MgCl_2(aq) + H_2(g)$$

(ii) Reaction between dilute hydrochloric acid and sodium hydroxide is a neutralisation reaction. Sodium chloride salt and water are formed.

$$NaOH(aq) + HCl(aq) \rightarrow NaCl(aq) + H_2O(l)$$

(iii) Egg shells are made of calcium carbonate, $CaCO_3$. Dilute hydrochloric acid dissolves the $CaCO_3$ and makes the shell soft.

$$\underset{\text{(Crushed egg shell)}}{CaCO_3(s)} + HCl(aq) \rightarrow CaCl_2 + CO_2(g) + H_2O(l)$$

8. (i) (a) No change in colour.

(b) Red litmus turns blue.

(ii) (a) $\underset{\text{Sodium hydroxide}}{2NaOH(aq)} + \underset{\text{Ferrous sulfate}}{FeSO_4(aq)} \rightarrow \underset{\text{Ferrous sulfate}}{Fe(OH)_2(s)} + \underset{\text{Sodium sulfate}}{Na_2SO_4(aq)}$

Observation: (b) Dirty gelatinous green precipitate of $Fe(OH)_2$ will be formed.

(c) $\underset{\text{Sodium hydroxide}}{3NaOH(aq)} + \underset{\text{Aluminium chloride}}{AlCl_3(aq)} \rightarrow \underset{\text{Aluminium hydroxide}}{Al(OH)_3(s)} + \underset{\text{Sodium chloride}}{NaCl(aq)}$

Observation: (ii) A white precipitate of $Al(OH)_3$ will be formed.

9. (i) It is soluble in water.

(ii) It is lighter than air.

(iii) Test for H_2 gas: Bring a burning matchstick near the gas jar. It burns with a pop sound.

(iv) Hydrogen gas will evolve with a greater speed.

(v) Almost the same amount of gas is evolved.

(vi) If sodium hydroxide is taken, hydrogen gas will be evolved only on heating.

$$Zn + 2NaOH \rightarrow \underset{\text{Sodium zincate}}{Na_2ZnO_2} + H_2$$

10. (i) Universal indicator is a mixture of many different indicators (or dyes) which give different colours at different pH values of the entire pH scale. The colour produced by a universal indicator is used to find the pH value of an acid or base by matching the colour with the colours on the pH colour chart.

(ii) Solution A is acidic and will turn the litmus solution from blue to red.
Solution B is basic and will turn phenolphthalein from colourless to pink.

(iii) Green colour will be obtained.

11. (i) (a) The solution is an acid. The substance in egg shells is calcium carbonate.

(b) Effervescence is due to the evolution of CO_2 gas.

(c) $2HNO_3 + CaCO_3 \rightarrow 2Ca(NO_3)_2 + H_2O + CO_2$

(d) Lime stone.

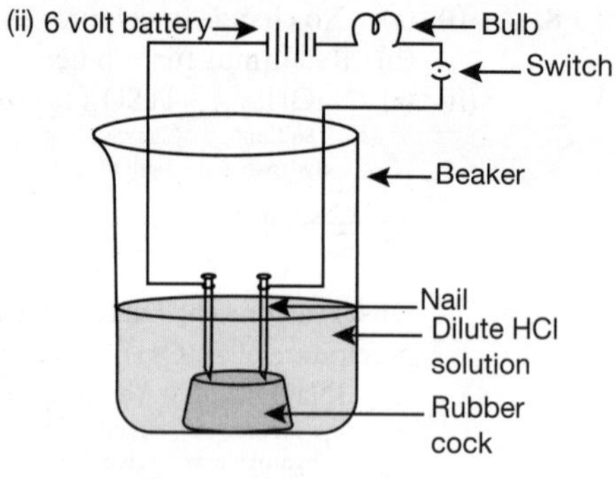

Acid solution in water conducts electricity.

12. (i) (a) Solution D (b) Solution C
(c) Solution B (d) Solution A
(e) Solution E
Increasing order C, E, D, A, B.

(ii) (a) Acid: CH_3COOH (Acetic acid)
Base: NaOH (Sodium hydroxide)
(b) Acid: HCl (Hydrochloric acid)
Base: NH_4OH (Ammonium hydroxide)

13. (i) Indicators are the substances which change their colour in different types of substances.
(ii) Lichen.
(iii) Examples of natural indicators:
(a) Beetroot extract shows the same colour in all sour substances and a different colour in bitter substances.
(b) Turmeric gives a yellow spot on a cloth but when washed with soap, the colour of the spot changes to red.

14. *(i) Pain and irritation in the stomach is due to excessive secretion of a strong acid, called hydrochloric acid. Baking soda is a mild base and neutralises the excessive acid, thereby giving relief from pain.
(ii) (a) Careful learning in the class.
(b) Remembering the teacher's knowledge and use of the right knowledge at the right place.

Section E: Case Study or Passage Based Questions

1. (b) **2.** (a) **3.** (b) **4.** (c)
5. (c)

High Order Thinking Skills (HOTS) Questions

1. When a few drops of acetic acid are added to:
(i) Phenolphthalein → Remains colourless/no change
(ii) Distilled water → No change, acetic acid dissolves in water
(iii) Universal indicator → Turns orange
(iv) Sodium hydrogen carbonate powder → Evolution of colourless, odourless gas with brisk effervescence

2. (i) None will turn blue litmus red. The solution P, that is, sodium chloride is neutral in nature and thus it will have no effect on litmus paper. The solution Q is a strong base.
(ii) The solution Q, that is, sodium hydroxide is a strong base and thus it will turn red litmus paper blue.

3. (i) After adding acid to water, dissociation of the acid and hydration of the ions takes place. As breaking of bonds needs some energy to proceed, it is an endothermic reaction.
(ii) When the pH of rain water is less than 5.6, it is called acid rain. Living organisms can survive only in a narrow range of pH change.
(iii) The pH of gastric acid is 1.5 to 3.5 in the human stomach.
(iv) 1 M HCl has a higher concentration of H^+ ions. This is because, when HCl dissolves in water it dissociates completely into ions while CH_3COOH is a weak acid and does not dissociate into ions completely.
(v) At a lower pH below 5.5, the calcium phosphate of the enamel of tooth gets corroded.

4. (i) Distilled water does not have free ions to conduct electricity. As rain water contains salts which ionise to form free mobile ions, they conduct electricity.
(ii) Copper metal slowly reacts with water, carbon dioxide and oxygen present in air to form basic copper carbonate which is green in colour. A layer slowly gets deposited on the surface of the metal. As lemon is acidic, when we rub the tarnished copper vessel with lemon, the basic copper

oxide or copper carbonate reacts with the acid present in lemon to form a salt which is washed away with water.

(iii) Sour substances like pickles and curd are acidic in nature, that is, they contain acids. So, when they are kept in brass or copper vessels, they react with them and form toxic chemicals which may prove harmful for the human body. So, they are not kept in metal containers.

5. (i) A dry strip of litmus and dry HCl do not contain any water, so they do not get ionised. But, in the presence of a drop of water, HCl is ionised and releases protons which turn blue litmus red.

(ii) HCl and HNO_3 produce H^+ ions in aqueous solution. Hence, they show acidic character.
Alcohol and glucose do not produce H^+ in aqueous solution, hence they do not show acidic character.

(iii) No, substances such as vinegar and tamarind contain acids which when accidentally fall on marble will react with the marble causing discoloration.

(iv) Solution Y is a stronger acid.

6.

Name of the solution	Colour change with phenolphthalein	Colour change with blue litmus
1. Sodium carbonate	turns pink	no change
2. Hydrochloric acid	no change	turns red
3. Sodium chloride	no change	no change

7. (i) HCl will give rise to more H^+ ions and CH_3COOH produces less H^+ ions on dissociation. The colour of the pH paper depends on the concentration of the H^+ ions. The colour becomes red for high H^+ concentration.

(ii) Aqueous solutions of acids have H^+ ions which carry electric current through the solution.

(iii) (a) Most acidic: A, Most basic: C.
(b) C, B, D, A / C < B < D < A
(c) In C: blue, in D: green

8. (i)

Indicators	Acid	Base
Litmus solution	Blue to red	Red to blue
Phenolphthalein	Colourless	Pink
Methyl orange	Yellow	Pink

(ii) Solution A: Basic solution because, $[H^+]$ is less than 1.0 × 10–7.
Solution B: Acidic solution because $[H^+]$ is greater than 1.0 × 10–7 m.

9. (i) If H^+ ion concentration is more, the pH will be less and vice versa. That is, the pH of a solution is inversely proportional to the H^+ ion concentration or pH = –log (H^+).

(ii) This solution is neutral. (iii) 1 M NaOH solution.

(iv) When the pH in the mouth is below 5.5, the bacteria present in the mouth produce acids by the degradation of sugar and corrode the tooth enamel. This can be prevented by using toothpastes (which are generally basic).

(v) As hydronium $[H_3O]^+$ ion.

10. (i) Soil B should be treated with powdered chalk which is basic in nature (as this soil is acidic), in order to adjust the pH.

(ii) (a) Formic acid (HCOOH)
(b) Formic acid (HCOOH)
The effect of these stings is neutralised by rubbing the skin with a mild base such as baking soda ($NaHCO_3$).

(iii) A pH change in the river water decreases the amount of oxygen dissolved in water since acidic water (pH value less than 5.6) is harmful for aquatic animals such as fish.

Hints

Section A: Multiple Choice Questions

8. Sodium hydroxide ionises in water and produces a large amount of hydroxide ions.

12. $Zn + 2NaOH \rightarrow Ma_2ZnO_2$ (Sodium zincate) + H_2

2.8 SALTS

Salts are the compounds formed by the partial or complete replacement of the hydrogen of an acid by metal atoms or positive radicals. For example, KCl, $CaCl_2$, Na_2SO_4, NH_4Cl, $(NH_4)_2SO_4$ are salts because they are formed by the complete replacement of hydrogen from the acids HCl or H_2SO_4. This phenomenon is called neutralisation. Salts are ionic compounds having two parts—one part has a positive charge (cation) and the other part has a negative charge (anion). In a salt, the total positive charge and total negative charge are equal which means a salt is electrically neutral. Here are some common salts (Table 2.5).

Table 2.5 Some common salts

S. No.	Name/ Formula/Ions	Crystal structures
1	Sodium chloride NaCl Na^+ and Cl^-	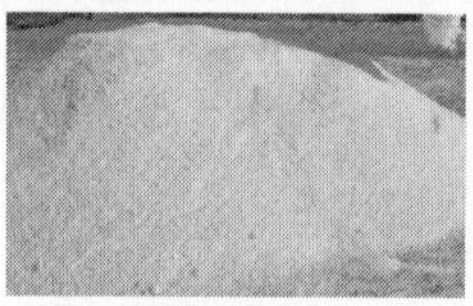 *Source:* Chemicalinterest, Public domain, via Wikimedia Commons
2	Potassium chloride KCl K^+ and Cl^-	 *Source:* Public Domain, https://commons.wikimedia.org/w/index.php?curid=95463
3	Calcium chloride $CaCl_2$ Ca^{2+} and Cl^-	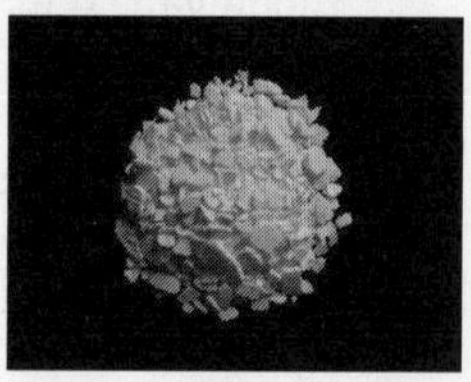 *Source:* No machine-readable source provided. Own work assumed (based on copyright claims)., CC BY-SA 3.0, https://commons.wikimedia.org/w/index.php?curid=342288

(Continued)

Table 2.5 (Continued)

S. No.	Name/ Formula/Ions	Crystal structures
4	Magnesium chloride $MgCl_2$ Mg^{2+} and Cl^-	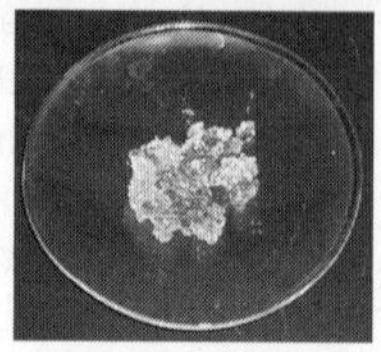 *Source:* Public Domain, https://commons.wikimedia.org/w/index.php?curid=95461
5	Magnesium oxide MgO Mg^{2+} and O^{2-}	 *Source:* Adam Rędzikowski, CC BY-SA 4.0 <https://creativecommons.org/licenses/by-sa/4.0>, via Wikimedia Commons
6	Aluminium oxide Al_2O_3 Al^{3+} and O^{2-}	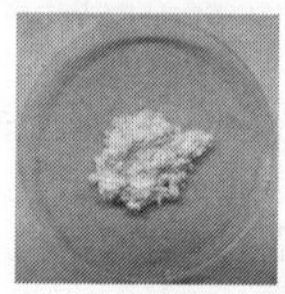 *Source:* By Aariuser I - https://www.flickr.com/photos/194184248@N08/51602553796/, CC BY-SA 2.0, https://commons.wikimedia.org/w/index.php?curid=114438757
7	Sodium hydroxide NaOH Na^+ and OH^-	 *Source:* By Walkerma - Own work, Public Domain, https://commons.wikimedia.org/w/index.php?curid=5190866
8	Copper sulfate $CuSO_4$ Cu^{2+} and SO_4^{2-}	 *Source:* By Stephanb - Own work, CC BY-SA 3.0, https://commons.wikimedia.org/w/index.php?curid=565837

(Continued)

Table 2.5 (Continued)

S. No.	Name/ Formula/Ions	Crystal structures
9	Sodium nitrate $NaNO_3$ Na^+ and NO_3^-	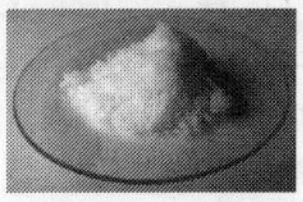 *Source:* By Ondřej Mangl - Own work, Public Domain, https://commons.wikimedia.org/w/index.php?curid=2899599
10	Sodium sulfate Na_2SO_4 Na^+ and SO_4^{2-}	 *Source:* Public Domain, https://commons.wikimedia.org/w/index.php?curid=292586
11	Iron sulfate $FeSO_4$ Fe^{2+} and SO_4^{2-}	 *Source:* By Benjah-bmm27 - Own work, Public Domain, https://commons.wikimedia.org/w/index.php?curid=2116160
12	Magnesium sulfate $MgSO_4$ Mg^{2+} and SO_4^{2-}	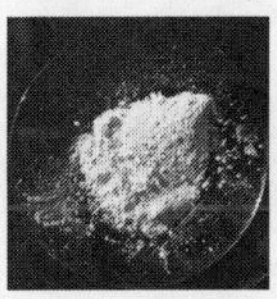 *Source:* Public Domain, https://commons.wikimedia.org/w/index.php?curid=292629
13	Potassium nitrate KNO_3 K^+ and NO_3^-	 *Source:* By w:User:Walkerma, Public Domain, https://commons.wikimedia.org/w/index.php?curid=184556
14	Ferric chloride $FeCl_3$ Fe^{3+} and Cl^-	 *Source:* Benjah-bmm27, Public domain, via Wikimedia Commons

2.8.1 Common Methods to Prepare Salts

Here are few important methods to prepare salts.

(i) Reaction of Acid and Base
An acid and a base react together to give a salt and water.

$$\text{Acid} + \text{Base} \rightarrow \text{Salt} + \text{Water} + \text{Heat}$$

$$HCl + NaOH \rightarrow NaCl + H_2O + \text{Heat}$$

$$HNO_3 + KOH \rightarrow KNO_3 + H_2O + \text{Heat}$$

(ii) Reaction of Acid and Metal
Metals like zinc react with an acid to give a salt and hydrogen gas.

$$\text{Metal} + \text{Acid} \rightarrow \text{Salt} + \text{Hydrogen gas}$$

$$\underbrace{Zn + H_2SO_4}_{\text{Reactants}} \rightarrow \underbrace{ZnSO_4 + H_2}_{\text{Products}}$$

(iii) Reaction of Metal Carbonate or Metal Hydroxide with Acid
Metal carbonate or metal hydroxide react with an acid to give a salt and water.

$$\underset{\substack{\text{Calcium carbonate}\\\text{(Limestone/chalk/}\\\text{marble)}}}{CaCO_3} + \underset{\substack{\text{Hydrochloric}\\\text{acid}}}{2HCl} \rightarrow \underset{\substack{\text{Calcium}\\\text{chloride}}}{CaCl_2} + \underset{\text{Water}}{H_2O} + \underset{\substack{\text{Carbon}\\\text{dioxide}}}{CO_2}$$

$$\underset{\substack{\text{Calcium}\\\text{hydroxide}}}{Ca(OH)_2(aq)} + \underset{\substack{\text{Carbon dioxide}\\\text{(From air)}}}{CO_2(g)} \longrightarrow \underset{\substack{\text{Calcium}\\\text{carbonate}}}{CaCO_3(s)} + \underset{\text{Water}}{H_2O(l)}$$

(iv) Reaction of Metal and Base
A few metals like zinc and tin react with alkali (NaOH) to give a salt and hydrogen.

$$Zn + 2NaOH \xrightarrow{\Delta} \underset{\text{Sodium zincate}}{Na_2ZnO_2} + H_2\uparrow$$

2.8.2 Classification of Salts

From what we have discussed, it is clear that the salts formed from different kinds of acids and bases can also be classified as follows.

Normal Salts

Salts formed by the complete replacement of the hydrogen atom/atoms of an acid by metal atoms or positive radicals of the base are known as normal salts. In such a salt, the first part is of the base and the second part is of the acid. So, during naming of this salt, the first part of the name is taken from the name of the base and the second part of the name is taken from the name of the acid. For example, the name of the salt known as potassium nitrate comes from potassium hydroxide base and nitric acid.

(i) **Classification based on the acids from which they are obtained:** It is important to note the following.
- The salts of hydrochloric acid are known as chlorides (Table 2.6).
- The salts of nitric acid are known as nitrates (Table 2.7).
- The salts of sulfuric acid are known as sulfates (Table 2.8).
- The salts of carbonic acid are known as carbonates (Table 2.9).
- The salts of acetic acid are known as acetates (Table 2.10).

Table 2.6 Chlorides formed by the reaction of hydrochloric acid with different bases

Formula	Name of the salt	Base involved	Name of the base
NaCl	Sodium chloride	NaOH	Sodium hydroxide
KCl	Potassium chloride	KOH	Potassium hydroxide
NH_4Cl	Ammonium chloride	NH_4OH	Ammonium hydroxide
$AlCl_3$	Aluminium chloride	$Al(OH)_3$ or Al_2O_3	Aluminium hydroxide or aluminium oxide
$MgCl_2$	Magnesium chloride	$Mg(OH)_2$ or MgO	Magnesium hydroxide or magnesium oxide
$CaCl_2$	Calcium chloride	$Ca(OH)_2$ or CaO	Calcium hydroxide or calcium oxide
$BaCl_2$	Barium chloride	$Ba(OH)_2$ or BaO	Barium hydroxide or barium oxide

Table 2.7 Nitrates formed by the reaction of nitric acid with different bases

Formula	Name of the salt	Base involved	Name of the base
$NaNO_3$	Sodium nitrate	NaOH	Sodium hydroxide
KNO_3	Potassium nitrate	KOH	Potassium hydroxide
$Ca(NO_3)_2$	Calcium nitrate	$Ca(OH)_2$ or CaO	Calcium hydroxide or calcium oxide
$Cu(NO_3)_2$	Copper(II) nitrate	$Cu(OH)_2$ or CuO	Copper(II) hydroxide or copper(II) oxide

Table 2.8 Sulfates formed by the reaction of sulfuric acid with different bases

Formula	Name of the salt	Base involved	Name of the base
Na_2SO_4	Sodium sulfate	NaOH	Sodium hydroxide
K_2SO_4	Potassium sulfate	KOH	Potassium hydroxide
$MgSO_4$	Magnesium sulfate	$Mg(OH)_2$ or MgO	Magnesium hydroxide or magnesium oxide
$CaSO_4$	Calcium sulfate	$Ca(OH)_2$ or CaO	Calcium hydroxide or calcium oxide
$CuSO_4$	Copper(II) sulfate	$Cu(OH)_2$ or CuO	Copper(II) hydroxide or copper(II) oxide
$ZnSO_4$	Zinc(II) sulfate	$Zn(OH)_2$ or ZnO	Zinc(II) hydroxide or zinc(II) oxide

Table 2.9 Carbonates formed by the reaction of carbonic acid with different bases

Formula	Name of the salt	Base involved	Name of the base
Na_2CO_3	Sodium carbonate	NaOH	Sodium hydroxide
K_2CO_3	Potassium carbonate	KOH	Potassium hydroxide
$MgCO_3$	Magnesium carbonate	$Mg(OH)_2$	Magnesium hydroxide
$CaCO_3$	Calcium carbonate	$Ca(OH)_2$	Calcium hydroxide

Table 2.10 Families of salts based on acids

Family	Acid involved	Example
Phosphates	Phosphoric acid (H_3PO_4)	$Ca_3(PO_4)_2$, Calcium phosphate $AlPO_4$, Aluminium phosphate
Acetates	Acetic acid (CH_3COOH)	CH_3COONa, Sodium acetate $(CH_3COO)_2Ca$, Calcium acetate
Oxalates	Oxalic acid [$(COOH)_2$]	$(COO)_2Ca$, Calcium oxalate $(COONa)_2$, Sodium oxalate

(ii) Classification based on the bases from which they are obtained: Salts obtained from different bases are classified on the basis of the cationic part of the salt. For example, salts of sodium hydroxide and potassium hydroxide are known as sodium and potassium salts respectively (Table 2.11).

Table 2.11 Families of salts on the basis of bases used

Family	Base involved	Example
Sodium salts	Sodium hydroxide (NaOH)	NaCl, $NaNO_3$, Na_2SO_4, Na_2CO_3, CH_3COONa, $(COONa)_2$
Potassium salts	Potassium hydroxide (KOH)	KCl, KNO_3, K_2SO_4, KBr, K_2CO_3, CH_3COOK, $(COOK)_2$
Magnesium salts	Magnesium hydroxide [$Mg(OH)_2$] or magnesium oxide (MgO)	$MgCl_2$, $MgSO_4$, $MgCO_3$
Calcium salts	Calcium hydroxide [$Ca(OH)_2$] or calcium oxide (CaO)	$CaCl_2$, $(COO)_2Ca$, $(CH_3COO)_2Ca$
Copper salts	Copper hydroxide [$Cu(OH)_2$] or copper oxide (CuO)	$CuSO_4$, $CuCO_3$, $CuCl_2$
Ammonium salts	Ammonium hydroxide (NH_4OH)	NH_4Cl, NH_4NO_3, $(NH_4)_2SO_4$, NH_4Br

(Continued)

Table 2.11 (Continued)

Acidic Salts

Salts formed by the partial replacement of the hydrogen atom/atoms of a polybasic acid by metal atoms or positive radicals are known as *acid salts*. For example,

Acid	**Acid salts**
H_2SO_4	$KHSO_4$, NH_4HSO_4
H_2CO_3	$NaHCO_3$, $KHCO_3$, $Ca(HCO_3)_2$, NH_4HCO_3
H_3PO_4	NaH_2PO_4 (sodium dihydrogenphosphate) $(NH_4)H_2PO_4$ (ammonium dihydrogenphosphate) Na_2HPO_4 (disodium hydrogenphosphate)

Basic Salts

Salts formed by the partial replacement of hydroxyl groups or the incomplete neutralisation of a base are known as *basic salts*. When a base having more than one hydroxyl (–OH) group reacts with an acid, a salt having one or more hydroxyl groups may be formed. For example,

$$Pb(OH)_2 + HCl \rightarrow Pb(OH)Cl + H_2O$$

2.8.3 General properties of Salts

- Salts are electrovalent (ionic) compounds. They conduct electricity in their molten states as well as in their aqueous solutions (due to their dissociation into ions).
- Salts are non-volatile solids that form crystals.
- Solubility of salts: Most of the salts are soluble in water. For example, all the compounds of ammonium, sodium and potassium are soluble in water, all chlorides are soluble in water.

2.9 Salt Hydrolysis

When a salt is added into water, the ions of the salt interact with water to cause acidity or basicity in the aqueous solution. This ionic interaction is called *salt hydrolysis*. Interaction of the cation is *cationic hydrolysis* and interaction of the anion is *anionic hydrolysis*.

2.9.1 Types of Salt Hydrolysis

(i) **Hydrolysis of a salt of a strong acid and a strong base:** This type of hydrolysis produces a strong acid and a strong base which are completely ionised, giving an equal number of H^+ and OH^- ions. Hence, the aqueous solution is neutral with a pH = 7. For example, KNO_3, K_2SO_4, NaCl.

$$Na^+ + Cl^- + H_2O \rightarrow Na^+ + OH^- + H^+ + Cl^-$$

(ii) **Hydrolysis of a salt of a weak acid and a strong base:** This produces a weak acid which does not ionise and a strong base which ionises and gives OH^- ions. This turns the aqueous solution basic and has a pH > 7. For example, Na_2CO_3, CH_3COONa, CH_3COOK.

$$CH_3COO^- + Na^+ + H_2O \rightarrow CH_3COOH + Na^+ + OH^-$$

As the anion of the salt reacts with water, it is also known as *anionic hydrolysis*.

(iii) **Hydrolysis of a salt of a strong acid and a weak base:** This type of hydrolysis produces a strong acid which ionises and a weak base which does not ionise much. This increases the concentration of H^+ ions, making the solution acidic with a pH < 7. For example, $CuSO_4$, NH_4NO_3, $FeCl_3$, $(NH_4)_2SO_4$.

$$Cu^{2+} + SO_4^{2-} + 2H_2O \rightarrow Cu(OH)_2 + 2H^+ + SO_4^{2-}$$

As only the cation of the salt reacts with water, it is known as cationic hydrolysis.

(iv) **Hydrolysis of salt of a weak acid and a weak base:** This produces a weak acid and weak base, which exist in the molecular form. Hence, there are no free H^+ or OH^- ions and the solution is neutral with a pH = 7. For example, CH_3COONH_4, $(NH_4)_2CO_3$.

$$2NH_4^+ + CO_3^{2-} + 2H_2O \rightarrow 2NH_4OH + H_2CO_3$$

TEST YOUR KNOWLEDGE

2.18 Fill in the blanks.

(i) Potassium nitrate has a pH value equal to ________.

(ii) The salts of a strong acid and weak base are ________ with a pH value ________ than 7.

(iii) ________ is the fixed number of water molecules chemically attached to each formula unit of a salt in its crystalline form.

(iv) Sodium bicarbonate is an ______ salt.

(v) Formation of a salt from acid and water is known as ________.

Solution: (i) 7, (ii) acidic, less, (iii) water of crystallisation, (iv) acidic, (v) neutralisation

2.19 Define neutralisation with an example, also tell whether it is exothermic or endothermic.

Solution: Neutralisation is a reaction in which an acid reacts with a base to give salt and water. It is an exothermic process.

$$\underset{\text{(Base)}}{NaOH} + \underset{\text{(Acid)}}{HNO_3} \rightarrow \underset{\text{(Salt)}}{NaNO_3} + H_2O$$

2.20 Write the names of salts formed by sodium with chloride, sulfate and nitrate.

Solution: NaCl, Na_2SO_4 and $NaNO_3$.

2.21 Explain the following.

(i) What is the nature of the solution obtained by dissolving ammonium sulfate in water?

(ii) Which of the following is/are acid salts?

Na_2CO_3, $NaHCO_3$, $FeCl_3$, Na_3PO_4, $KHSO_4$, $BaHPO_4$

Solution: (i) The solution obtained will be slightly acidic as it contains a strong acid and a weak base.

$$\underset{\text{Ammonium sulfate}}{(NH_4)_2SO_4} + 2H_2O \rightarrow \underset{\text{Strong acid}}{H_2SO_4} + \underset{\text{Weak base}}{2NH_4OH}$$

(ii Here, the acidic salts are $NaHCO_3$, $KHSO_4$ and $BaHPO_4$.

2.9.2 Important Salts

Caustic Soda or Sodium Hydroxide (NaOH)

Preparation

Electrolysis of brine solution (modern method): It is manufactured by the electrolysis of an aqueous solution of sodium chloride (brine solution). For the electrolysis of aqueous NaCl, electrolytic cells like Nelson, Castner–Kelner, Solvay are used. Here, NaOH, Cl_2, H_2 are formed. This process of electrolysis is known as the chlor-alkali process as the product formed here are chlorine and alkali.

Reactions: $NaCl \rightarrow Na^+ + Cl^-$ $\quad H_2O \rightarrow H^+ + OH^-$

At anode: $2Cl^- \rightarrow Cl_2 + 2e^-$

At cathode: $2Na^+ + 2e^- \rightarrow 2Na$

$2Na + 2H.OH \rightarrow 2NaOH + H_2\uparrow$

The overall reaction is given as,

$$\underset{\substack{\text{Sodium}\\\text{chloride}\\\text{(Brine)}}}{2NaCl(aq)} + \underset{\text{Water}}{2H_2O(l)} \xrightarrow[\text{(Electrolysis)}]{\text{Electricity}} \underset{\substack{\text{Sodium}\\\text{hydroxide}\\\text{(Caustic}\\\text{soda)}}}{2NaOH(aq)} + \underset{\text{Chlorine}}{Cl_2(g)} + \underset{\text{Hydrogen}}{H_2(g)}$$

Physical Properties

- It is a white crystalline, deliquescent soapy solid with an MP of 591.4 K (Fig. 2.24).

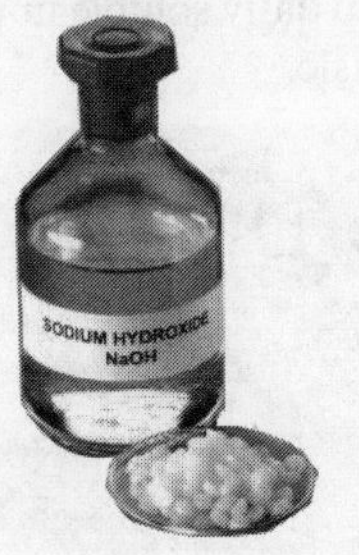

Fig. 2.24 NaOH

- It is highly soluble in water but less in alcohol.
- It is corrosive in nature and bitter in taste.

Uses of NaOH

- It is used in the hydrolysis of esters and saponification.
- It is used in the preparation of paper, soap, rayon, dyes, drugs and a number of chemicals.
- It is used as a reagent in the laboratory.
- It is used in the refining of bauxite (by Baeyer's method), petroleum and vegetable oils.
- It is used to prepare $H_2\uparrow$ (Uyeno method).
  $2Al + 2NaOH + 2H_2O \rightarrow 2NaAlO_2 + 3H_2\uparrow$
- It is used in making cotton unshrinkable; this process is called *mercerisation* of cotton.

Uses of Hydrogen

- It is used in the manufacture of ammonia which is further used for the production of fertilisers like urea, ammonium sulfate, calcium ammonium nitrate (CAN).
- It is used in the hydrogenation of vegetable oils to form solid fats such as Vanaspati ghee and margarine (artificial butter).
- It is used in metallurgy to reduce heavy metal oxides to metals.
- Liquid hydrogen mixed with liquid oxygen is used as a rocket fuel.

Uses of Chlorine

- It is used as a germicide and disinfectant for the sterilisation of water in the swimming pools.
- It is used in the manufacture of bleaching powder, chloroform, hydrochloric acid, DDT.
- It is used in the manufacture of chlorofluorocarbons (CFCs) used as refrigerants.
- It is used in the bleaching of wood pulp and cotton fabrics.
- It is used in the manufacture of PVC (polyvinyl chloride), used for making shoe soles.
- It is used in the manufacture of pesticides.

Uses of Hydrochloric Acid

- It is used as a reagent in the laboratory.
- It is used in making aqua regia (after mixing with nitric acid) for dissolving gold and platinum.
- It is used for cleaning steel.
- It is used in the textile industry.
- It is used in the manufacture of medicines and cosmetics.

Sodium Chloride (NaCl)

It is called common salt, rock salt, sea salt or table salt (Fig. 2.25). Sea water has about 2.95% NaCl. 28% aqueous NaCl solution is called *brine*. It is known as table salt as it is used as an important food material.

Normally table salt is not in the pure form but contains small amounts of other salts such as potassium iodide. That is why it is also called iodised salt and it is essential for us as it protects us from thyroid disorders.

Fig. 2.25 NaCl

Occurrence and Preparation

The major sources of common salt are the following.

(i) **Sea water:** Sea water also contains sodium chloride along with many other salts like $MgCl_2$ and $MgSO_4$. It is mainly manufactured from sea water by evaporation in the sun (Fig. 2.26). As it contains impurities of $MgCl_2$, $CaCl_2$ and $CaSO_4$, it is further purified by passing HCl gas; due to the common ion effect, pure NaCl gets precipitated.

Fig. 2.26 Salt pans

Source: Sandip Dey, CC BY-SA 4.0 <https://creativecommons.org/licenses/by-sa/4.0>, via Wikimedia Commons

(ii) **From underground deposits or rock salt:** Common salt is present in the form of solid deposits in many parts of the word—Himachal Pradesh in India and Khewra in Pakistan. The large crystals of common salt found in underground deposits are known as rock salt. Rock salt is brownish in colour as it has impurities. Rock salt can be dug out like coal.

(iii) **Inland lakes:** Common salt in large quantities can be obtained by the natural evaporation of the water of inland lakes. For example, Sambhar lake in Rajasthan (India), Great salt lake in Utah (USA).

Physical Properties

- It is a white crystalline solid and hygroscopic in nature.
- It melts at 1081 K and boils at 1713 K.
- It dissolves in water and the process of dissolution is endothermic.
- Its solubility is 36 g per 100 g of water at 273 K. The solubility does not increase much with increase in temperature.

Uses

- It is an essential constituent of our food.
- It is used in the manufacture of sodium, sodium hydroxide, washing soda, hydrogen chloride, chlorine.
- It is used in freezing mixture (in making of ice cream).
- It is used for the preparation of soap. When soap is manufactured it remains in the soluble form but when a little common salt is added, soap separates out as a solid mass. This known as salting out of soap.
- It is used for regenerating ion exchange resins.

Sodium Carbonate or Washing Soda ($Na_2CO_3.10H_2O$)

Na_2CO_3 is known as anhydrous sodium carbonate or soda ash while its hydrated form $Na_2CO_3.H_2O$ is known as crystal carbonate or sodium carbonate monohydrate and $Na_2CO_3.10H_2O$ is known as washing soda.

Methods of Preparation

Solvay ammonia soda process: It is prepared industrially mainly by the Solvay method using NaCl, $CaCO_3$ and NH_3.

$$2NH_3 + CO_2 + H_2O \rightarrow (NH_4)_2CO_3$$

$$2NaCl + (NH_4)_2CO_3 \rightarrow Na_2CO_3 + 2NH_4Cl$$

$$Na_2CO_3 + CO_2 + H_2O \rightarrow 2NaHCO_3$$

$$2NaHCO_3 \xrightarrow{\Delta} Na_2CO_3 + CO_2 + H_2O$$

Here Na_2CO_3, $NaHCO_3$, $CaCl_2$ are obtained as byproducts. $NaHCO_3$ is precipitated due to its low solubility product.

Properties:

- **Physical properties**: It is a white crystalline solid (Fig. 2.27), partially soluble in water and its aqueous solution is basic.

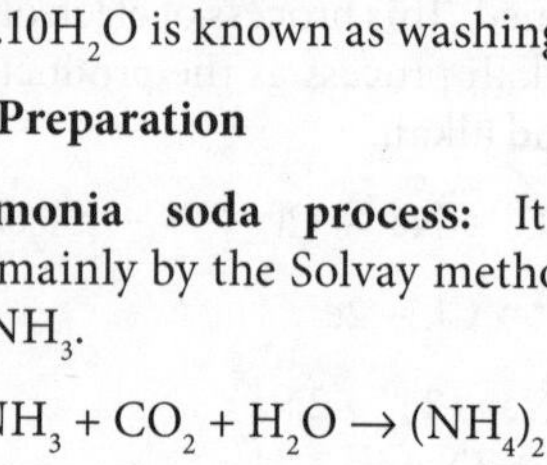

Fig. 2.27 Washing soda

Source: Chemicalinterest, Public domain, via Wikimedia Commons

- **Efflorescence**: It is an efflorescent substance and loses its water of crystallisation to form its monohydrate in open air.

$$Na_2CO_3.10H_2O \xrightarrow{\text{Air}} \underset{\substack{\text{Powder form} \\ \text{(loss of weight)}}}{Na_2CO_3.H_2O} + 9H_2O$$

- **Heating effect**: On heating, it changes into anhydrous sodium carbonate.

$$Na_2CO_3 \cdot 10H_2O \xrightarrow[-10H_2O]{} Na_2CO_3 \xrightarrow[\text{Strong}]{\Delta} \text{No reaction}$$

- **With acids**: It is easily decomposed by acids.

$$Na_2CO_3 + HCl \rightarrow NaHCO_3 + NaCl$$

$$NaHCO_3 + HCl \rightarrow NaCl + H_2O + CO_2 \uparrow$$

- **With silica**: When it is treated with silica it gives sodium silicate.

$$Na_2CO_3 + SiO_2 \rightarrow Na_2SiO_3 + CO_2 \uparrow$$

Sodium silicate is known as water glass or soluble glass as it dissolves in water.

Uses

- It is used for softening hard water, refining of petrol and in the manufacture of glass, borax.
- $Na_2CO_3 + K_2CO_3$ is a fusion mixture so it is used in quantitative and qualitative analysis.
- It is used for washing in laundries.
- It is used in paper, textile and paint industries.
- Na_2CO_3 gives pink colour with HPh, yellow colour with MeOH or methyl red and blue colour with red litmus.

Sodium Bicarbonate or Baking Soda ($NaHCO_3$)

$NaHCO_3$ is sodium bicarbonate and it is commonly known as baking soda as it is widely used in the baking industry.

Preparation

It is prepared by passing carbon dioxide through a saturated solution of sodium carbonate.

$$Na_2CO_3 + CO_2 + H_2O \rightarrow 2NaHCO_3$$

The sodium bicarbonate can be easily precipitated from here as it is less soluble in water.

Properties

- **Physical properties**: It is a white crystalline solid partially soluble in water and its aqueous solution is basic (Fig. 2.28). $NaHCO_3$ gives a yellow colour with methyl red or methyl orange but no colour with HPh, that is, phenolphthalein.

Fig. 2.28 Baking soda

Source: Rainer Z ..., CC BY-SA 3.0 <https://creativecommons.org/licenses/by-sa/3.0>, via Wikimedia Commons

$$HCO_3^- + H_2O \rightarrow H_2O + CO_2 + OH^-$$

- **Heating effect**: On heating at 100°C, it decomposes into Na_2CO_3 with the evolution of CO_2.

$$2NaHCO_3 \xrightarrow{100°C} Na_2CO_3 + CO_2 + H_2O$$

- With zinc sulfate, it forms normal zinc carbonate.

$$ZnSO_4 + 2NaHCO_3 \rightarrow ZnCO_3 + Na_2SO_4 + H_2O + CO_2$$

Uses

- It is used to remove acidity in the stomach so it is used as an ingredient of antacids.
- It is used in the making of baking powder (sodium bicarbonate + potassium hydrogen tartrate) which is used for creating softness and fluffiness in making food stuff like breads and cakes. It is possible since sodium bicarbonate on heating gives carbon dioxide which causes the bread or cake to rise and this makes them highly porous (fluffiness).
- It is used in soda acid fire extinguishers. In such extinguishers, a solution of sodium hydrogen carbonate displaced in an iron cylinder fitted with a nozzle (Fig. 2.29). A bottle having concentrated sulfuric acid is also fitted in the cylinder. When the knob fitted on the cylinder is pressed, the bottle breaks and the acid mixes into the solution of sodium hydrogen carbonate. Alternatively, if the fire extinguisher is inverted, the acid falls into the solution contained in the cylinder. Immediately, a brisk reaction starts giving out CO_2 along with a stream of liquid with pressure through the nozzle. On directing the nozzle towards the fire, CO_2 surrounds the fire. As a result, the supply of air is cut off and the fire is extinguished.

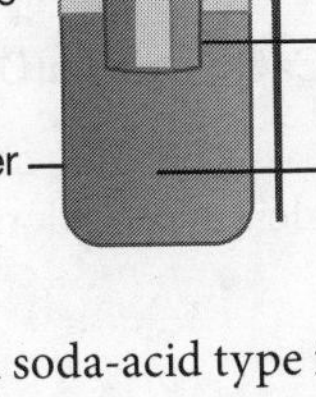

Fig. 2.29 A soda-acid type fire extinguisher

- It is used in medicines as a mild antiseptic for skin infections.
- It is used in making effervescent drinks.

Plaster of Paris ($CaSO_4.½H_2O$) or [$(CaSO_4)_2H_2O$]

It is also known as calcium sulfate hemihydrate as it contains half molecule of water of crystallisation.

Preparation

On heating gypsum (calcium sulfate dihydrate) at 120°C, plaster of Paris is formed as it loses ¾ of its water of crystallisation.

$$2CaSO_4 \cdot 2H_2O \xrightarrow[\text{Setting } (\Delta H = -Ve)]{120°C} \underset{\text{Plaster of Paris}}{(CaSO_4)_2 H_2O} + 3H_2O$$

Properties

- It is a white powder.
- **Setting of plaster of Paris**: When plaster of Paris is mixed with a sufficient amount of water, a paste is formed which on standing for some time changes into a hard mass with more volume. Here, gypsum is formed with the evolution of heat.

$$(CaSO_4)_2 \cdot H_2O \xrightarrow[H_2O]{\text{Setting}} \underset{\text{(Orthorhombic)}}{\underset{\text{Gypsum}}{CaSO_4 \cdot 2H_2O}} \xrightarrow{\text{Hardening}} \underset{\text{(Monoclinic)}}{\underset{\text{Gypsum}}{CaSO_4 \cdot 2H_2O}}$$

Setting of plaster of Paris may be catalysed with the help of NaCl while it is retarded by adding borax or alum.

- **Heating effect**: When plaster of Paris is heated at 200°C it gives anhydrous calcium sulfate (dead plaster or dead burnt) which does not have any setting property.

$$(CaSO_4)_2 H_2O \xrightarrow[200°C]{\Delta} \underset{\text{Dead burnt plaster}}{2CaSO_4} + H_2O$$

Uses

- Plaster of Paris is used as a cast to mend broken bones, making statues, toys, chalk.
- It is used as a building material.
- It is used for making moulds for casting.
- In the laboratory, it is used for making the apparatus air-tight by sealing the gaps at the stopper.
- In making fire-proof materials.

Bleaching Powder [$CaOCl_2$ or $Ca(OCl)Cl$]

It is also known as calcium chlorohypochlorite, chloride of lime or calcium oxychloride.

Preparation

It is prepared by passing a current of chlorine (obtained during manufacture of caustic soda) over dry slaked lime. The methods used are Bachmann method and Hasenclever method.

$$Ca(OH)_2 + Cl_2 \rightarrow \underset{\text{Bleaching powder}}{CaOCl_2} + H_2O$$

Properties

- It is a mixture of calcium hypochlorite ($CaOCl_2.4H_2O$) and basic calcium chloride [$CaCl_2.Ca(OH)_2.4H_2O$].
- It is a yellowish white powder (Fig. 2.30).

Fig. 2.30 Bleaching powder

Source: شهاب, CC BY-SA 4.0 <https://creativecommons.org/licenses/by-sa/4.0>, via Wikimedia Commons

- It is an oxidising agent and a bleaching agent (decolorise delicate fibres).
- Cl_2 (available chlorine) is generally obtained in less amount (35%–40%) from bleaching powder by using CO_2.

$$CaOCl_2 + CO_2 \rightarrow CaCO_3 + Cl_2\uparrow$$

- $CoCl_2$ acts as a catalyst to decompose bleaching powder to liberate oxygen.

$$2CaOCl_2 \xrightarrow{CoCl_2} 2CaCl_2 + O_2\uparrow$$

Uses

- It is used to purify water, to prepare $CHCl_3$, germicides.
- It is used for bleaching cotton, wood pulp.
- It is added in water as a disinfectant as chlorine released by it kills germs (sterilisation of drinking water).

Water of Crystallisation

Some of the salts, while crystallising out from their solution, unite with a definite quantity of water, which is called the *water of crystallisation* (Table 2.12). This water of crystallisation is in loose chemical combination with the salts and it can be driven out by heating the powdered crystals of these salts above 373 K.

Table 2.12 Some salts containing water of crystallisation

S. No	Common name	Chemical name	Formula
1.	Blue vitriol	Copper(II) sulfate pentahydrate	$CuSO_4.5H_2O$
2.	Epsom salt	Magnesium sulfate heptahydrate	$MgSO_4.7H_2O$
3.	Glauber's salt	Sodium sulfate decahydrate	$Na_2SO_4.10H_2O$
4.	Gypsum	Hydrated calcium sulfate	$CaSO_4.2H_2O$
5.	Plaster of Paris	Calcium sulfate semihydrate	$CaSO_4.\frac{1}{2}H_2O$
6.	Potash alum	Potassium aluminium sulfate	$K_2SO_4.Al_2(SO_4)_3.24H_2O$
7.	White vitriol	Zinc sulfate heptahydrate	$ZnSO_4.7H_2O$
8.	Washing soda crystals	Sodium carbonate decahydrate	$Na_2CO_3.10H_2O$

Experiment: To show that copper sulfate crystals contain water of crystallisation.

Procedure: Take some powdered copper sulfate crystals in a clean, dry test tube and clamp it in a tilted position (Fig. 2.31) so that the condensed water may not slip back and crack the test tube at its hot end. Heat the crystals. Drops of a colourless liquid will condense on the cooler parts of the test tube, and finally a little of it can be collected in a dish.

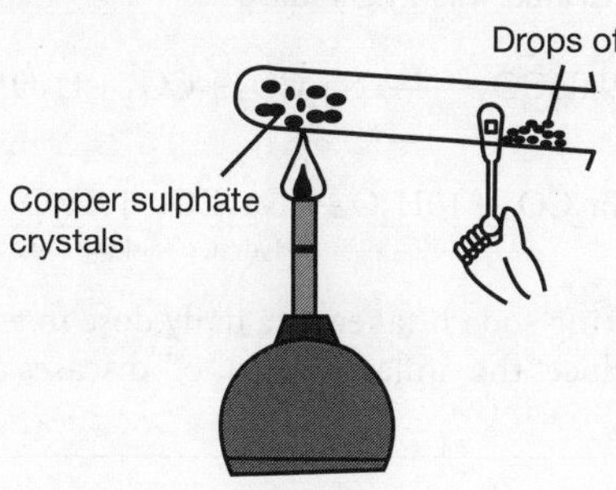

Fig. 2.31 Removal of water of crystallisation by heating

Reaction: $CuSO_4 \cdot 5H_2O \underset{\text{Cool}}{\overset{\text{Heat}}{\rightleftharpoons}} CuSO_4 + 5H_2O$
(Blue crystal) (White powder)

Test for condensed liquid

- It is neutral with respect to litmus.
- Spread some anhydrous copper sulfate (white) on a glass slide in a thin layer and add a few drops of the liquid in the dish. A blue patch will be formed, thereby showing that the liquid is water.
- The liquid turns blue $CoCl_2$ paper pale pink—almost colourless.

$$\underset{\text{Blue}}{CoCl_2} + 2H_2O \rightarrow \underset{\text{Pale Pink}}{CoCl_2 \cdot 2H_2O}$$

Crystalline solids not containing water of crystallisation

Common salt (NaCl), nitre (KNO_3), sugar ($C_{12}H_{22}O_{11}$), potassium permanganate ($KMnO_4$), ammonium chloride (NH_4Cl).

Experiment: Heat gently a few crystals of pure sodium chloride and potassium nitrate separately, in different test tubes.

Observation for sodium chloride: On heating, crystals of sodium chloride produce a crackling sound, called *decrepitation*. This is due to the bigger crystals breaking up into smaller ones.

Observation for potassium nitrate: The crystals of potassium nitrate melt into a colourless liquid and on further heating, produce oxygen, which rekindles a glowing splinter. Potassium nitrite, a pale yellow solid residue, is left behind.

$$\underset{\text{Potassium nitrate}}{2KNO_3} \rightarrow \underset{\text{Potassium nitrite}}{2KNO_2} + O_2$$

Note: These salts do not produce water.

Conclusion: All crystalline salts do not contain water of crystallisation.

Competition Edge

Double salts: Such salts are formed by the union of two simple salts that dissolve in water and crystallise. For example, Mohr's salt $FeSO_4.(NH_4)_2SO_4.6H_2O$, dolomite $CaCO_3.MgCO_3$, potash alum $K_2SO_4.Al_2(SO_4)_3.24H_2O$.

Mixed salts: Such salts are those salts that contain more than one basic or acid radical. For example, sodium potassium carbonate $NaKCO_3$, disodium potassium phosphate Na_2KPO_4, bleaching powder $CaOCl_2$.

Complex salts: Such salts are formed by mixing the saturated solutions of simple salts followed by the crystallisation of the mixture. For example, copper ammonium sulfate $[Cu(NH_3)_4]SO_4$, potassium ferrocyanide $K_4[Fe(CN)_6]$.

Efflorescence: It is the property of some substances to lose wholly or partly, their water of crystallisation when their crystals are exposed to dry air even for a short time. They become powdery. Such substances are known as *efflorescent substances.* For example, washing soda (hydrated sodium carbonate) when exposed to dry air, becomes a monohydrate.

$$Na_2CO_3 \cdot 10H_2O \xrightarrow{\text{Dry air}} Na_2CO_3 \cdot H_2O + 9H_2O$$

Hygroscopic substances: Such substances absorb moisture (water vapour) from the atmosphere without dissolving in it. They are called hygroscopic substances. For example, conc. sulfuric acid H_2SO_4 and phosphorus pentoxide (P_2O_5).

Deliquescent substances: Those water-soluble substances, when exposed to the atmosphere at ordinary temperature, absorb moisture from the atmospheric air to become moist and ultimately dissolve in the absorbed water, forming a saturated solution. These substances are known as *deliquescent substances,* and the phenomenon is called *deliquescence.* For example, caustic soda (NaOH), caustic potash (KOH), magnesium chloride ($MgCl_2$).

Drying or desiccating agent: Such substances remove moisture from other substances and are therefore known as *desiccants* or *desiccating agents* or drying agents. Almost all the hygroscopic substances are desiccating agents. For example, conc. sulfuric acid, phosphorus pentoxide, quicklime and silica gel.

Dehydrating agents: Dehydrating agents are the substances that can remove water molecules even from compounds. Concentrated sulfuric acid can remove water molecules from blue vitriol ($CuSO_4.5H_2O$), so it is a dehydrating agent as well.

TEST YOUR KNOWLEDGE

2.22 Fill in the blanks.

(i) ________ are the products obtained when bleaching powder reacts with dilute sulfuric acid.

(ii) ______ is one of the raw materials for the production of baking soda.

Solution: (i) $CaSO_4$, Cl_2, H_2O, (ii) sodium chloride

2.23 What happens when nitric acid is added to egg shells?

Solution: Egg shells contain calcium carbonate. When nitric acid is added to it, carbon dioxide gas is evolved. The reaction can be given as:

$$CaCO_3 + 2HNO_3 \rightarrow Ca(NO_3)_2 + H_2O + CO_2$$

2.24 Why does a bee sting cause pain and irritation? Rubbing of baking soda on the sting area gives relief. How?

Solution: A bee sting causes pain and irritation due to the presence of formic acid. Rubbing of baking soda (sodium hydrogen carbonate) neutralises the formic acid to provide relief.

2.25 (i) State the chemical properties on which the following uses of baking soda are based.

(a) As an antacid.
(b) As a soda fire extinguisher.
(c) To make bread and cake soft and spongy.

(ii) How is washing soda obtained from baking soda? Write a balanced chemical equation. **[CBSE 2015]**

Solution: (i) (a) An antacid is weakly basic in nature and hence neutralises hyperacidity.

(b) Baking soda liberates CO_2 with H_2SO_4 that extinguishes fire.

(c) Baking soda makes bread and cake soft and spongy as it liberates CO_2 on heating.

(ii) Baking soda on heating gives sodium carbonate which on crystallisation forms hydrated washing soda.

$$2NaHCO_3 \xrightarrow{\Delta} Na_2CO_3 + CO_2 + H_2O$$

$$Na_2CO_3 + 10H_2O \rightarrow \underset{\text{Hydrated washing soda}}{Na_2CO_3 \cdot 10H_2O}$$

Note: Baking soda if taken as a daily dose in water could reduce the inflammation of diseases like arthritis.

Quick Review

- **Salts** are the compounds formed on the partial or complete replacement of the hydrogen of an acid by metal atoms or positive radicals. For example, KCl, $CaCl_2$, Na_2SO_4.
- Salts formed by the complete replacement of the hydrogen atom/atoms of an acid by metal atoms or positive radicals are known as **normal salts**. For example, KCl, NH_4Cl.
- Salts formed by the partial replacement of the hydrogen atoms of a polybasic acid by metal atoms or positive radicals are called **acid salts**. For example, $KHSO_4$, NH_4HSO_4.
- Salts formed by the partial replacement of hydroxyl groups or the incomplete the neutralisation of a base are known as basic salts. For example, Pb(OH)Cl.
- **Double salts** are formed by the union of two simple salts that dissolve in water and crystallise. For example, Mohr's salt $FeSO_4.(NH_4)_2SO_4.6H_2O$, dolomite $CaCO_3.MgCO_3$, potash alum $K_2SO_4.Al_2(SO_4)_3.24H_2O$.
- **Mixed** salts are those salts that contain more than one basic or acid radical. For example, $NaKCO_3$.
- **Complex** salts are formed by mixing the saturated solutions of simple salts followed by the crystallisation of the mixture. For example, copper ammonium sulfate $[Cu(NH_3)_4]SO_4$, potassium ferrocyanide $K_4[Fe(CN)_6]$.
- Some of the salts, while crystallising out from their solution, unite with a definite quantity of water, which is called the **water of crystallisation**. For example, $CuSO_4.5H_2O$ (blue vitriol).
- Common salt or table salt is NaCl. It contains a little potassium iodide also.
- Table salt is used in the preparation of ice cream (freezing mixture) and in salting out soap.
- Caustic soda (NaOH) is manufactured by electrolysis of brine solution (chlor-alkali process) and the byproducts are H_2, Cl_2 and HCl.
- Caustic soda is used in making soaps, petroleum refining, mercerising cotton.
- Bleaching powder ($CaOCl_2$) is obtained by the reaction of slaked lime and chlorine. Bleaching power has available chlorine which is used as a bleaching agent, germicide.
- Sodium bicarbonate is baking soda ($NaHCO_3$) which is used as an antacid, in baking power and in fire extinguishers.
- Washing soda is hydrated sodium carbonate ($Na_2CO_3.10H_2O$) and obtained by Solvay process.
- Plaster of Paris is calcium sulfate hemihydrate ($CaSO_4.1/2\ H_2O$). It is used in making toys, chalk, ceiling, plaster.

Exercise 2

Section A: Multiple Choice Questions (1 Mark)

1. Which salt is a neutral salt?
(a) NH_4Cl (b) CH_3COONH_4
(c) CH_3COONa (d) Na_2CO_3

2. Which salt is acidic in nature?
(a) NH_4Cl (b) CH_3COONH_4
(c) NaCl (d) Na_2CO_3

3. Which of the following does not form an acidic salt?
(a) phosphoric acid (b) carbonic acid
(c) hydrochloric acid (d) sulfuric acid

4. Brine is an
(a) aqueous solution of sodium hydroxide
(b) aqueous solution of sodium carbonate
(c) aqueous solution of sodium chloride
(d) aqueous solution of sodium bicarbonate

5. $Na_2CO_3.10H_2O$ is known as
(a) baking soda
(b) baking powder
(c) washing soda
(d) bleaching powder

6. Farmers neutralise the effect of acidity on the soil by adding
(a) slaked lime (b) gypsum
(c) caustic soda (d) baking soda

7. $CuSO_4.5H_2O$. In this compound the water molecules are called
(a) pure water
(b) water of crystallisation
(c) soda water
(d) none of these

8. Which of the following salts does not contain water of crystallisation?
(a) blue vitriol (b) baking soda
(c) washing soda (d) gypsum

9. What is the chemical formula of POP (plaster of Paris)?
(a) $CaSO_4.2H_2O$ (b) $CaSO_4.3H_2O$
(c) $CaSO_4.1/2H_2O$ (d) $CaCO_3.1/2H_2$

10. At what temperature is gypsum heated to form plaster of Paris?
(a) 45°C (b) 70°C
(c) 90°C (d) 100°C

11. Sodium carbonate reacts with hydrochloric acid to produce
(a) NaCl (b) CO_2
(c) H_2O (d) all of the above

12. The function of quicklime in sodalime mixture is to
(a) absorb moisture present in soda lime
(b) increase the efficiency of soda lime
(c) increase the pH of soda lime
(d) take part in reaction with NaOH

13. The difference of molecules of water in gypsum and POP is
(a) 5/2 (b) 2
(c) 3/2 (d) 1/2

14. How many water molecules does hydrated calcium sulfate contain?
(a) 5 (b) 10
(c) 7 (d) 2

15. Lime water reacts with chlorine to give
(a) bleaching powder (b) baking powder
(c) baking soda (d) washing soda

16. Salt A commonly used in food products is a reactant to produce salt B, used in the kitchen for making tasty, crispy pakoras. Salt B on heating converts into another salt C which is used in the manufacturing of glass. Salts A, B and C respectively are
(a) Na_2CO_3, NaCl, $NaHCO_3$
(b) NaCl, $NaHCO_3$, Na_2CO_3
(c) $NaHCO_3$, NaCl, Na_2CO_3
(d) Na_2CO_3, $NaHCO_3$, NaCl **(CBSE 2021)**

17. Given below is a reaction showing the chlor-alkali process: **(CBSE 2021)**

$$\underset{}{2NaCl(aq) + 2H_2O(l)} \rightarrow \underset{(A)}{2NaOH(aq)} + \underset{(B)}{Cl_2(g)} + \underset{(C)}{H_2(g)}$$

The products A, B and C produced respectively are,
(a) at the anode, near the cathode, at the cathode
(b) at the cathode, near the cathode, at the anode
(c) near the cathode, at the anode, at the cathode
(d) at the anode, at the cathode, near the cathode

18. Which of the following chemicals is used in the soda–acid fire extinguishers? **(CBSE 2021)**
(a) sodium acetate
(b) sodium chloride
(c) ammonium sulfate
(d) sodium hydrogen carbonate

19. Study the following table and choose the correct option. **(CBSE 2021)**

	Salt	Parent acid	Parent base	Nature of salt
(a)	Sodium chloride	HCl	NaOH	Basic
(b)	Sodium acetate	CH_3COOH	NaOH	Basic
(c)	Sodium sulfate	H_2SO_4	NaOH	Acidic
(d)	Sodium carbonate	H_2CO_3	NaOH	Neutral

20. Study the experimental set up shown in the figure and choose the correct option from the following. **(CBSE 2021)**

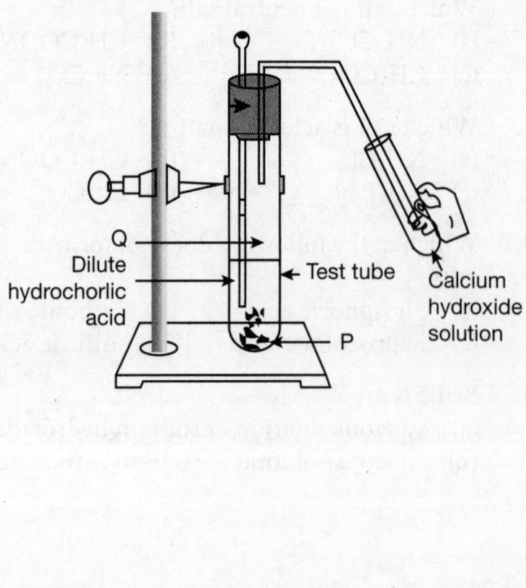

	P	Q	Change observed in calcium hydroxide solution
(a)	K_2CO_3	Cl_2 gas	No change
(b)	$KHCO_3$	H_2 gas	Turns milky
(c)	K_2CO_3	CO_2 gas	Turns milky
(d)	$KHCO_3$	CO_2 gas	No change

21. Which of the following salts do not contain water of crystallisation? **(CBSE 2021)**
   (i) bleaching powder
   (ii) plaster of Paris
   (iii) washing soda
   (iv) baking soda
   (a) (i) and (iii) (b) (ii) and (iv)
   (c) (i) and (iv) (d) (ii) and (iii)

22. Match the following.

Column I	Column II
1. Plaster of Paris	(i) $NaHCO_3$
2. Washing soda	(ii) $CaOCl_2$
3. Bleaching powder	(iii) $CaSO_4.1/2\ H_2O$
4. Baking soda	(iv) $Na_2CO_3.10H_2O$

   (a) (1)—(ii), (2)—(iii), (3)—(i), (4)—(iv)
   (b) (1)—(iii), (2)—(iv), (3)—(ii), (4)— (i)
   (c) (1)—(iii), (2)—(ii), (3)—(i), (4)—(iv)
   (d) (1)—(iii), (2)—(ii), (3)—(iv), (4)—(i)

23. Match the following.

Column I (Compound)	Column II (Uses)
1. Caustic soda	(i) Chalk making
2. Baking soda	(ii) Fire exinguisher
3. Washing soda	(iii) Mercerising of cotton
4. Plaster of Paris	(iv) In laundry

   (a) (1)—(ii), (2)—(iii), (3)—(i), (4)—(iv)
   (b) (1)—(iii), (2)—(iv), (3)—(ii), (4)—(i)
   (c) (1)—(iii), (2)—(ii), (3)—(i), (4)—(iv)
   (d) (1)—(iii), (2)—(ii), (3)—(iv), (4)—(i)

Assertion Reason Questions

Directions: In the following questions, two statements (Assertion) A and Reason (R) are given. Mark
(a) if A and R are both correct and R is the correct explanation of A;
(b) if A and R are both correct but R is not the correct explanation of A;
(c) A is true but R is false;
(d) A is false but R is true.

Assertion	Reason
1. Sodium hydrogen carbonate is used as an ingredient in antacids. **(CBSE 2021)**	**1.** $NaHCO_3$ is a mild non-corrosive basic salt.
2. The pH of an acidic solution is less than 7, the pH of a basic solution is greater than 7 whereas the pH of salt solutions is equal to 7.	**2.** Salts are formed by the neutralisation of an acid by a base or vice versa.
3. Plaster of Paris has to be stored in moisture-proof bags.	**3.** Plaster of Paris contains no water of crystallisation. On coming in contact with moisture, it forms the hydrated salt $CaSO_4.2H_2O$ (called gypsum).
4. Baking soda is used in making baking powder.	**4.** Baking soda is also known as soda ash.
5. Aqueous solution of sodium carbonate is basic in nature.	**5.** In aqueous solution, the strong base NaOH and the weak acid H_2CO_3 are present.
6. Bleaching powder can be used as a bleaching agent and germicide.	**6.** The formula of bleaching powder is CaOCl.
7. In gypsum, two water molecules are present.	**7.** Gypsum on heating gives plaster of Paris.
8. Washing soda is $Na_2CO_3.H_2O$.	**8.** It is manufactured by the Solvay method.

Section B: Very Short Answer Questions (2 Marks)

1. Define a salt.
2. Write the name of any two salts of sodium.
3. Differentiate between a basic radical and an acidic radical.
4. Differentiate between a normal salt and an acidic salt.
5. Out of the following salts which are acidic salts?

   NaCl, Na_2CO_3, $NaHCO_3$, Na_2S, NaHS
6. Give any two examples of salts of a strong acid and strong base.
7. The nature of aqueous solution of $NaNO_3$ is ______ and that of Na_2CO_3 is ______.
8. The nature of an aqueous solution of NH_4Cl is ______ and that of KNO_3 is __________.
9. Name one salt with a pH value more than 7 and one salt with a pH value less than 7.
10. What do you mean by hygroscopic nature of common salt?
11. What happens when salt and ice are mixed together?
12. What do you understand by 'available chlorine' as used in the case of bleaching powder?
13. Which compound of calcium is used for sterilisation of drinking water?
14. What do you understand by mercerisation of cotton? Which compound is generally used for doing so?
15. Which material is used for making chalk for writing on the blackboard?
16. Out of tap water, sea water and pure water, which is the best conductor of electricity and why?
17. Why table salt is labelled as iodised salt?
18. Why is potassium iodide added in common salt?
19. Write the names of the products formed by heating gypsum at 373 K. Write one use.
20. Name a salt which does not contain water of crystallisation.
21. Name two constituents of baking powder.
22. Define efflorescence with an example.
23. Name the sodium compound which is used for softening hard water.
24. Which byproduct of the chlor-alkali process is used for manufacturing bleaching powder?
25. Name the substance which on treatment with chlorine yields bleaching powder.
26. What would be the colour of litmus in a solution of sodium carbonate?
27. Write the chemical name and formula of the compound which is used as an antacid.

Case Study or Passage Based Questions

A salt is formed by the combination reaction between an acid and a base. Salts are normal, acidic and basic in nature. In a normal salt there is no H^+ or OH^- to donate while an acidic salt has one or more H^+ to donate and similarly a basic salt has one or more OH^- to donate. A normal salt has two parts—one cationic part and one anionic part.

28. Define basic salts with an example.
29. Out of the following salts which are normal salts?

    NaCl, Na_2CO_3, $NaHCO_3$, Na_2S, NaHS
30. Out of the following salts which are acidic salts?

    KCl, $CaCO_3$, $NaHCO_3$, Na_2S, NaHS, $Ca(HCO_3)_2$
31. The nature of an aqueous solution of KCl is ______ and that of CH_3COONa is ______.
32. Name one metal carbonate which is soluble in water and one which is insoluble in water.

Section C: Short Answer Questions (3 Marks)

1. Define normal and acidic salts with two examples of each.
2. Mention the products produced when an acid reacts with a base. Give the equation of an example of the reaction involved. What is this kind of reaction known as?

3. Classify the following salts as acidic, basic or neutral.
   (i) NaCl (ii) Na_2SO_4
   (iii) $CaCl_2$ (iv) K_2CO_3

4. Write the formulae of the acid and the base that formed from the following salts.
   (i) $NaClO_3$ (ii) $Al_2(SO_3)_3$
   (iii) $Ca(NO_2)$ (iv) NH_4NO_3

5. Classify the following salts into acidic, basic and neutral—potassium sulfate, ammonium chloride, sodium carbonate, sodium chloride.

6. (i) What is meant by the term hydrated salt?
   (ii) Give two examples of hydrated salts which are white and state their chemical formula.

7. Define water of crystallisation with two examples. How will you prove their existence in the examples given by you?

8. 'Sodium hydrogen carbonate is a basic salt.' Justify the statement. How is it converted into washing soda? Explain.

9. Name the three products of the chlor-alkali process. Write one commercially or industrially important material each that can be prepared from each of these products.

10. Why does bleaching powder smell strongly of chlorine and does not dissolve completely in water?

11. What is bleaching powder? How is it prepared? List two uses of bleaching powder.

12. A white coloured powder is used by doctors for supporting fractured bones.
    (i) Write chemical name and formula of the powder.
    (ii) When this white powder is mixed with water a hard solid mass is obtained. Write a balanced chemical equation for the change. **[CBSE 2015]**

13. What is baking powder? How does it make the cake soft and spongy?

14. How does bleaching powder help in disinfection of drinking water?

15. What is the role of toothpastes in preventing cavities?

16. Give two important uses of washing soda and baking soda.

17. Explain why an aqueous solution of sodium sulfate is neutral while an aqueous of sodium carbonate is basic in nature. **[Board Term-I Set-WDCXXOV, 2016]**

18. Write the chemical formula of baking soda. How is baking soda prepared? Write the reaction which takes place when it is heated during cooking.

19. Give the chemical equation to show the changes that occur when green coloured ferrous sulfate crystals are heated. Mention the change observed as well as reason for this change.

20. A few crystals of copper sulfate are heated in a dry boiling tube.
    (i) What is the colour before and after heating?
    (ii) What is the reason for the colour change?
    (iii) Can its original colour be restored? How?

21. Name the gas evolved when sodium hydrogen carbonate is made to react with dilute hydrochloric acid. How will you test the gas?

22. A white powder is added while baking breads and cakes to make them soft and fluffy. What is the name of the powder? What are the main ingredients in it? What are the functions of each ingredient?

23. A sanitary worker uses a white chemical having a strong smell of chlorine gas to disinfect the water tank.
    (i) Identify the chemical compound, write its chemical formula.
    (ii) Give the chemical equation for its preparation.
    (iii) Write two uses other than for disinfection. **[Board Term-I Set-L7ZSVLH, 2016]**

24. Explain the following giving the chemical equation in each case.
    (i) Baking soda is heated.
    (ii) Washing soda is heated
    (iii) Gypsum is heated at 373 K. **[CBSE 2017]**

25. How are sodium hydroxide and Cl_2 gas produced from common salt? What is this process called? **[CBSE 2017]**

Section D: Long Answer Questions (5 Marks)

1. Name the acids and bases that combine to form the following salts.
NaCl, $BaSO_4$, $Mg_3(PO_4)_2$, $Pb(NO_3)_2$, $(NH_4)_2SO_4$ and $ZnCO_3$

2. (i) Crystals of a substance changed their colour on heating in a closed test tube but regained it after sometime when they were allowed to cool down. Name the substance and write its formula and explain the phenomenon involved.
(ii) Name the acid and base that would be used to prepare the following salts.
(a) Potassium sulfate(b) Ammonium chloride

3. Give suitable reasons for the following statements.
(i) A tarnished copper vessel regains its shine when rubbed with lemon.
(ii) The crystals of washing soda change to a white powder on exposure to air.
(iii) An aqueous solution of sodium chloride is neutral but an aqueous solution of sodium carbonate is basic.
[Board Term-I Set-NS9SX1D; 2016 Set 2ZGOVVV; 2015 Set IN14KGB, 2014]

4. (i) Identify the acid and the base whose combination forms common salt (that you use in your food). Write its formula and chemical name. Name the source from where it is obtained.
(ii) What is rock salt? Mention its colour and the reason due to which it has this colour.
(iii) What happens when electricity is passed through brine? Write the chemical equation.
(iv) How is it connected to our struggle for freedom?

5. Write the chemical name of $Na_2CO_3.10H_2O$ and Na_2CO_3. Write the significance of $10H_2O$. Mention the term used for water molecules attached with a salt. With the help of a chemical equation, explain the method of preparation of both $Na_2CO_3.10H_2O$ and Na_2CO_3. Also list two uses of $Na_2CO_3.10H_2O$.

6. (i) Study the following chemical equation:

$$CaSO_4.2H_2O \xrightarrow{373K} CaSO_4.H_2O + H_2O$$

Name the reactant and the product and mention one use of the product.
(ii) The following salts are formed by the reaction of an acid with a base:
(a) sodium chloride
(b) ammonium nitrate
Identify the acid and the base and tabulate your answer in the format given below.

S. No	Salt	Acid	Base	Nature
(i)	Sodium chloride			
(ii)	Ammonium nitrate			

7. *We use colours dissolved in water during Holi and clothes get spoiled. Many colours used to celebrate Holi are oxidised metals or industrial dyes mixed with engine oil. Doctors say these are harmful colours which should be banned for usage. Manufacturers mix dyes with colours, and sell them for profits. Holi revellers can make their Holi colourful by using natural colours available in the market. Natural colours start fading when they come in contact with sunrays.
(i) Why do doctors say that synthetic colours are harmful?
(ii) What are natural colours?
(iii) As a student, what initiative will you take to motivate your classmates to use natural colours during Holi? Give any three suggestions.

8. *A group of students, while on an excursion is camping on the hills. One morning, they find themselves engulfed in a thick blanket of snow. One of the senior members of the group suggests that common salt should be sprinkled on the ice silt covering the pavement. Answer the following questions.
(i) What is the purpose of sprinkling common salt on ice silt?
(ii) Can we use any other substance in place of common salt?
(iii) What values are associated with the passage?

9. (i) Write the chemical formula of hydrated copper sulfate and anhydrous copper sulfate. Giving an activity illustrate how these are interconvertible.
(ii) Write chemical names and the formula of plaster of Paris and gypsum.
[CBSE 2015]

10. Define water of crystallisation. Give the chemical formula for two compounds as examples. How can it be proved that the water of crystallisation makes a difference in the state and colour of the compounds?

[CBSE 2020]

Section E: Case Study or Passage Based Questions (4 Marks)

One day Kamal saw that her mother was roasting peanuts in a pan in the kitchen and she had taken ordinary salt to roast the peanuts. She was surprised to observe that in spite of very high temperature, the salt does not melt and peanuts also get roasted.

(CBSE 2021)

1. The chemical name of common salt is
(a) sodium carbonate
(b) potassium chloride
(c) sodium chloride
(d) sodium hydrogen carbonate

2. Common salt is
(a) an ionic compound formed by sharing of electrons
(b) a covalent compound
(c) soluble in organic solvents
(d) an ionic compound formed by the transfer of electrons

3. Common salt is not a raw material for
(a) caustic soda
(b) baking soda
(c) plaster of Paris
(d) bleaching powder

4. Common salt is used as a raw material in the preparation of washing soda in which the number of molecules of water of crystallisation is
(a) 2 (b) 5
(c) 7 (d) 10

High Order Thinking Skills (HOTS) Questions

1. (i) While eating food, you happen to spill some curry on your white shirt. You immediately scrub it with soap. What happens to its yellow colour on scrubbing with soap? Why? What happens to this stain when the shirt is washed with plenty of water?

(ii) After drinking excess of a cold drink, a person suffered from acidity. Explain.

(iii) Why does a bee sting cause pain and irritation? Rubbing of baking soda on the sting area gives relief. How?

2. (i) Explain why an aqueous solution of sodium sulfate is neutral while an aqueous solution of sodium carbonate is basic in nature.

(ii) The colour of copper sulfate solution changes when an iron nail is dipped in it. State the reason giving the chemical equation for the reaction involved.

(iii) An aluminium can is used to store ferrous sulfate solution. It is observed that in few days, holes appeared in the can. Explain the cause for this observation and write the chemical reaction to support your answer.

3. (i) What is the colour of $FeSO_4.7H_2O$ crystals? How does this colour change upon heating? Give a balanced chemical equation for the changes.

(ii) Crystals of a substance changed their colour on heating in a closed vessel but regained it after sometime, when they were allowed to cool down.
(a) Name one such substance.
(b) Explain the phenomenon involved.

4. (i) (a) What would be the colour of the solution when copper oxide and dilute hydrochloric acid are mixed?
(b) Write a chemical equation which represents that the effect of a base is nullified by an acid and vice versa.

(ii) How will you prove that a given salt is a carbonate of a metal?

5. (i) How does the chloride of lime chemically differ from calcium chloride?

(ii) What happens when the chloride of lime reacts with sulfuric acid? Write the chemical equation involved.

(iii) Mention two uses of the chloride of lime.

6. A compound *x* on losing partial crystallisation of water gives a compound *y* which is used for making chalk and other pottery articles. But when *x* is strongly heated, it gives *z* which is used as a drying agent. Identify *x*, *y*, *z* and write the chemical equation.

7. A compound P forms the enamel of teeth. It is the hardest substance of the body. It doesn't dissolve in water but gets corroded when the pH is lowered below 5.5.
   (i) Identify the compound P.
   (ii) How does it undergo damage due to eating chocolate and sweets? What should we do to prevent tooth decay?

8. (i) A metal compound X reacts with dil. H_2SO_4 to produce effervescence. The gas evolved extinguishes a burning candle. If one of the compounds formed is calcium sulfate, then what is X and the gas evolved? Also write a balanced chemical equation for the reaction which occurred.
   (ii) (a) Name one antacid. How does it help to relieve indigestion in the stomach?
   (b) A farmer treats the soil with quicklime or calcium carbonate. What is the nature of the soil? Why does the farmer treat the soil with quicklime?

9. (i) Dry pellets of a base X when kept in the open absorbs moisture and turns sticky. The compound is also formed by the chlor-alkali process. Write the chemical name and formula of X. Describe the chlor-alkali process with a balanced chemical equation. Name the type of reaction that occurs when X is treated with dilute hydrochloric acid. Write the chemical equation.
   (ii) While diluting an acid, why is it recommended that the acid should be added to water and not water to the acid?

10. The pH of a salt used to make tasty and crispy pakoras is 14. Identify the salt and write a chemical equation for its formation. List its two uses. **[CBSE 2018]**

Answers

Section A: Multiple Choice Questions

1. (b)	**2.** (a)	**3.** (b)	**4.** (c)
5. (c)	**6.** (a)	**7.** (b)	**8.** (b)
9. (c)	**10.** (d)	**11.** (d)	**12.** (a)
13. (c)	**14.** (d)	**15.** (c)	**16.** (b)
17. (c)	**18.** (d)	**19.** (b)	**20.** (c)
21. (c)	**22.** (b)	**23.** (d)	

Assertion Reason Question

1. (a)	**2.** (d)	**3.** (c)	**4.** (c)
5. (a)	**6.** (c)	**7.** (b)	**8.** (d)

Section B: Very Short Answer Questions

1. See text (Section 2.8.1).
2. Sodium chloride (NaCl) and sodium sulfate (Na_2SO_4).
3. The positive part of a salt is the basic radical and the negative part of the salt is an acidic radical. For example, in NaCl, Na^+ is the basic radical while Cl^- is the acidic radical.
4. See text (Section 2.8.1).
5. $NaHCO_3$, NaHS
6. NaCl, Na_2SO_4.
7. Neutral, basic.
8. Acidic, neutral.
9. Sodium acetate and ammonium sulfate.
10. The *hygroscopic nature* of common salt means that it catches moisture from the air and becomes wet.
11. It becomes a freezing mixture.
12. Chlorine liberated when bleaching powder is treated with excess of dilute sulfuric acid is called available chlorine.
13. Bleaching powder ($CaOCl_2$).
14. Mercerisation of cotton means making cotton unshrinkable. NaOH is used for doing so.
15. Plaster of Paris is used for making chalk.
16. Sea water is the best conductor of electricity because it contains large amounts of salts which dissociate into ions.
17. Table salt contains potassium iodide so it is labelled as iodised salt.
18. Potassium iodide is added in common salt as iodide salts prevent thyroid disorders.

19. Plaster of Paris and water. **Use:** It is used for plastering fractured bones.

20. Baking soda.

21. Baking powder is a mixture of baking soda and a mild edible acid like tartaric acid.

22. It is the property due to which a crystalline substance loses water of crystallisation when exposed to air and becomes amorphous. For example,

$$Na_2CO_3 \cdot 10H_2O \xrightarrow{\text{Air}} Na_2CO_3 \cdot H_2O + 9H_2O$$

23. Sodium carbonate (Na_2CO_3).

24. Bleaching powder is produced by the action of chlorine on dry slaked lime.

25. Slaked lime $Ca(OH)_2$ (calcium hydroxide).

26. The colour of litmus in a solution of sodium carbonate is blue.

27. Sodium bicarbonate, $NaHCO_3$.

Case Study or Passage Based Questions

28. A basic salt has a minimum of one –OH group, that is, it is formed by incomplete neutralisation. For example, Mg(OH)Cl.

29. NaCl, Na_2CO_3, Na_2S

30. $NaHCO_3$, NaHS, $Ca(HCO_3)_2$.

31. Neutral, basic.

32. Na_2CO_3 (or K_2CO_3) is soluble in water whereas $CaCO_3$ (or $BaCO_3$) is insoluble in water.

Section C: Short Answer Questions

1. See text (Section 2.8.1).

2. The reaction between an acid and a base that gives a salt and water is known as a neutralisation reaction.

$$\text{Base} + \text{Acid} \rightarrow \text{Salt} + \text{Water}$$

$$NaOH(aq) + HCl(aq) \rightarrow NaCl(aq) + H_2O(l)$$

3. (i) Acidic salt: $CaCl_2$
(ii) Basic salt: K_2CO_3
(iii) Neutral salt: NaCl, Na_2SO_4

4. (i) NaOH and $HClO_3$
(ii) $Al(OH)_3$ and H_2SO_3
(iii) $Ca(OH)_2$ and HNO_2
(iv) NH_4OH and HNO_3

5. **Neutral salts:** Potassium sulfate, sodium chloride
Basic salts: Sodium carbonate
Acidic salts: Ammonium chloride

6. (i) Salts that contain water of crystallisation are known as hydrated salts.
(ii) Washing soda: $Na_2CO_3.10H_2O$
Gypsum: $CaSO_4.2H_2O$

Plaster of Paris: $CaSO_4.\frac{1}{2}H_2O$

7. Water of crystallisation is a fixed number of water molecules present in one formula unit of a salt. One formula unit of copper sulfate contains five water molecules ($5H_2O$). The water molecules which form part of the structure of a crystal are called water of crystallisation.

$$\underset{\substack{\text{(Hydrated copper}\\\text{sulfate)}\\\text{(Blue)}}}{CuSO_4.5H_2O} \xrightarrow{\Delta} \underset{\substack{\text{(Anhydrous}\\\text{coppersulfate)}\\\text{(White)}}}{CuSO_4} + \underset{\substack{\text{Water}\\\text{(Goes away)}}}{5H_2O}$$

* Due to loss of water of crystallisation.

$$\underset{\substack{\text{Anhydrous}\\\text{copper sulfate}\\\text{(blue)}}}{CuSO_4} + \underset{\text{Water}}{5H_2O} \rightarrow \underset{\substack{\text{Hydrated copper}\\\text{sulfate (white)}}}{CuSO_4.5H_2O}$$

* By adding water

8. It is a salt produced by the neutralisation reaction between a strong base (NaOH) and a weak acid (H_2CO_3), hence it is a basic salt. It is heated strongly to produce sodium carbonate.

$$2NaHCO_3 \rightarrow Na_2CO_3 + H_2O + CO_2$$

Sodium carbonate is recrystallised to produce washing soda.

$$Na_2CO_3 + 10H_2O \rightarrow Na_2CO_3.10H_2O$$

9. The three products are sodium hydroxide (NaOH), chlorine (Cl_2) and hydrogen (H_2).
NaOH: for soaps and detergents, paper making and artificial fibres.
Cl_2: PVC, pesticides, CFCs.
H_2: fuels, margarine, NH_3 for fertilisers.

10. When exposed to air, bleaching powder gives the smell of chlorine. This is because bleaching powder reacts with carbon dioxide from the atmosphere to produce calcium carbonate and chlorine gas.

11. Bleaching powder is calcium oxychloride. It is prepared by the action of chlorine on dry slaked lime.

$$Ca(OH)_2 + Cl_2 \rightarrow CaOCl_2 + H_2O$$

Bleaching power

Uses: (i) Used for bleaching cotton and linen in the textile industry and wood pulp in the paper industry.

(ii) It is used for disinfecting drinking water.

12. (i) **Chemical name:** Calcium sulfate hemihydrate.

Formula: $CaSO_4 \cdot \frac{1}{2}H_2O$

(ii) $CaSO_4 \cdot \frac{1}{2}H_2O + \frac{3}{2}H_2O \rightarrow \underset{\text{Gypsum}}{CaSO_4 \cdot 2H_2O}$

13. Baking powder is a mixture of sodium hydrogen carbonate and tartaric acid. On heating, it liberates CO_2 which makes the cake soft and spongy.

14. Bleaching powder dissolves in water to generate hypochlorous acid.

$$Ca(OCl)_2 + 2H_2O \rightarrow \underset{\text{Hypochlorous acid}}{2HOCl} + Ca(OH)_2$$

Hypochlorous acid has effective germicidal property which is responsible for the disinfecting action of bleaching powder.

15. Fluoride is a very safe and effective way to help prevent tooth decay, so it is an important ingredient in toothpaste and mouthwashes. It works by making the enamel more resistant to acids produced by the bacteria living on the teeth and gums.

16. Uses of washing soda: (i) In glass, soap and paper industry.

(ii) Manufacture of borax.

Uses of baking soda: (i) An ingredient in antacid.

(ii) Used in soda–acid fire extinguishers.

17. Sodium sulfate is a neutral salt because it is made from a strong acid and a strong base. Sodium carbonate is a basic salt because it is made from a strong base and a weak acid.

18. The chemical formula of baking soda is $NaHCO_3$. It is prepared by using sodium chloride.

$$NaCl + H_2O + CO_2 + NH_3 \rightarrow NH_4Cl + NaHCO_3.$$

On heating: $2NaHCO_3 \xrightarrow{\Delta} Na_2CO_3 + H_2O + CO_2$

19. $2FeSO_4 \xrightarrow{\Delta} Fe_2O_3 + SO_2 + SO_3$

Crystals lose their colour due to loss of water of crystallisation on heating.

20. (i) Blue and white respectively.

(ii) Water of crystallisation is lost on heating.

(iii) Yes, by adding water.

21. Carbon dioxide (CO_2) is evolved and when we pass this gas into lime water, it turns lime water milky.

22. Baking powder.

Baking soda and an edible acid like tartaric acid.

Baking soda ($NaHCO_3$) is used to release CO_2 gas when heated.

Tartaric acid is used to avoid the bitter taste by reacting with the Na_2CO_3 formed.

23. (i) Bleaching powder: $CaOCl_2$ (ii) $Ca(OH)_2 + Cl_2 \rightarrow CaOCl_2 + H_2O$

(iii) Two uses other than disinfection are—paper industry and chemical industries.

24. (i) Baking soda (sodium hydrogencarbonate, $NaHCO_3$) on heating gives Na_2CO_3 salt and CO_2 gas which will turn lime water milky.

$$2NaHCO_3 \xrightarrow{\Delta} Na_2CO_3 + H_2O + CO_2 \uparrow$$

(ii) Washing soda is sodium carbonate (Na_2CO_3) which occurs as a crystalline heptahydrate on heating, loses its water to form a white powder monohydrate. This process is called efflorescence.

$$Na_2CO_3 \cdot 7H_2O \xrightarrow{\Delta} Na_2CO_3 \cdot H_2O + 6H_2O$$

(iii) When gypsum is heated at 373 K, it loses water molecules and becomes calcium sulfate hemihydrate ($CaSO \cdot \frac{1}{2}H_2O$). This is called plaster of Paris.

$$\underset{\text{Gypsum}}{CaSO_4 \cdot 2H_2O} \xrightarrow{\Delta} \underset{\text{Plaster of paris}}{CaSO_4 \cdot \frac{1}{2}H_2O} + \frac{3}{2}H_2O$$

Note: Gypsum can be used as a dietary source of calcium as an ingredient in canned vegetables, flour, and so on.

25. Sodium hydroxide and Cl_2 gas are produced as products when electricity is passed through an aqueous solution of NaCl and it gets decomposed. This process is called the chlor-alkali process.

$$2NaCl(aq) + 2H_2O(l) \rightarrow 2NaOH(aq) + Cl_2(g) + H_2(g)$$

Here, Cl_2 gas is given off at the anode and H_2 gas at the cathode.

Section D: Long Answer Questions

1.

Salt	→	Acid used	+	Base used
NaCl	→	HCl	+	NaOH
$BaSO_4$	→	H_2SO_4	+	$Ba(OH)_2$
$Mg_3(PO_4)_2$	→	H_3PO_4	+	$Mg(OH)_2$
$Pb(NO_3)_2$	→	HNO_3	+	$Pb(OH)_2$
$(NH_4)_2SO_4$	→	H_2SO_4	+	NH_4OH
$ZnCO_3$	→	H_2CO_3	+	$Zn(OH)_2$

2. (i) Copper sulfate crystals ($CuSO_4.5H_2O$): 5 molecules of water are present in one formula unit of copper sulfate. When heated, it loses its water and the salt turns white.
   (ii) **Basic salt:** (a) Base: KOH (Potassium hydroxide); Acid: H_2SO_4 (Sulfuric acid)
   (b) Base: NH_4OH (Ammonium hydroxide); Acid: HCl (Hydrochloric acid)

3. (i) Copper vessels get tarnished due to the formation of basic copper carbonate which gets neutralised when rubbed with lemon and the copper vessel regains its shine.
   (ii) Washing soda is sodium carbonate decahydrate, which when exposed to air loses 10 molecules of water and changes to a white powder.
   (iii) Sodium chloride is a salt of the strong acid HCl and the strong base NaOH, so it is neutral.
   Sodium carbonate is a salt of the weak acid H_2CO_3 and the strong base NaOH, so it is basic.

4. (i) Hydrochloric acid and sodium hydroxide.
   $NaOH + HCl \rightarrow NaCl + H_2O$
   NaCl, sodium chloride, ocean water.
   (ii) Deposits of solid salt are found in several parts of the world. These large crystals are called rock salt.
   The colour of rock salts is brown, due to the presence of impurities.
   (iii) $2NaCl + 2H_2O \xrightarrow{\text{Electricity}} 2NaOH + Cl_2 + H_2$
   (iv) Mahatma Gandhi's Dandi March was for the procurement of common salt for the common man.

It is the most essential product of daily use.

5. $Na_2CO_3.10H_2O$: Sodium carbonate decahydrate. Na_2CO_3: Anhydrous sodium carbonate.
   $10H_2O$: Water of crystallisation which imparts shape and colour to the crystals.
   Preparation: $NaCl + H_2O + CO_2 + NH_3 \rightarrow NH_4Cl + NaHCO_3$
   $2NaHCO_3 \xrightarrow{\Delta} Na_2CO_3 + H_2O + CO_2$
   $Na_2CO_3 + 10H_2O \xrightarrow{\Delta} Na_2CO_3.10H_2O$
   Uses: (i) Used in glass, soap and paper industries.
   (ii) Used in the manufacture of borax.
   (iii) Used as a cleansing agent for domestic purpose.

6. (i) **Reactant:** Gypsum
   Product: Plaster of Paris
   Use: As cast (plaster) used for supporting fractured bones.

S. No	Salt	Acid	Base	Nature
(i)	Sodium chloride	Hydrochloric acid	Sodium hydroxide	Neutral
(ii)	Ammonium nitrate	Nitric acid	Ammonium hydroxide	Acidic

7. (i) Doctors say these colours are toxic and can cause skin allergies, eye irritation and blindness.
   (ii) The colours derived from vegetables and other organic sources are known as natural colours.
   (iii) **Suggestions:** (a) Doctor's consultancy in the class to satisfy the classmate.
   (b) Application of natural colours in class.
   (c) To make them understand the importance of playing safe Holi.

8. (i) When common salt is sprinkled on icy roads, ice/snow starts melting from the surface because the freezing point of

water is lowered. The softened ice/ snow can be easily cleared from the roads.

(ii) Calcium chloride can be used in the place of common salt. It can lower the freezing point up to 55°C.

(iii) Associated values are scientific knowledge, caring, helping and supporting nature.

9. (i) $CuSO_4.5H_2O$ is hydrated copper sulfate and $CuSO_4$ is anhydrous copper sulfate.

Activity:

Aim: To show that hydrated copper sulfate and anhydrous copper sulfate are interconvertible.

Apparatus required: $CuSO_4.5H_2O$ (blue vitriol), burner, boiling tube, delivery tube, test tube.

Procedure:

- Heat a few crystals of copper sulfate in a dry boiling tube.
- After heating, water is removed and the salt turns white.
- Add 2–3 drops of water on the sample of copper sulfate obtained after heating.
- If the crystals are moistened again with water, the blue colour reappears.

Water of crystallisation is the fixed number of water molecules present in one formula unit of a salt. Here, five water molecules are present in one formula unit of copper sulfate.

Chemical reaction:

$$\underset{\text{Blue vitriol}}{CuSO_4 \cdot 5H_2O} \xrightarrow{\Delta} \underset{\text{White}}{CuSO_4} + 5H_2O$$

Result: Crystalline substances have water of crystallisation which is lost on heating. When hydrated copper sulfate is heated, it gets converted into anhydrous copper sulfate. When water is added again to it, it gets converted into hydrated copper sulfate. This shows that both are interconvertible.

(ii) (a) **Plaster of Paris:**

Chemical name: Calcium sulfate dehydrate.

Formula: $CuSO_4.2H_2O$

10. Water of crystallisation is the number of molecules of water which are loosely bonded to one molecule of salt. For example, $BaCl_2.2H_2O$ and $CuSO_4.5H_2O$. We can prove that by heating the compound containing water of crystallisation, its colour or state changes due to the removal of water of crystallisation. For example, copper sulfate crystals containing water of crystallisation are blue but on heating they turn white. Similarly, plaster of Paris is a white powder and on mixing with water it changes to gypsum which forms a hard, solid mass.

Section E: Case Study or Passage Based Questions

1. (c) **2.** (d) **3.** (c) **4.** (d)

High Order Thinking Skills (HOTS) Questions

1. (i) The colour changes from yellow to reddish brown. Soap is basic in nature and the colour of turmeric changes from yellow to reddish brown in the basic medium. The stain turns yellow again.

(ii) Cold drinks contain carbon dioxide dissolved in them, under pressure. Some amount of the gas escapes when a bottle of cold drink is opened while a portion of it remains dissolved. Carbon dioxide dissolved in water forms a weak acid, carbonic acid. Consumption of excess of cold drink causes accumulation of carbonic acid in the stomach which is responsible for the development of acidity.

(iii) Bee sting has acid that causes pain and irritation. Baking soda being alkaline, neutralises the acid and gives relief.

2. (i) Sodium sulfate when dissolved in water forms a strong acid and a strong base so its aqueous solution is neutral. Sodium carbonate when dissolved in water forms an alkaline solution which turns red litmus blue. It shows that its aqueous solution is basic in nature.

(ii) $\underset{\text{(Blue)}}{CuSO_4} + Fe \rightarrow \underset{\text{(Green)}}{FeSO_4} + Cu$

Iron is more reactive than copper. Hence it displaces copper from copper sulfate and due to formation of $FeSO_4$, the colour changes from blue to green.

(iii) Aluminium is more reactive than iron. Aluminium from the walls can react

with ferrous sulfate to form its sulfate and iron is precipitated.
$2FeSO_4 + Al \rightarrow Al_2(SO_4)_3 + 3Fe$

3. (i) $FeSO_4.7H_2O$ is green in colour and loses its water of crystallisation when it is heated.
It is then decomposed to Fe_2O_3 (brown coloured), SO_2 and SO_3.

$$FeSO_4.7H_2O \xrightarrow{\Delta} FeSO_4 + 7H_2O$$

$$2FeSO_4(s) \xrightarrow{\Delta} Fe_2O_3(s) + SO_2(g) + SO_3(g)$$

(ii) (a) $CuSO_4.5H_2O$

$$\underset{\text{(Blue)}}{CuSO_4.5H_2O} \xrightarrow{\Delta} \underset{\text{(White)}}{CuSO_4} + 5H_2O$$

(b) **Explanation:** On heating, the crystal loses its water of crystallisation and regains it on cooling.

4. (i) (a) It is blue green.
(b) $NaOH + HCl \rightarrow NaCl + H_2O$

(ii) To the salt, add a few drops of dilute HCl or H_2SO_4, if a colourless gas is evolved with brisk effervescence and on passing it through lime water, the lime water turns milky, the gas is CO_2 and salt is a carbonate of a metal.

5. (i) Chemically, chloride of lime is known as calcium oxychloride.

(ii) $2CaOCl_2(s) + H_2SO_4(aq) \rightarrow CaSO_4(aq) + CaCl_2(aq) + 2HClO(aq)$

$\underset{\text{Hypochlorous acid}}{HClO} \rightarrow HCl + [O]$

Nascent oxygen: Coloured Substance + $[O] \rightarrow$ Colourless

(iii) (a) Bleaching powder is used for bleaching purposes in textile and paper industry and in laundries.
(b) It is also used as a disinfectant for water.

6. A commonly used compound for making chalk and other pottery articles is plaster of Paris. Its chemical formula is $CaSO_4.H_2O$. It is obtained from $CaSO_4.2H_2O$ on partial loss of water of crystallisation by controlled heating at 120–130°C. When $CaSO_4.2H_2O$ is strongly heated, it loses the entire of water of crystallisation and gives anhydrous calcium sulfate. Thus x is $CaSO_4.2H_2O$, y is $CaSO_4.H_2O$ and z is $CaSO_4$.

$$\underset{(x)}{CaSO_4.2H_2O} \rightarrow \underset{(y)}{CaSO_4.\frac{1}{2}H_2O} + \frac{3}{2}H_2O(g)$$

$$CaSO_4.2H_2O \rightarrow CaSO_4$$

7. (i) The compound P is calcium phosphate.

(ii) It undergoes damage when the bacteria present in the mouth work on the left over food particles and produce acid.
To prevent tooth decay, toothpastes (basic) are used which neutralises the excess acid.

8. (i) A → $CaCO_3$ (Limestone)
Gas → CO_2
Equation: $CaCO_3 + H_2SO_4 \rightarrow CaSO_4 + H_2O + CO_2$

(ii) (a) **Milk of magnesia:** It is basic in nature. The stomach produces dil. HCl. When suffering from indigestion, more acid is produced, this causes irritation. Antacid neutralises some acid and then gives relief.

(b) **Soil is acidic:** The farmer wants to neutralise some acid and hence adds basic substances like CaO or $CaCO_3$ to it.

9. (i) X is NaOH (sodium hydroxide). When electricity is passed through an aqueous solution of sodium chloride, it decomposes to form sodium hydroxide.
$2NaCl + 2H_2O \rightarrow 2NaOH + Cl_2 + H_2$
Neutralisation reaction: $NaOH + HCl \rightarrow NaCl + H_2O$

(ii) Acid should be added slowly to water with constant stirring. If water is added to a concentrated acid, the heat generated may cause the mixture to splash out and cause burns. The glass container may also break due to excessive local heating.

10. The salt is sodium hydrogen carbonate ($NaHCO_3$) which is used to make tasty and crispy pakoras.

Chemical equation:

$$NaCl + H_2O + CO_2 + NH_3 \rightarrow \underset{\text{Ammonium chloride}}{NH_4Cl} + \underset{\text{Sodium hydrogen carbonate}}{NaHCO_3}$$

Uses: (i) It is used as an ingredient in anti-acid medicine.

(ii) It is used in the form of baking powder for making various food items.

Hints

Multiple Choice Questions

14. The chemical formula of hydrated calcium sulfate or gypsum is $CaSO_4.2H_2O$.

15. $$\underset{\text{Slaked lime}}{Ca(OH)_2} + \underset{\text{Chlorine}}{Cl_2} \rightarrow \underset{\text{Bleaching powder}}{CaOCl_2} + H_2O$$

Activities with Discussion and Conclusion

Activity 2.1: To test acids and bases in the laboratory using colour change acid–base indicators.

- Collect the following samples from the science laboratory in your school: hydrochloric acid (HCl), sulfuric acid (H_2SO_4), nitric acid (HNO_3), acetic acid (CH_3COOH), sodium hydroxide (NaOH), potassium hydroxide (KOH), calcium hydroxide $[Ca(OH)_2]$, magnesium hydroxide $[Mg(OH)_2]$ and ammonium hydroxide (NH_4OH).
- Put a drop of each of the above solutions on a watchglass and test with a drop of the following indicators.

Red litmus solution, blue litmus solution, phenolphthalein solution, methyl orange solution.

Now answer:

- What change in colour do you observe with each of the above indicators?

Discussion: The changes observed are as follows:

Sample solution	Red litmus solution	Blue litmus solution	Phenolphthalein solution	Methyl orange solution
1. Hydrochloric acid (HCl)	No effect	Red	Colourless	Red
2. Sulfuric acid (H_2SO_4)	No effect	Red	Colourless	Red
3. Nitric acid (HNO_3)	No effect	Red	Colourless	Red
4. Acetic acid (CH_3COOH)	No effect	Red	Colourless	Red
5. Sodium hydroxide (NaOH)	Blue	No effect	Pink	Yellow
6. Calcium hydroxide $Ca(OH)_2$	Blue	No effect	Pink	Yellow
7. Potassium hydroxide (KOH)	Blue	No effect	Pink	Yellow
8. Magnesium hydroxide, $Mg(OH)_2$	Blue	No effect	Pink	Yellow
9. Ammonium hydroxide (NH_4OH)	Blue	No effect	Pink	Yellow

Conclusion: Acids turn blue litmus red but have no effect on red litmus. Bases turn red litmus blue but have no effect on blue litmus. Phenolphthalein is colourless in the acidic medium and turns pink in basic medium. Methyl orange is yellow in the basic medium while it is red in the acidic medium.

Activity 2.2: To test acids and bases using olfactory indicators.

- Take some finely chopped onions in a plastic bag along with some strips of clean cloth. Tie up the bag tightly and leave overnight in the fridge. The cloth strips can now be used to test for acids and bases.
- Take two of these cloth strips and check their odour.
- Keep them on a clean surface and put a few drops of dilute HCl solution on one strip and a few drops of dilute NaOH solution on the other.
- Rinse both cloth strips with water and again check their odour.
- Note your observations.
- Now take some dilute vanilla essence and clove oil and check their odour.
- Take some dilute HCl solution in one test tube and dilute NaOH solution in another. Add a few drops of dilute vanilla essence to both test tubes and shake well. Check the odour once again and record changes in odour, if any.

- Similarly, test the change in the odour of clove oil with dilute HCl and dilute NaOH solution and record your observations.

Discussion: On the basis of your observation, report which out of vanilla, onion and clove oil can be best used as an olfactory indicator.

Conclusion: A substance which has one odour in the acidic medium and a different odour in the basic medium can be used as an acid indicator known as olfactory indicator.

Activity 2.3: To study the reaction of acids with metals.

Caution: This activity needs a teacher's assistance.

- Set up the apparatus as shown in Fig. 2.12.
- Take 5 mL of dilute sulfuric acid in a test tube and add a few pieces of zinc granules to it.
- What do you observe on the surface of the zinc granules?
- Pass the gas being evolved through the soap solution.

Now answer:

- Why are bubbles formed in the soap solution?
- Take a burning candle near a gas-filled bubble.
- What do you observe when a burning candle is brought near the gas bubble?
- Repeat this activity with some more acids such as HCl, HNO_3 and CH_3COOH.
- Are the observations in all the cases the same or different?

Discussion: Bubbles are formed in the soap solution indicating that some gas is evolved during the reaction between the metal and acid. The gas burns with a popping sound showing that the gas evolved is hydrogen. The same observation is found in all the cases.

Conclusion: An active metal like zinc displaces hydrogen from all the acids and the remaining part of the acid combines with the metal to form the corresponding salt.

Metal + Acid → Salt + Hydrogen gas

Activity 2.4: To study the reaction of a base with a metal.

- Place a few pieces of granulated zinc metal in a test tube.
- Add 2 mL of sodium hydroxide solution and warm the contents of the test tube.
- Repeat the rest of the steps as in Activity 2.3 and record your observations.

Discussion: Metals like zinc react with strong bases like NaOH and KOH to liberate hydrogen gas and form the corresponding salt. For example,

$$\underset{\text{Zinc}}{Zn(s)} + \underset{\text{Sodium hydroxide}}{2NaOH(aq)} \rightarrow \underset{\text{Sodium zincate}}{Na_2ZnO_2(aq)} + \underset{\text{Hydrogen}}{H_2(g)}$$

Activity 2.5: To study the reaction of metal carbonates and metal hydrogen carbonates with acids.

- Take two test tubes. Label them as A and B.
- Take about 0.5 g of sodium carbonate (Na_2CO_3) in test tube A and about 0.5 g of sodium hydrogen carbonate ($NaHCO_3$) in test tube B.
- Add about 2 mL of dilute HCl to both the test tubes.
- Pass the gas produced in each case through lime water (calcium hydroxide solution) (Fig. 2.22).

Now answer:

- What do you observe?

Discussion: It is observed that carbon dioxide gas is evolved in each case which turns lime water milky.

$$\underset{\text{Lime water}}{Ca(OH)_2} + \underset{\text{Carbon dioxide}}{CO_2(g)} \rightarrow \underset{\substack{\text{Calcium carbonate} \\ \text{(White ppt.)}}}{CaCO_3(s)} + H_2O(l)$$

On passing excess CO_2, the solution becomes clear due to the conversion of insoluble $CaCO_3$ into soluble calcium hydrogen carbonate.

$$CaCO_3(s) + H_2O(l) + CO_2(g) \rightarrow \underset{\substack{\text{Calcium hydrogen carbonate} \\ \text{(Soluble in water)}}}{Ca(HCO_3)_2(aq)}$$

Conclusion: All metal carbonates and hydrogen carbonates react with acids to give the corresponding salt, carbon dioxide and water.

Metal carbonate/Metal hydrogen carbonate + Acid → Salt + Carbon dioxide + Water

Activity 2.6: To study the reaction between acids and bases, that is, neutralisation reaction.

- Take about 2 mL of dilute NaOH solution in a test tube and add two drops of phenolphthalein solution.
- Note the colour of the solution.
- Add dilute HCl solution to the above solution drop by drop.
- Note the change in colour of the reaction mixture.
- Now add a few drops of NaOH solution to the above mixture.

Now answer:

- What is the colour of the NaOH solution on adding phenolphthalein?
- What change in colour takes place on adding dilute HCl?

- Does the pink colour of phenolphthalein reappear on adding NaOH?
- Why do you think this has happened?

Discussion: On adding phenolphthalein to NaOH solution, the colour becomes pink. On adding dilute HCl solution dropwise, finally the pink colour disappears and the solution again becomes colourless. On adding NaOH, the pink colour reappears as the medium becomes basic again.

$$\text{Acid} + \text{Base} \rightarrow \text{Salt} + \text{Water}$$

The reaction is known as a neutralisation reaction.

Activity 2.7: To study the reaction of metal oxides with acids.

- Take a small amount of copper oxide in a beaker and add dilute hydrochloric acid slowly while stirring.
- Note the colour of the solution.

Now answer:

- What has happened to the copper oxide?

Discussion: It is observed that the colour of the solution becomes blue-green and the copper oxide dissolves. The blue-green colour of the solution is due to the formation of copper(II) chloride in the solution.

Conclusion: Metal oxides react with the acids to give salt and water.

For example,

$$\text{Metal oxide} + \text{Acid} \rightarrow \text{Salt} + \text{Water}$$

$$\underset{\text{Copper (II) oxide}}{CuO} + \underset{\text{Hydrochloric acid}}{2HCl} \rightarrow \underset{\substack{\text{Copper (II) chloride}\\ \text{(Blue-green)}}}{CuCl_2} + H_2O$$

This shows that metal oxides are basic or alkaline in nature.

Activity 2.8: To study the conduction of electricity by aqueous solutions of acid and bases.

- Take solutions of glucose, alcohol, hydrochloric acid, sulfuric acid, and so on.
- Fix two nails on a cork, and place the cork in a 100 mL beaker.
- Connect the nails to the two terminals of a 6 volt battery through a bulb and a switch, as shown in Fig. 2.15.
- Now pour some dilute HCl in the beaker and switch on the current.
- Repeat with dilute sulfuric acid.
- Repeat the experiment separately with glucose and alcohol solutions.

Now answer:

- What do you observe on passing electric current?
- What do you observe in the case of glucose and alcohol solution?
- Does the bulb glow in all cases?

Discussion: The bulb is found to glow in the case of the acids but not in the case of glucose and alcohol solutions.

Conclusion: As electric current is carried through the solution by ions, this shows that acids dissociate in the solution to give $H^+(aq)$ ions but substances like glucose and alcohol do not dissociate to give H^+ ions and so are not acidic. Hence, H^+ ions are responsible for the acidic properties.

Activity 2.9: To test that dry HCl gas is not acidic but its aqueous solution is acidic.

- Take about 1 g solid NaCl in a clean and dry test tube and set up the apparatus as shown in Fig. 2.16.
- Add some concentrated sulfuric acid to the test tube.
- Test the gas evolved successively with dry and wet blue litmus paper.

Now answer:

- In which case does the litmus paper change colour?
- On the basis of the above activity, what do you infer about the acidic character of:
  (i) dry HCl gas, (ii) HCl solution?

Discussion: Dry blue litmus paper does not change colour but wet blue litmus paper changes its colour to red.

Conclusion: Dry HCl gas is not acidic while HCl solution is acidic. This is because, in the presence of water, HCl dissociates to give H^+ ions.

$$HCl(g) + H_2O(l) \rightarrow \underset{\text{Hydronium ion}}{H_3O^+} \text{ or } H^+(aq) + \underset{\text{Chloride ions}}{Cl^-(aq)}$$

Activity 2.10: To test that dilution of a strong acid like H_2SO_4 and strong base like NaOH is exothermic.

- Take 10 mL water in a beaker.
- Add a few drops of concentrated H_2SO_4 to it and stir it slowly.
- Touch the base of the beaker.
- Repeat the above activity with sodium hydroxide pellets.

Caution: Add sulfuric acid into water for dilution and not water into sulfuric acid as otherwise the solution may splash out and cause burns.

Now answer:

- Is there a change in temperature in the two cases?
- Is this an exothermic or endothermic process?

Discussion: In both the cases, the temperature is found to rise.

Conclusion: Dilution of concentrated sulfuric acid with water as well as dissolution of NaOH in water is exothermic.

Activity 2.11: To test the pH values of the given solutions using the universal indicator.

- Test the pH values of the following solutions:
  1. Saliva (before meal) 2. Saliva (after meal)
  3. Lemon juice 4. Colourless aerated drink
  5. Carrot juice 6. Coffee
  7. Tomato juice 8. Tap water
  9. 1 M NaOH 10. 1 M HCl

Now answer:

- What is the nature of each substance on the basis of your observations?

Discussion: See table and colours of universal indicator.

Conclusion: Solution with pH < 7 but close to 7 are weakly acidic. Solutions with pH << 7 are strongly acidic. Solutions with pH > 7 but close to 7 are weakly basic. Solutions with pH >> 7 are strongly basic.

Activity 2.12: To test the pH of the soil in the backyard of your home.

- Put about 2 g soil in a test tube and add 5 mL water to it.
- Shake the contents of the test tube.
- Filter the contents and collect the filtrate in a test tube.
- Check the pH of this filtrate with the help of universal indicator paper.

Now answer:

- What can you conclude about the ideal soil pH for the growth of plants in your region?

Discussion: The ideal soil pH should be close to 7, that is, neither very acidic nor very alkaline.

Activity 2.13: To write the formulae of the salts and to identify their acids and bases and the families.

- Write down the formulae of the salts given below: (i) potassium sulfate, (ii) sodium sulfate, (iii) calcium sulfate, (iv) magnesium sulfate, (v) copper sulfate, (vi) sodium chloride, (vii) sodium nitrate, (viii) sodium carbonate, (ix) ammonium chloride.
- Identify the acids and bases from which the above salts may be obtained.
- Salts having the same positive or negative radicals are said to belong to a family. For example, NaCl and Na_2SO_4 belong to the family of sodium salts. Similarly, NaCl and KCl belong to the family of chloride salts.

Now answer:

- How many families can you identify among the salts given in this activity?

Discussion and conclusion:

Salt	Formula	Acid	Base
1. Potassium sulfate	K_2SO_4	H_2SO_4	KOH
2. Sodium sulfate	Na_2SO_4	H_2SO_4	NaOH
3. Calcium sulfate	$CaSO_4$	H_2SO_4	$Ca(OH)_2$
4. Magnesium sulfate	$MgSO_4$	H_2SO_4	$Mg(OH)_2$
5. Copper sulfate	$CuSO_4$	H_2SO_4	$Cu(OH)_2$
6. Sodium chloride	NaCl	HCl	NaOH
7. Sodium nitrate	$NaNO_3$	HNO_3	NaOH
8. Sodium carbonate	Na_2CO_3	H_2CO_3	NaOH
9. Ammonium chloride	NH_4Cl	HCl	NH_4OH

Families: (i) On the basis of common acids.
Sulfates: K_2SO_4, Na_2SO_4, $CaSO_4$, $MgSO_4$, $CuSO_4$
Chlorides: NaCl, NH_4Cl
Carbonates: Na_2CO_3
(ii) On the basis of common bases.
Sodium salts: Na_2SO_4, NaCl, $NaNO_3$, Na_2CO_3
Potassium salts: K_2SO_4
Calcium salts: $CaSO_4$
Magnesium salts: $MgSO_4$
Copper salts: $CuSO_4$
Ammonium salts: NH_4Cl

Activity 2.14: To find the pH of the given salt solutions.

- Collect the following salt samples: (i) sodium chloride, (ii) potassium nitrate, (iii) aluminium chloride, (iv) zinc sulfate, (v) copper sulfate, (vi) sodium acetate, (vii) sodium carbonate, (viii) sodium hydrogen carbonate.
- Check their solubility in water (use distilled water only).
- Check the action of these solutions on litmus and find the pH using a pH paper.
- Report your observations in a tabular form.

Now answer:

- Which of the salts are acidic, basic or neutral?

- Identify the acid or base used to form the salt.

Discussion and conclusion:

Salt	Formula	Solubility in water	Action on litmus	pH	Acid used	Base used
1. Sodium chloride	NaCl	Soluble	No action	7	HCl	NaOH
2. Potassium nitrate	KNO_3	Soluble	No action	7	HNO_3	KOH
3. Aluminium chloride	$AlCl_3$	Soluble	Turns red	<7	HCl	$Al(OH)_3$
4. Zinc sulfate	$ZnSO_4$	Soluble	Turns red	<7	H_2SO_4	$Zn(OH)_2$ or ZnO
5. Copper sulfate	$ZnSO_4$	Soluble	Turns red	<7	H_2SO_4	$Cu(OH)_2$ or CuO
6. Sodium acetate	CH_3COONa	Soluble	Turns blue	>7	CH_3COOH	NaOH
7. Sodium carbonate	Na_2CO_3	Soluble	Turns blue	>7	H_2CO_3	NaOH
8. Sodium hydrogen carbonate	$NaHCO_3$	Soluble	Turns blue	>7	H_2CO_3	NaOH

Activity 2.15: To test the presence of water of crystallisation in copper sulfate crystals.

- Heat a few crystals of copper sulfate in a dry boiling tube (Fig. 2.44).
- Note the colour of copper sulfate after heating.
- Observe the water droplets in the boiling tube.
- Add 2–3 drops of water on the sample of copper sulfate obtained after heating.

Now answer:

- What do you observe on heating the blue copper sulfate crystals?
- Is the blue colour of copper sulfate restored on adding water?

Discussion: Blue coloured copper sulfate crystals on heating leave behind white anhydrous copper sulfate and water droplets are seen in the upper cooler parts of the boiling tube. On adding 2–3 drops of water to the white residue, the blue colour reappears.

Conclusion: Copper sulfate crystals on heating lose water to form white anhydrous copper sulfate which combines with water to give blue coloured copper sulfate crystals.

Group Activities

Activity 2.1: To prepare your own acid–base indicator from beetroot extract.

- Crush beetroot in a mortar.
- Add sufficient amount of water to obtain the extract.
- Filter the extract.
- Collect the filtrate (which is cherry red in colour).
- Arrange four test tubes in a test tube stand and label them as A, B, C and D. Pour 2 mL each of lemon juice solution, soda water, vinegar and baking soda solution in them respectively.
- Put 2–3 drops of the beetroot extract in each test tube and note the colour change if any. Write your observations in the table as follows:

Substance	Lemon juice solution	Soda water	Vinegar	Baking soda solution
Colour in the presence of beetroot indicator				

Discussion: Beetroot extract contains a cherry red coloured dye which acts as an acid–base indicator.

Conclusion: The cherry red colour of the beetroot extract changes to one particular colour in the presence of acidic (sour) substances like lemon juice, vinegar and soda water, while it changes to a different particular colour in the presence of basic (bitter) substances like baking soda.

Activity 2.2: To prepare a soda–acid fire extinguisher.

- Take 20 mL of sodium hydrogen carbonate ($NaHCO_3$) solution in a wash bottle.
- Suspend an ignition tube containing dilute sulfuric acid in the wash bottle (check figure below).
- Close the mouth of the wash bottle.
- Tilt the wash bottle so that the acid from the ignition tube mixes with the sodium hydrogen carbonate solution below.
- It is noticed that an effervescence is coming out of the nozzle.
- Direct this gas on a burning candle.

Discussion: When sulfuric acid mixes into sodium hydrogen carbonate, a brisk reaction starts giving out CO_2 along with a stream of liquid with pressure through the nozzle.

$$2NaHCO_3(aq) + H_2SO_4(aq) \rightarrow Na_2SO_4(aq) + 2H_2O(l) + CO_2(g)$$

Conclusion: The fire (burning candle) is extinguished because CO_2 is heavier than air and is non-supporter of combustion. It surrounds the fire and cuts off the supply of air.

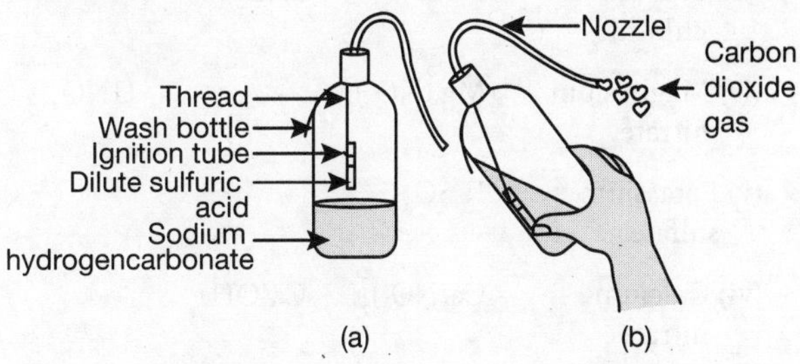

Exemplar Problems

Short Answer Questions

1. Match the acids given in Column I with their correct source given in Column II

Column I	Column II
(a) Lactic acid	(i) Tomato
(b) Acetic acid	(ii) Lemon
(c) Citric acid	(iii) Vinegar
(d) Oxalic acid	(iv) Curd

2. Match the important chemicals given in Column (A) with the chemical formulae given in Column (B)

Column I	Column II
(a) Plaster of Paris	(i) $Ca(OH)_2$
(b) Gypsum	(ii) $CaSO_4.1/2\ H_2O$
(c) Bleaching powder	(iii) $CaSO_4.2H_2O$
(d) Slaked lime	(iv) $CaOCl_2$

3. What will be the action of the following substances on litmus paper? Dry HCl gas, moistened NH_3 gas, lemon juice, carbonated soft drink, curd, soap solution.

4. Name the acid present in ant sting and give its chemical formula. Also give the common method to get relief from the discomfort caused by the ant sting.

5. What happens when nitric acid is added to eggshells?

6. A student prepared solutions of (i) an acid and (ii) a base in two separate beakers. She forgot to label the solutions and litmus paper is not available in the laboratory. Since both the solutions are colourless, how will she distinguish between the two?

7. How would you distinguish between baking powder and washing soda by heating?

8. Salt A, commonly used in bakery products, on heating gets converted into another salt B which itself is used for the removal of hardness of water and a gas C is evolved. The gas C when passed through limewater, turns it milky. Identify A, B and C.

9. In one of the industrial processes used for the manufacture of sodium hydroxide, a gas X is formed as a byproduct. The gas X reacts with limewater to give a compound Y which is used as a bleaching agent in the chemical industry. Identify X and Y giving the chemical equations of the reactions involved.

10. Fill in the missing data in the following table.

Name of the salt	Formula	Salt obtained from	
		Base	Acid
(i) Ammonium chloride	NH_4Cl	NH_4OH	—
(ii) Copper sulfate	—	—	H_2SO_4
(iii) Sodium chloride	NaCl	NaOH	—
(iv) Magnesium nitrate	$Mg(NO_3)_2$	—	HNO_3
(v) Potassium sulfate	K_2SO_4	—	—
(vi) Calcium nitrate	$Ca(NO_3)_2$	$Ca(OH)_2$	—

11. What are strong and weak acids? In the following list of acids, separate strong acids from weak acids. Hydrochloric acid, citric acid, acetic acid, nitric acid, formic acid, sulfuric acid.

12. When zinc metal is treated with a dilute solution of a strong acid, a gas is evolved, which is utilised in the hydrogenation of oil. Name the gas evolved. Write the chemical equation of the reaction involved and also write a test to detect the gas formed.

Long Answer Questions

1. In the following schematic diagram for the preparation of hydrogen gas as shown in Fig. 2.3, what would happen if the following changes are made?

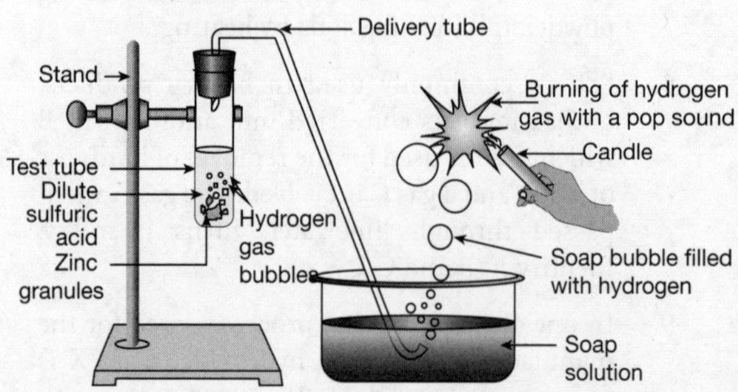

(a) In place of zinc granules, the same amount of zinc dust is taken in the test tube.

(b) Instead of dilute sulfuric acid, dilute hydrochloric acid is taken.

(c) In place of zinc, copper turnings are taken.

(d) Sodium hydroxide is taken in place of dilute sulfuric acid and the tube is heated.

2. For making cake, baking powder is used. If at home your mother uses baking soda instead of baking powder in cake,

(a) how will it affect the taste of the cake and why?

(b) how can baking soda be converted into baking powder?

(c) what is the role of tartaric acid added to baking soda?

3. A metal carbonate X on reacting with an acid gives a gas which when passed through a solution Y gives the carbonate back. On the other hand, a gas G that is obtained at the anode during electrolysis of brine is passed on dry Y, it gives a compound Z, used for disinfecting drinking water. Identity X, Y, G and Z.

4. A dry pellet of a common base B, when kept in the open absorbs moisture and turns sticky. The compound is also a byproduct of the chlor-alkali process. Identify B. What type of reaction occurs when B is treated with an acidic oxide? Write a balanced chemical equation for one such solution.

5. A sulfate salt of a group 2 element of the periodic table is a white, soft substance, which can be moulded into different shapes by making its dough. When this compound is left in the open for some time, it becomes a solid mass and cannot be used for moulding purposes. Identify the sulfate salt and why does it show such a behaviour? Give the reaction involved.

6. Identify the compound X on the basis of the reactions given below. Also, write the name and chemical formulae of A, B and C.

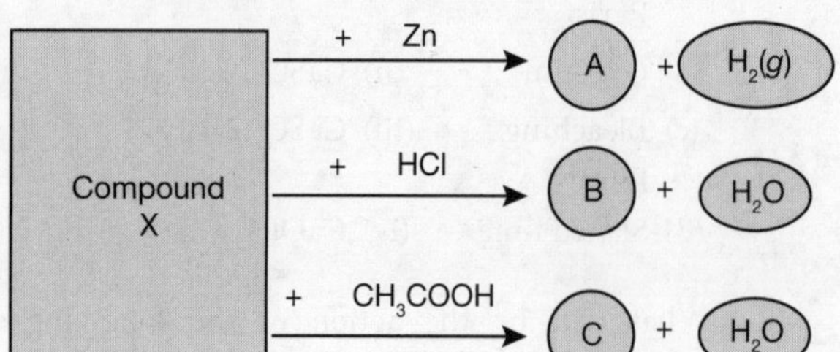

Answers and Solutions

Short Answer Questions

1. (a)—(iv), (b)—(iii), (c)—(ii), (d)—(i)
2. (a)—(ii), (b)—(iii), (c)—(iv), (d)—(i)
3. Substance action on litmus paper
   Dry HCl gas: No change
   Moistened NH_3 gas: Turns red to blue
   Lemon juice: Turns blue to red
   Carbonated soft drink: Turns blue to red
   Curd: Turns blue to red
   Soap solution: Turns red to blue
4. The acid present in ant sting is methanoic acid (formic acid). The chemical formula is HCOOH. To get relief, one should apply any available basic salt such as baking soda ($NaHCO_3$) on it.
5. Egg shells contain calcium carbonate. When nitric acid is added to it, carbon dioxide gas is evolved. The reaction can be given as $CaCO_3 + 2HNO_3 \rightarrow Ca(NO_3)_2 + H_2O + CO_2$
6. **Hint:** Using a chemical indicator like phenolphthalein or natural indicators like turmeric, china rose.
7. The chemical formula of baking powder is sodium hydrogen carbonate ($NaHCO_3$), whereas that of washing soda is sodium carbonate ($Na_2CO_3.10H_2O$). Sodium hydrogen carbonate on heating gives CO_2 gas which will turn limewater milky whereas no such gas is obtained from sodium carbonate.
   $$2NaHCO_3 \xrightarrow{\Delta} Na_2CO_3 + H_2O + CO_2$$
   $$Na_2CO_3.10H_2O \xrightarrow{\Delta} Na_2CO_3 + 10H_2O$$
8. Baking powder ($NaHCO_3$), salt A, is commonly used in bakery products. On heating, it forms sodium carbonate (Na_2CO_3), B and CO_2 gas, C is evolved. When CO_2 gas is passed through limewater it forms calcium carbonate ($CaCO_3$), which is slightly soluble in water making it milky.
   A: $NaHCO_3$
   B: Na_2CO_3
   C: CO_2 gas
9. In the manufacture of sodium hydroxide, hydrogen gas and chlorine gas (X) are formed as byproducts. When chlorine gas (X) reacts with lime water, it forms calcium oxychloride (bleaching powder) Y. The reactions are
   $$2NaCl(aq) + 2H_2O(l) \rightarrow 2NaOH(aq) + Cl_2(g) + H_2(g)$$
   $X \rightarrow Cl_2$ (Chlorine gas)
   $$Ca(OH)_2(s) + Cl_2(g) \rightarrow CaOCl_2(s) + H_2O$$
   Y: Calcium oxychloride (bleaching powder)
10.

Name of the salt	Formula	Salt obtained from	
		Base	Acid
(i) Ammonium chloride	NH_4Cl	NH_4OH	HCl
(ii) Copper sulfate	$CuSO_4$	$Cu(OH)_2$	H_2SO_4
(iii) Sodium chloride	NaCl	NaOH	HCl
(iv) Magnesium nitrate	$Mg(NO_3)_2$	$Mg(OH)_2$	HNO_3
(v) Potassium sulfate	K_2SO_4	KOH	H_2SO_4
(vi) Calcium nitrate	$Ca(NO_3)_2$	$Ca(OH)_2$	HNO_3

11. In aqueous solutions, strong acids ionise completely and provide hydronium ions. On the other hand, weak acids are partially ionised and an aqueous solution of the same molar concentration provides a much smaller concentration of H_3O^+ ions. Strong acids: hydrochloric acid, sulfuric acid, nitric acid; weak acids: citric acid, acetic acid, formic acid.
12. When zinc reacts with a dilute solution of a strong acid, it forms a salt and hydrogen gas is evolved. $Zn + 2HCl \rightarrow ZnCl_2 + H_2$
    When a burning splinter is brought near the mouth of the test tube, the gas burns with a pop sound.

Long Answer Questions

1. **Hint:** (a) Hydrogen gas will evolve with greater speed.
   (b) Almost the same amount of gas is evolved.
   (c) Hydrogen gas is not evolved.
   (d) If sodium hydroxide is taken, hydrogen gas will be evolved.

$$Zn + 2NaOH \rightarrow \underset{\text{Sodium zincate}}{Na_2ZnO_2} + H_2$$

2. (a) Baking soda is sodium hydrogen carbonate. On heating, it is converted into sodium carbonate which is bitter in taste.

$$2NaHCO_3 \xrightarrow{\text{Heat}} Na_2CO_3 + H_2O + CO_2$$

(b) Baking soda can be converted into baking powder by the addition of an appropriate amount of tartaric acid to it.

(c) The role of tartaric acid is to neutralise the sodium carbonate formed and the cake will not taste bitter.

3. The gas evolved at the anode during electrolysis of brine is chlorine (G). When chlorine gas is passed through dry $Ca(OH)_2$ (Y) produces bleaching powder (Z) used for disinfecting drinking water.

$$\underset{\text{Slaked lime}}{Ca(OH)_2} + Cl_2 \rightarrow \underset{\text{Bleaching powder}}{CaOCl_2} + H_2O$$

Since Y and Z are calcium salts, X is also a calcium salt and is calcium carbonate.

$$CaCO_3 + 2HCl \rightarrow CaCl_2 + CO_2 + H_2O$$
$$Ca(OH)_2 + CO_2 \rightarrow CaCO_3 + H_2O$$

4. Sodium hydroxide (NaOH) is a commonly used base and is hygroscopic, that is, it absorbs moisture from the atmosphere and becomes sticky. The acidic oxides react with the base to give a salt and water. The reaction between NaOH and CO_2 can be given as:

$$2\,NaOH + CO_2 \rightarrow Na_2CO_3 + H_2O$$

5. The substance which is used for making different shapes is plaster of Paris. Its chemical name is calcium sulfate hemihydrate ($CaSO_4 \cdot H_2O$). Two formula units of $CaSO_4$ share one molecule of water. As a result, it is soft. When it is left open for some time, it absorbs moisture from the atmosphere and forms gypsum, which is a hard solid mass.

$$\underset{\substack{\text{Plaster of Paris}\\ \text{(Soft)}\\ \text{(Sulfate salt)}}}{CaSO_4 \cdot \frac{1}{2}H_2O} + 1\frac{1}{2}H_2O \rightarrow \underset{\substack{\text{Gypsum}\\ \text{(Hard mass)}}}{CaSO_4 \cdot 2H_2O}$$

6. X: NaOH (Sodium hydroxide)
A: Na_2ZnO_2 (Sodium zincate)
B: NaCl (Sodium chloride)
C: CH_3COONa (Sodium acetate)

NCERT CORNER

In-Text Questions

1. You have been provided with three test tubes. One of them contains distilled water and the other two contain an acidic solution and a basic solution, respectively. If you are given only red litmus paper, how will you identify the contents of each test tube?

2. Why should curd and sour substances not be kept in brass and copper vessels?

3. Which gas is usually liberated when an acid reacts with a metal? Illustrate with an example. How will you test for the presence of this gas?

4. Metal compound A reacts with dilute hydrochloric acid to produce effervescence. The gas evolved extinguishes a burning candle. Write a balanced chemical equation for the reaction if one of the compounds formed is calcium chloride.

5. Why do HCl, HNO_3, and so on, show acidic character in aqueous solutions while solutions of compounds like alcohol and glucose do not show acidic character?

6. Why does an aqueous solution of acid conduct electricity?

7. Why does dry HCl gas not change the colour of the dry litmus paper?

8. While diluting an acid, why is it recommended that the acid should be added to water and not water to the acid?

9. How is the concentration of hydronium ions H_3O^+ affected when a solution of an acid is diluted?

10. How is the concentration of hydroxide ions OH^- affected when excess base is dissolved in a solution of sodium hydroxide?

11. You have two solutions, A and B. The pH of solution A is 6 and the pH of solution B is 8. Which solution has more hydrogen ion concentration? Which of this is acidic and which one is basic?

12. What effect does the concentration of $H^+(aq)$ ions have on the nature of the solution?

13. Do basic solutions also have $H^+(aq)$ ions? If yes, then why are these basic?

14. Under what soil condition do you think a farmer would treat the soil of his fields with quicklime (calcium oxide) or slaked lime (calcium hydroxide) or chalk (calcium carbonate)?

15. What is the common name of the compound $CaOCl_2$?

16. Name the substance which on treatment with chlorine yields bleaching powder.

17. Name the sodium compound which is used for softening hard water.

18. What will happen if a solution of sodium hydrogen carbonate is heated? Give the equation of the reaction involved.

19. Write an equation to show the reaction between plaster of Paris and water.

Textbook Exercises

1. A solution turns red litmus blue; its pH is likely to be:
   (a) 1 (b) 4
   (c) 5 (d) 10

2. A solution reacts with crushed eggshells to give a gas that turns limewater milky. The solution contains:
   (a) NaCl (b) HCl
   (c) LiCl (d) KCl

3. 10 mL of a solution of NaOH is found to be completely neutralised by 8 mL of a given solution of HCl. If we take 20 mL of the same solution of NaOH, the amount of HCl solution (the same solution as before) required to neutralise it will be:
   (a) 4 mL (b) 8 mL
   (c) 12 mL (d) 16 mL

4. Which of the following types of medicines is used for treating indigestion?
   (a) antibiotic (b) analgesic
   (c) antacid (d) antiseptic

5. Write word equations and then balanced equations for the reactions taking place when
   (i) dilute sulfuric acid reacts with zinc granules
   (ii) dilute hydrochloric acid reacts with magnesium ribbon
   (iii) dilute sulfuric acid reacts with aluminium powder
   (iv) dilute hydrochloric acid reacts with iron filings

6. Compounds such as alcohols and glucose also contain hydrogen but are not categorised as acids. Describe an activity to prove it.

7. Why does distilled water not conduct electricity, whereas rainwater does?

8. Why do acids not show acidic behaviour in the absence of water?

9. Five solutions A, B, C, D and E, when tested with universal indicator, showed the pH as 4, 1, 11, 7 and 9, respectively. Which solution is
   (i) neutral?
   (ii) strongly alkaline?
   (iii) strongly acidic?
   (iv) weakly acidic?
   (v) weakly alkaline?
   Arrange the pH in increasing order of hydrogen ion concentration.

10. Equal lengths of magnesium ribbons are taken in test tubes A and B. Hydrochloric acid (HCl) is added to test tube A, while acetic acid (CH_3COOH) is added to test tube B. Amount and concentration taken for both the acids are the same. In which test tube will the fizzing occur more vigorously and why?

11. Fresh milk has a pH of 6. How do you think the pH will change as it turns into curd? Explain your answer.

12. A milkman adds a very small amount of baking soda to fresh milk.
    (i) Why does he shift the pH of the fresh milk from 6 to slightly alkaline?
    (ii) Why does this milk take a long time to set as curd?

13. Plaster of Paris should be stored in a moisture-proof container. Explain why.

14. What is a neutralisation reaction? Give two examples.

15. Give two important uses of washing soda and baking soda.

Answers and Solutions

In-Text Questions

1. Few drops of each solution from the test tubes are added to red litmus paper separately.
   (i) The solution which turns red litmus to blue is the basic solution. Use this blue litmus paper to test the solutions in the other two test tubes.
   (ii) The solution which turns blue litmus paper to red will be the acidic solution.
   (iii) The solution which does not change the colour of either red or blue litmus paper contains distilled water.

2. Curd and sour substances contain acids. These acids can react with brass and copper vessels to form toxic compounds. This makes the substances unfit for human consumption. Hence, they are not kept in brass and copper vessels.

3. Hydrogen gas is usually liberated when an acid reacts with a metal. For example,

   $$Zn(s) + 2HCl(aq) \rightarrow ZnCl_2(aq) + H_2(i)$$

   The equation given above illustrates how zinc reacts with dilute hydrochloric acid resulting in the liberation of hydrogen gas and the formation of the salt, zinc chloride. A burning matchstick, when brought near the mouth of the test tube where H_2 gas is being released makes a pop sound. This confirms the presence of hydrogen gas.

4. Metal compound A is calcium carbonate ($CaCO_3$). When A reacts with dilute hydrochloric acid it produces an effervescence. The chemical equation is given as: $CaCO_3(s) + 2HCl(aq) \rightarrow CaCl_2(aq) + CO_2(g) + H_2O(l)$
   The gas evolved is CO_2. CO_2 extinguishes a burning candle.

5. When HCl or HNO_3 are mixed with water, they dissolve in water to form H^+ or H_3O^+ ions which show their acidic character. For example,

   $$HCl(aq) \rightarrow H^+(aq) + Cl^-(aq)$$

   $$H^+(aq) + H_2O(l) \rightarrow H_3O + (aq)$$

   Solutions of compounds like alcohol and glucose do not dissolve to form any such ions. Hence, they do not show any acidic character.

6. An acid dissolves in water and dissociates to form H^+ or H_3O^+ ions. Electricity is conducted through these moving ions.

7. Dry HCl gas does not dissociate into H^+ ions and hence, the acidic property of the gas is not observed. So, dry HCl gas does not change the colour of the dry litmus paper.

8. The process of adding water to acid is highly exothermic. If not done very carefully, the acid may splash, cause burns and even the bottle/container can break due to the vigorous reaction. To avoid this and to allow the heat evolved to be absorbed by water slowly, the acid is added to water for diluting it. This significantly reduces the risk of splashing.

9. When a solution of acid is diluted, the concentration of the hydronium ions H_3O^+ per unit volume decreases. On adding water, the H^+ ions of the acid and hydroxyl ions of water react to form water molecules and the concentration of the hydronium ions decreases.

10. When the base is mixed with sodium hydroxide solution, there is an increase in the number of hydroxide ions whereas the volume remains almost the same. This leads to an increase in the concentration of OH^- ions per unit volume.

11. Solutions with pH < 7 are said to be acidic. Solutions with pH = 7 are neutral and pH >7 are basic.
    Therefore, Solution A with pH = 6 is acidic and Solution B with pH = 8 is basic.
    So A has more hydrogen ion concentration.

12. The higher the concentration of $H^+(aq)$ ions, the more is its acidic nature.

13. Yes, basic solutions also have H^+ ions, but they are basic in nature due to a greater number of OH^- ions than H^+ ions.

14. Quicklime (calcium oxide), slaked lime (calcium hydroxide) and chalk (calcium carbonate) are all basic in nature. When the soil becomes more acidic in nature, the farmer would add quicklime (CaO), slaked lime ($CaOH_2$) or chalk ($CaCO_3$) to make it neutral. This helps the farmer to make the soil fit to grow his crops.

15. $CaOCl_2$ (chemical name is calcium oxychloride) is commonly called bleaching powder.

16. Bleaching powder is prepared by treating calcium hydroxide ($CaOH_2$) with chlorine. The chemical equation for the reaction is:

$$Ca(OH)_2 + Cl_2 \rightarrow CaOCl_2 + H_2O$$

17. Sodium carbonate is commonly used for softening hard water.

18. When sodium hydrogen carbonate ($NaHCO_3$) is heated, sodium carbonate, water and carbon dioxide gas are obtained. The chemical equation for the reaction is

$$2NaHCO_3 + heat \rightarrow Na_2CO_3 + H_2O + CO_2$$

19. Plaster of Paris is $CaSO_4\frac{1}{2}H_2O$. The reaction between the Plaster of Paris and water is given as

$$\underset{\text{Plaster of Paris (white powder)}}{CaSO_4\frac{1}{2}H_2O} + \frac{3}{2}H_2O \rightarrow \underset{\text{Gypsum (solid mass)}}{CaSO_4 \cdot 2H_2O}$$

Textbook Exercises

1. (d) A solution which turns the colour of red litmus to blue must be basic in nature. A basic solution has a pH value of greater than 7. The only option with a pH greater than 7 is 10 which is the correct answer.

2. (b) Eggshells are made up of calcium carbonate, which on reaction with HCl liberates CO_2 gas. When CO_2 reacts with lime water [$Ca(OH)_2$], it forms calcium carbonate which gives the solution a milky appearance.

$$CaCO_3 + 2HCl \rightarrow CaCl_2 + H_2O + CO_2$$

$$Ca(OH)_2 + CO_2 \rightarrow CaCO_3 + H_2O$$

3. (d) The number of hydroxide ions and hydrogen ions is directly proportional to the volume of their respective solutions. Hence, the problem can be solved by the unitary method.
10 mL NaOH reacts with = 8 mL HCl
Therefore, 20 mL NaOH reacts with = $(8/10)\times20$
= 16 mL HCl
∴ 16 mL of HCl solution will be required to completely neutralise 20 mL of NaOH solution.

4. (c) An antacid is used for the treatment of indigestion which is caused due to excessive acidity in the stomach. The antacid is basic in nature and thus, it neutralises the acidity in the stomach.

5. (i) Sulfuric acid + Zinc → Zinc sulfate + Hydrogen
$H_2SO_4(aq) + ZnS(s) \rightarrow ZnSO_4(aq) + H_2(g)$
(ii) Hydrochloric acid + Magnesium → Magnesium chloride + Hydrogen
$2HCl(aq) + MgS(s) \rightarrow MgCl_2(aq) + H_2(g)$
(iii) Sulfuric acid + Aluminium → Aluminium sulfate + Hydrogen
$3H_2SO_4(aq) + 2Al(s) \rightarrow Al_2(SO_4)3(aq) + 3H_2(g)$
(iv) Hydrochloric acid + Iron (Ferrous) chloride + Hydrogen
$6HCl(aq) + 2Fe(s) \rightarrow 2FeCl_2(aq) + 3H_2(g)$

6. Two nails are fitted on a cork and are kept in a 100 mL beaker. The nails are then connected to the two terminals of a 6-volt battery through a bulb and a switch. Some dilute HCl is poured into the beaker and the current is switched on. The same experiment is then performed with a glucose solution and an alcohol solution.
Observations: It is observed that the bulb glows in the HCl solution and does not glow in the glucose or alcohol solution.

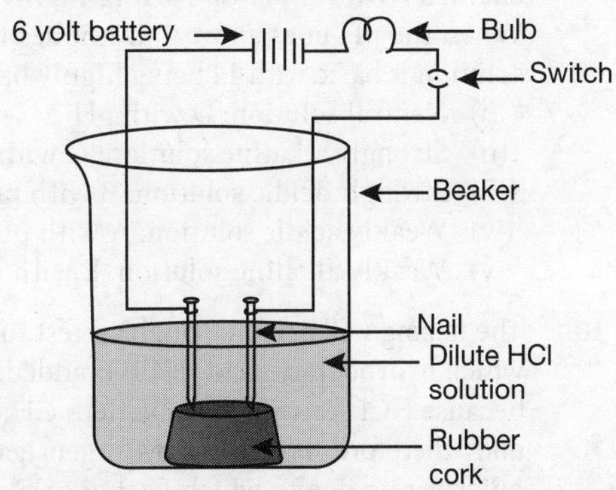

Result: HCl dissociates into H^+ and Cl^- ions. These ions conduct electricity in the solution resulting in the glowing of the bulb. On the other hand, neither the glucose solution nor the alcohol solution dissociates into ions. Therefore, these two solutions do not conduct electricity.

Conclusion: From this activity, it can be concluded that all acids generate hydrogen ions but not all compounds containing hydrogen are acids. That is why, though alcohols and glucose contain hydrogen, they are not categorised as acids.

7. Distilled water is a pure form of water, its pH is 7 and it is devoid of any ionic species. Therefore, it does not conduct electricity. Rainwater, being an impure form of water, contains CO_2, SO_2, NO. These oxide gases react with water and get chemically converted into acids which are responsible for the release of ions that conduct electricity. It contains many ionic species (acids) whose pH is less than 7 and therefore it conducts electricity.
$SO_2 + H_2O \rightarrow H_2SO_3$ (Sulfuric acid)
$H_2O + 2NO \rightarrow HNO_3 + O_2$ (Nitrous acid)
$CO_2 + H_2O \rightarrow H_2CO_3$ (Carbonic acid)

8. A substance is said to be acidic when it can generate hydrogen ions (H^+). The hydrogen ions generally come from the acid which dissociates in the presence of water. Thus, for the acid to dissociate into the hydrogen ion and the respective anion, water must be present. Hence, an acid fails to show its acidic behaviour in the absence of water.

9. At 25°C, the pH of neutral solutions = 7
As the pH falls below 7, it denotes acidic character with a pH of 1 being highly acidic. When the pH goes above 7, it implies that the solution is basic with 14 being highly basic.
   (i) Neutral solution: D with pH 7
   (ii) Strongly alkaline solution: C with pH 11
   (iii) Strongly acidic solution: B with pH 1
   (iv) Weakly acidic solution: A with pH 4
   (v) Weakly alkaline solution: E with pH 9

10. The fizzing will occur strongly in test tube A in which hydrochloric acid (HCl) is added. This is because HCl is a stronger acid than CH_3COOH and therefore produces hydrogen gas at a quicker rate due to which fizzing occurs more vigorously in test tube A.
$Mg + 2HCl \rightarrow MgCl_2 + H_2\uparrow$ (faster)
$Mg + 2CH_3COOH \rightarrow Mg(CH_3COO)_2 + H_2$ (slower)

11. The pH of milk is 6. As it gets converted to curd (pH between 4.5 to 5.5) the pH will reduce because curd has lactic acid which is more acidic in nature. Thus, the pH drops below 6 when milk gets converted to curd.

12. (i) The milkman adds a small amount of baking soda to shift the pH of the fresh milk from 6 to slightly alkaline because in the alkaline condition, milk does not set as curd easily and this helps in preventing the milk from getting converted to curd.
   (ii) Since this milk is slightly more basic than usual milk, the acids produced to set the curd are neutralised by the base. Only when the alkalinity of milk is neutralised, the process of curd formation begins. Therefore, it takes a longer time for the curd to set.

13. Plaster of Paris (POP) should be stored in a moisture-proof container because plaster of Paris, a powdery mass, absorbs water (moisture) to form a hard solid known as gypsum. $CaSO_4.H_2O_{\text{Plaster of Paris}} + H_2O_{\text{Water}} \rightarrow CaSO_4.2H_2O_{\text{Gypsum [Hard solid]}}$

14. A reaction in which an acid and a base react with each other to give a salt and water is termed as a neutralisation reaction. In this reaction, energy is evolved in the form of heat and thus, such reactions are usually exothermic by nature.
Example:
   (i) NaOH (Base) + HCl (Acid) → NaCl (Salt) + H_2O (Water)
   (ii) During indigestion (caused due to the production of excess of hydrochloric acid in the stomach), we administer an antacid (generally milk of magnesia, $MgOH_2$, which is basic in nature). The antacid neutralises the excess acid and thus gives relief from indigestion.
   $Mg(OH)_2 + 2HCl \rightarrow MgCl_2 + 2H_2O$

15. Two important uses of washing soda are:
   - it is used in glass, soap and paper industries, and
   - it is used to remove the permanent hardness of water.

   Two important uses of baking soda are the following.
   - It is used in baking. Baking powder is a mixture of baking soda and a mild acid known as tartaric acid. When it is heated or mixed with water, it releases CO_2 that makes bread or cake fluffy.
   - It is used in soda–acid fire extinguishers.

CHAPTER 3

Metals and Non-metals

Learning Objectives

After studying this unit, you will be able to understand:

- the characteristic and distinguishing properties of metals and non-metals
- the importance of the reactivity series of metals
- why atoms combine
- chemical bond formation and types of bonds
- electron dot representation, properties of ionic and covalent compounds
- the terms mineral, ore, concentration, calcination, roasting, refining
- corrosion and its prevention
- the composition, properties and uses of important alloys

3.1 Introduction

Today 118 chemical elements are known to exist. Many of them are found to occur in nature while some of them have been artificially prepared. Such elements obtained by artificial methods are known as synthetic elements. For example, elements with atomic number above 92 are synthetic and commonly known as transuranium elements. These elements have similarities as well as dissimilarities in their properties. On the basis of the properties of the elements they can be further divided into two main categories—metals and non-metals. Both types of elements have a great importance in our daily lives. For example (Fig. 3.1),

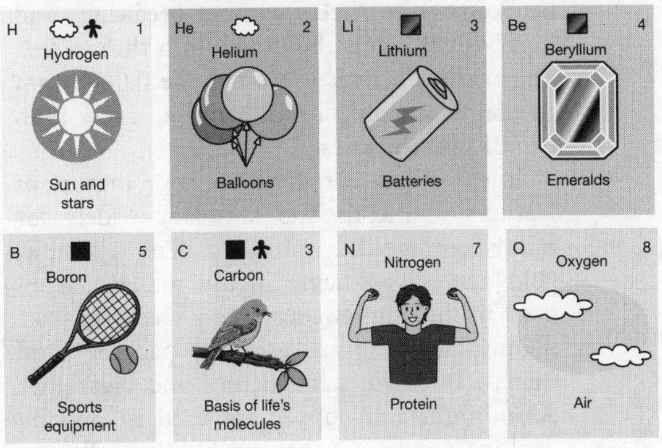

Fig. 3.1 Uses of the first eight elements

3.2 Comparative Study of Metals and Non-Metals

For conveniently studying them, elements can be roughly divided into non-metals, metalloids and metals. The majority of elements are metals and they occupy the largest portion of the periodic table. In general, metals are malleable, ductile, conductors and lose electrons, while non-metals are non-malleable, non-ductile, non-conductors and accept electrons.

3.2.1 Metals

Those elements which have 1, 2 or 3 electrons in their valence shells to lose, are lustrous, malleable, ductile (can be drawn into sheets or pulled into wires) good conductors of heat and electricity and have high densities and tensile strength are metals. For example, except hydrogen, the rest of the *s*-block elements are metals. All elements of the *d*- and *f*-block are metals. The elements at the bottom of the *p*-block are also metals such as Pb, Bi, Po. Some common and highly used metals in our daily life are aluminium (Al), iron (Fe), copper (Cu), silver (Ag), gold (Au), platinum (Pt) and tin (Sn).

The metals can be arranged on the basis of their electronic configuration as in Table 3.1.

Table 3.1 Position of metals in the periodic table

Group	Metals
1 (IA)	The alkali metals
2 (IIA)	The alkaline earth metals
3–10 (IIIB–VIIIB)	The transition metals
11 (IB)	The coinage metals (called so because they were used to make coins earlier)
12 (IIB)	The zinc group metals
Lower part of 13–16 (IIIA–VIA)	Metals like Al, Sn, Pb …

Physical Properties

Metals are characterised by some specific physical properties which differentiate them from non-metals.

(i) **Physical state:** Metals exist in the solid state as strong metallic bonds are present in the metal lattices. Metallic bonds are strong interactions which hold the metal atoms together. More the strength of the metallic bond, the closer are the metal atoms and more will be the solid nature. Metallic bond strength depends upon the number of valence electrons; as the number of valence electrons increase, the metallic bond strength and solid nature increase. For example, alkali metals (Li, La) have only one valence electron and therefore they are soft metals (cuttable with knife) while metals like iron and tin are hard. Mercury and gallium are in the liquid state and have low melting points due to weak metallic bonds.

(ii) **Colour:** Metals are usually silver or grey in colour with the exception of copper (reddish-brown) and gold (yellow).

(iii) **Lustrous nature:** Metals are shiny with a bright metallic lustre because of the to and fro oscillation of the mobile (free) electrons. They can be polished to give a highly reflective surface. The shiny look of metals makes them highly useful in making jewellery and decorative articles. For example, gold and silver are very widely used in making jewellery. As they are quite shiny they are good reflectors of light and so are used in making mirrors (silver mirror). It is important to note that these metals lose their brightness on keeping in the air for a long duration and they become dull due to the formation of a thin layer of oxide, sulfide or carbonate on the metal surface, that is, the metal surface has been corroded. However, on rubbing the metal surface with a sand paper, the layer is removed and metal becomes shiny again.

(iv) **Sonorous nature:** Metals are sonorous, so a metal makes a ringing sound when it is struck. Due to their sonorous nature they are used for making wires or strings for musical instruments such as violins, guitars and sitars, and plate-type musical instruments such as cymbals.

(v) **Hardness, density, melting point and boiling point:** Metals are generally strong, hard and have high values of densities, melting points and boiling points due to strong metallic bonds (interaction between metal atoms).

Density, MP, BP $\propto$ Strength of metallic bond $\propto$ Number of valence electrons

For example, metals like iron and copper are quite hard, that is, they cannot be cut with a knife and the melting points of iron and copper are 1808 K and 1356 K respectively. Metals like iron (steel) can hold large weights without snapping, which means they are very strong and can be used in the construction of bridges, railway lines, grinding machines.

The exceptions are the alkali metals (Na, K) which are soft, cuttable with a knife and have low densities and low melting points due to weak metallic bonds (as they have only 1 valence electron). Gallium and cesium also have very low melting points so they can melt even in our hands. *Iron has a very high density of 7.8 g/cm^3 while sodium has a very low density of 0.97 g/cm^3.

(vi) **Malleable and ductile nature:** Metals (except Zn, Hg and Ga) are highly malleable (can be beaten into sheets without breaking) and highly ductile (can be drawn into thin wires). For example, silver wires may be 0.0008 mm in diameter and gold foil can be of less than 0.0001 mm thickness.

These two properties are important features of metals and are used widely for many commercial purposes. For example, gold and silver, being highly malleable, are used in making jewellery, thin sheets or foils. Aluminium foils are used for packing food items like biscuits, medicines and cigarettes. Aluminium and copper are used in making cooking utensils as they are highly malleable. Copper and aluminium are very widely used in making wires as they are highly ductile and conducting also. Iron, magnesium and tungsten are quite ductile—iron is used for making wire gauzes, magnesium wires are used in science experiments in labs and tungsten is used in making the filaments of electric bulbs.

(vii) **Tensile strength:** Metals have a very high tensile strength. Tensile strength is the resistance of any metal (such as steel) to breaking under tensile tension.

Tensile strength $\propto$ Metallic bond strength

Due to high tensile strength, steel is used in making wire bridges.

(viii) **Thermal and electrical conductivity:** Metals (except tungsten) are good conductors of heat, electricity and sound due to the presence of free electrons in the metal lattices.

Thermal conduction: Metals are good conductors of heat, that is, they allow heat to pass through them easily. When a metal is heated, its atoms gain heat energy and start vibrating. This energy is transferred to the electrons present in the atoms due to which the movement of the electrons starts. They can transfer energy to other electrons and atoms of the metal to cause thermal or heat conduction. For example, silver has the highest thermal conductivity followed by copper. That is why these metals are used in making cooking utensils and water boilers. Lead and mercury are the poorest conductors of heat.

Electrical conduction: Metals are good conductors of electricity, that is, they allow electricity to pass through them easily. When a metal foil (aluminium foil) is connected with a bulb and battery (Fig. 3.2), the electric bulb lights up showing that the metal foil conducts electric current. For example, silver has the highest electrical conductivity followed by copper. That is why these metals are used in making electrical wires. These electric wires that carry current are covered by a plastic-like substance PVC (polyvinyl chloride) which is an insulator, that is, it does not allow electric current to pass through it. PVC coating is used to prevent any electric shock.

Metals like lead and iron are the poorest conductors of electricity.

Uses of metals: Metals and their alloys (homogenous mixture of metals) are very widely used in our daily lives in making cooking utensils, electric appliances, vehicles, and so on (Fig. 3.3).

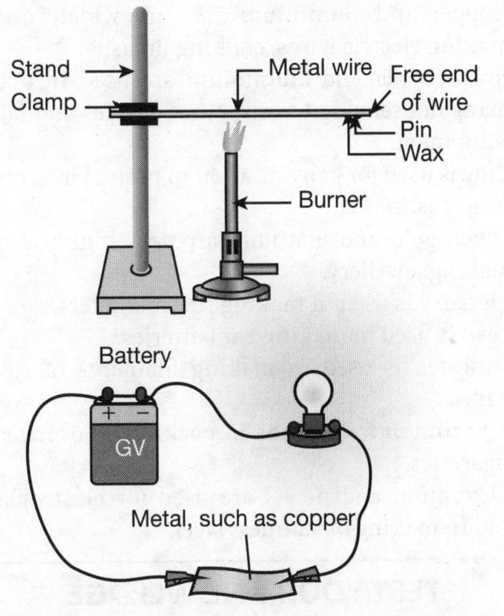

Fig. 3.2 Thermal and electrical conductance

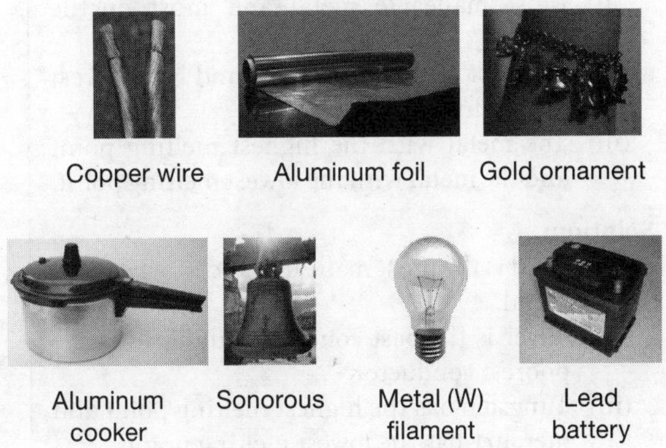

Fig. 3.3 Metals are very widely used in making cooking utensils, wires, electric appliances, vehicles, and so on

Source: Copper wire – By RobertKuhlmann - Own work, CC BY-SA 3.0, https://commons.wikimedia.org/w/index.php?curid=17659439; Aluminum foil – By MdeVicente - Own work, CC0, https://commons.wikimedia.org/w/index.php?curid=23122025; Gold ornament – By Mabsal - Own work, Public Domain, https://commons.wikimedia.org/w/index.php?curid=8852013; Aluminum cooker – Phrontis, CC BY-SA 3.0 <https://creativecommons.org/licenses/by-sa/3.0>, via Wikimedia Commons; Sonorous – CC BY-SA 2.0 fr, https://commons.wikimedia.org/w/index.php?curid=353742; Metal (W) filament – KMJ, alpha masking by Edokter, CC BY-SA 3.0 <http://creativecommons.org/licenses/by-sa/3.0/>, via Wikimedia Commons ; Lead battery – Cjp24, CC BY-SA 4.0 <https://creativecommons.org/licenses/by-sa/4.0>, via Wikimedia Commons

- Copper and aluminium are very widely used in making electric wires, cooking utensils.
- Iron, copper and aluminium are used in making many household utensils, other devices and factory equipment.
- Zinc is used for galvanisation to protect iron objects from rusting.
- Silver, gold and platinum are very highly used in making jewellery.
- Mercury is used in making thermometers.
- Lead is used in making car batteries.
- Tungsten is used in making filaments of electric bulbs.
- Aluminium foil is used in packing food materials, cigarettes.
- Chromium and nickel are used for electroplating and in making of stainless steel.

TEST YOUR KNOWLEDGE

3.1 Write one example of each of the following:

(i) Most malleable metal and most ductile metal.

(ii) The best conductor of heat and the poorest conductor of heat.

(iii) The metal with the highest melting point and the metal with the lowest melting point.

Solution:

(i) Gold is the most malleable and most ductile metal.

(ii) Silver is the best conductor and lead is the poorest conductor.

(iii) Tungsten has the highest melting point and mercury has the lowest melting point.

3.2 Name any two metals which lack malleability and ductility.

Solution:

Zinc and mercury lack in malleability and ductility.

3.3 Mention the names of the metals for the following.

(i) Two metals which are alloyed with iron to make stainless steel.

(ii) Two metals which are used to make jewellery.

Solution:

(i) Nickel and chromium are alloyed with iron to make stainless steel.

(ii) Gold and platinum are used to make jewellery.

3.4 Why are aluminium and copper used for making cooking vessels?

Solution:

Aluminium and copper are used for making cooking vessels because they are good conductors of heat and they have high melting points.

3.3 The Reactivity or Activity Series of Metals

All metals are not chemically equally reactive. For example, a few metals such as alkali metals and alkaline earth metals are more reactive, a few metals such as Al, Zn, Fe are moderately reactive and a few metals such as like Cu, Ag, Au, Pt are least reactive. When these metals are arranged in the decreasing order of their reactivity, an activity series of metals is formed. Thus, the activity series or reactivity series of metals represents a list of metals arranged in order of their decreasing chemical activity. Hence, the most reactive metal is potassium and is placed at the top of the list and the least active metal is gold which is placed at the bottom of the list (Fig. 3.4).

	Element	Symbol	
Metals more reactive than hydrogen	Potassium	K	←Most reactive
	Sodium	Na	Reactivity decreases downwards
	Barium	Ba	
	Calcium	Ca	
	Magnesium	Mg	
	Aluminium	Al	
	Zinc	Zn	
	Iron	Fe	
	Nickel	Ni	
	Tin	Sn	
	Lead	Pb	
	Hydrogen	H	
Metals less reactive than hydrogen	Copper	Cu	
	Mercury	Hg	
	Silver	Ag	
	Gold	Au	
	Platinum	Pt	←Least reactive

Fig. 3.4 Activity series of metals

3.3.1 Main Features of the Activity Series

- Any metal placed higher up in the series displaces any metal below it from the salt solution of the latter.
- The larger the difference in the positions of the metals in the series, the more rapidly the displacement occurs.
- Metals above hydrogen can displace hydrogen from water and from dilute acid but the metals below hydrogen cannot do it.
- Oxides of metals K, Na, Ca and Mg cannot be reduced by H_2, CO or C.
- Oxides and nitrates of less reactive metals such as Hg, Ag, Au decompose to give the metals on being strongly heated.
- Metals below copper do not rust easily.
- Hydrogen, though a non-metal has been included in this series. It would occupy the position based on the formation of positive ions.

3.4 Chemical Properties

Most of the metals show reactions with air, water, hydrogen, chlorine, dilute acids and metal salt solutions.

3.4.1 Reaction with Oxygen

Most of the metals form metal oxides on burning in air or reaction with oxygen. Metals oxides formed here are basic in nature and may also react with water to form metal hydroxides (alkalies). These oxides also turn red litmus solution blue.

$$\text{Metal} + \text{Oxygen or Air} \xrightarrow{\Delta} \text{Metal oxide}$$

All metals do not react equally with air as the tendency of metals to lose electrons differs. For example, highly reactive metals (group 1, Na, K) readily tarnish in air even at room temperature, so they are placed under kerosene to avoid this reaction, as this reaction is vigorous and explosive.

$$4Na + O_2 \rightarrow \underset{\text{Sodium oxide}}{2Na_2O} \qquad 4K + O_2 \rightarrow \underset{\text{Potassium oxide}}{2K_2O}$$

Both Na_2O and K_2O are strongly basic in nature and react with water to form alkalies, that is, NaOH and KOH respectively.

$$Na_2O + H_2O \rightarrow 2NaOH \qquad K_2O + H_2O \rightarrow 2KOH$$

Metals of group 2, that is, alkaline earth metals such as Mg, Ca react with oxygen only on heating and not at room temperature to form oxides.

$$2Mg + O_2 \rightarrow \underset{\text{Magnesium oxide}}{2MgO} \qquad 2Ca + O_2 \rightarrow \underset{\text{Calcium oxide}}{2CaO}$$

Both MgO and CaO are strongly basic in nature and react with water to form alkalies, that is, $Mg(OH)_2$ and $Ca(OH)_2$ respectively.

$$MgO + H_2O \rightarrow \underset{\text{Magnesium hydroxide}}{Mg(OH)_2} \qquad CaO + H_2O \rightarrow \underset{\text{Calcium hydroxide}}{Ca(OH)_2}$$

Many other metals (group 2, Zn, Al) tarnish in air after some time as they react slowly with O_2 because they are covered with a thin layer of oxide (protective oxide layer) which prevents further oxidation. Both zinc and aluminium react with oxygen on heating to give their oxides. Their oxides are amphoteric in nature, that is, they can react with both acids and bases.

$$4Al(s) + 3O_2(g) \rightarrow \underset{\substack{\text{Aluminium oxide} \\ \text{(Amphoteric oxide)}}}{2Al_2O_3(s)}$$

$$\underset{\substack{\text{Aluminium oxide} \\ \text{(Base)}}}{Al_2O_3(s)} + \underset{\substack{\text{Hydrochloric acid} \\ \text{(Acid)}}}{6HCl(aq)} \rightarrow \underset{\substack{\text{Aluminium chloride} \\ \text{(Salt)}}}{2AlCl_3(aq)} + 3H_2O(l)$$

$$\underset{\substack{\text{Alu min ium} \\ \text{oxide (Acid)}}}{Al_2O_3(s)} + \underset{\substack{\text{Sodium} \\ \text{hydroxide (Base)}}}{2NaOH(aq)} \rightarrow \underset{\substack{\text{Sodium} \\ \text{aluminate (salt)}}}{2NaAlO_2(aq)} + H_2O(l)$$

$$2Zn(s) + O_2(g) \rightarrow \underset{\substack{\text{Zinc oxide} \\ \text{(Amphoteric oxide)}}}{2ZnO(s)}$$

$$\underset{\text{Zinc oxide (base)}}{ZnO(s)} + \underset{\text{Hydrochloric acid (Acid)}}{2HCl(aq)} \rightarrow \underset{\text{Zinc chloride (Salt)}}{ZnCl_2(aq)} + H_2O(l)$$

$$\underset{\text{Zinc oxide (Acid)}}{ZnO(s)} + \underset{\text{Sodium hydroxide (Base)}}{2NaOH(aq)} \rightarrow \underset{\text{Sodium zincate (Salt)}}{Na_2ZnO_2(aq)} + H_2O(l)$$

Iron and copper do not burn in air even on heating strongly; however, iron filings can react with air or oxygen on heating strongly and copper on prolonged heating.

$$3Fe + 2O_2 \rightarrow \underset{\text{Iron (II, III) oxide}}{Fe_3O_4(s)} \qquad 2Cu + O_2 \rightarrow \underset{\substack{\text{Copper (II) oxide} \\ \text{(Black)}}}{2CuO}$$

Iron and copper react with moist air to give red-brown coloured rust which is hydrated ferric oxide ($Fe_2O_3.xH_2O$) and green coloured basic copper sulfate [$3Cu(OH)_2.CuSO_4$] respectively.

$$4Fe + 3O_2 + 2xH_2O \rightarrow 2Fe_2O_3.xH_2O$$

Some unreactive metals like Ag and Au do not tarnish in air. In air, the reactivity of metals decreases as follows:

$$\underbrace{\text{IA, IIA}}_{\substack{\text{Highly reactive}\\\text{Bright flame}}}, \underbrace{\text{Al, Zn, Fe, Pb, Cu}}_{\text{Reacts on heating}}, \underbrace{\text{Ag, Au}}_{\text{No reaction}}$$

3.4.2 Reaction with Water

Metals react with water to form metal oxides and hydrogen gas. Metal oxides are soluble in water and easily form metal hydroxides. All metals do not react equally with water and the intensity of the reaction is decided by the chemical reactivity of the metal.

$$\text{Metal + Water} \rightarrow \text{Metal hydroxide + Hydrogen}$$

or $$\text{Metal + Water} \rightarrow \text{Metal oxide + Hydrogen}$$

$$\text{Metal oxide + Water} \rightarrow \text{Metal hydroxide}$$

Alkali metals react quickly and vigorously and the reactivity increases down the group. Alkaline earth metals have a lower reactivity than group 1 metals. For example, magnesium cannot react with cold water but reacts with hot water; however, calcium reacts with cold water also (Fig. 3.5).

Fig. 3.5 Reaction of Ca, Na and K with water

$2Na(s) + 2H_2O(l) \rightarrow 2NaOH(aq) + H_2(g)$ + heat energy

$2K(s) + 2H_2O(l) \rightarrow 2KOH(aq) + H_2(g)$ + heat energy

$Ca(s) + 2H_2O(l) \rightarrow Ca(OH)_2(aq) + H_2(g)$

Sodium and potassium react with water very vigorously and violently. As a huge amount of heat is formed here, hydrogen catches fire at once while in the case of calcium, very less heat is released and so hydrogen does not catch fire.

Metals like Mg, Al, Fe and Zn do not react with cold water or hot water; however, they react with steam to form metal oxide and hydrogen. The reactivity order for these metals is Mg > Al > Zn > Fe. A few metals like Cu, Ag, Au and Pb do not react with cold or hot water or steam (Fig. 3.6).

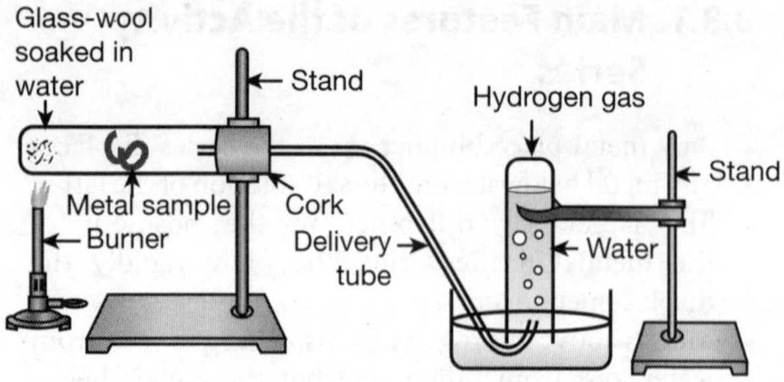

Fig. 3.6 Reaction of metals with steam

$$\text{Metal + steam} \rightarrow \text{Metal oxide + Hydrogen}$$

$Mg(s) + H_2O(s) \rightarrow MgO(s) + H_2(s)$

$2Al(s) + 3H_2O(g) \rightarrow Al_2O_3(s) + 3H_2(g)$

$3Fe(s) + 4H_2O(g) \rightarrow Fe_3O_4(s) + 4H_2(g)$

3.4.3 Reaction of Metals with Hydrogen

Most of the metals do not combine with hydrogen. Like the metals, hydrogen also prefers to form compounds by losing an electron. That is, generally, hydrogen does not accept electrons lost by a metal. However, highly electropositive metals such as sodium, potassium, calcium and magnesium can react with hydrogen to form a metal hydride. Here, hydrogen accepts the electron lost by the metal atom. These metal hydrides are ionic compounds in which the metal is in the cationic form and hydrogen is in the anionic form, that is, hydride (H^-).

$$2Na + H_2 \rightarrow \underset{\text{Sodium hydride}}{2NaH}$$

$$Ca + H_2 \rightarrow \underset{\text{Calcium hydride}}{CaH_2}$$

Less reactive metals like iron, zinc and copper cannot react with hydrogen to form these hydrides.

3.4.4 Reaction of Metals with Chlorine

Metals like sodium, potassium, magnesium, calcium, aluminium, zinc, iron react with chlorine to form metal chlorides which are ionic compounds. Here, the metal atom loses electrons to form the metal cation (M^{n+}) while the chlorine atoms accept these electrons to form chloride ions (Cl^-). Sodium, potassium, calcium and zinc directly combine with chlorine.

$$2Na + Cl_2 \rightarrow \underset{\text{Sodium chloride}}{2NaCl}$$

$$Ca + Cl_2 \rightarrow \underset{\text{Calcium chloride}}{CaCl_2}$$

$$Zn + Cl_2 \rightarrow \underset{\text{Zinc chloride}}{ZnCl_2}$$

Iron, aluminium and copper on heating with chlorine form their metal chlorides.

$$2Al(s) + 3Cl_2(g) \rightarrow \underset{\text{Aluminium chloride}}{2AlCl_3(s)}$$

$$2Fe(s) + 3Cl_2(g) \rightarrow \underset{\substack{\text{Iron (III) chloride} \\ \text{(or Ferric chloride)}}}{2FeCl_3(s)}$$

$$Cu(s) + Cl_2(g) \rightarrow \underset{\substack{\text{Copper (II) chloride} \\ \text{(Or Cupric chloride)}}}{CuCl_2(s)}$$

3.4.5 Reaction with Acids

Metals react with acids to form the metal salt along with the release of hydrogen, that is, the metal displaces hydrogen from dilute acids. More reactive metals like Na, K and Mg react quickly and violently while aluminium, zinc and iron react slowly. Less reactive metals like copper, silver and gold do not displace hydrogen from dilute acids as their reactivity is less than that of hydrogen.

$$\text{Metal} + \text{Acid} \rightarrow \text{Salt} + \text{Hydrogen}$$

Reaction with dil. HCl or dil. H_2SO_4: Alkali metals, alkaline earth metals, zinc, aluminium and iron can react readily with dilute HCl or H_2SO_4, however, the reactions of Zn, Al and Fe are slow compared to that of the alkali metals. Cu, Ag and Au cannot react with dil. HCl or dil. H_2SO_4. When metals react with dil. HCl, the metal chloride and bubbles of hydrogen are formed. The reactivity of some metals for this reaction is, Na > Mg > Al > Zn > Fe.

For example, $2Na + 2HCl \rightarrow 2NaCl(aq) + H_2$

$$Ca + H_2SO_4 \rightarrow CaSO_4(aq) + H_2$$

$$Zn + H_2SO_4 \rightarrow ZnSO_4(aq) + H_2$$

$$2Al + 6HCl \rightarrow 2AlCl_3(aq) + 3H_2$$

$$Fe + 2HCl \rightarrow FeCl_2(aq) + H_2$$

Reaction with dil. nitric acid: When metals react with HNO_3, H_2 is not evolved as HNO_3, being a strong oxidising agent, oxidises H_2 into H_2O and get reduced into its oxides such as N_2O, NO, NO_2. Only with Mg and Mn, hydrogen is evolved.

$$Mg + 2HNO_3 \rightarrow Mg(NO_3)_2 + H_2$$

$$Mn + 2HNO_3 \rightarrow Mn(NO_3)_2 + H_2$$

When dil. nitric acid is added into a mixture having Cu, Zn and Al, only Cu reacts with HNO_3.

$$Cu + 2HNO_3 + 2H^+ \rightarrow Cu^{2+} + 2NO_2 + 2H_2O$$

Al, Fe, Zn do not react with conc. HNO_3 due to the formation of a stable oxide layer on these metals.

> **Competition Edge**
>
> **Aqua regia:** It is a freshly prepared mixture of one part concentrated HNO_3 (nitric acid) and three parts of concentrated HCl (hydrochloric acid). It can dissolve even unreactive or noble metals like gold (Au) and platinum (Pt).

3.4.6 Reaction with Metal Salt Solution

When a more reactive metal is taken in the salt solution of a less reactive metal, the more reactive metal displaces the less reactive metal from its salt solution. This is a displacement reaction. Let us see a few examples.

- When a strip or nail of iron metal is taken in copper sulfate solution, iron being more reactive than copper, displaces copper from copper sulfate. During the reaction, the blue colour disappears and the iron metal acquires a pink-brown colour due to deposition of copper [Fig. 3.7(a)].

$$Fe(s) + \underset{\text{Blue}}{CuSO_4(aq)} \rightarrow \underset{\text{Pale green}}{FeSO_4(aq)} + Cu(s)$$

- When a strip of zinc metal is taken in copper sulfate solution, zinc being more reactive than copper displaces copper from copper sulfate. Here during the reaction, the blue colour fades due to the formation of colourless zinc sulfate and red-brown copper metal is deposited on the zinc strip [Fig. 3.7(b)].

$$Zn(s) + \underset{\text{Blue}}{CuSO_4(aq)} \rightarrow \underset{\text{Colourless}}{ZnSO_4(aq)} + Cu(s)$$

- When a strip of copper metal is taken in silver nitrate solution, copper being more reactive than silver, displaces silver from silver nitrate. During the reaction, the colour of the solution becomes blue due to the formation of copper nitrate. The copper wire or strip becomes greyish as silver gets deposited on it [Fig. 3.7(c)].

$$Cu(s) + \underset{\text{Colourless}}{2AgNO_3(aq)} \rightarrow \underset{\text{Blue}}{Cu(NO_3)_2(aq)} + 2Ag(s)$$

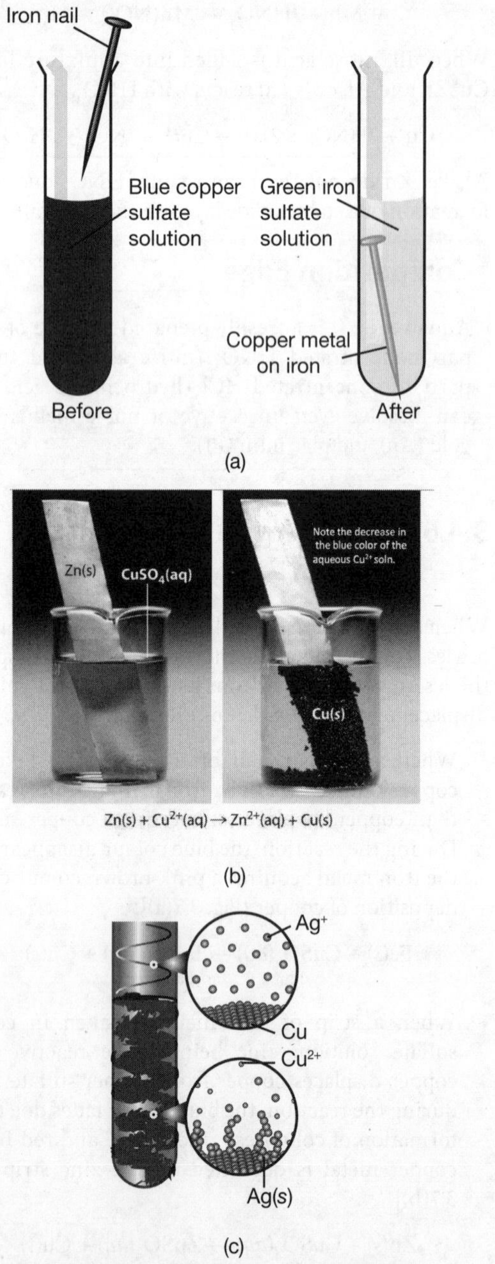

Fig. 3.7 (a) Iron nail taken in copper sulfate, (b) Zinc strip in copper sulfate and (c) Copper strip or wire in silver nitrate solution

It is important to note that when the added metal is less reactive than the metal to be displaced from the metal salt, the reaction is not possible. Some examples are discussed below.

- When an iron strip is put in magnesium sulfate solution, no reaction occurs as Fe has less reactivity than Mg; so, Fe cannot displace Mg.

$$Fe + MgSO_4 \rightarrow \text{No reaction}$$

- When a copper strip is put in lead nitrate solution, no reaction occurs as Cu has less reactivity than Pb; so, Cu cannot displace Pb.

$$Cu + Pb(NO_3)_2 \rightarrow \text{No reaction}$$

TEST YOUR KNOWLEDGE

3.5 Name any three metals present below copper in the activity series of metals.

Solution: Mercury, silver, gold, platinum.

3.6 Between Mg and Zn, which burns with bright flame to form the oxide?

Solution: Magnesium reacts quickly and vigorously with air to form an oxide with a bright flame while Zn reacts slowly with air.

$$2Mg + O_2 \rightarrow 2MgO$$

$$2Zn + O_2 \rightarrow 2ZnO$$

3.7 Explain why lead on reaction with dil. HCl does not displace hydrogen.

Solution: Lead on reaction with dil. HCl does not displace hydrogen because a thin film of $PbCl_2$ is formed which protects the metal from the action of the acid.

3.8 When calcium metal is added to water, the gas evolved does not catch fire but the same gas evolved on adding potassium metal to water catches fire. Explain why.

Solution: In both cases, the gas evolved is H_2. When calcium reacts with water, the heat evolved is not sufficient for hydrogen to catch fire. On the other hand, potassium reacts with water violently and a lot of heat is evolved which is sufficient for the hydrogen to catch fire.

3.9 (i) Why do cesium and gallium melt in our palm?

(ii) Why does magnesium ribbon start floating in hot water?

Solution:

(i) Some metals are more like liquids. A chip of gallium or cesium melts on our palm because the melting point of these metals is just below thirty degrees celsius (86°F). They have very low melting points.

(ii) Magnesium reacts with hot water and starts floating due to the bubbles of hydrogen gas sticking to its surface.

3.5 Non-Metals

A non-metal is an element that is neither malleable nor ductile and cannot conduct electricity [with the exception of graphite (C)]. All the non-metals are gases or solids, except bromine which is a liquid non-metal at room temperature. Examples are, hydrogen (H), nitrogen (N), carbon (C), oxygen (O), phosphorus (P), sulfur (S), chlorine (Cl), bromine (Br), iodine (I), helium (He), neon (Ne), argon (Ar), krypton (Kr) and xenon (Xe) (Fig. 3.8).

(a) Carbon (b) Sulfur (c) Iodine

(d) Bromine (e) Chlorine

Fig. 3.8 Some non-metals

Source: (a) Carbon – By Robert M. Lavinsky CC BY-SA 3.0, https://commons.wikimedia.org/w/index.php?curid=39385953; (b) Sulfur – Jamain, CC BY-SA 4.0 <https://creativecommons.org/licenses/by-sa/4.0>, via Wikimedia Commons; (c) Iodine – Benjah-bmm27, Public domain, via Wikimedia Commons; (d) Bromine – Hi-Res Images ofChemical Elements, CC BY 3.0 <https://creativecommons.org/licenses/by/3.0>, via Wikimedia Commons; (e) Chlorine – Larenmclane, CC BY-SA 4.0 <https://creativecommons.org/licenses/by-sa/4.0>, via Wikimedia Commons

Position in periodic table: Except hydrogen, all other non-metals are placed in the *p*-block in groups 13 to 18. They are 24 in number: **solids (12)**: B, C, Si, Ge, P, As, Sb, S, Se, Te, I, At; **liquids (1)**: Br and **gases (11)**: H, N, O, F, Cl, He, Ne, Ar, Kr, Xe, Rn.

3.5.1 Physical Properties of Non-metals

The physical properties of non-metals are just the opposite of the physical properties of metals.

1. **Physical state:** At room temperature they may be solids (carbon, sulfur, phosphorus), liquids (bromine) or gases (hydrogen, nitrogen, oxygen, inert gases) depending upon the magnitude of the force of attraction between the constituent atoms. With an increase in molecular weight, the magnitude of the force of attraction (van der Waals force) increases, so solid nature is favoured. For example, F_2, Cl_2 are gases, bromine is a liquid and iodine is a solid.

2. **Colour:** They may be colourless or may have different colours. For example, hydrogen and oxygen are colourless, phosphorus is white, sulfur is yellow, chlorine is greenish-yellow, bromine is brownish-red.

3. **Appearance:** They are not lustrous or shiny but are dull in appearance as they do not have a shining surface. For example, sulfur, phosphorus. Iodine is an exception and has a lustrous appearance.

4. **Non-sonorous nature:** They are not sonorous as they do not make any ringing sound on striking.

5. **Non-ductile and non-malleable nature:** They are not ductile (cannot be drawn into wires) and they are non-malleable (cannot be beaten into sheets on hammering). However, solid non-metals are brittle as on hammering they split into pieces. The property of solid non-metals breaking easily into pieces is known as brittleness. For example, pencil lead breaks easily into pieces.

6. **BP, MP and density:** They have comparatively low melting points and low boiling points. They can be easily melt and vaporise at low temperature. For example, sulfur can be melted at 119°C. Due to their low boiling points, most of the non-metals exist as gases at room

temperature. However, graphite has a very high melting point (of nearly 3700°C). Non-metals have low densities, that is, they are light substances. For example, sulfur has a density of 2 g/cm^3. However, iodine has a higher value of density.

7. **Soft and low tensile strength:** They are generally soft, not strong and have a low tensile strength. Being soft, they are easily cuttable with a knife with the exception of diamond (allotrope of carbon) which is one of the hardest natural substances.

8. **Non-conductors:** They are non-conductors of heat and electricity; that is, heat and electricity cannot pass through them as they do not have free electrons for conduction. However, graphite can show electrical conductance due to the presence of free electrons in its hexagonal sheet-like structure and that is why it is used in making electrodes (Fig. 3.9).

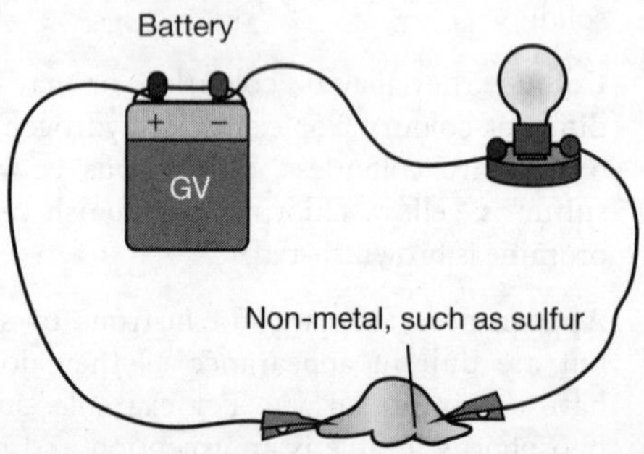

Fig. 3.9 No conduction in non-metals

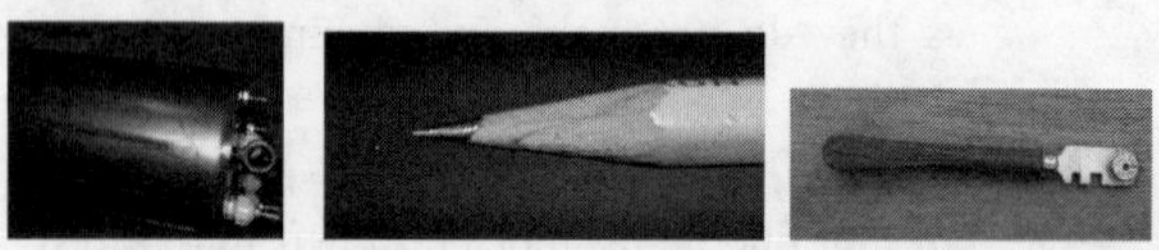

Fig. 3.10 Electrode of graphite, lead pencil having graphite, diamond as glass cutter

Source: Phil mc2, CC BY 3.0 <https://creativecommons.org/licenses/by/3.0>, via Wikimedia Commons; Juliancolton, Public domain, via Wikimedia Commons; Rasbak, CC BY-SA 3.0 <http://creativecommons.org/licenses/by-sa/3.0/>, via Wikimedia Commons

Uses of non-metals: Non-metals also have a lot of applications in our daily lives.

- All life on earth is made up of carbon compounds such as carbohydrates, proteins, fats and vitamins which are essential for the growth and maintenance of all living things.
- Hydrogen is used in the hydrogenation of oils into ghee, formation of ammonia which is used as a refrigerant and in making fertilisers.
- Oxygen is essential for breathing to sustain life and also for all combustion (burning) processes.
- Graphite is used in making electrodes, pencils, and diamond is used as a glass cutter and in making jewellery (Fig. 3.10).

TEST YOUR KNOWLEDGE

3.10 Name any two non-metals which are lustrous.

Solution:
Iodine and graphite are quite lustrous non-metals.

3.11 Name any two non-metals which are mono-atomic.

Solution:
Helium (He) and neon (Ne) are mono-atomic.

3.12 A non-metal X exists in two different forms Y and Z. Y is the hardest natural substance, whereas Z is a good conductor of electricity. Identify X, Y and Z.

Solution:
X is carbon (non-metal). X exists in the form of allotropes, Y is diamond and Z is graphite.

3.6 Chemical Properties

All non-metals do not have same the chemical activity, a few non-metals such as F_2, Cl_2, O_2, C, P, S are more reactive while a few are less reactive such as N_2 and a few are almost unreactive such as the inert gases (He, Ne).

3.6.1 Reaction with Air

Most of the non-metals can react with air on ignition. However, phosphorus can react with air even at room temperature to give acidic oxides which dissolve in water to give acids. The oxides of non-metals turn blue litmus into red. For example,

$$C(s)+O_2(g)\xrightarrow{\text{Ignition}}CO_2(g)$$

$$P_4(s)+5O_2(g)\xrightarrow{\text{Ignition}}P_4O_{10}(s)$$

$$S(s)+O_2(g)\rightarrow SO_2(g)$$

These non-metallic oxides readily dissolve in water to give oxyacids.

$$CO_2+H_2O(l)\rightarrow \underset{\text{Carbonic acid}}{H_2CO_3(aq)}$$

$$P_4O_{10}+6H_2O(l)\rightarrow \underset{\text{Phosphoric acid}}{4H_3PO_4(aq)}$$

$$SO_2(g)+H_2O(l)\rightarrow \underset{\text{Sulfurous acid}}{H_2SO_3(aq)}$$

A few non-metals like carbon and hydrogen form neutral oxides in air.

$$2C(s)+O_2(g)\xrightarrow{\text{Ignition}}2CO(g)$$

$$2H_2(g)+O_2(g)\xrightarrow[\text{Popping sound}]{\text{Ignition}}2H_2O(g)$$

Both CO (carbon monoxide) and H_2O (water) are neutral oxides.

3.6.2 Reaction with Water

Most of the non-metals do not react with water or steam to release hydrogen gas as non-metals cannot provide electrons to reduce the hydrogen ions of water into hydrogen gas. However, chlorine dissolves in water.as follows:

$$2Cl_2(g)+2H_2O(g)\xrightarrow[\text{or heat}]{\text{Sunlight}}2HCl(g)+\underset{\text{(Hypochlorous acid)}}{2HOCl(l)}$$

When steam is passed over red hot coke, water gas is formed.

$$\underset{\text{(Red hot)}}{C(s)}+\underset{\text{(Steam)}}{H_2O(l)}\rightarrow \underset{\text{(Water gas)}}{CO(g)+H_2(g)}$$

3.6.3 Reaction with Hydrogen

Non-metals on heating with hydrogen form covalent hydrides.

$$N_2(g)+3H_2(g)\xrightleftharpoons{500^\circ C}\underset{\text{Ammonia}}{2NH_3(g)}$$

$$H_2(g)+S(l)\rightarrow \underset{\substack{\text{Hydrogen sulfide}\\ \text{(Rotten egg smell)}}}{H_2S(g)}$$

3.6.4 Reaction with Chlorine

Non-metals on reaction with chlorine form covalent chlorides. ,

$$H_2(g)+Cl_2(g)\rightarrow 2HCl(g)$$

$$2P(s)+3Cl_2(g)\rightarrow \underset{\text{Phosphorus trichloride}}{2PCl_3(l)}$$

$$2P(s)+5Cl_2(g)\rightarrow \underset{\text{Phosphorus pentachloride}}{2PCl_5(l)}$$

3.6.5 Reaction with Acids

Most of the non-metals not react with acids. However, on heating with concentrated sulfuric acid or concentrated nitric acid, some non-metals do react.

$$S(s)+2H_2SO_4(conc.)\xrightarrow{\text{Heat}}3SO_2(g)+2H_2O(l)$$

$$S(l)+6HNO_3(conc.)\xrightarrow{\text{Heat}}H_2SO_4(aq)+6NO_2(g)+2H_2O(l)$$

3.6.6 Displacement of One Non-metal by Another from Salt Solution

A more reactive non-metal can displace a less reactive non-metal from its salts. For example, chlorine can displace bromine or iodine as it is more reactive than bromine and iodine.

$$Cl_2+2NaBr\rightarrow 2NaCl+Br_2$$

$$Cl_2+2NaI\rightarrow 2NaCl+I_2\,.$$

TEST YOUR KNOWLEDGE

3.13 Name any two oxides other than water which are neutral in nature.

Solution:
CO and NO are neutral in nature.

3.14 The compounds formed between non-metals are of which type?

Solution:
Compounds formed by non-metals are covalent in nature. For example, CO_2, PCl_5.

3.15 Define amphoteric oxides. Give two examples of such oxides.

Solution:
Oxides that show both acidic and basic behaviour and oxides that react with both acids as well as bases to produce salts and water are known as amphoteric oxides. For example, Al_2O_3 and ZnO.

3.7 Differences Between Metals and Non-Metals

Table 3.2 Differences in physical properties of metals and non-metals

Property	Metals	Non-metals
Physical state	They are mostly solids and are non-volatile in nature.	They may be gas, solid or liquid and are volatile in nature.
Density	They have high densities except sodium and potassium.	They have low densities.
Metallic lustre	They have metallic lustre and can be polished.	They do not have metallic lustre so cannot be polished except graphite and iodine.
Malleability and ductility	They are malleable and ductile, except Bi.	They are neither malleable nor ductile in nature.
Thermal and electrical conductivity	Metals are good conductors of heat and electricity except lead. For example, Ag and Cu are the best conductors.	They are poor conductors of heat and electricity except graphite.
Alloy formation	Due to similar sizes metals can readily form alloys.	They do not form alloys except phosphorus, carbon and silicon.

Table 3.3 Differences in chemical properties of metals and non-metals

Property	Metals	Non-metals
Number of valence electrons	Metals have 1, 2 or 3 valence electrons.	They have 4 to 7 valence electrons except H and He which have 1 and 2 valence electrons respectively.
Electrochemical nature	They are electropositive by nature; so, form cations and during electrolysis they are liberated at the cathode.	They are electronegative by nature; so, form anions and during electrolysis they are liberated at the anode.
Solubility	Mostly they dissolve by chemical reactions.	Most of the non-metals dissolve without any chemical reactions.
Nature of oxides	Their oxides are mainly basic in nature. $Na_2O + H_2O \rightarrow 2NaOH$	Their oxides are mainly acidic in nature. $2NO_2 + H_2O \rightarrow HNO_2 + HNO_3$
Action of acids	They dissolve in mineral acids and form salts and release H_2. $Zn + 2HCl \rightarrow ZnCl_2 + H_2$	Most of the non-metals do not dissolve in acids; however, S, P, I_2 can react with concentrated HNO_3. $I_2 + 10HNO_3 \rightarrow 2HIO_3 + 10NO_2 + 4H_2O$
Hydrides	Only a few metals can form stable hydrides, such as LiH, CaH_2.	Most of the non-metals form stable hydrides such as NH_3, CH_4.

Quick Review

- **Metals** occupy the largest portion of the periodic table. Those elements which have 1, 2 or 3 electrons in their valence cells, are lustrous, malleable, ductile (can be drawn into sheets or pulled into wires) and have high densities and tensile strength are metals.
- Density, MP, BP $\propto$ Strength of metallic bond $\propto$ Number of valence electrons.
- Metals (except Zn, Hg and Ga) are highly **malleable** (beaten into sheets) and highly **ductile** (drawn into wires) as the metallic bond is omnidirectional.
- **Metals** have a high tensile strength, show thermal and electrical conductivity and magnetic properties.
- Most of the metals form metal oxides on burning in air. Metals react with water to form metal oxides and hydrogen gas. Metal oxides are soluble in water and easily form metal hydroxides.
- The **reactivity** of different metals with water in **decreasing order**:
  K > Na > Ca > Mg > Al > Zn > Fe > Pb > Cu > Ag > Au
- A more reactive metal can displace a less reactive metal from its salt solution. It is a case of displacement reaction.
- Those elements which in general are non-lustrous, electronegative, bad conductors of heat and electricity are **non-metals**. These elements have a tendency to accept electrons to form anions due to their electronegative nature. **Non-metals** are soft with low density and non-sonorous nature.
- Most of the non-metals can react with air on ignition; however, phosphorus can react with air even at room temperature.
- Most of the non-metals do not react with water but chlorine dissolves in water to give hypochlorous acid.
- Non-metals on heating with hydrogen form hydrides like HX, NH_3, H_2S.
- A more reactive non-metal can displace a less reactive non-metal from its salts.
- Certain elements which show properties of both metals and non-metals are called **metalloids**. They are, boron (B), silicon (Si), germanium (Ge), arsenic (As), antimony (Sb) and tellurium (Te).

Exercise-1

Section A: Multiple Choice Questions (1 Mark)

1. Which of the following non-metals is lustrous?
(a) fluorine (b) iodine
(c) sulfur (d) oxygen

2. Select the statement that is correct about the non-metals.
(a) Bromine is an example of a liquid non-metal.
(b) Graphite is a good conductor of electricity.
(c) Most of the non-metal oxides are acidic.
(d) All are correct.

3. A non-metal used to preserve food material is:
(a) sulfur (b) phosphorus
(c) carbon (d) nitrogen

4. When a non-metal is allowed to react with water,
(a) CO_2 gas is formed
(b) the product formed depends on temperature
(c) H_2 gas is formed
(d) no products are formed

5. The poorest conductor of heat among metals is
(a) calcium (b) mercury
(c) lead (d) sodium

6. Which property of metals is used for making bells and strings of musical instruments like the sitar and violin?
(a) sonorous-ness (b) malleability
(c) ductility (d) conductivity

7. The metals that float when put into water are:
(a) manganese and sodium
(b) magnesium and calcium
(c) magnesium and sodium
(d) sodium and calcium

8. An element X is soft and can be cut with a knife. This is very reactive to air and cannot be kept open in air. It reacts vigorously with water. Identify the element from the following.
(a) P (b) Na
(c) Mg (d) Ca

9. The electronic configurations of three elements X, Y and Z are X = 2, 8; Y = 2, 8, 7 and Z = 2, 8, 2. Which of the following is correct?
(a) Y is a metal
(b) X is a metal
(c) Z is a non-metal
(d) Y is a non-metal and Z is a metal

10. What happens when a pellet of sodium is dropped in water?
(a) it catches fire and forms the oxide
(b) it catches fire and forms the hydroxide
(c) it absorbs heat and forms the oxide
(d) it absorbs heat and forms the hydroxide

11. When calcium oxide is added to water, it completely dissolves in water without forming bubbles. What products are formed in this reaction?
(a) Ca and H_2 (b) $Ca(OH)_2$
(c) Ca and H_2O_2 (d) CaH_2

12. Which of the following can undergo a chemical reaction?
(a) $ZnSO_4$ + Fe (b) $MgSO_4$ + Fe
(c) $MgSO_4$ + Pb (d) $CuSO_4$ + Fe

13. $Al_2O_3 + 2NaOH \rightarrow$ _______ + H_2O. What is the missing product?
(a) Na_2O (b) $Al(OH)_3$
(c) $AlNaO_2$ (d) $NaAlO_2$

14. Which of the following is the correct arrangement of the given metals in ascending order of their reactivity?
zinc, iron, magnesium, potassium
(a) potassium > zinc > magnesium > iron
(b) potassium> magnesium > zinc > iron
(c) zinc > iron > magnesium > potassium
(d) potassium > magnesium > iron > zinc

15. An example of an amphoteric oxide is:
(a) Al_2O_3 (b) K_2O
(c) Na_2O (d) MgO

16. Which one among the following is an acidic oxide?
(a) Al_2O_3 (b) CO
(c) CO_2 (d) Na_2O

17. A student placed an iron nail in copper sulfate solution. He observed the reddish-brown coating on the iron nail which is
(a) soft and dull
(b) rough and granular
(c) smooth and shining
(d) hard and fading

18. Aluminium is used for making cooking utensils. Which of the following properties of aluminium are responsible for the same?
(i) ductility
(ii) high melting point
(iii) good thermal conductivity
(iv) good electrical conductivity
(a) (i) and (ii) (b) (i) and (iii)
(c) (ii) and (iii) (d) (i) and (iv)

19. A student writes two incomplete chemical reactions.

$$P_4(s) + 5O_2(g) \rightarrow X$$

$$2Ca(s) + O_2(g) \rightarrow Y$$

Which option completes the reactions to form a balanced chemical equation?
(a) X: $P_5O_4(s)$; Y: $(CaO)_2(s)$
(b) X: $5P_4O_2(s)$; Y: $Ca_2O_2(s)$
(c) X: $P_4O_{10}(s)$; Y: 2CaO(s)
(d) X: $4PO_{10}(s)$; Y: 4CaO(s)

20. Select the correct option in the following table in connection with the given chemical reaction:
$CuSO_4 + Fe \rightarrow FeSO_4 + Cu$ **[CBSE 2021]**

	Initial colour of solution	Final colour of solution	Final colour of iron nail	Type of reaction
(a)	Pale green	Blue	Grey	Displacement
(b)	Blue	Pale green	Brownish	Double displacement
(c)	Blue	Light blue	Grey	Double displacement
(d)	Blue	Pale green	Brownish	Displacement

21. In the following diagram, what would happen if some amount of sodium hydroxide is taken in place of sulfuric acid and the test tube is heated. **[CBSE 2021]**

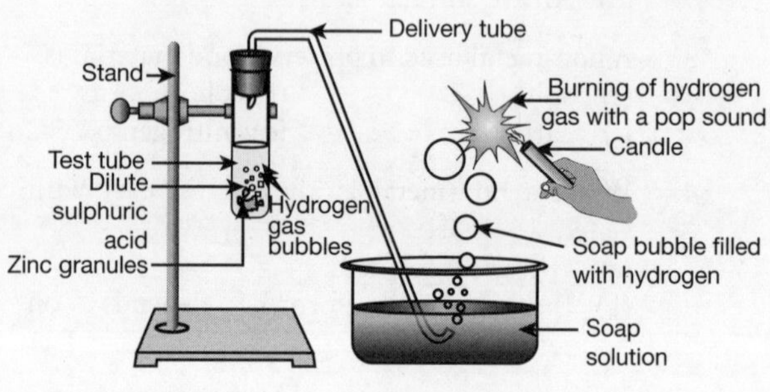

(a) same amount of H_2 gas is evolved
(b) H_2 gas is not evolved
(c) the amount of H_2 gas evolved is much less
(d) in place of H_2 gas, O_2 gas is evolved

22. Four test tubes A, B, C and D are taken. In test tube A, an iron nail is dipped in copper sulfate solution. In test tube B, copper wire is dipped in ferrous sulfate solution. In test tube C, zinc metal is dipped in ferrous sulfate solution and in test tube D, an iron nail is dipped in zinc sulfate solution. The reactivity order has been found to be Zn > Fe > Cu. In which test tubes was a colour change observed? **[CBSE 2021]**

(a) A and C (b) A and B
(c) B and C (d) B and D

23. The pair(s) which will show displacement reaction is/are **[CBSE 2021]**

(i) NaCl solution and copper metal
(ii) $AgNO_3$ solution and copper metal
(iii) $Al_2(SO_4)_3$ solution and magnesium metal
(iv) $ZnSO_4$ solution and iron metal

(a) (ii) only (b) (ii) and (iii)
(c) (iii) and (iv) (d) (i) and (ii)

24.

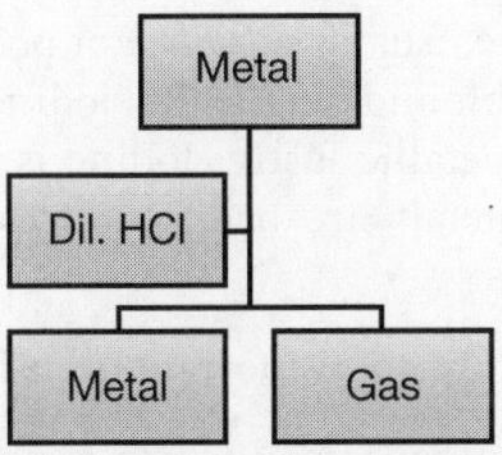

[CBSE 2021]

Which of the following combinations are correct?

Metal	Gas evolved
(i) Copper	Yes
(ii) Iron	Yes
(iii) Magnesium	No
(iv) Zinc	Yes

(a) (i) and (iii) (b) (i) and (iv)
(c) (ii) and (iii) (d) (ii) and (iv)

25. Match the metal (Column I) with its reaction with oxygen (Column II). **[CBSE 2021]**

Column I	Column II
A. Potassium	(i) Does not react even at high temperature
B. Zinc	(ii) Gets coated with a black-coloured layer of oxide
C. Copper	(iii) Does not burn at ordinary temperature
D. Silver	(iv) Burns vigorously

(a) (A)—(iv), (B)—(iii), (C)—(ii), (D)—(i)
(b) (A)—(iv), (B)—(ii), (C)—(i), (D)—(iii)
(c) (A)—(iii), (B)—(ii), (C)—(i), (D)—(iv)
(d) (A)—(iv), (B)—(ii), (C)—(iii), (D)—(i)

Assertion Reason Questions

Direction: In the following questions, two statements (Assertion) A and Reason (R) are given. Mark

(a) if A and R are both correct and R is the correct explanation of A;
(b) if A and R are both correct but R is not the correct explanation of A;
(c) A is true but R is false;
(d) A is false but R is true.

Assertion	Reason
1. Non-metals are elements that conduct heat and electricity and are malleable.	**1.** Non-metals are non-shiny and have a dull appearance.
2. Metals form cations.	**2.** Metals are electropositive, so they can lose electron/s.
3. Metals can be hammered into thin sheets.	**3.** Gold and silver are highly malleable metals.
4. Aluminium foils are used for packing food items.	**4.** Aluminium is quite malleable and convertible into thin sheets.
5. Cooking utensils are made of copper and aluminium.	**5.** Both copper and aluminium are good conductors of heat.
6. Sodium and potassium are cuttable with a knife.	**6.** Most of the metals are quite hard.
7. Metals on burning in air form acidic oxides.	**7.** Metal oxides turn red litmus into blue.
8. Zinc can displace hydrogen from dil. acids like HCl.	**8.** Zinc is more reactive than hydrogen and is placed above hydrogen.

Section B: Very Short Answer Questions (2 Marks)

1. Name two metals which are both ductile as well as malleable.
2. Name one metal which is a poor conductor of heat.
3. Give one example of a non-metal which is a conductor of electricity. Also explain why.
4. Name one non-metal and one metal which are in the liquid state at room temperature.
5. Metals are good conductors of electricity whereas non-metals are not. Why?
6. Name a non-metal which is lustrous and a metal which is non-lustrous.
7. Name a metal that offers higher resistance to the passage of electricity than copper.
8. Name the metal which has very low melting point and can melt with heat of your palm?
9. Why are metals highly electropositive in nature?
10. Name any two metals which can displace hydrogen from dilute HCl.
11. State two physical properties of gold which are of extreme use to jewellers.
12. Why is tungsten used almost exclusively for filaments of electric lamps?
13. What kind of oxides are formed when non-metals react with oxygen?
14. Differentiate between the oxides of magnesium and sulfur.
15. Why is sodium kept immersed in kerosene?
16. Which metals do not react with water at all?
17. Which one is more reactive, Cu or Pt?
18. An element X forms an oxide which turns red litmus blue. Identify whether X is a metal or a non-metal.
19. Between Zn and Ca, which burns with a bright flame to form an oxide?
20. Make a distinction between metals and non-metals with respect to the nature of their oxides.
21. Why are cutting tools usually made up of metals?
22. Name the metal which reacts with very dilute HNO_3 to evolve hydrogen gas.
23. Why do silver articles become black after sometime, when exposed to air?
24. An element forms an oxide A_2O_3 which is acidic in nature. Identify A as a metal or non-metal.
25. Arrange copper, silver and aluminium in the increasing order of reactivity.
26. A solution of $CuSO_4$ was kept in an iron pot. After few days the iron pot was found to have a number of holes in it. Explain the reason in terms of reactivity. Write the equation of the reaction involved.
27. What are strategic metals? Give a few examples.

Case Study or Passage Based Questions

There are 118 elements known till today. Some of them are metals and some are non-metals. Metals are usually hard, malleable and ductile and have a metallic lustre. Non-metals are usually soft, do not possess lustre and are not malleable and ductile. But iodine is a non-metal which has a metallic lustre. Iodine is also important for us. Non-metals are bad conductors of heat and electricity.

28. Why iodine is important for us?
29. Name a non-metal which is a good conductor of heat and electricity.
30. As a student, what initiative will you take to comment on the statement that 'Iodised salt is good for health.' Give any two suggestions.
31. Which are the most malleable metal and most ductile metal?
32. Name the best conductor of heat and the poorest conductor of heat.
33. Name the metal with the highest melting point and the metal with the lowest melting point. **[Board 2016]**

Section C: Short Answer Questions (3 Marks)

1. Name two solid metals and two solid non-metals along with their symbols.

2. Differentiate between metals and non-metals based on:
   (i) malleability (ii) electrical conduction

3. Write one example of each of:
   (i) a metal which is so soft that it can be cut with knife and a non-metal which is the hardest substance
   (ii) a metal and a non-metal which exist as liquids at room temperature

4. Name two metals that start floating after some time when immersed in water and explain why they do so.

5. Give two examples each of the metals that are good conductors and poor conductors of heat, respectively.

6. Name one metal and one non-metal that exist in the liquid state at room temperature. Also name two metals having a melting point less than 310 K (37°C).

7. Name the following.
   (i) A metal which is preserved in kerosene.
   (ii) A lustrous coloured non-metal.
   (iii) A metal which can melt while kept on the palm.
   (iv) A metal which is a poor conductor of heat.

8. State three reasons for the following facts.
   (i) Sulfur is a non-metal.
   (ii) Magnesium is a metal.
   One of these reasons must be supported with a chemical equation. **[CBSE 2015]**

9. Explain the meanings of malleable and ductile.

10. Which of the following listed metals can displace zinc from its salt solution? Give reasons for your answer along with a chemical equation. Copper, lead, magnesium, silver.

11. Write one example of each of the following.
    (i) Most malleable metal and most ductile metal.
    (ii) The best conductor of heat and the poorest conductor of heat.
    (iii) A metal with the highest melting point and a metal with the lowest melting point. **[CBSE 2016]**

12. What type of oxides are formed when non-metals combine with oxygen?

13. *A non-metal A is an important constituent of our food and forms two oxides B and C. Oxide B is toxic whereas C causes global warming.
    (i) Identify A, B and C.
    (ii) To which group of the periodic table does A belong?

14. Give reasons for the following.
    (i) School bells are made up of metals.
    (ii) Electric wires are made up of copper.

15. Mercury is the only metal found in the liquid state. It is largely used in thermometers to measure the temperature. But mercury is a very dangerous metal as its density is very high. What two precautions would you take while handling equipment containing mercury?

16. Explain why calcium metal after reacting with water starts floating on its surface. Write the chemical equation for the reaction. Name one more metal that starts floating after some time when immersed in water.

17. (i) What are amphoteric oxides? Choose the amphoteric oxides from amongst the following oxides: Na_2O_3, ZnO, Al_2O_3, CO_2, H_2O.
    (ii) Why do non-metals not displace hydrogen from dilute acids?

18. Write balanced chemical equations for the following reactions.
    (i) Dilute sulfuric acid reacts with aluminium powder.
    (ii) Dilute hydrochloric acid reacts with sodium carbonate.
    (iii) Carbon dioxide is passed through lime water.

19. Suggest a method of reduction for the following metals during their metallurgical processes.
    (i) Metal A which is at one of the last, second or third position in the reactivity.
    (ii) Metal B which gives a vigorous reaction even with water and air.
    (iii) Metal C which is kept in the middle of the activity series.

20. An iron nail is dipped in a solution of copper sulfate for about 30 minutes. State the change in colour observed. Give the reason for the change. **[CBSE 2015]**

21. Which of the following listed metals can displace zinc from its salt solution? Give

reasons for your answer along with the chemical equation. Copper, lead, calcium, gold. **[CBSE 2016]**

22. Explain the following statements.
    (i) Most metal oxides are insoluble in water but some of these dissolve in water. What are these oxides and their solutions in water called?
    (ii) At ordinary temperature, the surfaces of metals such as magnesium, aluminium and zinc, is covered with a thin layer. What is the composition of this layer? State its importance.
    (iii) Some alkali metals can be cut with a knife. **[CBSE 2016]**

Section D: Long Answer Questions (5 Makrs)

1. (i) Write two properties of gold which make it the most suitable metal for ornaments.
   (ii) Name two metals which are the best conductors of heat.
   (iii) Name two metals which melt when you keep them on your palm.
2. State the reason for the following.
   (i) Non-metals cannot displace hydrogen from the acids.
   (ii) Hydrogen is not a metal, yet it is placed in the activity series of metals.
   (iii) Aluminium is more reactive than iron, yet its corrosion is less that of iron. **[Board 2015]**
   (iv) Reaction of nitric acid with metals generally does not evolve hydrogen gas.
   (v) For making gold ornaments, 22 carat gold is preferred to 24 carat gold.
3. *(i) Of the three metals X, Y and Z, X reacts with cold water, Y with hot water and Z with steam only. Identify X, Y and Z and also arrange them in order of increasing reactivity.
   (ii) An element P burns with a golden flame in air. It reacts with another element Q with an atomic number 17 to give a product R. An aqueous solution of product R on electrolysis gives a compound S and liberates hydrogen. Identify P, Q, R and S. Also write down the equations for the reactions involved.
4. *Samples of four metals A, B, C and D were taken and added to the following solutions one by one. The results obtained have been tabulated as follows.

Metal	Iron(IIII) sulfate	Copper(IIII) sulfate	Zinc sulfate	Silver nitrate
A	No reaction	Displacement		
B	Displacement		No reaction	
C	No reaction	No reaction	No reaction	Displacement
D	No reaction	No reaction	No reaction	No reaction

   Use the table above to answer the following questions about metals A, B, C and D.
   (i) Select the most reactive metal.
   (ii) What are the colour changes observed when B reacts with copper sulphate?
   (iii) Write the decreasing order of reactivity of the metals A, B, C, D.
5. Complete the crossword given in the figure with the help of the clues.

Across

1. Which is generally hard, ductile, malleable and sonorous.
2. A metal is called so, if it can be drawn into wires.
3. Metal bells are used because of this property.

Down

4. A metal generally used for making jewellery.
5. A metal which is liquid at room temperature.
6. A metal which reacts with an acid as well as a base to form hydrogen gas.
7. Substances used to enhance the growth of plants.
8. Property by virtue of which metals can be beaten into thin sheets.

6. (i) What happen when—
(a) zinc reacts with copper sulfate solution
(b) aluminium reacts with steam
(c) sodium reacts with water
Give balanced equations for each.
(ii) Complete the following chemical reactions—
(a) $3Fe + 4H_2O \rightarrow$
(b) $Ca + H_2O \rightarrow$
(c) $K + H_2O \rightarrow$

7. (i) Write the balanced chemical equations for the following reactions.
(a) Copper is heated in air.
(b) Aluminium is heated in air.
(c) Aluminium oxide reacts with sodium hydroxide.
(ii) (a) Arrange the following metals in the order of their decreasing activities: aluminium, gold, sodium, copper.
(b) Give the chemical equation for the reaction of aluminium powder with manganese dioxide on heating.

8. Compose an activity to arrange Ca, Mg and Fe metals in the decreasing order of reactivity with water. Write suitable balanced chemical equation and draw diagrams.

9. *Shashi bought new silver anklets. After a month, she found that the anklets turned black. She doubts that it is not pure silver. Her neighbour Rakhi told her that it is pure silver and advised her to wash it with tamarind or lemon juice to get the lustre back. She followed the advice and got the lustre of her anklets back.
Answer the following questions based on the above passage.
(i) Why did the anklets turn black after some time?
(ii) Why did the anklets become lustrous on washing with tamarind or lemon juice?
(iii) What values are exhibited by Rakhi?

10. A metal (E) is stored under kerosene. When a small piece of it is left open in the air, it catches fire. When the product formed is dissolved in water, it turns red litmus to blue.
(i) Name the metal (E).
(ii) Write the chemical equation for the reaction when it is exposed to air and when the product is dissolved in water.
(iii) Explain the process by which the metal is obtained from its molten chloride.

11. What is meant by rusting? With labelled diagrams, describe an activity to find out the conditions under which iron rusts.

12. Compare in tabular form, the reactivities of the following metals with cold and hot water.
(i) Sodium
(ii) Calcium
(iii) Magnesium **[CBSE 2020]**

Section E: Case Study or Passage Based Questions (4 Makrs)

All metals are not equally reactive, that is, some metals are highly reactive while some are less reactive and some metals are almost unreactive. For example, Na, K are highly reactive and react with cold water while Mg, Ca are reactive and react with hot water. On the other hand, Fe and Zn are less reactive and can react with steam. When different metals are arranged in the decreasing order of their reactivity, an activity series of metals is obtained. A more reactive metal can displace a less reactive metal from its salt while a less reactive metal cannot do that.

1. The arrangement for copper, tin, lead and mercury, according to the reactivity series is:
(a) copper > mercury > tin > lead
(b) lead > copper > mercury > tin
(c) tin > lead > copper > mercury
(d) mercury> tin> lead> copper

2. Which of the following pairs will give displacement reactions?
(a) NaCl solution and copper metal
(b) $AgNO_3$ solution and copper metal
(c) $CuSO_4$ solution and silver metal
(d) $FeSO_4$ solution and copper metal

3. Which of the following oxide(s) of iron would be obtained on prolonged reaction of iron with steam?
   (a) Fe_2O_3 (b) FeO
   (c) Fe_2O_3 and Fe_2O_4 (d) Fe_3O_4

4. Copper objects lose their shine and form a green coating of
   (a) copper oxide
   (b) copper carbonate
   (c) basic copper carbonate
   (d) copper hydroxide and copper oxide

5. An aluminium strip is kept immersed in freshly prepared ferrous sulfate solution taken in a test tube, the change observed is that
   (a) a colourless gas with the smell of burning sulfur is observed
   (b) the lower end of the test tube becomes slightly warm
   (c) the green solution slowly turns brown
   (d) the light green solution changes to blue

High Order Thinking Skills (HOTS) Questions

1. *Shanvi treated a lustrous, divalent element M with sodium hydroxide. She observed the formation of bubbles in the reaction mixture. She made the same observations when this element was treated with hydrochloric acid. Suggest how she can identify the produced gas. Write chemical equations for both the reactions.

2. You are given a hammer, a battery, a bulb, wires and a switch.
   (i) How could you use them to distinguish between samples of metals and non-metals?
   (ii) Assess how useful these tests are in distinguishing between metals and non-metals.

3. Explain the following statements:
   (i) Most metal oxides are insoluble in water but some dissolve in water. What are these oxides and their solutions in water called?
   (ii) At ordinary temperature, the surface of metals such as magnesium, aluminium, zinc, and so on is covered with a thin layer. What is the composition of this layer? State its importance.
   (iii) Some alkali metals can be cut with a knife.

4. (i) Out of the two metals P and Q, P is less reactive than Q. Suggest an activity to arrange these metals in the order of their decreasing reactivity. Support your answer with a suitable chemical equation.
   (ii) A metal is found in the liquid state. It is widely used in instruments for measuring blood pressure. In what form does it occur in nature? How can we extract this metal from its ore?

5. (i) A metal M does not liberate hydrogen from acids but reacts with oxygen to give a black coloured product. Identify M and the black coloured product; also explain the reaction of M with oxygen.
   (ii) Explain why calcium metal after reacting with water starts floating on its surface. Write the chemical equation for the reaction.

6. An alkali metal A gives a compound B (molecular mass = 40) on reacting with water. The compound B gives a soluble compound C on treatment with aluminium oxide. Identify A, B and C give the reaction involved.

7. A student performed the following displacement reaction.
   $Fe(s) + CuSO_4(aq) \rightarrow FeSO_4(aq) + Cu(s)$
   $Zn(s) + FeSO_4(aq) \rightarrow ZnSO_4(aq) + Fe(s)$
   $2Al(s) + 3ZnSO_4(aq) \rightarrow Al_2(SO_4)_3(aq) + 3Zn(s)$
   Arrange Fe, Zn, Al and Cu in the decreasing order of reactivity on the basis of the above reaction.

8. A man went door to door posing to be a goldsmith. He promised to bring back the glitter of old and dull gold ornaments. An unsuspecting lady gave a set of gold bangles to him which he dipped in a particular solution. The bangles sparkled like new but their weight was reduced drastically. The lady was upset but after a futile argument, the man beat a hasty retreat.
   (i) Can you play the detective to find out the nature of the solution he had used?
   (ii) What is 24 carat gold?

9. (i) Why are electrical wires coated with PVC (polyvinyl chloride)?
   (ii) Why do we apply oil on iron hooks kept in storage?

10. You are given samples of three metals—sodium, magnesium and copper. Suggest any two activities to arrange them in the order of decreasing reactivity.

11. P, Q and R are 3 elements which undergo chemical reactions according to the following equations.
(a) $P_2O_3 + 2Q \rightarrow Q_2O_3 + 2P$
(b) $3RSO_4 + 2Q \rightarrow Q_2(SO_4)_3 + 3R$
(c) $3RO + 2P \rightarrow P_2O_3 + 3R$
Answer the following based on these.
(i) Which element is more reactive?
(ii) Which element is least reactive?
(iii) State the type of reaction listed above.

12. A student added a piece of zinc metal to four test tubes I, II, III, IV which respectively contain aqueous solutions of aluminium sulfate, zinc sulfate, ferrous sulfate and copper sulfate. In which solution will the reaction take place?

13. A student performed the following four experiments.

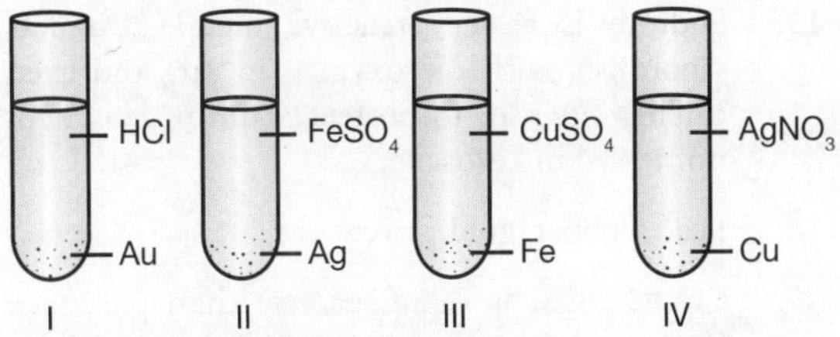

In which experiment did the reaction occur?

14. A student added few pieces of aluminium metal to two test tubes A and B, containing aqueous solutions of iron sulfate and copper sulfate. In the second part of her experiment, she added iron metal to another set of test tubes C and D, containing aqueous solutions of aluminium sulfate and copper sulfate. In which test tube or test tubes will she observe a change in colour? On the basis of this experiment, state which one is the most reactive metal and why. **[CBSE 2018]**

Answers

Section A: Multiple Choice Questions

1. (b)	**2.** (d)	**3.** (d)	**4.** (d)
5. (c)	**6.** (a)	**7.** (b)	**8.** (b)
9. (d)	**10.** (b)	**11.** (b)	**12.** (d)
13. (d)	**14.** (b)	**15.** (a)	**16.** (c)
17. (d)	**18.** (c)	**19.** (c)	**20.** (d)
21. (a)	**22.** (a)	**23.** (b)	**24.** (d)
25. (a)			

Assertion Reason Questions

1. (d)	**2.** (a)	**3.** (d)	**4.** (a)
5. (a)	**6.** (b)	**7.** (d)	**8.** (a)

Section B: Very Short Answer Questions

1. Gold and silver are ductile and malleable.

2. Lead (Pb) or mercury (Hg).

3. Graphite is a non-metal but a conductor of electricity. This is because it has free electrons.

4. Bromine (non-metal) and mercury (metal) are liquids at room temperature.

5. Metals have free electrons that move freely, so they conduct electricity.

6. Non-metal (lustrous) = iodine (I)
Metal (non-lustrous) = sodium (Na)

7. Iron or aluminium.

8. Gallium.

9. Metals are highly electropositive elements as they can lose electrons very easily to form positively charged ions. For example,

$$Na \rightarrow Na^+ + e^- \qquad Mg \rightarrow Mg^{2+} + 2e^-$$

10. Zinc and iron can displace H_2 from HCl.

11. Malleability, ductility, lustrous **(any two)**.

12. Tungsten has a high melting point and high resistance.

13. Acidic oxides are formed when non-metals react with oxygen.

$$C + O_2 \rightarrow CO_2$$

14. Oxide of magnesium is basic in nature. Oxide of sulfur is acidic in nature.

15. Sodium is a very reactive metal. It reacts vigorously with the oxygen (in air) and even catches fire. So, to protect sodium, it is kept immersed in kerosene.

16. Lead, copper, gold, silver.

17. Copper (Cu) is more reactive than platinum (Pt) as copper is above Pt in the reactivity series.

18. Metal, because metal oxides are basic in nature.

19. Ca reacts and burns with a bright flame on heating in air while Zn reacts slowly.

20. Metallic oxides are basic, a few are amphoteric. Non-metallic oxides are acidic, a few are neutral.

21. Cutting tools are usually made up of metals as metals are hard and strong.

22. Manganese or magnesium.

23. They get tarnished by reacting with atmospheric air to form silver sulfide.

24. The oxide of the element is acidic in nature, therefore, A will be a non-metal.

25. Ag < Cu < Al.
Silver < Copper < Aluminium.

26. Fe is more reactive as compared to Cu. Therefore, Fe displaces Cu from $CuSO_4$ and forms $FeSO_4$.
$Fe + CuSO_4 \rightarrow FeSO_4 + Cu$

27. Metals used for a country's defence are called strategic metals. For example, titanium, chromium, manganese and zirconium.

Case Study or Passage Based Questions

28. Deficiency of iodine leads to goitre. It also leads to hypothyroidism, a common disease among women above 40 years of age.

29. Graphite is a non-metal and a good conductor of heat and electricity.

30. **Suggestions:**
(a) To start an awareness campaign in localities, homes and schools.
(b) To understand the importance of iodine for people.

31. Gold and silver are the most malleable metal and most ductile metal.

32. Silver is the best conductor of heat and lead is the poorest conductor of heat.

33. Metals with the highest melting point—tungsten, iron. Metals with the lowest melting point—gallium, cesium.

Section C: Short Answer Questions

1. Two solid metals are copper (Cu) and sodium (Na).
Two solid non-metals are iodine (I) and silicon (Si).

2.

(i)	Malleability	Metals can be beaten into thin sheets.	Non-metals are non-malleable.
(ii)	Electrical conductivity	Metals are good conductors of heat and electricity.	Non-metals are poor conductors of heat and electricity.

3. (i) Sodium is a soft metal and carbon (diamond) is the hardest non-metal.
(ii) Mercury is a metal (liquid) and bromine is a non-metal (liquid).
Note: Mercury is a metal as it has free electrons as do all metals.

4. The two metals are calcium (Ca) and magnesium (Mg). They start floating because the bubbles of hydrogen gas which are formed during the reaction stick to the surface of the metal.

5. (i) Good conductors: Ag and Cu.
(ii) Poor conductors: Pb and Hg.

6. Metal: mercury (Hg); non-metal: bromine (Br). Two metals with melting points less than 310 K are cesium (Cs) and gallium (Ga).

7. (i) Sodium is preserved in kerosene.
(ii) Iodine is a lustrous coloured non-metal.
(iii) Gallium can melt while kept on the palm.
(iv) Lead is a poor conductor of heat.

8. (i) Sulfur is a non-metal.
(a) It is a poor conductor of heat and electricity.
(b) It is neither malleable nor ductile.
(c) It forms acidic oxides.
$S + O_2 \rightarrow SO_2$

$SO_2 + H_2O \rightarrow H_2SO_3$ (sulfurous acid)
SO_2 is an acidic oxide.
Note: Sulfur is an ingredient of gunpowder and is used in making matches and as an antiseptic and fungicide.

(ii) Magnesium is a metal.
   (a) It is a good conductor of heat and electricity.
   (b) It is malleable and ductile.
   (c) It forms basic oxides.
   $2Mg + O_2 \rightarrow 2MgO$

   $MgO + H_2O \rightarrow \underset{\text{Magnesium hydroxide}}{Mg(OH)_2}$

   MgO is basic in nature.

9. **Malleable:** A metal is said to be malleable if it can be beaten into thin sheets. For example, silver and gold are highly malleable metals.
Ductile: A metal is said to be ductile when it can be drawn into thin wires. For example, platinum and gold are highly ductile metals.

10. **Metals** that are more reactive than zinc can displace it from its salt solution. So, magnesium can displace Zn from its salt solution.

$$Mg(s) + ZnSO_4(aq) \rightarrow MgSO_4(aq) + Zn(s)$$

11. (i) Gold is the most malleable and most ductile metal.
(ii) Silver is the best conductor and lead is the poorest conductor.
(iii) Tungsten has the highest melting point and mercury has the lowest melting point.

12. Non-metals combine with oxygen to form neutral and acidic oxides.

$C + \frac{1}{2}O_2 \rightarrow CO$ (neutral oxide)

$N_2 + O_2 \rightarrow NO$ (neutral oxide)

$C(s) + O_2 \rightarrow CO_2$ (acidic oxide)

$N_2 + 2O_2 \rightarrow 2NO_2$ (acidic oxide)

$S(s) + O_2(g) \rightarrow SO_2(g)$ (acidic oxide)

13. (i) A is carbon, B is carbon monoxide and C is carbon dioxide.
(ii) A belongs to group 14 of the periodic table.

14. (i) School bells are made up of metals, as metals produce sound when hit (that is, they are sonorous).
(ii) Copper, being a good conductor of electricity, is used for making electric wires.

15. Equipment containing mercury should be handled carefully.
If broken or discarded, their disposal should be done carefully.

16. Calcium metal after reacting with water starts floating on its surface because the bubbles of hydrogen gas formed will stick to the surface of metal.

$$Ca(s) + 2H_2O(l) \rightarrow Ca(OH)_2 + H_2(g)$$

Another metal is magnesium. It reacts with hot water and starts floating after some time when immersed in water.

17. (i) Those oxides which are both acidic as well as basic are called amphoteric oxides. For example, Al_2O_3, ZnO.
(ii) Except carbon, all other non-metals are below hydrogen in the activity series. So, they cannot displace hydrogen from dilute acids.

18. (i)

$$2Al(s) + 3H_2SO_4(dil.) \rightarrow Al_2(SO_4)_3(aq) + 3H_2(g)$$

(ii) $Na_2CO_3(s) + 2HCl(dil.) \rightarrow 2NaCl(aq) + H_2O(l) + CO_2(g)$

(iii)

$$\underset{\text{Lime water}}{Ca(OH)_2(aq)} + CO_2(g) \rightarrow \underset{\text{White ppt.}}{CaCO_3(s)} + H_2O(l)$$

Note: If excess carbon dioxide is allowed to pass, the turbidity disappears due to the formation of calcium bicarbonate which is soluble in water.

19. (i) A can be obtained by chemical reduction using carbon or carbon monoxide as the reducing agent.
(ii) B can be obtained by electrolytic reduction.
(iii) C can be reduced through the use of a reducing agent such as Al.

20. Iron is more reactive than copper. So, it displaces Cu from copper sulfate solution. Hence, the blue colour of $CuSO_4$ solution fades and it turns green due to the formation of ferrous sulfate solution.

$$\underset{\text{Blue}}{CuSO_4(aq)} + Fe(s) \rightarrow \underset{\text{Green}}{FeSO_4(aq)} + Cu(s)$$

21. Metals that are more reactive than zinc can displace it from its salt solution. So, calcium can displace Zn from its salt solution.

$$Ca(s) + ZnSO_4(aq) \rightarrow CaSO_4(aq) + Zn(s)$$

22. (i) Solution of a metal oxide in water is called an alkali. For example, NaOH (sodium hydroxide).

(ii) This layer is called the oxide layer. It makes the metal less reactive and due to this layer, the metal cannot get corroded and remains intact.

(iii) Some alkali metals are very soft so they can be cut with a knife. For example, sodium metal.

Note: Due to the presence of weak metallic bonding, alkali metals are very soft.

Section D: Long Answer Questions

1. (i) Gold is an inert metal and it has a shiny surface. It is also malleable and ductile. Therefore, it is used for making ornaments.

(ii) Copper and aluminium.

(iii) Cesium and gallium.

2. (i) Non-metals are electron acceptors; they cannot supply electrons to convert H^+ ions to $H_2(g)$.

(ii) Like metals, hydrogen can lose an electron to form the positive H^+ ion.

(iii) Aluminium is covered with a strong protective layer of oxide which protects the metal from further corrosion.

(iv) Reaction of metals with nitric acid does not evolve hydrogen gas. It is because HNO_3 is a strong oxidising agent. It oxidises H_2 to produce water and itself gets reduced to any of its oxides such as N_2O, NO or NO_2.

(v) Pure gold is 24 carat and is very soft. Therefore, it is not suitable for making jewellery. It is alloyed with Cu or Ag to make it hard. Generally, 22 carat gold is used for making ornaments, which contains 22 parts of pure gold and 2 parts of either Cu or Ag.

3. (i) X is an alkali metal, Na or K.
Y is an alkaline earth metal, Mg or Ca.
Z is Fe.
Increasing reactivity series: Na > Mg > Fe

(ii) P = Na; Q = Cl_2; R = NaCl; S = NaOH

$2Na + Cl_2 \rightarrow 2NaCl$

$2NaCl(aq) + 2H_2O(l) \rightarrow 2NaOH(aq) + Cl_2(g) + H_2(g)$

4. From the given table we can write the following observations:

- A is less reactive than Fe but more reactive than Cu.
- B is more reactive than Fe but less reactive than Zn.
- C is more reactive than Ag but less reactive than Fe, Cu and Zn.
- D is less reactive than Fe, Cu, Zn, Ag.

Hence: (i) B is the most reactive.

(ii) The blue colour of copper sulfate solution disappears, and reddish-brown copper metal is deposited on the metal B.

(iii) B > A > C > D is the order of reactivity.

5.

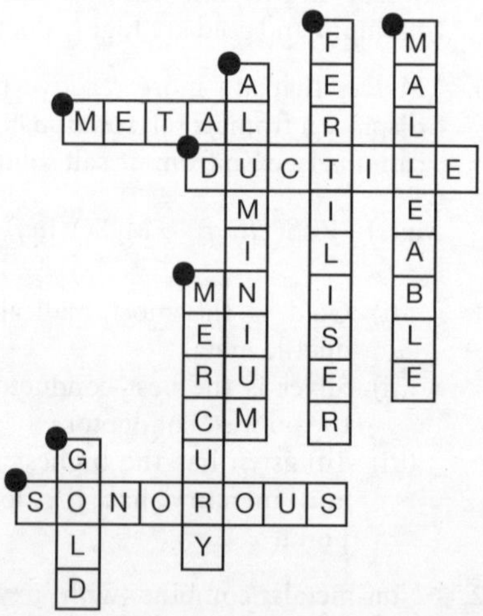

Across

1. Metal
2. Ductile
3. Sonorous

Down

4. Gold
5. Mercury
6. Aluminium
7. Fertiliser
8. Malleable

6. (i) (a) $Zn + CuSO_4 \rightarrow ZnSO_4 + Cu$

(b) $2Al + 3H_2O \rightarrow Al_2O_3 + 3H$

(c) $2Na + 2H_2O \rightarrow 2NaOH + H_2$

(ii) (a) $3Fe + 4H_2O \rightarrow Fe_2O_3 + 4H_2$

(b) $Ca + H_2O \rightarrow Ca(OH)_2 + H_2$

(c) $K + H_2O \rightarrow 2KOH + H_2$

7. (i) (a) $2Cu + O_2 \rightarrow 2CuO$

(b) $4Al + 3O_2 \rightarrow 2Al_2O_3$

(c) $Al_2O_3 + 2NaOH \rightarrow 2NaAlO_2 + H_2O$

(ii) (a) sodium (Na) > aluminium (Al) > copper (Cu) > gold (Au).

(b) $3MnO_2(s) + 4Al(s) \rightarrow 3Mn(l) + 2Al_2O_3(s) + Heat$

(ii) (a) $4Na + O_2 \rightarrow 2Na_2O$

(b) $Na_2O + H_2O \rightarrow 2NaOH$

(iii) **Electrolytic reduction:** Electric current is passed through molten NaCl. Sodium is formed at the cathode and $Cl_2(g)$ is liberated at the anode.

$$2NaCl \xrightarrow{\text{Molten}} 2Na + Cl_2(g)$$

8. Take some metals like Ca, Mg and Fe. Place separately small pieces of these metals in beakers with a small amount of water in them. If the metals react with the cold water, then the reaction is vigorous. If metal does not react with cold water, then treat it with hot water. Record the observations and arrange the metals in decreasing order of reactivity.

Metal	Reaction	Observation	Chemical reaction
Fe	Steam	Fe_3O_4 and $H_2(g)$	$3Fe(s) + 4H_2O(g) \rightarrow Fe_3O_4(s) + H_2(g)$
Mg	Hot water	$Mg(OH)_2(aq)$ and $H_2(g)$	$Mg(s) + 2H_2O(hot) \rightarrow Mg(OH)_2(aq) + H_2(g)$
Ca	Cold water	$Ca(OH)_2(aq)$ and $H_2(g)$	$Ca(s) + 2H_2O(l) \rightarrow Ca(OH)_2(aq) + H_2(g)$

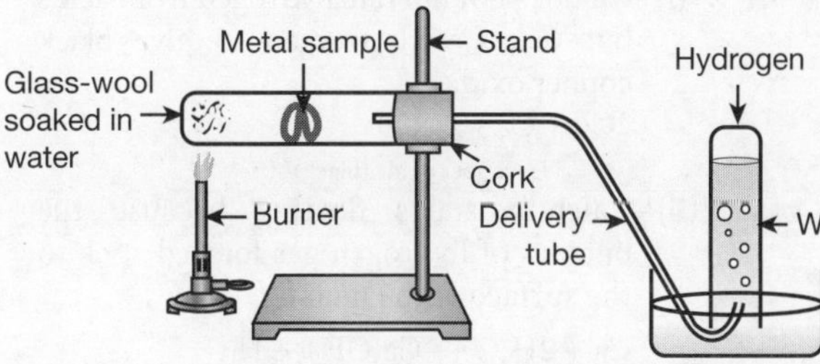

9. (i) Anklets turned black due to the reaction of silver with atmospheric H_2S, which results in the formation of black coloured silver sulfide (Ag_2S).

(ii) Ag_2S dissolves in acids present in tamarind or lemon juice, so the anklets got their lustre back.

(iii) The values exhibited by Rakhi are awareness, helpful, knowledgeable and scientific temperament.

10. (i) E is sodium. It catches fire in the presence of moisture.

11. **Rusting:** The process in which iron reacts with oxygen in the presence of moisture to form a reddish-brown coating of hydrated ferric oxide [$Fe_2O_3.xH_2O$] is called rusting.

Activity: (i) Take three boiling tubes A, B and C.

(ii) Pour some water in test tube A. Put iron nails in it and cork it.

(iii) Pour boiled, distilled water in another test tube B and put iron nails in it. Add 1 mL of oil over it such that oil floats over it prevent the air from entering in.

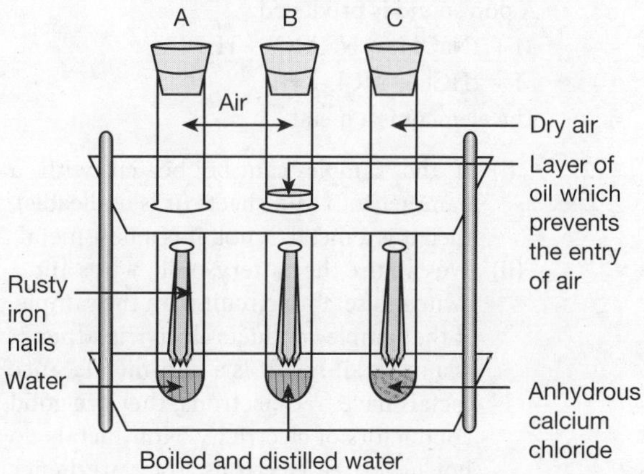

(iv) Take some more iron nails in test tube C and put some anhydrous calcium chloride in it and cork it.

(v) Leave all the three test tubes for a day and then observe the changes.

Observation: Iron nails get rusted in test tube A because both air and water are present in it. Iron nails do not get rusted in test tube B because there is water but no air. In test tube C, rusting will not take place because there is neither air nor water.

Conclusion: Iron gets rusted in the presence of air and water.

12.

Metal	Reactivity	Reactions
Sodium (Na)	It reacts vigorously with cold water.	$2Na(s) + 2H_2O(l) \rightarrow 2NaOH(aq) + H_2(g)$
Calcium (Ca)	It reacts slowly with cold water and moderately with hot water.	$Ca(s) + 2H_2O(l) \rightarrow Ca(OH)_2 + H_2(g)$ (Hot)
Magnesium (Mg)	Magnesium does not react with cold water. It reacts slowly with hot water and vigorously with steam.	$Mg(s) + 2H_2O(l) \rightarrow Mg(OH)_2 + H_2(g)$ (Hot) $Mg(s) + H_2O(l) \rightarrow MgO + H_2(g)$ (Steam)

Section E: Case Study or Passage Based Questions

1. (c) 2. (b) 3. (d) 4. (c)
5. (d)

High Order Thinking Skills (HOTS) Questions

1. The produced gas can be identified by bringing a burning match stick near the reaction vessel. A pop sound is produced.
$M + 2NaOH \rightarrow Na_2MO_2 + H_2$
$M + 2HCl \rightarrow MCl_2 + H_2$
The element is a metal.

2. (i) If the sample can be beaten with a hammer into thin sheets (it is malleable), then it is a metal, if not, it is a non-metal.
(ii) We can use the battery, bulb, wires and a switch to set up a circuit with the sample. If the sample conducts electricity, then it is a metal, if not, it is a non-metal. Since metals have free electrons, they are good conductors of electricity. Non-metals do not have free electrons; they are hence bad conductors of electricity.
The above tests are useful in distinguishing between metals and non-metals as these are based on the physical properties. No chemical reactions are involved in these tests.

3. (i) A solution of a metal oxide in water is called an alkali; for example, NaOH (sodium hydroxide).
(ii) This layer is called an oxide layer. It makes the metal less reactive and due to this layer, the metal cannot get corroded and remains intact.
(iii) Some alkali metals are very soft so they can be cut with a knife; for example, sodium metal.

Note: Due to the presence of weak metallic bonding, alkali metals are very soft.

4. (i) **Activity:** A metal Q is added to the salt solution of P. Q being more reactive, displaces the metal P from its salt solution.
Chemical equation: Metal Q + Salt solution of P $\rightarrow$ Salt solution of Q + Metal P.
(ii) Mercury is the metal which is found in liquid state and is used in instruments for measuring blood pressure. It occurs in nature as its sulfide ore in the form of cinnabar (HgS). Mercury is obtained from purified cinnabar by roasting in the presence of air.
$$HgS \xrightarrow[\text{Heat}]{\text{Air}} Hg + SO_2$$

5. (i) Cu does not liberate hydrogen from acids but reacts with oxygen to give black copper oxide.
$2Cu + O_2 \rightarrow \underset{\text{Copper oxide (black)}}{2CuO}$
(ii) Calcium starts floating because the bubbles of hydrogen gas formed stick to the surface of the metal.
$Ca + 2H_2O \rightarrow Ca(OH)_2 + H_2$

6. Let the atomic weight of the alkali metal A be x. When it reacts with water, it forms a compound B having a molecular mass of 40. Let the reaction be:
$2A + 2H_2O \rightarrow 2AHO + H_2\uparrow$
According to the question, $x + 16 + 1 = 40$
$x = 40 - 17 = 23$
23 is the atomic weight of Na (sodium) Therefore, the alkali metal (A) is Na and the reaction is:
$2Na + 2H_2O \rightarrow 2NaOH(aq) + H_2(g)$
So, the compound B is sodium hydroxide which reacts with aluminium oxide to give sodium aluminate.
$Al_2O_3 + \underset{(B)}{2NaOH} \rightarrow \underset{(C)}{2NaAlO_2} + H_2O$

7. According to the reactivity series, Al > Zn > Fe > Cu.

8. (i) The nature of the solution is acidic. It is a mixture of conc. HNO_3 and conc. HCl in the ratio of 1 : 3. It can dissolve gold as it is an oxidising agent.
   (ii) Pure gold is 24 carat gold which is very soft.

9. (i) Electric wires and cables are coated with substance like PVC to prevent electric shocks since PVC is an insulator (bad conductor of electricity) and it does not conduct electricity. It is therefore an excellent material to use for electrical applications such as insulation sheathing for cables.
   (ii) Oils help to prevent corrosion by excluding moisture and oxygen from the metal surface. All metals will form a thin layer of oxide on the surface and oil can penetrate into this, helping to stabilise it and prevent destructive corrosion.

10. **Activity 1:** Sodium reacts with cold water vigorously to form sodium hydroxide and hydrogen gas.

$$2Na(s) + \underset{\text{Cold}}{2H_2O(l)} \rightarrow 2NaOH(aq) + H_2(g)$$

Magnesium does not react with cold water; but it reacts with hot water to form magnesium hydroxide and hydrogen gas.

$$Mg(s) + \underset{\text{Hot}}{2H_2O} \rightarrow Mg(OH)_2(aq) + H_2(g)$$

Thus, sodium is more reactive than magnesium.
Activity 2: Magnesium reacts with copper sulfate to give magnesium sulfate and copper which gives no further reaction.

$$Mg(s) + CuSO_4(aq) \rightarrow MgSO_4(aq) + Cu(s)$$

$$Cu(s) + MgSO_4(aq) \rightarrow \text{No reaction}$$

Hence, magnesium is more reactive than copper. So, the order of decreasing reactivity is: Na > Mg > Cu.

11. (i) Most reactive metal is Q as it has replaced both P and R from their compounds.
   (ii) Element R is least reactive as it has been replaced by both R and Q.
   (iii) Displacement reaction.

12. The reaction will take place in solutions III and IV. Zn will react with ferrous sulfate and copper sulfate.
$Zn + CuSO_4 \rightarrow ZnSO_4 + Cu$: Zn > Cu
$Zn + FeSO_4 \rightarrow ZnSO_4 + Fe$: Zn > Fe

13. Reaction does not occur in I, II.
Reason: The reaction occurred in (III) because Fe is more reactive than Cu, so it will displace Cu from $CuSO_4$. It occurred in (IV) because Cu is more reactive than Ag, so it will displace Ag from $AgNO_3$.

14. When aluminium is added to the test tubes A and B containing the aqueous solutions of iron sulfate and copper sulfate, then aluminium will displace both iron and copper from their salt solutions. This is because aluminium is more reactive than both iron and copper.
   Again, when iron metal is added to test tube C containing aluminium sulfate, it will not react but when added to copper sulfate in test tube D, it will displace copper from its salt solution. So, a colour change will be observed in test tubes A, B and D but no colour change will take place in test tube C. Aluminium is the most reactive metal among all as it can displace both iron and copper from their salt solutions.

Hints

Section A: Multiple Choice Questions

3. Nitrogen is used during packing of food materials like chips to protect them from rancidity.

9. X = 2, 8; so it is neon with atomic number 10 and is a non-metal.
Y = 2, 8, 7; so it is chlorine with atomic number 17 and is a non-metal.
Z = 2, 8, 2; so it is magnesium with atomic number 12 and a is metal.

10. When a pellet of sodium is dropped in water, it catches fire and forms sodium hydroxide.

11. $CaO + H_2O \rightarrow Ca(OH)_2$

12. Iron being more reactive than copper displaces copper from copper sulfate.

$CuSO_4 + Fe \rightarrow FeSO_4 + Cu$

13. $Al_2O_3 + 2NaOH \rightarrow \underset{\text{Sodium meta aluminate}}{2NaAlO_2} + H_2O$

19. $P_4(s) + 5O_2(g) \rightarrow P_4O_{10}(s)$

$2Ca(s) + O_2(g) \rightarrow 2CaO(s)$

Case Study or Passage Based Questions

2. Copper being more reactive than silver displaces Ag from $AgNO_3$

$2AgNO_3 + Cu \rightarrow Cu(NO_3)_2 + 2Ag$

3. $3Fe(s) + 4H_2O(g) \rightarrow Fe_3O_4(s) + 4H_2(g)$

5. As aluminium is more reactive than iron, it will displace iron from the solution. That is, when an aluminium strip is kept immersed in freshly prepared light green coloured ferrous sulfate solution, aluminium displaces iron from iron sulfate and forms blue colored aluminium sulfate.

$2Al(s) + \underset{\text{Green}}{3FeSO_4} \rightarrow \underset{\text{Blue}}{Al_2(SO_4)_3} + 3Fe(s)$

3.8 Why Do Metals and Non-Metals Combine?

Metals on reaction with non-metals form ionic compounds having ionic bonds while non-metals on reaction with other non-metals form covalent compounds having covalent bonds. Now a question arises—why does a metal not react with other metals? In order to understand this, we have to know what a chemical bond is. When atoms of elements combine together they form molecules and a force of attraction is developed between them, this force of attraction that holds atoms or ions together in a molecule is known as a *chemical bond.* In order to understand the formation of a chemical bond we have to understand why inert gases are inert or why atoms combine.

We know that different elements have different atomic numbers and electronic configurations. The properties of atoms depend upon their electronic configurations. Some atoms are more reactive than others. Noble gas (He, Ne, Ar, Kr, Xe and Rn) atoms are not reactive at all; they are inert and stable. Then the question arises, why do noble gases not react to form compounds, while other elements do so? This can be answered by comparing the electronic configurations of noble gases with those of other elements. Also, it is important to understand how and why atoms react to form molecules and compounds. Atoms gain electrons in their outermost shells or lose them from their outermost shells, or share electrons with other atoms in such a way that their outermost shells become filled to capacity. They can do this by reacting with other atoms. As long as the outermost shell can accommodate more electrons, that is, it is not full, an atom tends to combine with other atoms in order to fill its outermost shell. When the outermost shell is filled to capacity, the atom becomes stable.

The atoms of all other elements (elements other than the noble gases) have in their outermost shells less than 8 electrons, that is, their outermost shells are not filled to capacity. Therefore, the atoms of these elements combine with other atoms to achieve stable configurations like those of the noble gases. It is the tendency on the part of an atom to achieve a stable configuration (noble gas configuration, Table 3.4) which is responsible for its chemical reactivity. An atom can obtain nearest inert gas configuration in the following three ways.

Table 3.4 Electronic configurations of noble gases (inert gases)

Noble gas (inert gas)	Symbol	Atomic number	Electronic configuration K L M N O P	Number of electrons in outermost shell (valence shell)
1. Helium	He	2	2	2
2. Neon	Ne	10	2, 8	8
3. Argon	Ar	18	2, 8, 8	8
4. Krypton	Kr	36	2, 8, 18, 8	8
5. Xenon	Xe	54	2, 8, 18, 18, 8	8
6. Radon	Rn	86	2, 8, 18, 32, 18, 8	8

(i) By losing one or more electrons to any other atom.
(ii) By gaining one or more electrons from any other atom.
(iii) By sharing one or more electrons with any other atom.

3.9 Ions

Before we start discussing the reactions between metals and non-metals by the transfer of valence electrons, it is quite necessary to understand about ions, their types and how they are formed.

An atom is formed of sub-atomic particles—electrons (negatively charged), protons (positively charged) and neutrons (neutral) and in an atom, the number of protons and electrons are equal, so it is neutral. If electrons are removed or added, the neutrality of the atom is lost as the number of protons and electrons are not equal and the new species that is formed and possesses a charge is known as an *ion*. That is, an ion is formed by the loss or gain of electrons. Now we can say that compounds composed of metals and non-metals can contain charged species which are known as *ions*. An ion is a positively or negatively charged atom (or group of atoms). It is formed by the loss or gain of electrons by an atom, so it contains an unequal number of electrons and protons.

Types of Ions

There are two types of ions—cations and anions.

3.9.1 Cations and their Formation

In case any element has 1, 2 or 3 valence electrons in the outermost orbital or valence shell of its atoms, then it prefers to lose these electrons to get the nearest inert gas configuration or octet state. In such a case, the addition of 7 or 6 or 5 electrons is not favourable at all due to energy considerations. Now it is clear that such an element or metal atom loses its electron(s) to form a positively charged ion or cation. A positively charged ion is known as a cation and it is formed by the loss of one or more electrons by an atom. The ions of all the metal elements are cations. As we know, in a neutral atom, the number of electrons and protons are always equal; however, in the case of a cation, the number of electrons become less than those of protons. The number of electrons lost by an atom is equal to the positive charge assigned on the ion formed. For example, a sodium atom loses one electron to form a sodium ion, Na^+ (cation), a magnesium atom loses two electrons to form a magnesium ion (Mg^{2+}) cation and an aluminium atom loses three electrons to form an aluminium ion (Al^{3+}) cation.

Formation of a few cations

(i) **Formation of lithium ion (Li^+):** As a lithium atom has only one valence electron, it prefers to lose this electron and form the Li^+ cation with the stable inert gas configuration of He.

$$Li \xrightarrow{-1\text{ electron}} Li^+$$

Lithium atom	Lithium ion
Protons = 3 (+ charge)	Protons = 3 (+charge)
Electrons = 3 (– charge)	Electrons = 2 (–charge)
Overall charge = 0	Overall charge = 1+

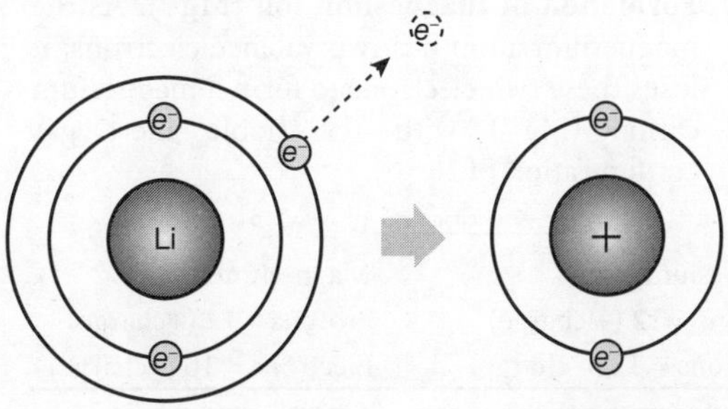

Lithium   Lithium ion

Both Li and Li^+ have 3 protons; however, the number of electrons in Li and Li^+ are 3 and 2 respectively.

(ii) **Formation of sodium ion (Na^+):** As the sodium atom has only one valence electron, it loses this one electron to form the sodium cation (Na^+) with the stable inert gas configuration of Ne.

$$Na \xrightarrow{-1\text{ electron}} Na^+$$

Sodium atom	Sodium ion
Protons = 11 (+ charge)	Protons = 11 (+charge)
Electrons = 11 (– charge)	Electrons = 10 (–charge)
Overall charge = 0	Overall charge = 1+
Elec. conf. 2, 8, 1	Elec. conf. 2, 8

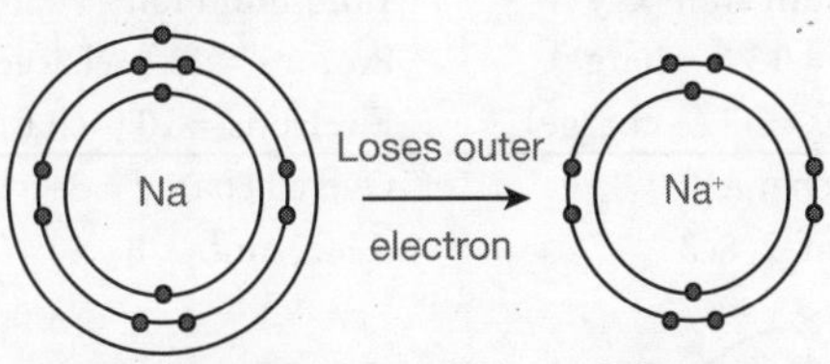

Sodium atom   Sodium ion

(iii) **Formation of potassium ion (K^+):** As the potassium atom has only one valence electron, it loses one electron to form a potassium cation (K^+) with the stable inert gas configuration of Ar.

K $\xrightarrow{-1\ \text{electron}}$	K^+
Potassium atom	Potassium ion
Protons = 19 (+ charge)	Protons = 19 (+charge)
Electrons = 19 (– charge)	Electrons = 18 (–charge)
Overall charge = 0	Overall charge = 1+
Elec. conf. 2, 8, 8, 1	Elec. conf. 2, 8, 8

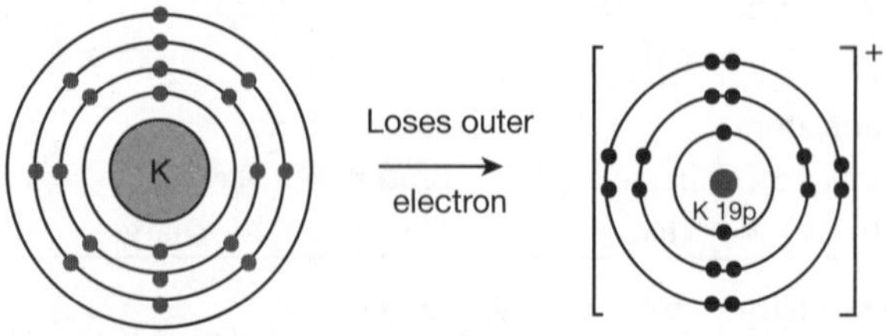

(iv) Formation of magnesium ion (Mg^{2+}): As the magnesium atom has two valence electrons, it loses these two electrons to form a magnesium cation (Mg^{2+}) with the stable inert gas configuration of Ne.

Mg $\xrightarrow{-2\ \text{electron}}$	Mg^{2+}
Magnesium atom	Magnesium ion
Protons = 12 (+ charge)	Protons = 12 (+charge)
Electrons = 12 (– charge)	Electrons = 10 (–charge)
Overall charge = 0	Overall charge = 2+
Elec. conf. 2, 8, 2	Elec. conf. 2, 8

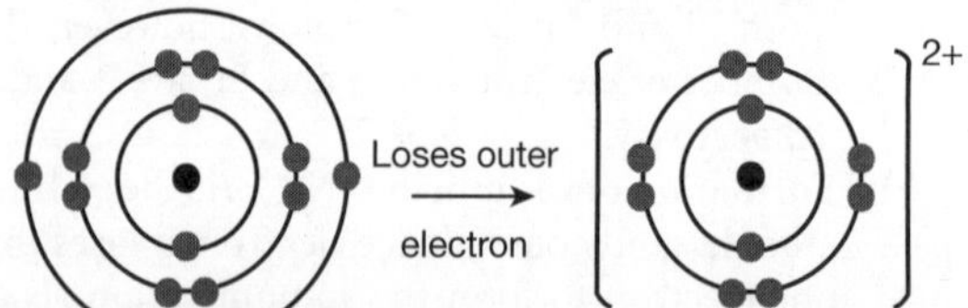

(v) Formation of aluminium ion (Al^{3+}): As an aluminium atom has three valence electrons, it loses these three electrons to form an aluminium cation (Al^{3+}) with the stable inert gas configuration of Ne.

Al $\xrightarrow{-3\ \text{electron}}$	Al^{3+}
Aluminium atom	Aluminium ion
Protons = 13 (+ charge)	Protons = 13 (+charge)
Electrons = 13 (– charge)	Electrons = 10 (–charge)
Overall charge = 0	Overall charge = 3+
Elec. conf. 2, 8, 3	Elec. conf. 2, 8

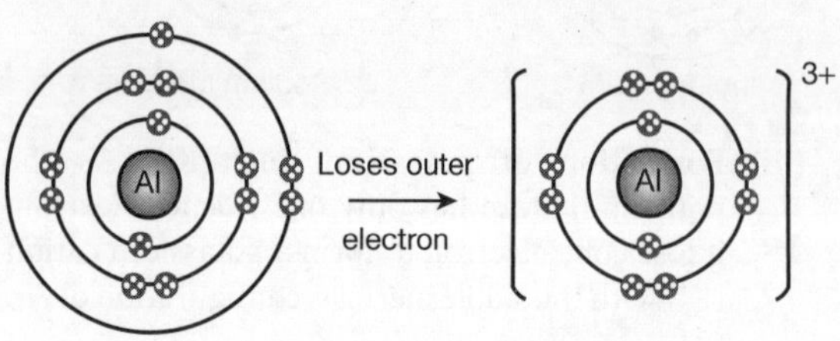

(vi) Formation of calcium ion (Ca^{2+}): As a calcium atom has only two valence electrons, it loses two electrons to form a calcium cation (Ca^{2+}) with the stable inert gas configuration of Ar.

Ca $\xrightarrow{-2\ \text{electron}}$	Ca^{2+}
Calcium atom	Calcium ion
Protons = 20 (+ charge)	Protons = 20 (+charge)
Electrons = 20 (– charge)	Electrons = 18 (–charge)
Overall charge = 0	Overall charge = 2+
Elec. conf. 2, 8, 8, 2	Elec. conf. 2, 8, 8

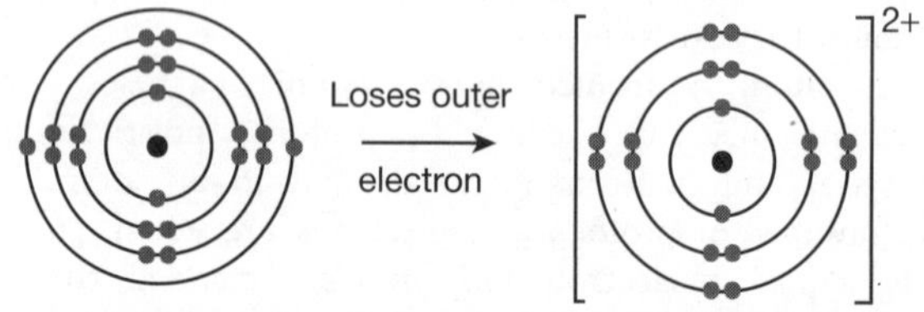

Competition Edge

Lithium ion battery: In mobile phones, there is a rechargeable lithium ion battery.

Fig. 3.11 Lithium battery used in mobiles

Source: Phrontis, CC BY-SA 3.0 <https://creativecommons.org/licenses/by-sa/3.0>, via Wikimedia Commons

3.9.2 Anions and their Formation

In case any element has 5, 6 or 7 valence electrons in the outermost orbit or valence shell of its atoms, then it prefers to gain 3, 2 or 1 electrons respectively to attain the nearest inert gas configuration or octet state. In such a case, addition of 3, 2 or 1 electrons is favourable and the removal of 5, 6 or 7 electrons is almost impossible due to energy considerations. A negatively charged ion is known as an anion and is formed by the gain of one or more electrons by an atom. The ions of all the non-metal elements are anions. A normal or a neutral atom contains an equal number of protons and electrons but an anion contains more electrons than protons since it is formed by the addition of one or more electrons to an atom. The number of electrons gained by an atom is equal to the negative charge assigned to the ion formed. For example, a chlorine atom gains one electron to form a chloride ion, Cl^- (anion), an oxygen atom gains two electrons to form an oxide ion (O^{2-}) anion and a nitrogen atom gains three electrons to form a nitride ion (N^{3-}) anion.

Formation of a few anions

(i) **Formation of chloride ion (Cl^-):** As a chlorine atom has seven valence electrons, it prefers to gain one electron and forms the Cl^- anion having the stable inert gas configuration of Ar.

$Cl \xrightarrow{+1\ \text{electron}} Cl^-$

Chlorine atom	Chlorine ion
Protons = 17 (+ charge)	Protons = 17 (+charge)
Electrons = 17 (– charge)	Electrons = 18 (–charge)
Overall charge = 0	Overall charge = 1–
Elec. conf. 2, 8, 7	Elec. conf. 2, 8, 8

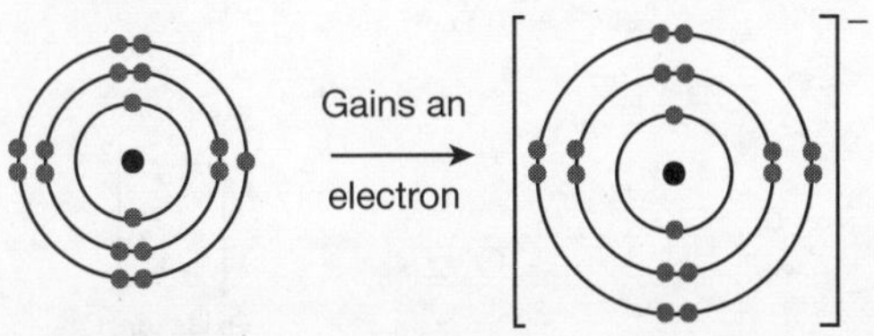

(ii) **Formation of oxide ion (O^{2-}):** As an oxygen atom has six valence electrons, it prefers to gain two electrons and forms an O^{2-} anion having the stable inert gas configuration of Ne.

$O \xrightarrow{+2\ \text{electron}} O^{2-}$

Oxygen atom	Oxide ion
Protons = 8 (+ charge)	Protons = 8 (+charge)
Electrons = 8 (– charge)	Electrons = 10 (–charge)
Overall charge = 0	Overall charge = 2–
Elec. conf. 2, 6	Elec. conf. 2, 8

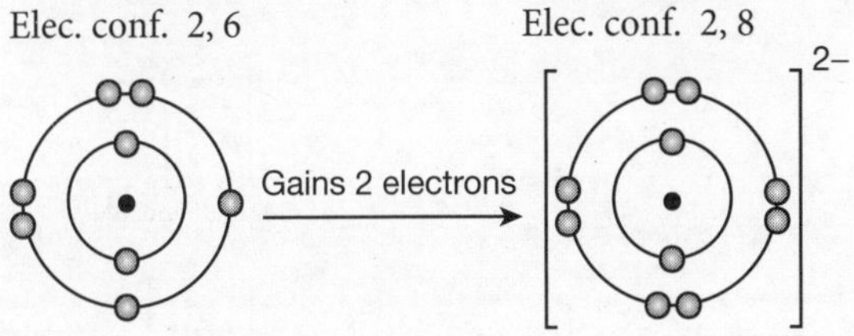

(iii) **Formation of nitride ion (N^{3-}):** As a nitrogen atom has five valence electrons, it prefers to gain three electrons and forms a N^{3-} anion having the stable inert gas configuration of Ne.

$N \xrightarrow{+3\ \text{electron}} N^{3-}$

Nitrogen atom	Nitride ion
Protons = 7 (+ charge)	Protons = 7 (+charge)
Electrons = 7 (– charge)	Electrons = 10 (–charge)
Overall charge = 0	Overall charge = 3–
Elec. conf. 2, 5	Elec. conf. 2, 8

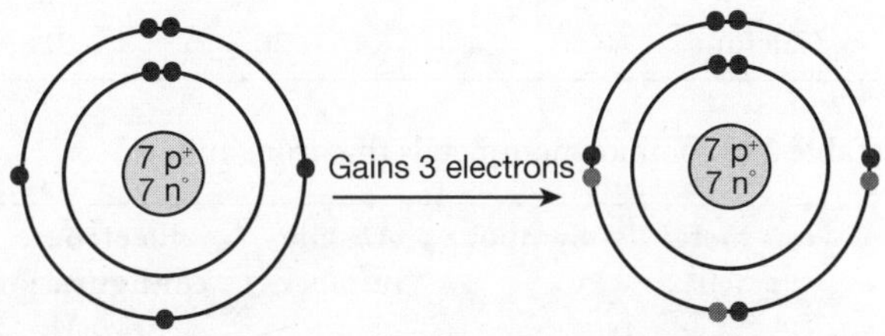

3.10 Electron Dot Representation

Only the electrons of the outermost shell of an atom take part in chemical bonding and these valence electrons are represented by putting dots (•) on the symbol of the element. It is important to note that one dot is used to represent one valence electron. For example, as sodium and magnesium atoms have 1 and 2 valence electrons respectively, the electron dot representation is Na• and Mg:. In order to write the correct electron dot structures, it is important to write the electronic configuration of the element to find the exact number of valence electrons. The electron dot representations of some common metals and non-metals are given in Tables 3.5 and 3.6.

3.10.1 Octet Rule

This rule was introduced by Lewis and Kossel. According to the octet rule, any atom or ion having the octet state (8 valence electrons) is quite stable; and, every atom tries to attain the octet state (that is, the eight valence electron state) by chemical combination (chemical bond formation) to gain stability (nearest inert gas configuration).

Table 3.5 Common metals that form cations

Metal element	Symbol	Atomic number	Electronic configuration *K L M N*	Number of outermost electrons	Electron dot structure	Cation formed
1. Lithium	Li	3	2, 1	1	Li•	Li^+
2. Sodium	Na	11	2, 8, 1	1	Na•	Na^+
3. Magnesium	Mg	12	2, 8, 2	2	Mg:	Mg^{2+}
4. Aluminium	Al	13	2, 8, 3	3	•Al:	Al^{3+}
5. Potassium	K	19	2, 8, 8, 1	1	K•	K^+
6. Calcium	Ca	20	2, 8, 8, 2	2	Ca:	Ca^{2+}

Table 3.6 common non-metals that from anions

Non-metal element	Symbol	Atomic number	Electronic configuration *K L M*	Number of outermost electrons	Electron dot structure	Anion formed
1. Nitrogen	N	7	2, 5	5	•N: (5 dots)	$[:N:]^{3-}$ (8 dots) Nitride ion, N^{3-}
2. Oxygen	O	8	2, 6	6	:O: (6 dots)	$[:O:]^{2-}$ (8 dots) Oxide ion, O^{2-}
3. Fluorine	F	9	2, 7	7	•F: (7 dots)	$[:F:]^{-}$ (8 dots) Fluoride ion, F^-
4. Phosphorus	P	15	2, 8, 5	5	•P: (5 dots)	$[:P:]^{3-}$ (8 dots) Phosphide ion, P^{3-}
5. Sulfur	S	16	2, 8, 6	6	:S: (6 dots)	$[:S:]^{2-}$ (8 dots) Sulphide ion, S^{2-}
6. Chlorine	Cl	17	2, 8, 7	7	•Cl: (7 dots)	$[:Cl:]^{-}$ (8 dots) Chloride ion, Cl^-

3.11 Chemical Bond

During chemical combination between two atoms, to attain the nearest inert noble gas configuration for stability by redistribution of electron/s, a chemical bond is formed between them. A chemical bond is defined as *the force of attraction that binds two atoms together.* A chemical bond is formed to:

- attain the octet state,
- minimise energy,
- gain stability, and
- decrease reactivity.

When two atoms come closer to each other, forces of attraction and repulsion operate between them. The distance at which the forces of attraction overcome the forces of repulsion is called the bond distance. Here, the potential energy for the system is the lowest, so the bond is formed.

Types of Bonds

A chemical bond is formed by the complete transfer of electron/s (ionic bond), by equal or mutual sharing of electrons (covalent bond), or by donating and accepting electron pairs (coordinate bond). Thus, the three types of chemical bonds are—ionic, covalent and coordinate bonds. Here, we shall discuss only the ionic and covalent bonds.

3.11.1 Electrovalent Bond or Ionic Bond

The ionic bond was introduced by Kossel. It is formed by the complete transfer of valence electrons from the metal to the non-metal or from one atom to another. One atom (metallic or electropositive having 1, 2, 3 valence electrons) donates its excess electrons to another atom (more electronegative non-metal having 5, 6, 7 valence electrons) so that both the atoms may acquire stable noble gas configurations. The atom which loses the electron/s becomes positively charged and is called the cation. The atom which takes up the electron/s lost by the first atom becomes negatively charged and is called the anion. These two oppositely charged ions are now held together by an electrostatic force of attraction or Coulombic force. This force of attraction binding the two ions together is known as an electrovalent or ionic bond. It is a non-directional bond. An electrovalent bond is polar, that is, the positive and negative charges are separated.

Electrovalency: When an element forms an electrovalent bond, its valency is known as electrovalency. The number of electrovalent bonds an atom can form is equal to its electrovalency. In other words, the electrovalency of an element is equal to the number of electrons lost or gained by the atom to form an ion. Elements which lose electrons show positive electrovalency and those which gain electrons show negative electrovalency. For example, in the formation of sodium chloride (Na^+Cl^-), the electrovalency of sodium (Na) is +1, while that of chlorine (Cl) is –1.

Ionic compounds: The compounds that are composed of ions are known as ionic compounds. In an ionic compound, the positively charged ions (cations) and negatively charged ions (anions) are held together by strong electrostatic forces of attraction (Coulombic forces). These forces that hold the ions together in an ionic compound are called as *ionic bonds* or *electrovalent bonds.* An ionic compound has an equal number of positive ions and negative ions so that the overall charge on an ionic compound is always zero. For example, NaCl, Na_2O, MgO, CaO, $CaCl_2$, $MgCl_2$ (Table 3.7).

> **Competition Edge**
>
> In the formula of an ionic compound, the positive ion is written first. Charges on the ions of an ionic compound are usually not shown with the formula. So, sodium chloride is usually expressed as NaCl, not as Na^+Cl^-.

Formation of Ionic Bonds in NaCl, $MgCl_2$, MgO, CaO

Formation of Sodium Chloride (NaCl)

The combination of sodium (Na, metal) and chlorine (Cl, non-metal) atoms give rise to formation of sodium chloride (NaCl) by ionic bond formation.

The atomic number of sodium is 11 so its electronic configuration is 2, 8, 1. As it has only one electron in its outermost shell, the sodium atom transfers this electron and becomes a positively charged sodium ion (Na^+).

$$\underset{2,8,1}{Na} \xrightarrow{-1e} \underset{2,\,8}{[Na]^+}$$

Thus, the electronic configuration of the Na^+ ion is the same as that of neon which is the noble gas nearest to sodium in the periodic table.

Table 3.7 Common ionic compounds

S. No.	Name/ Formula/Ions	Crystal structures
1	Sodium chloride NaCl Na^+ and Cl^-	 *Source:* Chemicalinterest, Public domain, via Wikimedia Commons
2	Potassium chloride KCl K^+ and Cl^-	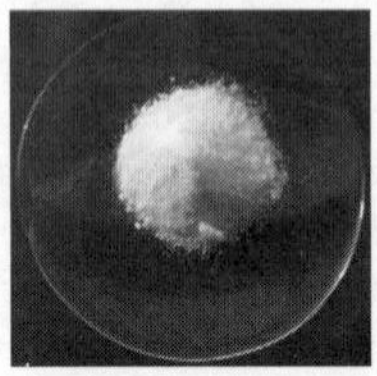 *Source:* Public Domain, https://commons.wikimedia.org/w/index.php?curid=95463
3	Calcium chloride $CaCl_2$ Ca^{2+} and Cl^-	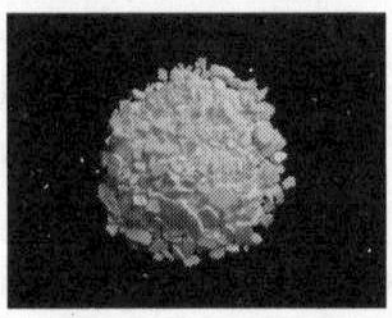 *Source:* By No machine-readable author provided. Firetwister assumed (based on copyright claims). - No machine-readable source provided. Own work assumed (based on copyright claims)., CC BY-SA 3.0, https://commons.wikimedia.org/w/index.php?curid=342288
4	Magnesium chloride $MgCl_2$ Mg^{2+} and Cl^-	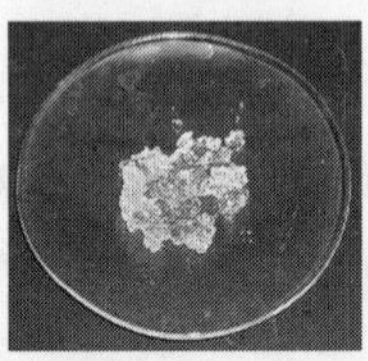 *Source:* Public Domain, https://commons.wikimedia.org/w/index.php?curid=95461
5	Magnesium oxide MgO Mg^{2+} and O^{2-}	 *Source:* Adam Rędzikowski, CC BY-SA 4.0 <https://creativecommons.org/licenses/by-sa/4.0>, via Wikimedia Commons
6	Calcium oxide CaO Ca^{2+} and O^{2-}	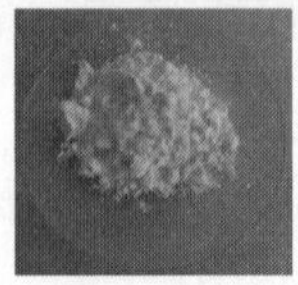 *Source:* Leiem, CC BY-SA 4.0 <https://creativecommons.org/licenses/by-sa/4.0>, via Wikimedia Commons
7	Aluminium oxide Al_2O_3 Al^{3+} and O^{2-}	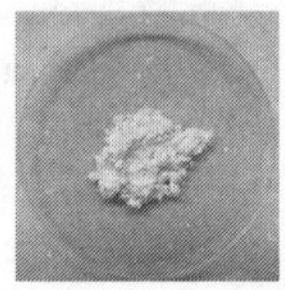 *Source:* By Aariuser I - https://www.flickr.com/photos/194184248@N08/51602553796/, CC BY-SA 2.0, https://commons.wikimedia.org/w/index.php?curid=114438757
8	Sodium hydroxide NaOH Na^+ and OH^-	 *Source:* By Walkerma - Own work, Public Domain, https://commons.wikimedia.org/w/index.php?curid=5190866
9	Copper sulfate $CuSO_4$ Cu^{2+} and SO_4^{2-}	 *Source:* By Stephanb - Own work, CC BY-SA 3.0, https://commons.wikimedia.org/w/index.php?curid=565837
10	Calcium sulfate $CaSO_4$ Ca^{2+} and SO_4^{2-}	 *Source:* Martin Walker, Public domain, via Wikimedia Commons

(Continued)

Table 3.7 (Continued)

S. No.	Name/ Formula/Ions	Crystal structures
11	Sodium sulfate Na_2SO_4 Na^+ and SO_4^{2-}	 *Source:* Public Domain, https://commons.wikimedia.org/w/index.php?curid=292586
12	Iron sulfate $FeSO_4$ Fe^{2+} and SO_4^{2-}	 *Source:* By Benjah-bmm27 - Own work, Public Domain, https://commons.wikimedia.org/w/index.php?curid=2116160
13	Magnesium sulfate $MgSO_4$ Mg^{2+} and SO_4^{2-}	 *Source:* Public Domain, https://commons.wikimedia.org/w/index.php?curid=292629
14	Potassium nitrate KNO_3 K^+ and NO_3^-	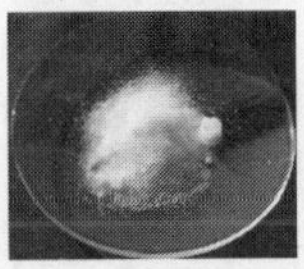 *Source:* By w:User:Walkerma, Public Domain, https://commons.wikimedia.org/w/index.php?curid=184556
15	Sodium nitrate $NaNO_3$ Na^+ and NO_3^-	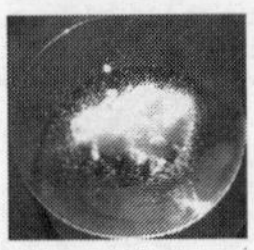 *Source:* No machine-readable author provided. Walkerma assumed (based on copyright claims)., Public domain, via Wikimedia Commons
16	Ferric chloride $FeCl_3$ Fe^{3+} and Cl^-	 *Source:* Benjah-bmm27, Public domain, via Wikimedia Commons

Now let us consider the chlorine atom (Cl). The atomic number of chlorine is 17 so its electronic configuration is 2, 8, 7. As it has 7 electrons in its outermost shell, that is, it lacks 1 electron to acquire a stable noble gas configuration, the chlorine atom takes the 1 electron transferred by the sodium atom and becomes a negatively charged chloride ion (Cl^-).

$$\underset{(2,\,8,\,7)}{:\ddot{\underset{..}{Cl}}\cdot} + 1e \longrightarrow \underset{(2,\,8,\,8)}{[:\ddot{\underset{..}{Cl}}:]^-}$$

Thus, the chloride ion (Cl^-) attains the configuration of the nearest noble gas, argon. The two ions (Na^+ and Cl^-) being oppositely charged, are now held together by the electrostatic force of attraction as Na^+Cl^-.

$$\underset{(2,\,8,\,1)}{Na} + \underset{(2,\,8,\,7)}{Cl} \longrightarrow \underset{(2,\,8)}{Na^+}\ \underset{(2,\,8,\,8)}{Cl^-}$$

$$\dot{Na} + :\ddot{\underset{..}{Cl}}\cdot \longrightarrow Na^+\underset{(2,\,8,\,8)}{[:\ddot{\underset{..}{Cl}}:]^-} \text{ or } Na^+Cl^-$$

The formation of sodium chloride and electron dot structure can be shown diagrammatically (Fig. 3.12).

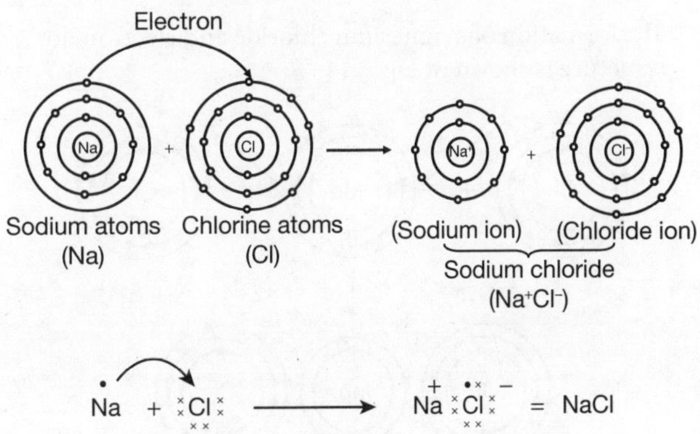

Fig. 3.12 Sodium chloride

It is important to note that the formation of potassium chloride is similar to that of sodium chloride (Fig. 3.13).

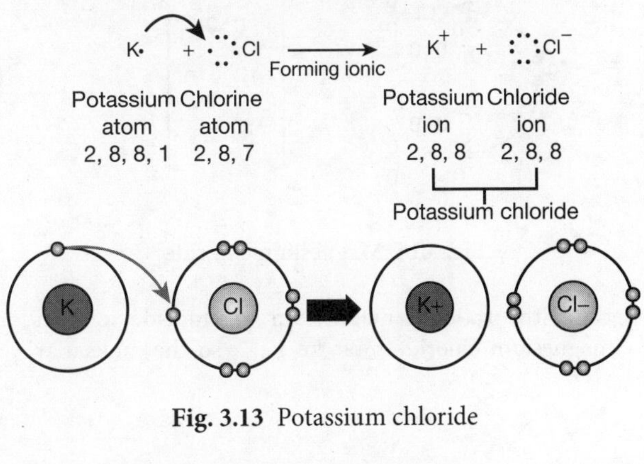

Fig. 3.13 Potassium chloride

Formation of Magnesium Chloride ($MgCl_2$)

The number of valence electrons in magnesium (atomic number 12) is 2 and that in chlorine (atomic number 17) is 7. The magnesium atom acquires a stable configuration of 8 electrons by losing two electrons from its outermost shell (one each to each atom of chlorine) and so becomes a positive magnesium ion, Mg^{2+}.

$$\underset{\substack{(2,\,8,\,2)\\ \text{Atom}}}{Mg} - 2e^- \rightarrow \underset{\substack{(2,\,8)\\ \text{Cation}}}{Mg^{2+}}$$

However, each chlorine atom, which contains 7 valence electrons in its outermost shell, can accept only 1 of the 2 electrons donated by a magnesium atom to attain the nearest inert gas configuration. So, for each magnesium atom forming a magnesium ion (Mg^{2+}), there must be two chlorine atoms to form two chloride ions.

$$\underset{\substack{(2,\,8,\,7)\\ \text{2 Cl atoms}}}{2Cl} + 2e^- \rightarrow \underset{\substack{(2,\,8,\,8)\\ \text{2 Chloride anions}}}{2Cl^-}$$

The formation of magnesium chloride and electron dot structure is shown in Fig. 3.14.

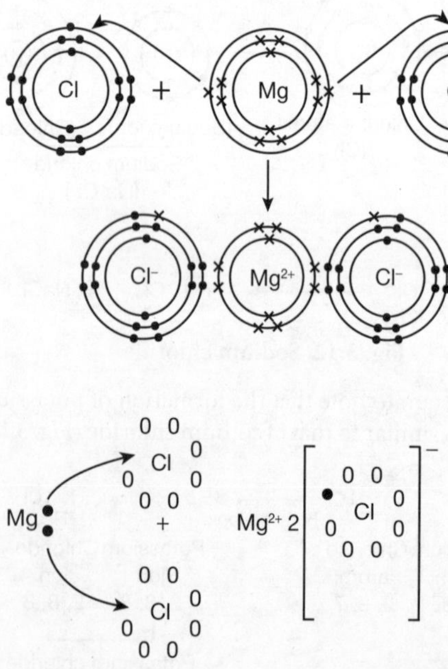

Fig. 3.14 Magnesium chloride

Here, the ratio of magnesium to chloride ions in magnesium chloride must be 1 : 2, so the molecular formula of the compound magnesium chloride is $MgCl_2$.

It is important to note that the formation of calcium chloride is similar to that of magnesium chloride (Fig. 3.15).

$$\underset{\substack{(2,\,8,\,8,\,2)\\ \text{Atom}}}{Ca} - 2e^- \rightarrow \underset{\substack{(2,\,8,\,8)\\ \text{Atom}}}{Ca^{2+}},\quad \underset{\substack{(2,\,8,\,7)\\ \text{2 Cl atoms}}}{2Cl} + 2e^- \rightarrow \underset{\substack{(2,\,8,\,8)\\ \text{2 Chloride anions}}}{2Cl^-}$$

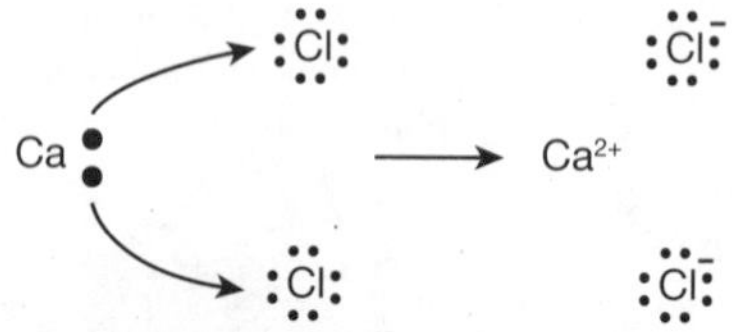

Fig. 3.15 Calcium chloride

Formation of Magnesium Oxide (MgO)

The number of valence electrons of a magnesium atom (atomic number 12) is 2, and that of an oxygen atom is 6, that is, the oxygen atom requires 2 electrons to attain the octet to get the nearest inert gas (neon) configuration. In the presence of oxygen, each magnesium atom loses its 2 valence electrons to one oxygen atom (Fig. 3.xx). As a result, the magnesium atom forms a magnesium ion with a charge of +2 (Mg^{2+}), and the oxygen atom forms an oxide ion with a charge of –2 (O^{2-}). Since only one oxygen atom is needed to accept the 2 valence electrons donated by a magnesium atom, the formula of magnesium oxide is MgO.

$$\underset{\substack{(2,\,8,\,2)\\ \text{Atom}}}{Mg} - 2e^- \rightarrow \underset{\substack{(2,\,8)\\ \text{Atom}}}{Mg^{2+}},\ \underset{\substack{(2,\,6)\\ \text{Atom}}}{O} + 2e^- \rightarrow \underset{\substack{(2,\,8)\\ \text{Atom}}}{O^{2-}}$$

The formation of magnesium oxide and the electron dot structure is shown in Fig. 3.16.

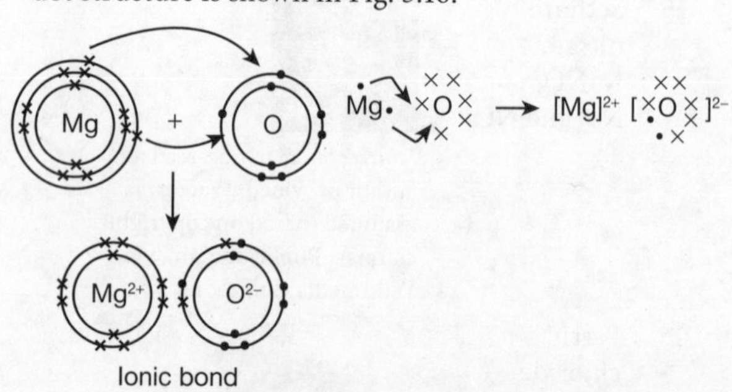

Fig. 3.16 Magnesium oxide

It is important to note that the formation of calcium oxide is similar to that of magnesium oxide (Fig. 3.17).

$$\underset{\substack{(2,\,8,\,8,\,2)\\ \text{Atom}}}{Ca} - 2e^{-} \rightarrow \underset{\substack{(2,\,8,\,8)\\ \text{Atom}}}{Ca^{2+}},\ \underset{\substack{(2,\,6)\\ \text{Atom}}}{O} + 2e^{-} \rightarrow \underset{\substack{(2,\,8)\\ \text{Atom}}}{O^{2-}}$$

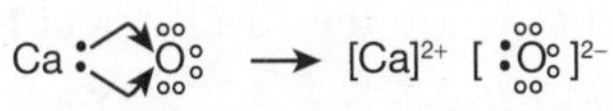

Fig. 3.17 Calcium oxide

Features of Electrovalent (ionic) Compounds

The following are the important features of ionic compounds.

(i) **Physical state:** Ionic compounds have solid crystalline structures (flat surfaces) with definite geometry due to strong electrostatic forces of attraction as the constituents are arranged in a definite pattern of positively and negatively charged ions. The crystals of such compounds are quite hard and brittle. For example, sodium chloride (NaCl) is made up of Na^+ and Cl^- ions arranged in a definite order in three dimensions to form crystals.

(ii) **Melting and boiling points:** Electrovalent compounds have high melting and boiling points. This is due to the presence of strong electrostatic forces of attraction between the positive and negative ions which means a large amount of heat energy is required to break this force of attraction. Hence, the melting and boiling points of these compounds are very high (Table 3.8).

BP, MP α Electrostatic force of attraction

For example, the boiling and melting points of sodium chloride are 1413°C (1686 K), 800°C (1073 K) respectively.

Table 3.8 Melting and boiling points of some ionic compounds

Ionic compound	Melting point (K)	Boiling point (K)
NaCl	1073	1686
LiCl	887	1600
$CaCl_2$	1045	1900
CaO	2850	3120
$MgCl_2$	981	1685

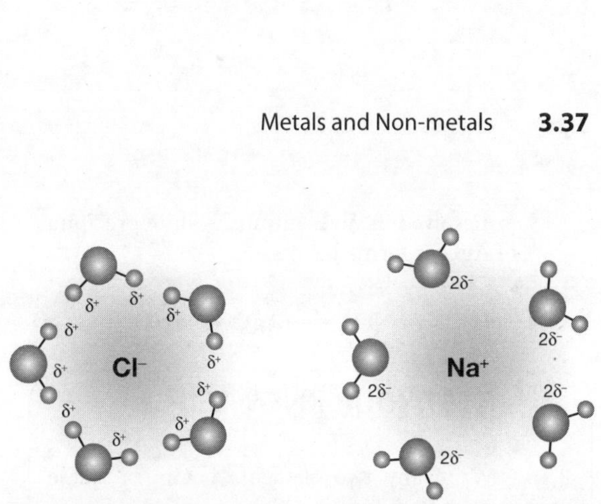

Slightly positive hydrogen are attracted to chlorine anions

Slightly negative oxygen are attracted to sodium cations

Fig. 3.18 Hydrated Na^+ and Cl^- ions

(iii) **Solubility:** Electrovalent compounds are usually soluble in water but insoluble in organic solvents such as benzene, acetone, carbon disulfide and carbon tetrachloride. Ionic compounds are soluble in polar solvents such as water, due to the high dielectric constant of these solvents. The force of attraction between ions gets destroyed and ionic compounds get dissolved in them. In fact, the ions of ionic compounds get hydrated or solvated on dissolving in water or a solvent. For example, NaCl dissolved in water.

Activity: In order to prove that ionic compounds are soluble in water but insoluble in organic solvents like benzene, take some water in a test tube and add a little sodium chloride into it. On shaking the test tube, sodium chloride dissolves in water. Now take benzene or ether in another test tube and try to dissolve a little sodium chloride in it. On shaking the test tube, it is observed that sodium chloride does not dissolve in it.

(iv) **Electrical conductance:** Electrovalent compounds can conduct electricity in the molten or fused state and in their aqueous solutions but not in the solid state. In the solid state, the ions of the electrovalent compounds are held together in fixed positions and cannot move freely so they do not conduct electricity. When an electrovalent compound is dissolved in water or is melted, the crystal structure breaks down and as a result the ions now become free to move and can, therefore, conduct electricity easily.

(v) **Nature of reactions:** These compounds can show fast ionic reactions as no additional external energy is needed for the ions to undergo the reaction. For example, when NaCl is dissolved in silver nitrate, a white precipitate of AgCl is formed at once.

$$\overset{+}{Na}\overset{-}{Cl} + \overset{+}{Ag}\overset{-}{NO_3} \longrightarrow \overset{+}{Ag}\overset{-}{Cl} + \overset{+}{Na}\overset{-}{NO_3}$$

TEST YOUR KNOWLEDGE

3.16 Write any two elements having a stable octet state configuration.

Solution:
Neon and argon.

3.17 Explain why chemical bond formation takes place.

Solution:
See text (Section 3.13).

3.18 Write any two cations and anions having octet state configuration.

Solution:
Cation: Na^+, Mg^{2+}
Anion: O^{2-}, N^{3-}

3.19 Write the type of bond and formula of the compound when potassium and calcium combine with chlorine.

Solution:
Ionic bond in both cases and the compounds formed are KCl and $CaCl_2$.

3.20 Ionic compounds have ________ boiling and melting points and they are ________ of electricity in the solid state.

Solution:
High, non-conductors.

3.11.2 Covalent Bond

The chemical bond formed when two atoms share electrons between them is known as a *covalent bond.* It is formed by the equal sharing of electrons between two similar or different atoms. The sharing of electrons between the two atoms takes place in such a way that both the atoms acquire the stable electronic configurations of their nearest noble gases. Atoms of non-metals having 4, 5, 6, 7 valence electrons prefer to form covalent bonds and not ionic bonds because the atoms of such elements do not favour the loss of valence electrons due to energy considerations which prevents the transfer of electrons. The molecule or ion formed due to mutual sharing of electrons is known as a covalent molecule. Here, the number of electrons shared or the number of covalent bonds represents covalency. For example,

- in ammonia (NH_3), the covalency of the nitrogen atom is 3,
- in methane (CH_4), the covalency of the carbon atom is 4.

Types of Covalent Bonds

Covalent bonds are of three types—single covalent bond, double covalent bond and triple covalent bond.

(i) **Single covalent bond:** A single covalent bond is formed when one pair of electrons is shared between two atoms. That is, each atom contributes one electron during sharing. It is shown by putting a short line (–) between two atoms. Examples are, H_2, X_2, HX, CH_4, CCl_4, NH_3, BF_3, BCl_3, H_2O, H_2S.

- **Formation of a hydrogen molecule (H_2):** A molecule of hydrogen has two hydrogen atoms. Each hydrogen atom has one electron. When two atoms of hydrogen combine, one electron of each hydrogen atom takes part in sharing. Hence, two electrons (one pair of electrons) are shared between the two hydrogen atoms. The shared electron pair always exists between these two atoms. The two dots between the two H atoms represent the pair of shared electrons. One pair of shared electrons gives a single bond. Such a bond is represented by a short line between the two atoms. Hence, a hydrogen molecule may be shown as in Fig. 3.19.

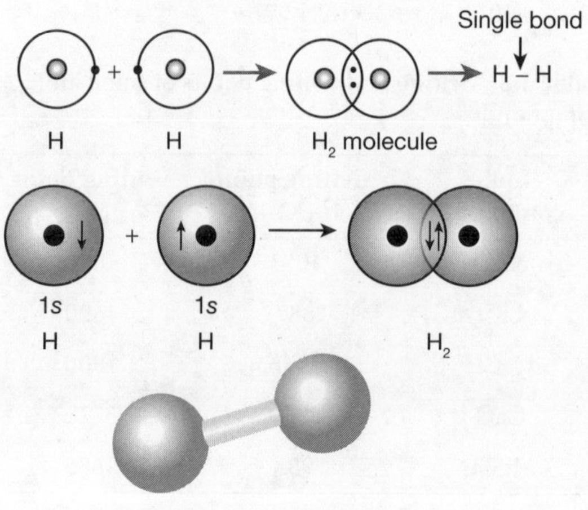

Fig. 3.19 Hydrogen

Once the covalent bond is formed, both H atoms have the stable configuration of the noble gas helium ($1s^2$).

- **Formation of a chlorine molecule (Cl_2):** A molecule of chlorine has two chlorine atoms. Each chlorine atom has seven valence electrons and needs one electron to get to the nearest inert gas (Ar) configuration. When two atoms of chlorine combine, one electron of each chlorine atom takes part in sharing. Hence, two electrons (one pair of electrons) are shared between the two chlorine atoms. The shared electron pair always exists between these two atoms. The two dots between the two Cl atoms represent the pair of shared electrons. One pair of shared electrons gives a single bond. Such a bond is represented by a short line between the two atoms. Hence, a chlorine molecule may be shown as in Fig. 3.20.

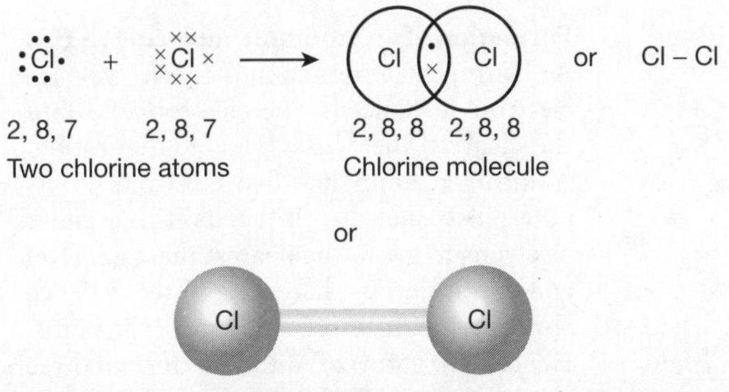

Fig. 3.20 Chlorine

Once the covalent bond is formed, both Cl atoms have the stable configuration of the noble gas, argon.

- **Formation of a hydrogen chloride molecule (HCl):** The hydrogen atom contains only one valence electron, so it needs one electron to get to the nearest inert gas (He) configuration. The chlorine atom contains seven valence electrons so it needs one more electron to get to the nearest inert gas (Ar) configuration. As both the H atom and the Cl atom need one electron to attain stable electronic configurations, only one electron is shared by each atom, that is, hydrogen and chlorine form a single bond between them (Fig. 3.21).

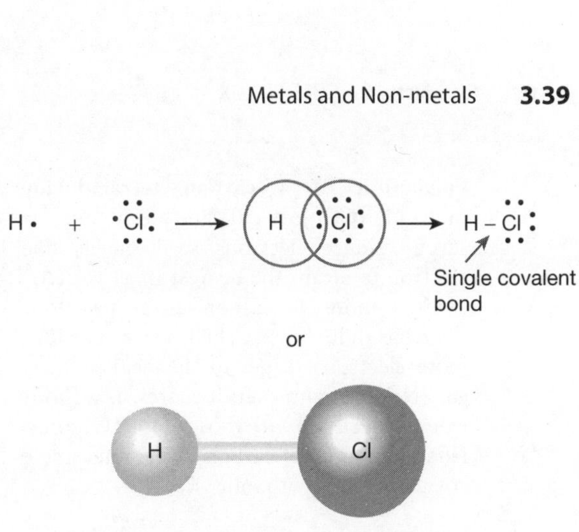

Fig. 3.21 Hydrogen chloride

- **Formation of a methane molecule (CH_4):** A hydrogen atom contains only one valence electron so it needs one electron to attain the nearest inert gas (He) configuration. A carbon atom has four electrons in its valence shell, so it needs four more electrons to get to the nearest inert gas (Ne) configuration. It shares all its four valence electrons with those of four H atoms. Hence, an atom of carbon forms four single covalent bonds with four H atoms.

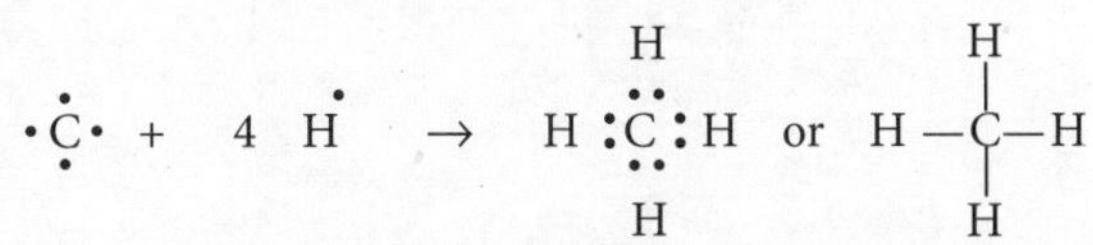

Pictorially, a methane molecule can be shown as in Fig. 3.22.

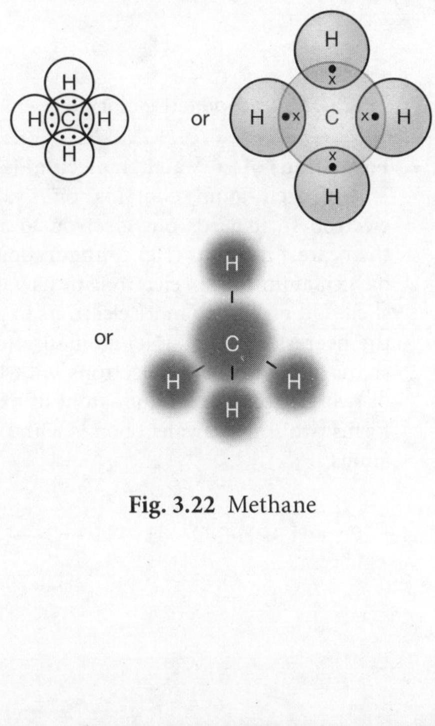

Fig. 3.22 Methane

- **Formation of a carbon tetrachloride molecule (CCl_4):** A chlorine atom contains seven valence electrons so it needs one electron to attain the nearest inert gas (Ar) configuration. A carbon atom has four electrons in its valence shell, so it needs four more electrons to get to the nearest inert gas (Ne) configuration. It shares its all four valence electrons with those of four C atoms. Hence, an atom of carbon forms four single covalent bonds with four Cl atoms.

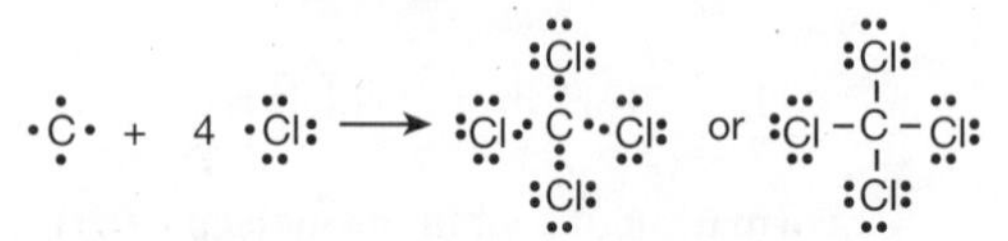

Pictorially, a carbon tetrachloride molecule can be shown as in Fig. 3.23.

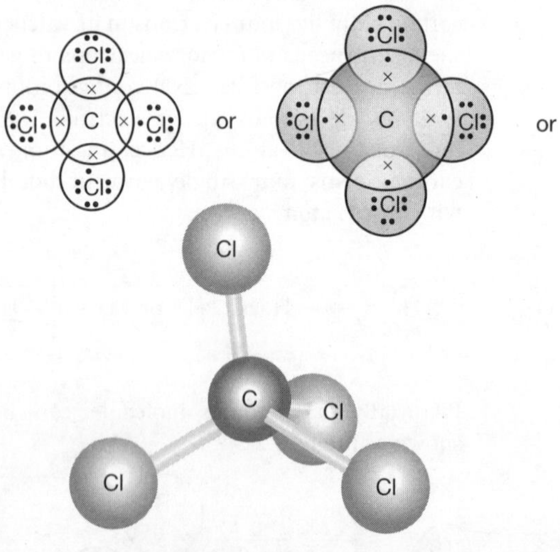

Fig. 3.23 Carbon tetrachloride

- **Formation of a water molecule (H_2O):** A hydrogen atom contains one valence electron so it needs one electron to attain the nearest inert gas (He) configuration. An oxygen atom has six electrons in its valence shell so it needs two more electrons to get to the nearest inert gas (Ne) configuration. It shares its two valence electrons with those of two H atoms. Hence, an atom of oxygen forms two single covalent bonds with two H atoms.

H•+ •O• +•H → H• •O• •H → H•–•O–•H

Pictorially, a water molecule can be shown as in Fig. 3.24.

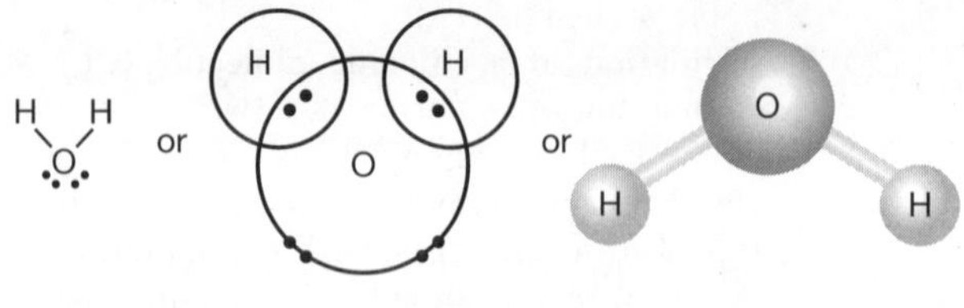

Fig. 3.24 Water

It is important to note that the formation of hydrogen sulfide (H_2S) is similar to that of H_2O.

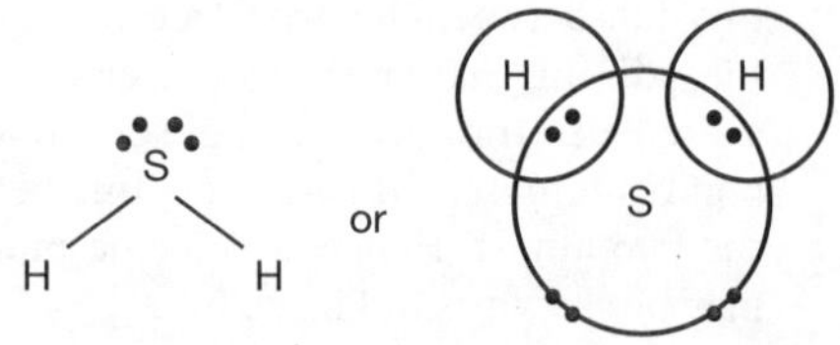

- **Formation of an ammonia molecule (NH_3):** A hydrogen atom contains one valence electron so it needs one electron to attain the nearest inert gas (He) configuration. A nitrogen atom has five electrons in its outermost shell so it needs three more electrons to get to the nearest inert gas (Ne) configuration. It shares its three valence electrons with those of three H atoms. Hence, an atom of nitrogen forms three single covalent bonds with three H atoms.

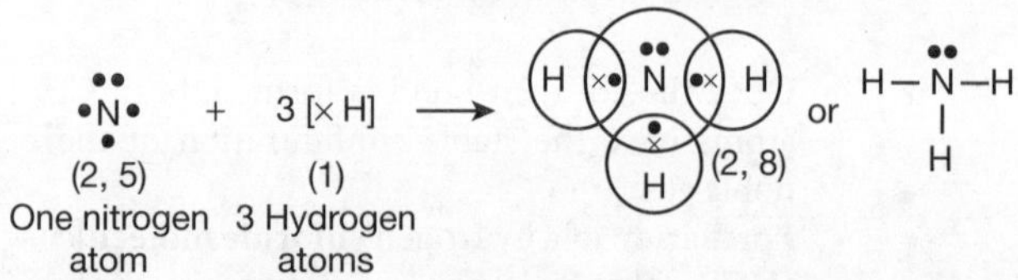

Pictorially, an ammonia molecule can be shown as in Fig. 3.25.

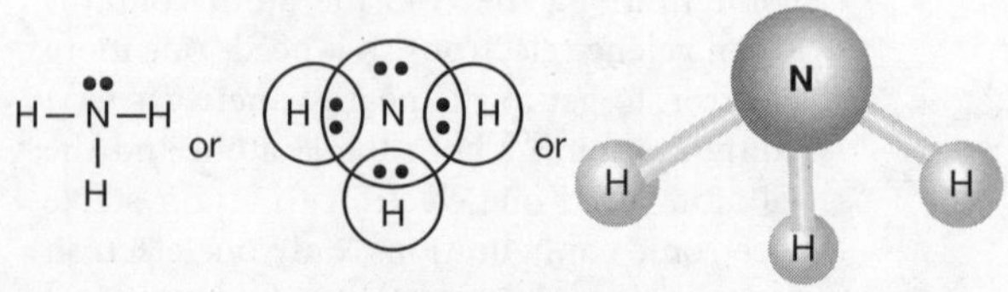

Fig. 3.25 Ammonia

It is important to note that the formation of phosphine (PH_3) is similar to that of NH_3 (Fig. 3.26).

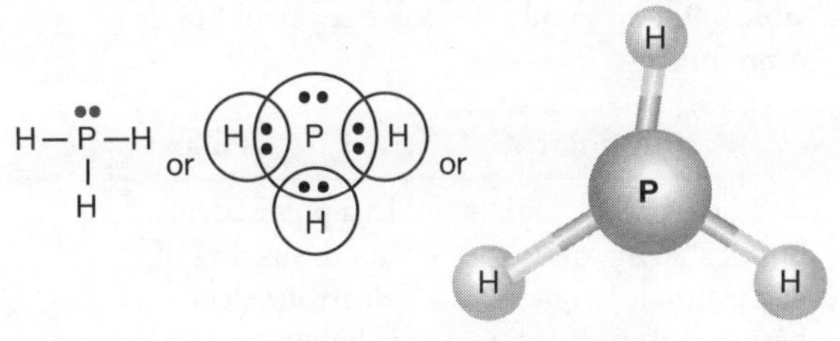

Fig. 3.26 Phosphine

(ii) Double covalent bond (=): A double covalent bond is formed when two pairs of electrons are shared between the two combining atoms. The sharing of two pairs of electrons is shown by marking two short lines (=) between the symbols of the two atoms. Examples are O_2, CO_2, CS_2, C_2H_4, C_6H_6.

- **Formation of an oxygen molecule (O_2):** An atom of oxygen contains six electrons in its valence shell. It needs two more electrons to attain a stable eight electron configuration (octet) like inert gas (neon). This is achieved when each of the two oxygen atoms shares its two electrons with the other, resulting in the formation of a stable oxygen molecule.

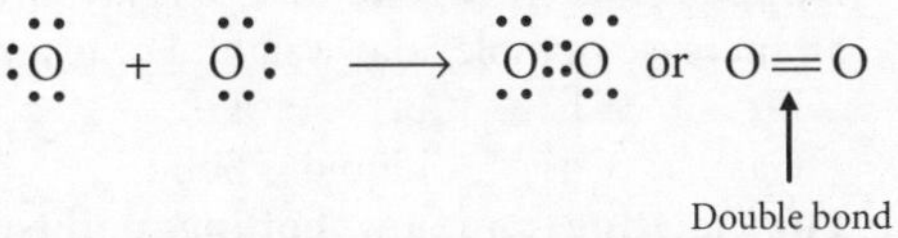

Pictorially, the oxygen molecule can be shown as in Fig. 3.27.

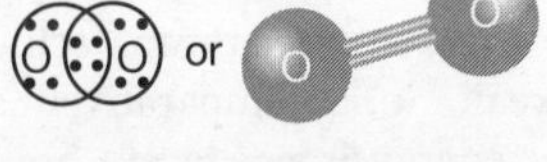

Fig. 3.27 Oxygen

- **Formation of a carbon dioxide molecule (CO_2):** In a carbon dioxide molecule, there are one carbon and two O atoms. A carbon atom has four valence electrons so it needs four more electrons to attain the nearest inert gas (Ne) configuration. An O atom has six valence electrons so it needs two more electrons to get to the nearest inert gas (Ne) configuration. Hence, in the formation of carbon dioxide (CO_2), one C atom combines with two O atoms to form two double covalent bonds by sharing two electrons with each oxygen atom.

Pictorially, a molecule of carbon dioxide (CO_2) can be shown as in Fig. 3.28.

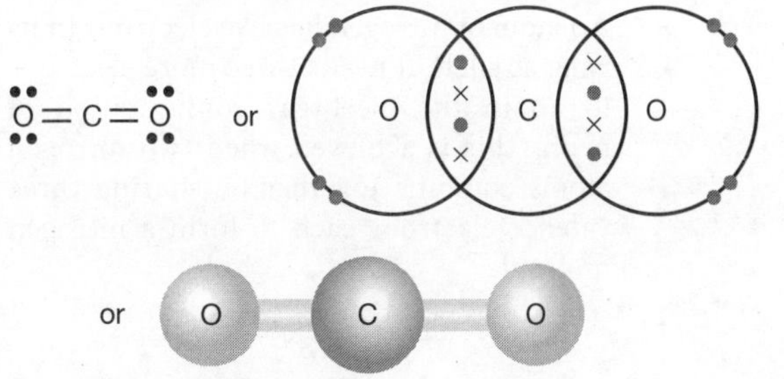

Fig. 3.28 Carbon dioxide

- **Formation of an ethylene molecule (C_2H_4):** A hydrogen atom contains only one valence electron so it needs one electron to attain the nearest inert gas (He) configuration. A carbon atom has four electrons in its valence shell, so it needs four more electrons to get to the nearest inert gas (Ne) configuration. In the formation of an ethylene (ethene) molecule (C_2H_4), each of the two C atoms combines with two H atoms to form two single covalent bonds. The remaining two electrons of each C atom form a double bond between the two C atoms.

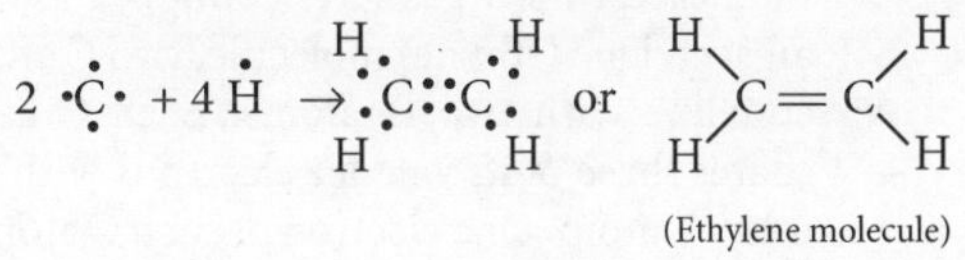

Pictorially, a molecule of ethylene (C_2H_4) can be shown as in Fig. 3.29.

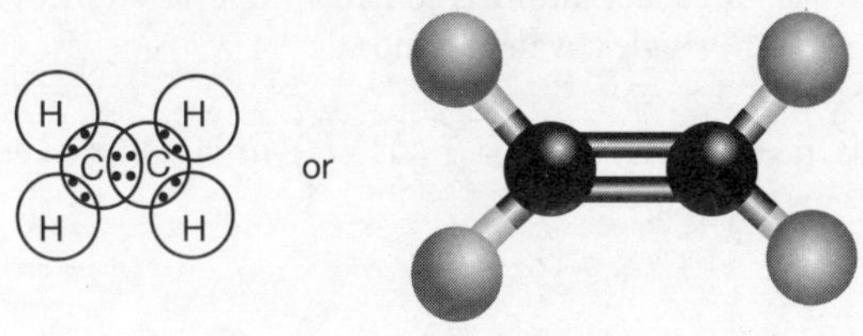

Fig. 3.29 Ethylene

(iii) Triple covalent bond (≡): A triple covalent bond is formed when three pairs of electrons (six electrons) are shared between the two combining atoms. A triple bond is shown by marking three short lines (≡) between the two symbols of the atoms. Examples are N_2, C_2H_2, HCN.

- **Formation of a nitrogen molecule (N_2):** An atom of nitrogen has five electrons in its valence shell. It needs three more electrons to attain the inert gas configuration of neon. This is achieved when two nitrogen atoms combine together by sharing three valence electrons each to form a nitrogen molecule.

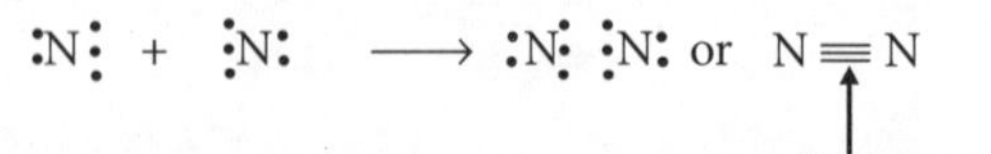

Pictorially, a nitrogen molecule can be shown as in Fig. 3.30.

Fig. 3.30 Nitrogen

- **Formation of an acetylene molecule (C_2H_2):** A hydrogen atom contains only one valence electron so it needs one electron to get the stable nearest inert gas (He) configuration. A carbon atom has four electrons in its valence shell so it needs four more electrons to attain the nearest inert gas (Ne) configuration. In an acetylene (ethyne) molecule, two C atoms combine with two H atoms. Each C atom shares three of its valence electrons with the other C atom. One electron of each C atom is shared with one electron of a H atom. Hence, in a molecule of acetylene, there is a triple covalent bond between the two C atoms and each C atom is joined to one H atom by a single covalent bond.

Pictorially, a molecule of acetylene can be shown as in Fig. 3.31.

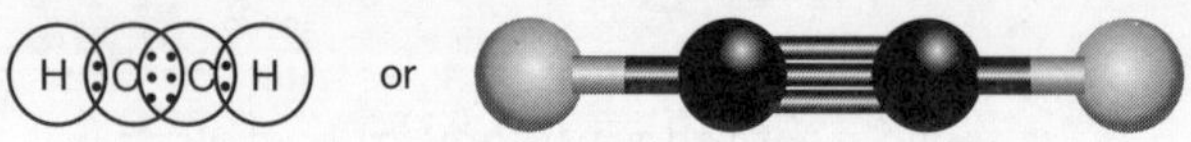

Fig. 3.31 Acetylene

Table 3.9 Polar and non-polar covalent bonds or compounds

Non-polar	Polar
In a non-polar covalent bond, sharing of electron/ electrons is between atoms of the same element or atoms with nearly the same electronegativity. For example, X–X, O=O, N≡N.	In a polar covalent bond, sharing of electron/ electrons is between atoms of different elements or atoms with different electronegativities. For example, H_2O, NH_3, HCl, HF.

Features of Covalent Compounds

The following are the important features of covalent compounds.

(i) Physical state: Covalent compounds are made up of neutral atoms. So, the forces of attraction (van der Waals force) between the molecules are weaker than those (Coulombic force) found in ionic compounds. Therefore, covalent compounds are usually volatile liquids or gases. Covalent compounds are mostly liquids or gases. But, if the molecular weight is high, they may be solids since solid nature is favoured by an increase in molecular weight. For example,

F_2	Cl_2	Br_2	I_2
Gas	Gas	Liquid	Solid

(ii) Low melting and low boiling points: The melting and boiling points of covalent compounds are generally low. As covalent compounds are made up of neutral atoms or molecules, the forces of attraction (van der Waals force) between the molecules or atoms are very weak. So, a comparatively small amount of heat energy is needed to break these weak intermolecular forces of attraction. Hence, they have low melting and boiling points. Covalent compounds have lower boiling and melting point values than ionic compounds since the covalent bond is held by weak van der Waals forces while the ionic bond is held by strong Coulombic forces. For example,

KOH	>	HX
Strong force of attraction		Weak van der Waals forces

BP, MP $\propto$ Molecular weight; for example, HI > HBr > HCl.

(iii) Solubility: Solubility in covalent compounds follows the 'like dissolves like' concept. That is, a non-polar molecule dissolves in a non-polar solvent while a polar molecule dissolves in a polar solvent. For example, CH_4 or CCl_4 (non-polar) dissolves in an organic solvent like benzene, HCl or NH_3 (polar) dissolves in H_2O.

(iv) **Electrical conductance:** Covalent compounds do not conduct electricity. This is because they are made up of neutral molecules or atoms and not of ions and do not produce ions in the molten state or aqueous solutions. Graphite is an exception and is a conductor of electricity since free electrons are available in its hexagonal sheet-like slippery structure. In the case of diamond, the structure is tetrahedral so free electrons are not available and hence it is a non-conductor.

(v) **Nature of reactions:** They show slow molecular reactions as they need external energy. For example, hydrogen and chlorine can react in the presence of sunlight to give hydrogen chloride.

$$H_2 + Cl_2 \xrightarrow{\text{Sunlight}} 2HCl$$

Table 3.10 Differences between ionic and covalent compounds

Ionic compound	Covalent compound
1. They are mostly crystalline solids.	1. They are mostly gases or liquids but may be solid if molecular weight is high.
2. They have ionic bonds and electrostatic forces of attraction.	2. They have covalent bonds and van der Waals forces of attraction.
3. They have high values of boiling points and melting points.	3. They have low values of boiling points and melting points.
4. They are soluble in polar solvents like water but insoluble in non-polar solvents like benzene.	4. Polar covalent compounds dissolve in polar solvents while non-polar covalent compounds dissolve in non-polar solvents.
5. They show electrical conductance in the molten state and in the aqueous solution state as they have free ions.	5. They do not show electrical conductance as they do not have free ions.
6. They show fast ionic reactions.	6. They show slow molecular reactions.

Activity: Effect of electricity on electrovalent and covalent compounds

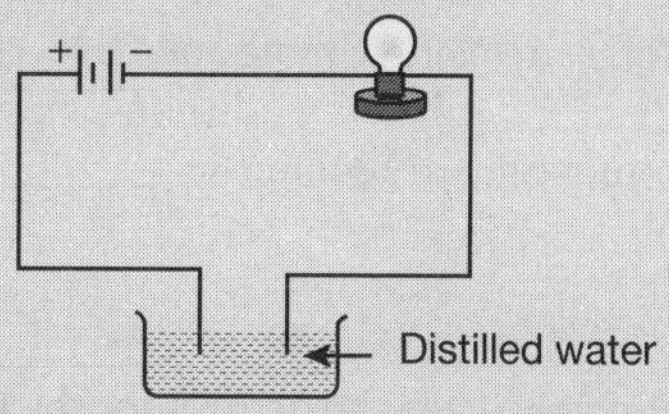

Experiment: Arrange an electrolytic cell (Fig. 3.xx). The electric circuit contains a 6-volt battery, an ammeter bulb and platinum electrodes connected in series.

Take the following electrovalent and covalent compounds in solution in separate beakers and dip the platinum electrodes in them for a moment, one by one, and note the change in the ammeter reading.

Observations

Electrovalent compounds: When the current is passed in the solutions of electrovalent compounds (magnesium chloride, sodium chloride, sodium hydroxide and copper sulfate) the bulb glows. That is, they allow electric current to pass through them. Hence, ionic compounds are good conductors of electricity in the molten or aqueous states due to free mobile ions.

Covalent compounds: When the current is passed through the solutions of covalent compounds (distilled water, sugar solution, alcohol, chloroform, benzene and petrol) the bulb does not glow. That is, they do not conduct electricity. This happens because solutions of covalent compounds contain only molecules and no ions.

TEST YOUR KNOWLEDGE

3.21 Define a covalent bond and write any two compounds having only single bonds.

Solution:

A covalent bond is formed by equal sharing of valence electrons between two atoms. For example, H_2, HCl.

3.22 In the formation of the oxygen molecule ______ electrons are shared by each oxygen atom while in the formation of the nitrogen molecule ________ electrons are shared by each nitrogen atom.

Solution:
2, 3.

3.23 Polar covalent compounds are soluble in ________ solvents while non-polar covalent compounds are soluble in ________ solvents.

Solution:
Polar, non-polar.

3.24 Draw the Lewis structures for the following molecules and ions: H_2S, $SiCl_4$, BeF_2, CO_3^{2-}.

Solution:

H_2S = 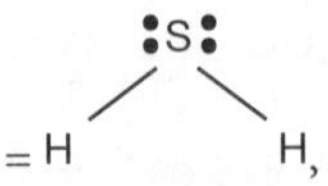

$SiCl_4$ = 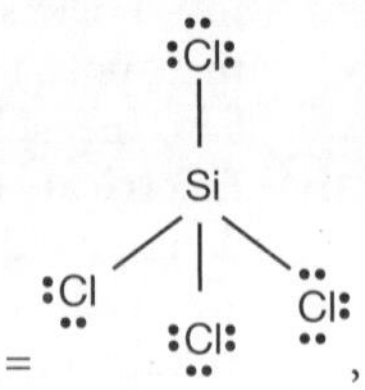

BeF_2 = F–Be–F

CO_3^{2-} = 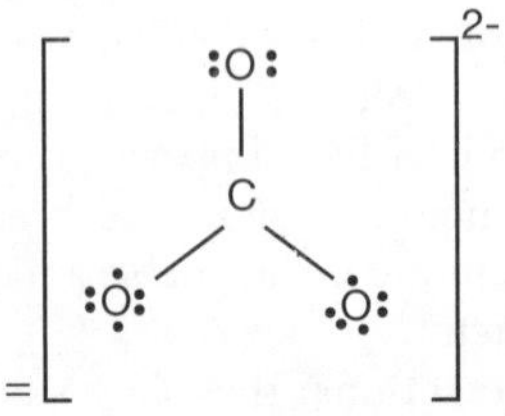

Quick Review

- **Valency:** The combining capacity of an element is known as its valency. It is measured in terms of the number of valence electrons (present in outer orbit) or number of H atoms or O atoms or Cl atoms linked with one atom of an element. The number of electrons shared by an atom is known as its valency.
- **Octet rule:** any atom or ion having an octet state (8 valence electron) is quite stable and every atom tries to attain the octet state (eight valence electron state) by chemical combination (chemical bond formation) to gain stability (nearest inert gas configuration).
- **Chemical bond:** It is the force of attraction that binds two atoms together. A chemical bond implies that the force of attraction and force of repulsion are balanced at a particular distance. It is formed to attain the octet state, to minimise energy, to gain stability, to decrease reactivity.
- **Ionic bond** is formed by the complete transfer of valence electrons from a metal to a non-metal or from one atom to another. The number of electrovalent bonds an atom can form is equal to its electrovalency. That is, the electrovalency of an element is equal to the number of electrons lost or gained by the atom to form an ion.
- **Ionic compounds** have solid, crystalline structures, high BP, MP, are soluble in water, show electrical conductance in molten or aqueous solution state and show ionic reactions.
- The chemical bond formed when two atoms share electrons between them is known as a **covalent bond**. It is formed by equal sharing of electrons between two similar or different atoms.
- Covalent bonds are of three types—single, double and triple covalent bonds.
- **Non-polar:** In a non-polar covalent bond, sharing of electron/ electrons is between atoms of the same element or atoms with nearly the same electronegativity.
- **Polar:** In a polar covalent bond, sharing of electron/ electrons is between atoms of different elements or atoms with different electronegativities.
- **Covalent compounds** may be gas, liquid or solid with low BP, MP values, the polar compounds dissolve in polar solvents and the non-polar compounds dissolve in non-polar solvents, are non-conductors and show slow molecular reactions.

Exercise 2

Section A: Multiple Choice Questions (1 Mark)

1. A chemical bond is formed to attain ______
   (a) the octet state (b) stability
   (c) minimum energy (d) all of these

2. Two elements with atomic number 12 and 13 have valences respectively,
   (a) 1, 3 (b) 2, 3
   (c) 3, 3 (d) 3, 2

3. Which one of the following structures correctly depicts the compound $CaCl_2$? **[CBSE 2021]**

(a) 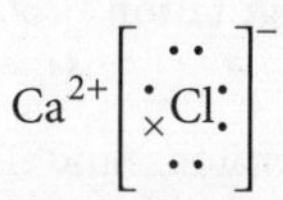

(b) 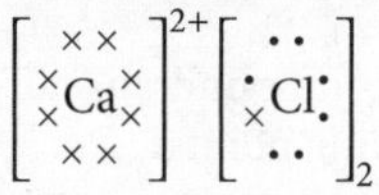

(c) 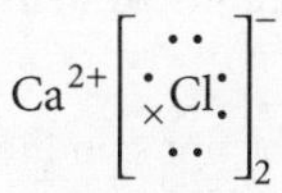

(d) 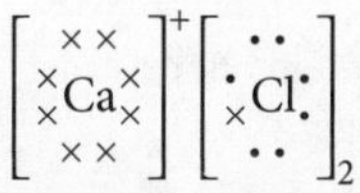

4. In which of the following molecules does transfer of electron(s) take place?
(a) PCl_3 (b) CsCl
(c) BCl_3 (d) CCl_4

5. During a covalent bond formation ________.
(a) transfer of electrons takes place
(b) unequal sharing of electrons takes place
(c) equal sharing of electrons takes place
(d) all of these occur

6. An element reacts with oxygen to give a compound with a high melting point. The compound is soluble in water. The element is likely to be
(a) carbon (b) nitrogen
(c) sodium (d) silicon

7. Non-metals form covalent chlorides because
(a) they can give electrons to chlorine atoms to form chloride ions
(b) they can share electrons with chlorine
(c) they can give electrons to chlorine
(d) they cannot share electrons with chlorine atoms

8. Which pair has only covalent bonds?
(a) NaCl, HI (b) HI, CCl_4
(c) NaCl, $MgCl_2$ (d) Na_2O, CO_2

9. Which pair contains only single bonds?
(a) O_2, N_2 (b) H_2, O_2
(c) CH_4, NH_3 (d) CH_4, C_2H_4

10. Ionic compounds can show conductance in ________
(a) solid state (b) molten state
(c) aq. solution state (d) both (b) and (c)

11. Which of thre following are not ionic compounds?
(i) NaCl (ii) HCl
(iii) CCl_4 (iv) KCl
(a) (i) and (ii) (b) (ii) and (iii)
(c) (iii) and (iv) (d) (i) and (iii)

12. The electronic configuration of three elements X, Y and Z are as follows:
X = 2, 4, Y = 2, 7, Z = 2, 1. Which two elements will combine to form an ionic compound? Write the correct formula.
(a) Y_2Z (b) XZ_3
(c) ZY (d) X_2Y

13. The reaction between X and Y forms compound Z. X loses electrons and Y gains electrons. Which of the following properties is not shown by Z?
(a) has a high melting point
(b) occurs as a solid
(c) conducts electricity in the molten state
(d) has a low melting point

14. Select the set having correct statements.
(i) Both O_2 and CO_2 are non-polar and have double bonds.
(ii) NaCl has a higher boiling point than HCl.
(iii) NH_3 contains a double bond.
(iv) Both NaCl and HCl are soluble in water.
(a) (i), (ii) (b) (iii), (iv)
(c) (i), (ii), (iii) (d) (i), (ii), (iv)

15. Match the following.

Column I	Column II
A. $MgCl_2$	(i) Polar covalent
B. CCl_4	(ii) Non-polar covalent
C. HF	(iii) Chemically inactive
D. Argon	(iv) Ionic compound

(a) (A)—(iv), (B)—(iii), (C)—(ii), (D)—(i)
(b) (A)—(iv), (B)—(ii), (C)—(i), (D)—(iii)
(c) (A)—(iii), (B)—(ii), (C)—(i), (D)—(iv)
(d) (A)—(iv), (B)—(ii), (C)—(iii), (D)—(i)

Assertion Reason Questions

Directions: In the following questions, two statements (Assertion) A and Reason (R) are given. Mark
(a) if A and R are both correct and R is the correct explanation of A;
(b) if A and R are both correct but R is not the correct explanation of A;
(c) A is true but R is false;
(d) A is false but R is true.

Assertion	Reason
1. Noble gases are chemically unreactive.	1. Noble gases have a highly stable octet state electronic configuration.
2. Every atom tries to attain the octet state for stability.	2. Octet state is attainable by losing or gaining or sharing of electrons.
3. An ionic bond is formed by the transfer of valence electrons.	3. An ionic bond is directional in nature.
4. A covalent bond is formed by equal sharing of electrons.	4. A covalent bond is an electrostatic force by nature.
5. A chlorine molecule is covalent in nature.	5. It contains a single bond between two chlorine atoms.
6. Ionic compounds are soluble in water.	6. Water has a high dielectric constant value.
7. Covalent compounds can conduct electricity.	7. Graphite can show electric conductance.
8. The boiling and melting points of ionic compounds are high.	8. Ionic compounds are generally crystalline solids.

Section B: Very Short Answer Questions (2 Marks)

1. The number of electrons in the valence shell of noble gases is ______ except for helium which has __________.
2. Define a chemical bond.
3. Name the bond formed by complete transfer of valence electrons from one atom to another atom.
4. What type of force is an ionic bond?
5. Sharing of two electron pairs between two atoms results in the formation of a ________ bond.
6. Ionic compounds are generally soluble in which type of solvents?
7. Sodium chloride, being an ionic solid, has a _______ melting point.
8. The atomic number of potassium atom is 19. What is its electrovalency?
9. An atom has a configuration of 2, 8, 4. What will be its covalency?
10. Element X has 10 protons and 10 electrons. Will it be reactive?
11. Why do ionic compounds usually have high melting and boiling points?
12. What is the difference in covalent bonding in CCl_4 and CH_4?
13. Why does solid NaCl not conduct electricity but aqueous or molten NaCl solution does?
14. How many covalent bonds are there in a molecule of nitrogen?
15. What type of covalent bond is present in a nitrogen molecule?
16. What is the number of single bonds present in a methane molecule (CH_4)?
17. Which of the following molecules have at least one double bond?
    O_2, F_2, HCl, CO_2, CH_4
18. Which of the following molecules have at least one triple bond?
    O_2, N_2, CO_2, C_2H_4, C_2H_2

Case Study or Passage Based Questions

Valence electrons are involved in chemical bond formation. Chemical bonds are of two types—ionic and covalent. Compounds having ionic bonds are ionic compounds while compounds having covalent bonds are known as covalent compounds. Ionic and covalent compounds differ in their physical state and many other properties like solubility, boiling and melting points, conduction.

19. The number of electrons lost or gained by the atom of an element is known as ________.
20. How do metals and non-metals combine?
21. What is the general physical state of an ionic compound?
22. Between an ionic and a covalent compound, which can show conductance in the molten state?
23. Name any two polar covalent compounds which are soluble in water.

Section C: Short Answer Questions (3 Marks)

1. What is a covalent bond? What type of bond exists in (i) CCl_4, (ii) $CaCl_2$?

2. What are covalent compounds? Why are they different from ionic compounds? List three characteristic properties of covalent compounds. **[CBSE 2016]**

3. Two neon atoms do not form a covalent bond to give a neon molecule Ne_2. Why?

4. Elements forming ionic compounds attain the noble gas electronic configuration by either gaining or losing electrons from their valence shells. Explain giving reasons why carbon cannot attain such a configuration in this manner to form its compounds. Name the type of bonds formed in ionic compounds and in the compounds formed by carbon. Also explain with reason why carbon compounds are generally poor conductors of electricity. **[CBSE 2015]**

5. A metal X combines with a non-metal Y by the transfer of electrons to form a compound Z.
   (i) State the type of bond in compound Z.
   (ii) What can you say about the melting point and boiling point of compound Z?

6. Explain why ionic bond formation takes place between potassium and chlorine but covalent bond formation takes place between hydrogen and chlorine.

7. What are ionic compounds? List two properties of these compounds.

8. Ionic compounds conduct electricity only in the molten state and not in the solid state. Why?

9. Give reasons for the following.
   (i) Covalent compounds are bad conductors of electricity.
   (ii) Covalent compounds have low melting and boiling points.

10. Illustrate the formation of bonds in:
    (i) sodium chloride
    (ii) magnesium chloride
    Identify the ions present in these compounds.

11. Write the symbols of the cation and anion present in MgO. Why do ionic compounds have higher melting points?

12. Describe the nature of bond between the following.
    (i) sodium and chlorine
    (ii) carbon and chlorine
    (iii) hydrogen and chlorine

13. From the list of compounds – Cl_2, N_2, CO_2, C_2H_2 and O_2 – choose the molecule that contains:
    (i) only a single bond
    (ii) only a triple bond
    (iii) only a double bond
    (iv) two double bonds
    (v) single and triple bonds

Section D: Long Answer Questions (5 Marks)

1. Define cations and anions. Discuss with suitable examples.

2. What is an electrovalent bond and how is it formed? Explain with an example.

3. Show the formation of covalent bonds in HCl, CCl_4, CH_4, H_2, O_2 and Cl_2.

4. Write any four differences between ionic and covalent compounds.

5. (i) Write the electron dot diagram for chlorine (atomic number 17) and calcium (atomic number 20). Show the formation of calcium chloride by the transfer of electrons.
   (ii) Identify the nature of the above compound and explain three physical properties of such a compound. **[CBSE 2015]**

6. Explain the formation of the ionic compound CaO with the electron dot structure. Atomic numbers of calcium and oxygen are 20 and 8 respectively. **[CBSE 2020]**

7. Explain the nature of the covalent bond using the bond formation in CH_3Cl.

Section E: Case Study or Passage Based Questions (4 Marks)

Chemical bonds are of two types—ionic and covalent. Compounds having ionic bonds are ionic compounds while compounds having covalent bonds are known as covalent compounds. Covalent compounds can be further divided into polar and non-polar covalent compounds depending upon the difference in electronegativities between atoms undergoing bond formation. Ionic and covalent compounds differ in their physical state and many other properties like solubility, boiling and melting points, conduction.

1. During the formation of which of the following molecules does hydrogen gain an electron?
(a) NH_3 (b) HCl
(c) CH_4 (d) NaH

2. Which of the following molecules is non-polar?
(a) NH_3 (b) ClF_3
(c) CH_4 (d) PCl_3

3. Which of the following molecules is polar?
(a) CH_4 (b) C_2H_4
(c) NaCl (d) $CHCl_3$

4. Among the following compounds, identify the one in which the metal and non-metal show different electrovalencies.
(a) calcium oxide
(b) sodium chloride
(c) aluminium oxide
(d) aluminium nitride

5. Select the set having correct statements.
(i) Both O_2 and N_2 are non-polar covalent molecules.
(ii) C_2H_2 contains a double bond.
(iii) CH_4 and CCl_4 have 4 single bonds.
(iv) NaCl and $CaCl_2$ have ionic bonds.
(a) (i), (ii) (b) (iii), (iv)
(c) (i), (ii), (iii) (d) (i), (iii), (iv)

High Order Thinking Skills (HOTS) Questions

1. Write the names and symbols of two most reactive metals. Explain by drawing the electronic structure, how any one of the two metals react with a halogen. State any four physical properties of the compound formed.

2. In the formation of a compound XY_2, atom X donates one electron to each Y atom. Show the electron dot structure of X and Y and the formation of XY_2. What is the nature of the bonds in XY_2? Write any three properties of XY_2. The electronic configurations of three elements X and Y are as follows:
X: 2, 8, 2 Y: 2, 7

3. The electronic configuration of three elements PQR are as follows:
P: 2, 8, 8, 1 Q: 2, 8, 7 R: 2, 5
(i) Which one of these elements is a non-metal?
(ii) Which of these elements is a metal?
(iii) When P combines with Q which type of bond is formed and what is the formula of the compound?
(iv) What is the formula of the compound when R reacts with Q?
(v) Which of the following elements can get argon (inert gas) configuration?

4. The elements A and B have the following configurations:
A: 2, 6
B: 2, 8, 2
(i) What is the nature of the bond between A and B?
(ii) Which inert gas configuration can A and B achieve during bond formation?

5. (i) Write the formula of the chloride and nature of bonding formed by 2 elements P and Q having atomic numbers 11 and 6 respectively.
(ii) Are the properties of both the chlorides formed here the same or different?
(iii) Write the formula of the oxides of P and Q.

Answers

Section A: Multiple Choice Questions

1. (d)	2. (b)	3. (c)	4. (b)
5. (c)	6. (c)	7. (b)	8. (b)
9. (c)	10. (d)	11. (b)	12. (c)
13. (d)	14. (d)	15. (b)	

Assertion Reason Questions

1. (a)	2. (a)	3. (c)	4. (c)
5. (b)	6. (a)	7. (d)	8. (b)

Section B: Very Short Answer Questions

1. 8, 2.
2. A chemical bond is the force of attraction that holds two atoms together.
3. Ionic bond.
4. An ionic bond is an electrostatic force of attraction.
5. Covalent bond.
6. Ionic compounds are generally soluble in highly polar solvents.
7. High.
8. Its electronic configuration is 2, 8, 8, 1, so its electrovalency is one.
9. As the atom has a configuration 2, 8, 4, its covalency is 4.
10. No, as it is an inert gas (Ne), it is inactive.
11. A high amount of energy is required to break the strong interionic attraction in ionic compounds so their melting and boiling points are high.
12. In CCl_4, all C–Cl bonds are polar covalent bonds while in CH_4 all C–H bonds are non-polar covalent bonds.
13. In molten or aqueous NaCl solution, ions are free to move so electric conduction is possible; but it is not possible in solid NaCl as the ions are not free.
14. There are 3 covalent bonds in a molecule of nitrogen.
15. The type of covalent bond present in a nitrogen molecule is a triple bond ($N \equiv N$).
16. The methane molecule has 4 single bonds.
17. O_2 and CO_2 have double bonds.
18. N_2 and C_2H_2 have triple bonds.

Case Study or Passage Based Questions

19. Valency.
20. Metals and non-metals combine by the transfer of electrons from metals to non-metals.
21. An ionic compound is in the solid state.
22. An ionic compound can show conductance in the molten state due to the presence of free ions.
23. Ammonia (NH_3) and hydrogen chloride (HCl) are soluble in water.

Section C: Short Answer Questions

1. The chemical bonds formed between two atoms by the sharing of electrons between them is known as a covalent bond. The sharing of electrons between the two atoms takes place in such a way that both the atoms acquire a stable electronic configuration of the nearest noble gas.
   (i) CCl_4: Covalent bond
   (ii) $CaCl_2$: Ionic bond
2. Covalent compounds are those compounds which are formed by the sharing of electrons. They are different from ionic compounds because they do not contain ions. The characteristic properties of these compounds are:
   (i) They have low melting and boiling points.
   (ii) They are insoluble in water but are soluble in organic solvents.
   (iii) They do not conduct electricity in molten state or in the aqueous solution.
3. Two neon atoms do not form a covalent bond to give a neon molecule Ne_2. This is because neon is an inert gas with an octet configuration (2, 8); so, neon atoms do not undergo any type of chemical combination.
4. Carbon cannot lose four electrons because a very high amount of energy is required to remove four electrons from its outermost shell. Carbon cannot gain four electrons easily because six protons cannot hold ten electrons.
   - Ionic compounds form electrovalent bonds, whereas carbon can share four electrons to form covalent bonds.
   - Carbon compounds are poor conductors of electricity because they do not form ions in aqueous solution.
5. (i) Ionic bond (ii) High.
6. Potassium belongs to group 1 and is highly electropositive. Chlorine belongs to group 17 and is highly electronegative. Potassium has very low IP value and can form the K^+ ion easily. Chlorine

has very high IP value and high EA value and can form the Cl^- ion easily. Thus, an ionic bond results between potassium and chlorine. IP value of hydrogen is not as low as that of the alkali metals (K) so the tendency to form H^+ ion is very less. Sharing of electrons takes place between hydrogen and chlorine resulting in the formation of a polar covalent bond.

7. The compounds which have bonds formed by the loss and gain of electrons are called ionic compounds or electrovalent compounds and the bond formed is known as an ionic bond or electrovalent bond.
Two properties of ionic compounds: (i) Ionic compounds are hard and brittle. (ii) Ionic compounds have high melting and boiling points.

8. Ionic compounds in the solid state do not conduct electricity because movement of ions in the solid is not possible due to their rigid structure; but, in the molten state ions move freely so conduction is possible.

9. (i) Covalent compounds are bad conductors of electricity due to lack of free ions or electrons.
(ii) Covalent compounds are formed by covalent bonds and it has been found that the intermolecular forces of attraction (van der Waals forces) in covalent compounds are weak so less energy is needed to break them. Hence, their melting and boiling points are quite low.

10.

(i) Na → $Na^+ + e^-$
2, 8, 1   2, 8
$Cl + e^-$ → Cl^-
2, 8, 7   2, 8, 8

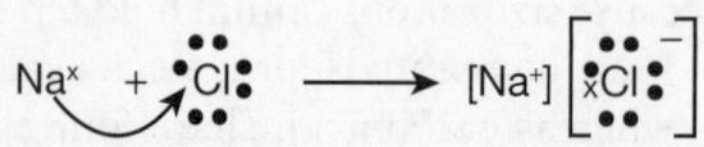

(ii) Mg → $Mg^{2+} + 2e^-$
2, 8, 2   2, 8
$Cl + e^-$ → Cl^-
2, 8, 7   (2, 8, 8)

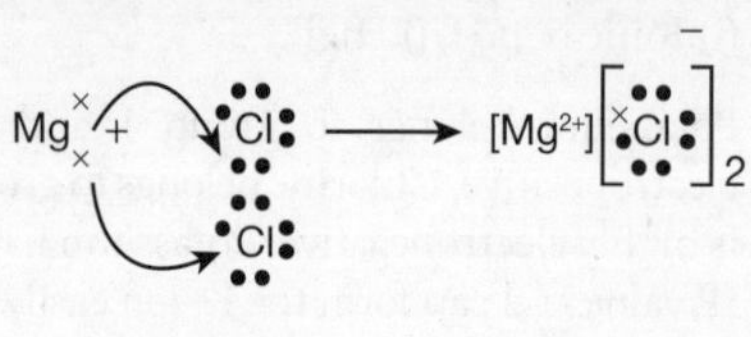

11. **Cation:** Mg++
Anion: O– –
Ionic compounds have high melting point because cations and anions are held by strong electrostatic forces of attraction and a considerable amount of energy is required to break the strong interionic forces.

12. (i) Sodium and chlorine combine to form NaCl so an ionic bond is formed between them.
(ii) Carbon and chlorine combine to form CCl_4 so four covalent bonds are formed between them.
(iii) Hydrogen and chlorine combine to form HCl so a covalent bond is formed between them.
(For structure of these compounds see text, Section 3.13.2.)

13. (i) Only a single bond: Cl_2
(ii) Only a triple bond: N_2
(iii) Only a double bond: O_2
(iv) Two double bonds: CO_2
(v) Single and triple bonds: C_2H_2

Section D: Long Answer Questions

1. See text (Section 3.13.1).
2. See text (Section 3.13.1).
3. See text (Section 3.13.2).
4. See text (Section 3.13.2).
5. (i) Cl: 2, 8, 7   Ca: 2, 8, 8, 2
$Ca \rightarrow Ca^{2+} + 2e^-$ $2Cl + 2e^- \rightarrow 2Cl^-$

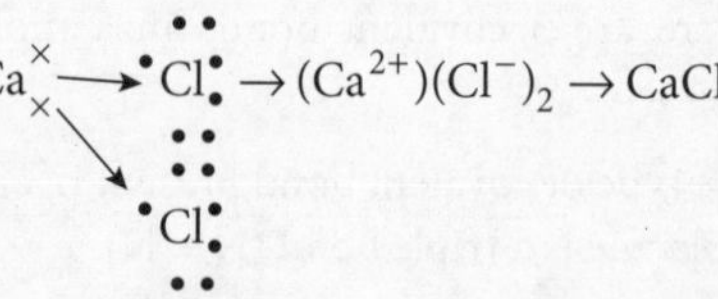

(ii) The above compound is an ionic compound. The physical properties of this compound are:
- it is hard and solid,
- it has high melting and boiling points, and
- it is soluble in water.

6. The atomic numbers of calcium and oxygen are 20 and 8 respectively. The electronic configuration of calcium will be 2, 8, 8, 2; thus,

it has 2 electrons in its outermost shell. On the other hand, oxygen (2, 6) has 6 electrons in its outermost shell. By losing two electrons, the calcium atom forms a calcium ion (Ca^{2+}) while by gaining 2 electrons, the oxygen atom forms an oxide ion (O^{2-}). Calcium reacts with oxygen to form an ionic compound calcium oxide (CaO).

The electron dot representation for the formation of calcium oxide is given below.

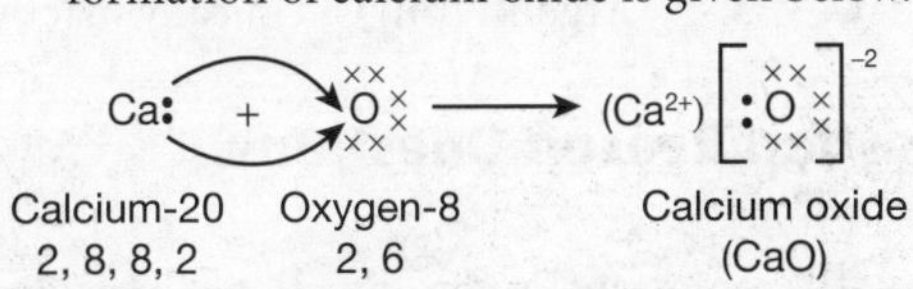

7. Carbon has four electrons in its valence shell. To complete its octet, it shares its four electrons with other atoms to create four bonds, thus becoming stable. The bonds that are formed by sharing electrons are known as covalent bonds.

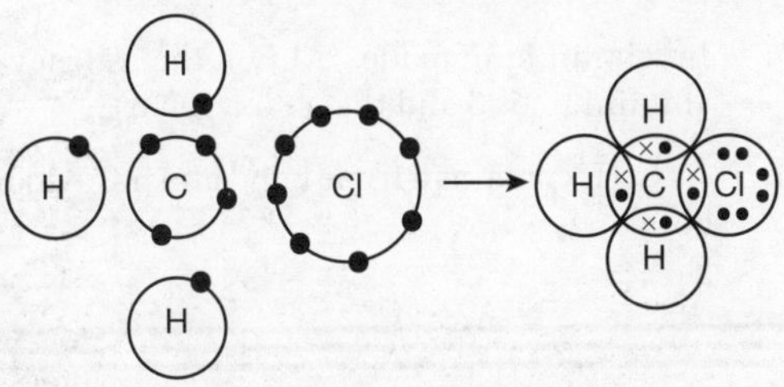

In CH_3Cl, carbon requires 4 electrons to complete its octet, while each hydrogen atom requires one electron to complete its duplet. Also, chlorine requires an electron to complete the octet. Therefore, all of these share the electrons and as a result, carbon forms 3 bonds with hydrogen and one with chlorine.

Section E: Case Study or Passage Based Questions

1. (d) **2.** (c) **3.** (d) **4.** (c)
5. (d)

High Order Thinking Skills (HOTS) Questions

1. The two reactive metals are sodium (Na) and potassium (K). Sodium reacts with chlorine.

$\underset{2,8,1}{Na} \rightarrow \underset{\substack{2,8\\ \text{(Sodium cation)}}}{Na^+} + e^-$

$\underset{2,\,8,\,7}{Cl} + e^- \rightarrow \underset{\substack{2,\,8,\,8\\ \text{(Chloride anion)}}}{Cl^-}$

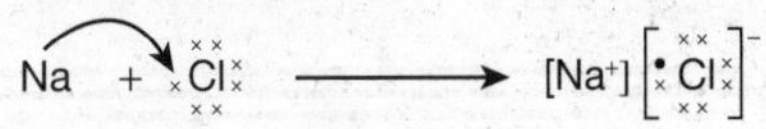

Ionic compounds have high melting and boiling points.
They are good conductors of electricity.
They are generally soluble in water and insoluble in organic solvents.
They are solid, hard and brittle.

2. X: 2, 8, 2, Y: 2, 7

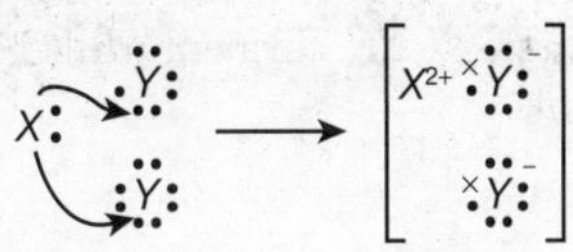

Nature of bond in XY_2 is ionic.
Properties: (i) Crystalline solids.
(ii) High melting and boiling point.
(iii) Soluble in water, insoluble in organic solvents.
(iv) Conductor of electricity. **(Any three)**

3. (i) Q and R are non-metals which are chlorine (17) and nitrogen (7) respectively.
(ii) P is a metal and it is potassium (19).
(iii) PQ is formed by ionic bond as P transfers one electron and Q gains that electron.
(iv) The compound formed is RQ_3 (NCl_3) as R will share three valence electrons with three Q atoms.
(v) P can get argon (inert gas) configuration by losing one electron while Q can get argon (inert gas) configuration by gaining one electron.

4. (i) Ionic bond is formed between A and B as A needs 2 electrons to complete the octet and B can lose 2 electrons to attain its octet state. The B atom will lose 2 electrons and A will get these 2 electrons (see example of MgO formation in text, Section 13.3.1).
(ii) Both A and B can achieve neon gas configuration.

5. (i) Here element P is sodium (11) so it will form PCl (NaCl) which is formed by ionic bonding. Here Q is carbon (6) so it will form QCl_4 (CCl_4) which is formed by covalent bonding.

(ii) Both PCl_5 and QCl_4 differ in properties as PCl_5 is an ionic compound while QCl_4 is a covalent compound.

(iii) P will form P_2O (Na_2O) and Q will form QO_2 (CO_2) by ionic bond and covalent bond formation respectively.

Hints

Section A: Multiple Choice Questions

2. Valency is equal to the number of valence electrons which are 2 and 3 for atomic number 12 and 13 respectively.
12: 2, 8, 2
13: 2, 8, 3

11. HCl and CCl_4 are covalent while NaCl and KCl are ionic.

12. As Z (3): 2, 1
Y (9): 2, 7
Here Z will lose one electron and Y will gain it to get stability and to form ionic bond to give ZY compound.

13. Z is an ionic compound. It has a high melting point.

14. As NH_3 does not have any double bond.

15.

A.	$MgCl_2$	(iv) Ionic compound
B.	CCl_4	(ii) Non-polar covalent
C.	HF	(i) Polar covalent
D.	Argon	(iii) Chemically inactive

Assertion Reason Questions

3. The ionic bond is non-directional in nature.

4. A covalent bond is van der Waals force.

Section E: Case Study or Passage Based Questions

4. In aluminium oxide (Al_2O_3) the valency of aluminium is 3 and that of oxygen is 2.

5. As C_2H_2 contains a triple bond and not a double bond.

3.12 Occurrence and Extraction of Metals

A few elements like carbon, sulfur, gold and noble gases, occur in the free state while others occur in combined forms in the earth's crust. The extraction and isolation of an element from its combined form involves various principles of chemistry. A particular element may occur in a variety of compounds. The process of metallurgy and isolation should be such that it is chemically feasible and commercially viable. Still, some general principles are common to all the extraction processes of metals. For obtaining a particular metal, first we look for **minerals** which are naturally occurring chemical substances in the earth's crust obtainable by mining. Out of many minerals in which a metal may be found, only a few are viable to be used as sources of that metal. Such minerals are known as **ores**. Rarely, an ore contains only a desired substance. It is usually contaminated with earthy or undesired materials known as **gangue**. The extraction and isolation of metals from ores involve the following major steps:

- concentration of the ore,
- isolation of the metal from its concentrated ore, and
- purification of the metal.

The entire scientific and technological process used for isolation of the metal from its ores is known as **metallurgy**.

3.12.1 Occurrence

The earth's crust is the major source of metals. Sea water also contains some soluble salts such as sodium chloride and magnesium chloride. Elements can occur as follows.

Elements in the atmosphere: The atmosphere mainly contains nitrogen (78.09%), oxygen (20.95%) and other gases (about 1%).

Elements in the sea: The sea is the major source of elements like Br, I, Ni, Cu, Zn, Sn, Au.

Elements in the earth's crust (lithosphere): Elements occur in two states in the earth crust

- **Free or native state**: Less reactive metals or noble metals with least electropositive nature are present in the free or native state, such as, copper, silver, gold, platinum.
  The most abundant elements in the lithosphere are:

	O,	Si,	Al,	Fe,	Ca,	Na
%	48.60	26.30	7.73	4.75	3.45	2.74

- **Combined state (minerals)**: Reactive metals occur in the combined state and are known as minerals. Those minerals from which metals can be profitably extracted are called *ores* (Table 3.11). These metals are generally associated with rocky materials, sand and clay, known as *gangue* or *matrix*.

Table 3.11 Important ores

Metal	Name of the ore	Name of main compound in ore	Formula of the ore
1. Sodium (Na)	Rock salt	Sodium chloride	$NaCl$
2. Aluminium (Al)	Bauxite	Aluminium oxide	$Al_2O_3.2H_2O$
3. Iron (Fe)	Hematite	Iron(III) oxide (Ferric oxide)	Fe_2O_3
4. Copper (Cu)	Copper glance Cuprite	Copper(I) sulfide Copper(I) oxide	Cu_2S Cu_2O
5. Silver (Ag)	Silver glance	Silver(I) sulfide	Ag_2S
6. Zinc (Zn)	Zinc blende Calamine	Zinc sulfide Zinc carbonate	ZnS ZnS
7. Lead (Pb)	Galena	Lead(II) sulfide	PbS
8. Mercury (Hg)	Cinnabar	Mercury(II) sulfide	HgS

The chemical form in which a metal occurs: The form in which a metal is present in its ore is extremely important as the process of extraction has to be done accordingly.

- The least active metals occur in the free state.
- Highly active metals such as the alkali metals, occur as salts—halides, carbonates, sulfates and nitrates.
- With the decrease in activity of metal oxides and sulfides, these forms become more common and dominant.

TEST YOUR KNOWLEDGE

3.25 Name the two most abundant metals in the earth's crust.

Solution:
Al and Fe are the two most abundant metals in the earth's crust.

3.26 Differentiate between mineral and ore.

Solution:
See text (Section 3.14.1).

3.27 What are the impurities associated with minerals known as?

Solution:
Matrix or gangue.

3.28 Which of these metals can have both sulfide and oxide ores?
Ag, Zn, Fe, Al, Cu.

Solution:
Here Zn, Cu, Fe can have both sulfide and oxide ores.

3.13 General Metallurgical Processes

Extraction of Metals

We have learnt about the reactivity series of metals. With this knowledge, we can easily understand how a metal is extracted from its ore. Some metals are found in the earth's crust in the free state. Some are found in the form of their compounds. The metals at the bottom of the activity series are the least reactive. They are often found in the free state. For example, gold, silver, platinum and copper are found in the free state. Copper and silver are also found in the combined state as their sulfide or oxide ores. The metals at the top of the activity series like K, Na, Ca, Mg and Al are so reactive that they are never found in nature as free elements. The metals in the middle of the activity series like Zn, Fe and Pb, are moderately reactive. They are found in the earth's crust mainly as oxides, sulfides or carbonates. The ores of many metals are oxides. This is because oxygen is a very reactive element and is very abundant on the earth.

Thus, on the basis of reactivity, we can group the metals into the following three categories (Fig. 3.9)— (i) Metals of low reactivity; (ii) Metals of medium reactivity; (iii) Metals of high reactivity. Different techniques are to be used for obtaining the metals falling in each category.

3.13.1 Extracting Metals Towards the Top of the Activity Series

Metals are electropositive and can form metal oxides. The strength of M–O in these oxides is directly proportional to the electropositivity of the metal. A reducing agent can break this M–O by removing oxygen from it. This breakage is decided by the strength of the M–O bond or upon the position of the metal in the activity series. The metals high up in the reactivity series are very reactive. For example, metals like K, Na, Ca, Mg and Al can be obtained by the electrolysis of their fused halides or oxides, that is, by electrolytic reduction, as their oxides cannot be reduced by common reducing agents like carbon, carbon monoxide and hydrogen. This is because these metals have more affinity for oxygen than carbon.

$$K_2O \text{ or } CaO \xrightarrow{C, H_2, CO} \text{No reaction}$$

These metals are obtained by electrolytic reduction of their molten chlorides or oxides. For example, sodium, magnesium and calcium are obtained by the electrolysis of their molten chlorides.

Electrolytic reduction: Electrolytic reduction is carried out by passing electric current through the molten salts of highly reactive metal chloride or oxides. Here electricity causes splitting or electrolysis. In electrolysis, the negatively charged electrode (cathode) acts like a powerful reducing agent by providing electrons to reduce the metal cations to the metal. The metals are deposited at the cathode. The positively charged electrode (anode) acts as a powerful oxidising agent by accepting electrons from non-metal ions and oxidises them into neutral atoms or molecules. For example, chlorine or oxygen is liberated at the anode. Here are few examples of electrolysis.

(i) Extraction of sodium metal: Sodium metal is extracted by the electrolytic reduction of molten sodium chloride by passing electric current through it. Due to electrolysis, NaCl splits into free sodium ions (Na^+) and free chloride ions (Cl^-). Here sodium ions are attracted towards the cathode and get reduced to form sodium atoms or sodium metal while the Cl^- ions are attracted to the anode and get oxidised into chlorine gas (Fig. 3.32). The reactions are:

$$\underset{\text{Sodium chloride (Molten)}}{2NaCl(l)} \xrightarrow{\text{Electrolysis}} \underset{\text{Sodium metal}}{2Na(s)} + \underset{\text{Chlorine gas}}{Cl_2(g)}$$

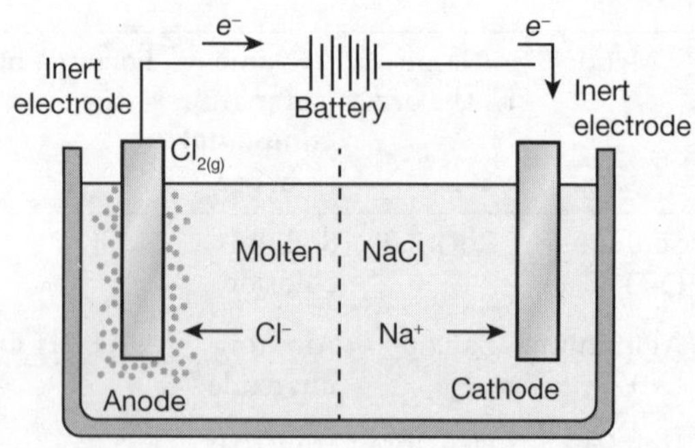

Fig. 3.32 Electrolysis of molten NaCl

At cathode: $2Na^+ + 2e^- \rightarrow 2Na$

At anode: $2Cl^- \rightarrow Cl_2 + 2e^-$

It is important to note that here, an aqueous solution of sodium chloride (brine solution) cannot be used for electrolysis as the sodium ion formed here will react with the OH^- ions obtained from water to form sodium hydroxide (NaOH).

$$Na^+ + OH^- \rightarrow NaOH$$

Similarly, potassium metal is obtained by the electrolysis of molten KCl, calcium metal is obtained by the electrolysis of molten $CaCl_2$ and magnesium metal is obtained by the electrolysis of molten $MgCl_2$.

(ii) Extraction of aluminium metal: Aluminium metal is extracted by the electrolytic reduction of molten aluminium oxide (Al_2O_3) by passing electric current through it. Due to electrolysis, Al_2O_3 splits into free aluminium ions (Al^{3+}) and free oxide ions (O^{2-}). The aluminium ions are attracted towards the cathode and get reduced to form aluminium atoms or aluminium metal while the O^{2-} ions are attracted to the anode and get oxidised into oxygen gas (Fig. 3.33). The reactions are:

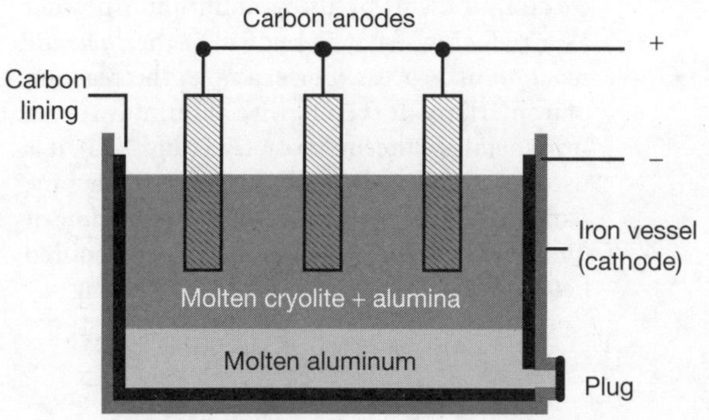

Fig. 3.33 Electrolysis of molten Al_2O_3

$$\underset{\text{Aluminium oxide (Molten)}}{2Al_2O_3(l)} \xrightarrow{\text{Electrolysis}} \underset{\text{Aluminium metal}}{4Al(s)} + \underset{\text{Oxygen gas}}{3O_2(g)}$$

At cathode: $Al^{3+} + 3e^- \rightarrow Al$

At anode: $2O^{2-} \rightarrow O_2 + 4e^-$

In the electrolysis of alumina, cryolite (Na_3AlF_6) and CaF_2 are also used for the easy and lower temperature reduction of alumina.

3.13.2 Extracting Metals in the Middle of the Activity Series

The metals in the middle of the activity series such as iron, zinc, lead and copper, are moderately reactive. These are usually present as sulfides or carbonates in nature. The metals present below Al like iron, lead and copper are obtained by the reduction of their oxides with carbon, carbon monoxide and hydrogen; however, zinc oxide can be reduced by carbon only. Oxides of mercury and silver can be reduced by heating above 300°C. It is easier to obtain a metal from its oxide, as compared to its sulfides and carbonates. Therefore, prior to reduction, the metal sulfides and carbonates must be converted into metal oxides. The process of extraction of pure metal from its ore is called metallurgy.

Metallurgy involves the following processes:

Crushing (pulverisation) of ore: Big lumps of ore obtained from the earth's crust are crushed into smaller pieces with the help of jaw crushers and grinders. This process is known as *crushing of ore*.

Concentration of the ore: Removal of impurities like sand, rocky substances, lime stone and mica from the crushed ore is known as *concentration* or *benefaction* or *dressing of ore*. The methods that can be used for removing the matrix from the ore depends on some difference in either physical properties or chemical properties of the ore and the matrix. These are the following methods—hand picking (selected pieces of ore are picked up), levigation (by washing so that impurities are washed away), magnetic separation (magnetic and non-magnetic substances are separated).

Calcination: Calcination is the process of heating the ores below their melting points in the absence of air or in a very limited amount of air, to remove volatile impurities like water, CO_2 and organic matter. During calcination the ore becomes porous, volatile impurities are removed and carbonate ore decomposes into oxides. Calcination is mainly used for carbonate ores like limestone ($CaCO_3$) and calamine ($ZnCO_3$). For example,

$$\underset{\text{Limestone}}{CaCO_3} \xrightarrow{\Delta} \underset{\text{Calcium oxide}}{CaO} + CO_2\uparrow$$

$$\underset{\text{Bauxite}}{Al_2O_3.2H_2O} \xrightarrow{\Delta} \underset{\text{Alumina}}{Al_2O_3} + 2H_2O\uparrow$$

Roasting: Roasting is the process of heating the ore in excess of air in order to convert the metal ores into their oxides and water insoluble sulfides into water soluble sulfates. During roasting, volatile impurities are removed. S, As, Sb are removed as SO_2, As_2O_3, Sb_2O_3 respectively and sulfide ores are converted into their oxides. Both calcination and roasting are carried out in the reverberatory furnace (Fig. 3.34). For example,

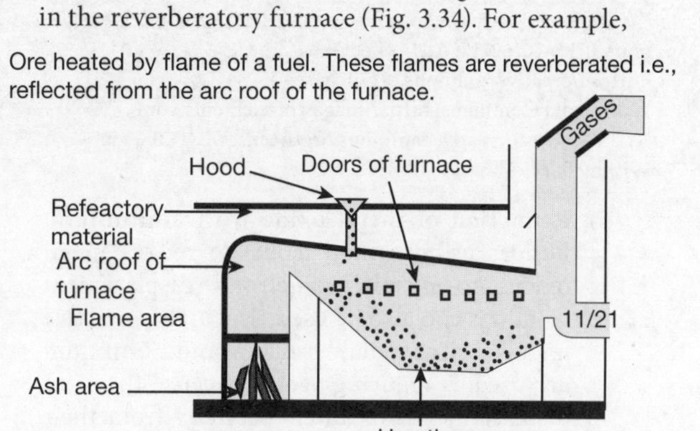

Fig. 3.34 Reverberatory furnace

$$\underset{\text{Pyrite}}{2FeS} + 3O_2 \xrightarrow{\Delta} 2FeO + 2SO_2$$

$$\underset{\text{Cinnabar}}{2HgS} + 3O_2 \xrightarrow{\Delta} 2HgO + 2SO_2$$

$$\underset{\text{Zinc sulfide}}{ZnS} + 2O_2 \xrightarrow{\Delta} ZnSO_4$$

$$\underset{\text{Lead sulfide}}{PbS} + 2O_2 \xrightarrow{\Delta} PbSO_4$$

Reduction: The calcinated or roasted metal oxides are reduced to the corresponding metals by using suitable reducing agents such as carbon, carbon monoxide and aluminium. The selection of the reducing agent depends upon the chemical reactivity of the metal to be extracted.

(i) **Reduction of metal oxide with carbon:** The oxides of comparatively less active metals like zinc, lead, tin, iron and copper are mainly reduced by using carbon as a reductant. Here the metal oxide is mixed with carbon (coke) and heated in a furnace for the reduction process.

$$\underset{\text{Metal oxide}}{MO(s)} + \underset{\text{Reductant}}{C(s)} \xrightarrow{\Delta} M(s) + \underset{\text{Carbon monoxide}}{CO(g)}$$

For example, when zinc oxide is heated with carbon, it is reduced to metallic zinc (Fig. 3.35).

$$ZnO(s) + C(s) \rightarrow Zn(s) + CO(g)$$

Fig. 3.35 Calamine, zinc oxide and zinc metal

Source: By Rob Lavinsky, iRocks.com – CC-BY-SA-3.0, CC BY-SA 3.0, https://commons.wikimedia.org/w/index.php?curid=10162115; Adam Rędzikowski, CC BY-SA 4.0 <https://creativecommons.org/licenses/by-sa/4.0>, via Wikimedia Commons; http://images-of-elements.com/, CC BY 1.0 <https://creativecommons.org/licenses/by/1.0>, via Wikimedia Commons

(ii) **Reduction of metal oxide with aluminium:** Besides using carbon (coke) to reduce metal oxides to metals, sometimes displacement reactions can also be used. The highly reactive metals such as sodium, calcium and aluminium are used as reducing agents because they can displace metals of lower reactivity from their compounds. For example, when manganese dioxide is heated with aluminium powder, the following reaction takes place.

$$3MnO_2(s) + 4Al(s) \rightarrow 3Mn(l) + 2Al_2O_3(s) + \text{Heat}$$

These displacement reactions are highly exothermic and the amount of heat evolved is so large that the metals are produced in the molten state. The reduction of a metal oxide into metal by using aluminium powder as a reducing agent is known as the *thermite reaction* or process. For example, the reaction of iron(III) oxide (Fe_2O_3) with aluminium to get iron metal is a thermite reaction (Fig. 3.36). It is used to join railway tracks or cracked machine parts. This process is called *thermite welding* or *aluminothermy* in which molten iron is poured between the broken iron pieces for welding.

$$Fe_2O_3(s) + 2Al(s) \rightarrow 2Fe(l) + Al_2O_3(s) + \text{Heat}$$

Fig. 3.36 Pyrolusite, alumina, hematite and thermite welding

Source: Rob Lavinsky, iRocks.com – CC-BY-SA-3.0, CC BY-SA 3.0 <https://creativecommons.org/licenses/by-sa/3.0>, via Wikimedia Commons; Aariuser I, CC BY-SA 2.0 <https://creativecommons.org/licenses/by-sa/2.0>, via Wikimedia Commons; : Rob Lavinsky, iRocks.com – CC-BY-SA-3.0, CC BY-SA 3.0 <https://creativecommons.org/licenses/by-sa/3.0>, via Wikimedia Commons; PetrS., CC BY-SA 3.0 <https://creativecommons.org/licenses/by-sa/3.0>, via Wikimedia Commons

3.13.3 Extracting Metals Low in the Activity Series

Metals low in the activity series like Cu, Ag and Hg are very unreactive. The oxides of these metals can be reduced to metals by heating alone. This type of reduction is known as self or auto-reduction. Some examples are discussed.

(i) **Extraction of mercury:** Cinnabar (HgS) is an ore of mercury. When it is heated in air, it is first converted into mercuric oxide (HgO). Mercuric oxide is then reduced to mercury on further heating.

$$2HgS(s) + 3O_2(g) \xrightarrow{\text{Heat}} 2HgO(s) + 2SO_2(g)$$

$$2HgO(s) \xrightarrow{\text{Heat}} 2Hg(l) + O_2(g)$$

(ii) Extraction of copper: Copper glance (Cu_2S) is an ore of copper. When it is heated in air, it is first converted into cuprous oxide (Cu_2O). Cuprous oxide is then reduced to copper on further heating.

$$2Cu_2S + 3O_2(g) \xrightarrow{\text{Heat}} 2Cu_2O(s) + 2SO_2(g)$$

$$2Cu_2O + Cu_2S \xrightarrow{\text{Heat}} 6Cu(s) + SO_2(g)$$

Fig. 3.37 Cinnabar, mercury, copper glance, cuprite and copper

Source: Rob Lavinsky, iRocks.com – CC-BY-SA-3.0, CC BY-SA 3.0 <https://creativecommons.org/licenses/by-sa/3.0>, via Wikimedia Commons; Tavo Romann, CC BY 4.0 <https://creativecommons.org/licenses/by/4.0>, via Wikimedia Commons; Rob Lavinsky, iRocks.com – CC-BY-SA-3.0, CC BY-SA 3.0 <https://creativecommons.org/licenses/by-sa/3.0>, via Wikimedia Commons; Rob Lavinsky, iRocks.com – CC-BY-SA-3.0, CC BY-SA 3.0 <https://creativecommons.org/licenses/by-sa/3.0>, via Wikimedia Commons; By Native_Copper_Macro_Digon3.jpg: "Jonathan Zander (Digon3)"derivative work: Materialscientist (talk) - Native_Copper_Macro_Digon3.jpg, CC BY-SA 3.0, https://commons.wikimedia.org/w/index.php?curid=7223304

Competition Edge

Reduction with carbon monoxide: Carbon monoxide produced by heating coke in limited supply of oxygen, is also used as a reducing agent.

$$\underset{\text{Ferric oxide}}{Fe_2O_3} + 3CO \rightarrow \underset{\text{Iron}}{2Fe} + 3CO_2$$

$$\underset{\text{Lead monoxide}}{PbO} + CO \rightarrow \underset{\text{Lead}}{Pb} + CO_2$$

3.13.4 Refining of Metals

The metals obtained by the various reduction processes described above are not very pure. They contain various impurities, which must be removed to obtain pure metals. The most widely used method for refining impure metals is electrolytic refining.

Electrolytic refining: Many metals like copper, zinc, nickel, tin, silver and gold, are refined electrolytically. In this process, the impure metal is made the anode and a thin strip of pure metal is made the cathode. A solution of the metal salt is used as an electrolyte. The apparatus is set up as shown in Fig. 3.38. On passing the current through the electrolyte, the pure metal from the anode dissolves into the electrolyte. An equivalent amount of pure metal from the electrolyte is deposited on the cathode. The soluble impurities go into the solution, whereas, the insoluble impurities settle down at the bottom of the anode and are known as anode mud. For example, in the electrolytic refining of copper, the electrolyte is a solution of acidified copper sulfate. The anode is impure copper, whereas, the cathode is a strip of pure copper. On passing electric current, pure copper is deposited on the cathode.

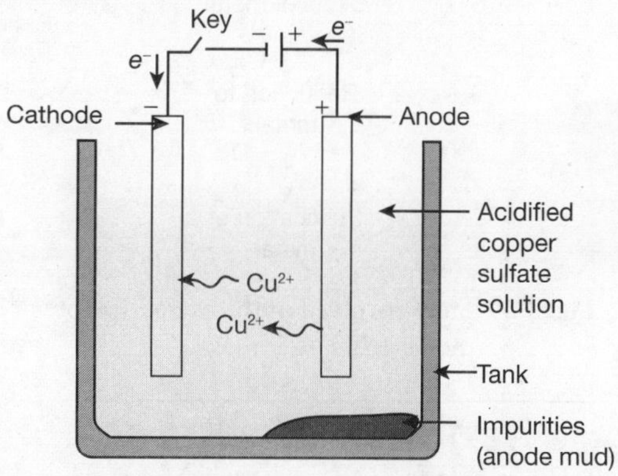

Fig. 3.38 Electrolytic refining of copper

At cathode:

$$\underset{\substack{\text{Copper ion}\\\text{(From electrolyte)}}}{Cu^{2+}} + \underset{\substack{\text{Electrons}\\\text{(From cathode)}}}{2e^-} \rightarrow \underset{\substack{\text{Copper atom}\\\text{(Deposited on cathode)}}}{Cu}$$

At anode:

$$\underset{\substack{\text{Copper atom}\\\text{(From impure anode)}}}{Cu} + \underset{\substack{\text{Electrons}\\\text{(Given to anode)}}}{2e^-} \rightarrow \underset{\substack{\text{Copper ion}\\\text{(Goes into electrolyte)}}}{Cu^{2+}}$$

Here, Cu^{2+} ions are taken from copper sulfate solution at the cathode and again put into the solution at the anode. As the process proceeds, the pure cathode becomes thicker and thicker while the anode becomes thinner and thinner—pure copper is collected at the cathode.

Electrolytic refining not only purifies the concerned metal but also recovers other valuable metals like silver and gold which are present as impurities.

Summary of Metallurgy

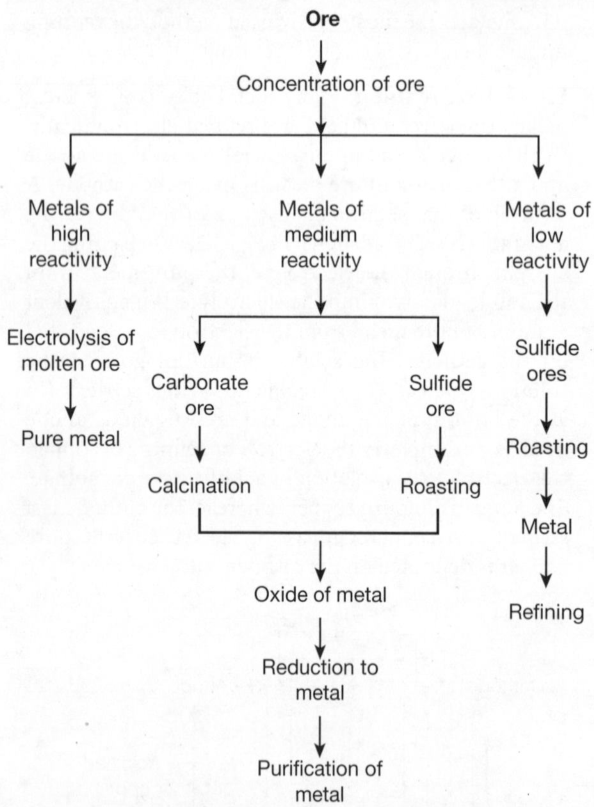

Fig. 3.39 Steps involved in the extraction of metals from ores

TEST YOUR KNOWLEDGE

3.29 Differentiate between calcination and roasting.

Solution:

Calcination	Roasting
1. Ore is heated in the absence of air.	1. Ore is heated in excess of air.
2. Carbonate and hydrated ores are calcinated.	2. Mostly sulfide ores are heated.
3. Moisture and CO_2 are removed	3. Volatile impurities of P, S are removed as P_2O_5, SO_2.

3.30 Why cannot oxides of sodium calcium and aluminium be reduced by C or CO or H_2?

Solution:
As the oxides of these metals have great affinity towards oxygen, they cannot be reduced by C or CO or H_2.

3.31 What is the significance of using cryolite in the metallurgy of aluminium?

Solution:
Cryolite not only lowers the fusion temperature of alumina from 2050°C to 950°C but also increases conductivity.

3.32 Name any two metals obtained by self reduction.

Solution:
Copper and mercury.

3.14 Corrosion

You would have noticed that in the case of a reactive metal, its surface may be attacked slowly by air and moisture (water) present in the atmosphere. Such a metal reacts with air and water vapour forming compounds on its surface due to which the surface of the metal appears dull as these compounds tarnish the metal. Such compounds formed on the surface of the metal are porous by nature and gradually fall off from the metal surface which exposes more metal to attack by air and water vapour. As this process continues, the whole metal is eaten up or converted into compounds. This eating up of metals by the action of air, water vapour (moisture) and chemicals like acid is known as *corrosion*. For example:

- Silver articles become black after some time when exposed to air. This is because it reacts with the sulfur in the air to form a coating of silver sulfide.
- Copper reacts with moist carbon dioxide in the air and slowly loses its shiny brown surface and gains a green coat. This green substance is basic copper carbonate.
- Iron when exposed to moist air for a long time acquires a coating of a brown flaky substance called rust.

3.14.1 Rusting of Iron and Its Prevention

Rusting: When iron objects are left in damp air or moisture for a considerable time, they get covered with a red-brown flaky substance - rust - and this phenomenon is called *rusting* (Fig. 3.40). During rusting, iron metal reacts with the oxygen of air in the presence of moisture to give a red-brown powder of hydrated ferric oxide ($Fe_2O_3.xH_2O$), that is, rust.

Fig. 3.40 Rusted iron object

Source: Angie from Sawara, Chiba-ken, Japan, CC BY 2.0 <https://creativecommons.org/licenses/by/2.0>, via Wikimedia Commons

Conditions needed for rusting: Rusting of iron requires the presence of air or oxygen and the presence of moisture or water.

Prevention: The rusting of iron can be prevented by painting, oiling, greasing, galvanising, chrome plating, anodising or alloying.

Painting: The most easy and common method to prevent rusting is to coat the surface of the iron object with a paint or varnish. On applying paint on the surface of the iron object the action of moisture and air is prevented as they are not in contact with the object. It is suitable for railings, iron pipes, railway coaches, body parts of vehicles and so on.

Grease or oil: The common method to prevent rusting is to coat the surface of the iron object with grease, oil or antirust resins. On applying grease on the surface of iron object, the action of moisture and air is prevented as they are not in contact with the iron object.

Galvanisation: Galvanisation is a method of protecting steel and iron from rusting by coating them with a thin layer of zinc. Zinc is more reactive than iron, so air or moisture attacks the zinc to form zinc oxide. As the zinc oxide coating is hard and also prevents action of air and moisture on the iron object, rusting of iron is prevented. It is interesting to note that the galvanised article is protected against rusting even if the zinc coating is broken as zinc is more easily oxidised than iron so corrosion of iron is not possible. For example, galvanisation is used for iron pipes used for water supply (Fig. 3.41).

Fig. 3.41 Galvanised pipe

Source: Tomas Castelazo, CC BY-SA 3.0 <https://creativecommons.org/licenses/by-sa/3.0>, via Wikimedia Commons

Fig. 3.42 Tin plated utensil

Source: By Sun Ladder - Own work, CC BY-SA 3.0, https://commons.wikimedia.org/w/index.php?curid=5746428

Tin plating and chromium plating: As both tin and chromium metals are resistant to corrosion, when a plating or electroplating of these metals is done on iron objects, rusting can be prevented. For example, tiffin boxes made of steel are tin plated to prevent rusting (Fig. 3.42). Chromium plating is mainly used on taps, handles of bicycles and so on.

Mixing iron with chromium and nickel: When iron is alloyed with chromium and nickel, stainless steel is obtained which is quite resistant to rusting. It is mainly used in making cooking utensils.

Corrosion of aluminium: As aluminium is an even more reactive metal than iron, it can undergo corrosion when it comes in contact with air and moisture. In air or moisture, a thin layer of aluminium oxide is formed on the surface of aluminium metal due to which the aluminium objects loses its shine and becomes dull. It is interesting to note that corrosion is fast in fresh aluminium; while in the older form, there is a protective layer of oxide so further corrosion is slow. The layer of aluminium oxide on the surface of aluminium objects can be made thicker by electrolysis to prevent corrosion. It is called anodising. It is possible by placing aluminium objects as the anode in an electrolytic tank with sulfuric acid. During electrolysis of dilute sulfuric acid, oxygen is released at the anode which reacts with the aluminium object to form a layer of aluminium oxide. Anodised aluminium utensils do not undergo corrosion easily.

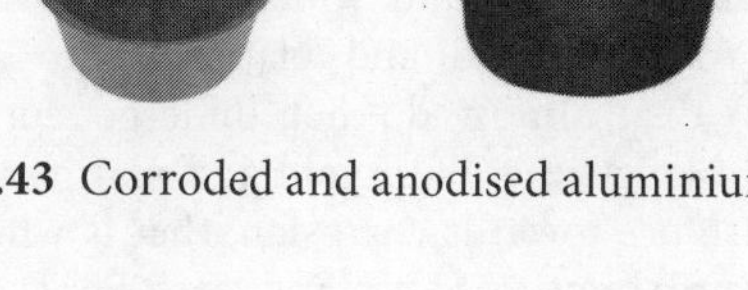

Fig. 3.43 Corroded and anodised aluminium utensils

Corrosion of copper: When a copper object remains in air and moisture for a considerable time then it undergoes corrosion and loses the shine and becomes dull. Copper reacts with the carbon dioxide and water present in air and moisture to form a green coating of basic copper carbonate [$CuCO_3.Cu(OH)_2$] (Fig. 3.44). As copper is a less reactive metal, the rate of corrosion is slow and the corroded copper object is easily cleanable with dil. acid solution such as acetic acid (vinegar) as basic copper carbonate readily dissolves in acid and the copper object becomes shiny and red brown again.

Fig. 3.44 Corroded copper object

Source: The Portable Antiquities Scheme/ The Trustees of the British Museum, CC BY-SA 4.0 <https://creativecommons.org/licenses/by-sa/4.0>, via Wikimedia Commons

Corrosion of silver: When a silver object is placed in air it gets tarnished and finally turns black. The silver metal undergoes corrosion with hydrogen sulfide present in air to give black coloured silver sulfide (Ag_2S); however, the rate of corrosion is slow as silver is very less reactive. Due to this corrosion of silver and formation of Ag_2S, silver ornaments get black. As silver has a low corrosion rate and it has a shiny surface, it is used in making silver utensils, jewellery and coins (Fig. 3.45).

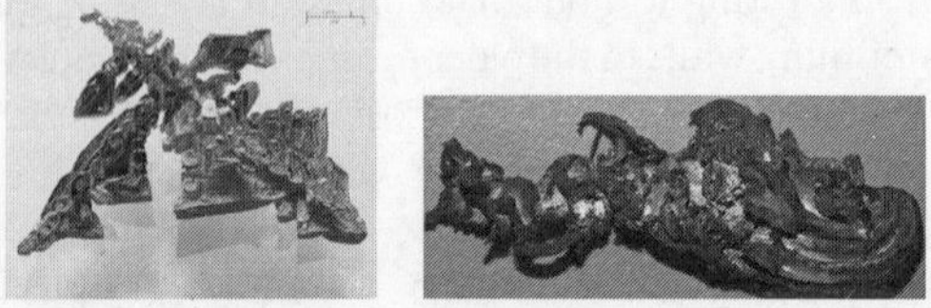

Fig. 3.45 Non-corroded and corroded silver objects

Source: By Alchemist-hp (talk) (www.pse-mendelejew.de) - Own work (additional processed by Waugsberg), CC BY-SA 3.0 de, https://commons.wikimedia.org/w/index.php?curid=7394995; James St. John, CC BY 2.0 <https://creativecommons.org/licenses/by/2.0>, via Wikimedia Commons

Corrosion of gold and platinum: Gold and platinum are both shiny metals and widely used in making jewellery. As both gold and platinum are almost unreactive, they almost do not undergo corrosion even after a long time. Both gold and platinum have a high resistance towards corrosion, that is why their ornaments are shiny even for many years. For example, Golden temple (Fig. 3.46).

Fig. 3.46 Golden temple and platinum ring

Source: txd, CC BY 2.0 <https://creativecommons.org/licenses/by/2.0>, via Wikimedia Commons; 1791Rings, CC BY-SA 3.0 <https://creativecommons.org/licenses/by-sa/3.0>, via Wikimedia Commons

3.14.2 Alloys

Alloying is a very good method of improving the properties of a metal. We can get the desired properties by this method. For example, iron is the most widely used metal, but it is never used in its pure state. This is because pure iron is very soft and stretches easily when hot. However, if it is mixed with a small amount of carbon (about 0.05%), it becomes hard and strong. When iron is mixed with nickel and chromium, we get stainless steel, which is hard and does not rust. Thus, if iron is mixed with some other substance, its properties change. In fact, the properties of any metal can be changed if it is mixed with some other substance. The substance added may be a metal or a non-metal. An *alloy* can be prepared by mixing various metals in the molten state in suitable proportions and then cooling the mixture to the room temperature. In the case of a metal and a non-metal alloy, the non-metal is added into the molten metal followed by cooling. An alloy is a homogeneous mixture of two or more metals, or a metal and a non-metal. It is prepared by first melting the primary metal, and then, dissolving the other elements in it in definite proportions. It is then cooled to room temperature. Alloys can be considered as solid solutions in which the metal with more concentration is the solvent and other metals in low concentration are the solutes (Fig. 3.47). For example, in brass, copper is the solvent and zinc is solute.

Fig. 3.47 Some alloys and metals (beryllium-copper, inconel, steel, titanium, aluminium, magnesium)

Source: By Bill Abbott - Metal samples: Cu, Ni, Fe, Ti, Al, Mg, CC BY-SA 2.0, https://commons.wikimedia.org/w/index.php?curid=28099343

If one of the metals is mercury, then the alloy is known as an amalgam; for example, sodium amalgam. An amalgam having silver, tin and zinc is used for dental fillings.

The properties of an alloy are different from the properties of the constituent metals. For example,

- alloys are stronger than the constituent metals from which they are formed,
- alloys are harder than the constituent metals from which they are formed,
- the electrical conductivity and melting point of an alloy is less than that of constituent metals, and
- alloys are more resistant to corrosion than the constituent metals.

Purpose of making alloys

The purpose of an alloy is to improve the specific usefulness of the primary component and not to adulterate or degrade it. Alloys are made to change the property of their major constituents to achieve a specific objective (Table 3.12). Some examples are given below.

(i) Brass, an alloy of copper and zinc (Cu and Zn), and bronze, an alloy of copper and tin (Cu and Sn), are not good conductors of electricity whereas copper is used for making electrical circuits.

(ii) Brass is stronger than its components—copper and zinc.

(iii) Pure gold (24 carat) is too soft to be used alone for making jewellery, so a small percentage of copper or silver is added in it to make it harder which is more suitable for making jewellery (22 carat gold has 22 parts of pure gold and 2 parts of silver or copper).

(iv) Steel having up to 10% of elements such as Cr, Mo and Ni with a low percentage of carbon is known as alloy steel. These added elements improve hardness and toughness.

(v) Solder, an alloy of lead and tin (Pb and Sn), has a low melting point and is used for welding electrical wires together.

Table 3.12 Some important alloys

Principal metal	Name of alloy	Composition	Properties	Uses
Aluminium	1. Magnalium	90%–95% Al 10%–5% Mg	1. Strong 2. Resists corrosion 3. Light	1. Scientific instruments 2. Aircraft 3. Metal mirrors 4. Beams of balance 5. Light tools 6. Household appliances
	2. Duralumin	95% Al 4% Cu 0.5% Mg 0.5% Mn	1. Highly ductile 2. Light but as strong as steel 3. Hard and resistant to corrosion	1. Pressure cookers 2. Bodies of aircraft, buses and tube trains 3. Light tools
Zinc and copper	Bronze	80% Cu 18% Sn 2% Zn	1. Can take up polish 2. Hard and easily cast	1. Coins 2. Bearing 3. Utensils 4. Statues 5. Medals
	Brass	40%–30% Zn 60%–70% Cu	1. Can be easily cast 2. Malleable and ductile 3. Yellow in colour 4. Resists corrosion	1. Electrical goods 2. Screws and handles 3. Musical instruments 4. Parts of watches 5. Cartridge, containers 6. Decorative hardware, utensils

(Continued)

Table 3.12 (Continued)

Principal metal	Name of alloy	Composition	Properties	Uses
	German silver	50% Cu 30% Zn 20% Ni	1. White and light like silver 2. Malleable and ductile 3. High electrical resistance	1. Decorative articles 2. Electric heaters 3. Rheostats 4. Resistors
	Gun metal	88% Cu, 8% Sn, 1% Zn, 1% Pb	1. Hard and brittle 2. Easily cut	1. Barrels of cannons 2. Bearings, gears
Iron	1. Stainless steel	73% Fe, 18% Cr, 8% Ni, 1% C	1. Resists corrosion 2. Lustrous, hard 3. Resistant to acids and alkalis	1. Utensils 2. Cutlery 3. Ornamental pieces 4. Surgical instruments
	2. Manganese steel	85% Fe, 1% C, 14% Mn	Durable, tough and hard	1. Safe 2. Rock drills and 3. Armour plates
	3. Tungsten steel	84% Fe, 5% W 1% C	Very hard	1. Cutting tools for high speed lathes
	4. Nickel steel	95%–98% Fe, 5%–3% Ni	1. Hard and elastic 2. Resistant to corrosion	1. Electric wire cables 2. Automobile parts
	5. Invar	63% Fe, 36% Ni, 1% C	1. Negligible expansion	1. Metre scales 2. Scientific instruments
Lead	1. Solder or fuse metal 2. Type metal	Pb, Sn 75% Pb, 15% Sb, 10% Sn	1. Low MP, high tensile strength 1. Low MP easily cast	1. Welding, 2. Fuse 1. Printing blocks

Competition Edge

The iron pillar near Qutab Minar in Delhi is more than 2000 years old but is still rust resistant as it has a film of magnetic oxide of iron (Fe_3O_4) and other salts.

TEST YOUR KNOWLEDGE

3.33 Write the formula of rust of iron and name any two methods to prevent rusting.

Solution:
$Fe_2O_3.xH_2O$. Galvanisation, tin plating.

3.34 Why are gold and platinum used in making jewellery?

Solution:
See text (Section 3.15.2).

3.35 What type of gold (carats) is used in making gold jewellery and which metal is added to gold?

Solution:
22 carats. Copper or silver.

3.36 (i) Why does silver jewellery become black when exposed to air?
(ii) What is this process called?

Solution:
(i) Silver becomes black when exposed to air because a coating of silver sulfide (which is black) is formed on it.
(ii) The process is called corrosion.

Quick Review

- Reactive metals occur in the combined state and are known as **minerals**. Those minerals from which metals can be profitably extracted are called **ores**.
- These metals are generally associated with rocky materials, sand and clay which is known as **gangue** or **matrix**.
- The process of extraction of pure metal from its ore is called **metallurgy**.
- Removal of impurities from the crushed ore is called **concentration** or **benefaction** or **dressing** of ore.
- In **levigation,** the crushed ore is washed in a stream of water. The lighter impurities are swept away while the heavier ore particles settle down.
- **Magnetic separation** is meant for the separation of magnetic ore from impurities.
- **Calcination** is the process of heating the ores below their melting points in the absence of air to remove volatile impurities like water, CO_2 and organic matter.
- **Roasting** is the process of heating ore in an excess of air in order to convert metals into their oxides and water-insoluble sulfides into water soluble sulfates.
- **Smelting** is the reduction of oxides of less electropositive metals like Pb, Fe, Zn, Sb and Cu. It is carried out by heating them with coal or coke in a blast furnace in the presence of limited air.
- Sulfide ores of less electropositive metals like Hg, Cu, Pb and Sb, undergo **self-reduction** without using any reducing agent.
- Highly electropositive metals like Cu, Ag, Au, Al, Zn, Sn, Pb are purified by **electrorefining**. Here, pure metal gets deposited at the cathode while impurities get deposited at the anode after electrolysis.
- An **alloy** is a homogeneous mixture of two or more metals or of one or more metals with certain non-metallic elements. The properties of alloys are mostly different from those of the components.
- **Aluminium, copper and iron** are extracted from alumina (Al_2O_3), copper glance (Cu_2S) and hematite (Fe_2O_3) respectively. Zinc is extracted from zinc blende.

Exercise 3

Section A: Multiple Choice Questions (1 Mark)

1. The most abundant metal in the earth's crust is
   (a) calcium (b) aluminium
   (c) iron (d) sodium

2. The second most abundant metal in the earth's crust is
   (a) oxygen (b) aluminium
   (c) silicon (d) iron

3. The highly reactive metals like sodium, potassium and magnesium are extracted by the
   (a) electrolysis of their molten oxides
   (b) electrolysis of their molten chloride
   (c) reduction by carbon
   (d) reduction by aluminium

4. The process in which a carbonate ore is heated strongly in the absence of air to convert it into the metal oxide is called
   (a) reduction (b) roasting
   (c) calcination (d) smelting

5. Which option gives the process of extraction of mercury from its ore cinnabar?
   (a) heating cinnabar in the presence of limited air and then adding a small amount of water
   (b) cooling cinnabar to convert it into mercuric oxide and then heating it
   (c) heating cinnabar to convert it into mercuric oxide and then heating it again
   (d) cooling cinnabar in the presence of excess air

6. During smelting, an additional substance is added which combines with impurities to form a fusible product known as
   (a) flux (b) mud
   (c) gangue (d) slag

7. Oxides of moderately reactive metals like zinc, iron, nickel, tin and copper are reduced by using
   (a) aluminium as a reducing agent
   (b) calcium as a reducing agent
   (c) carbon as a reducing agent
   (d) sodium as a reducing agent

8. The image shows the electrolytic refining of copper.

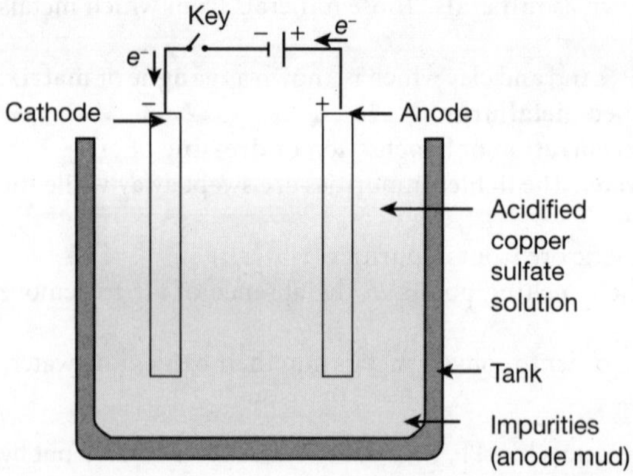

Which option explains the process to obtain pure copper?
(a) When current is passed, pure copper from the electrolytic solution deposits on the anode.
(b) When current is passed, pure copper from the anode deposits in the electrolytic solution.
(c) When current is passed, pure copper from the anode deposits on the cathode.
(d) When current is passed, pure copper from the electrolytic solution deposits on the cathode.

9. Which of the following metals is present in the anode mud during the electrolytic refining of copper?
(a) aluminium (b) sodium
(c) gold (d) iron

10. An alloy of Zn and Cu is dissolved in dil. HCl. Hydrogen gas is evolved. In this evolution of the gas
(a) only zinc reacts with dil. HCl
(b) only copper reacts with water
(c) both zinc and copper react with dil. HCl
(d) only copper reacts with dil. HCl

11. A greenish coating develops on copper utensils due to formation of
(a) $CuCO_3$
(b) $Cu(OH)_2.CuCO_3$
(c) $Cu(OH)_2$
(d) CuO

12. Rusting of iron takes place in
(a) distilled water
(b) ordinary water
(c) both ordinary and distilled water
(d) none of the above

13. The bronze medals are made up of
(a) Zn and Ni (b) Cu and Zn
(c) Cu and Sn (d) Cu, Zn, Tn

14. Silver articles become black on prolonged exposure to air. This is due to the formation of
(a) Ag_2O (b) Ag_2S
(c) Ag_2O and Ag_2S (d) AgCN

15. In thermite welding, a mixture of _____ and _____ is ignited with a burning magnesium ribbon which produces molten iron metal as large amount of heat is evolved.
(a) iron(III) sulfate and aluminium powder
(b) iron(II) oxide and aluminium powder
(c) iron(III) chloride and aluminium powder
(d) iron(III) oxide and aluminium powder

16. Galvanisation is a method of protecting iron from rusting by coating with a thin layer of
(a) aluminium (b) copper
(c) zinc (d) silver

17. An amalgam is an alloy of
(a) copper and zinc (b) mercury
(c) lead and tin (d) copper and tin

18. Which of the following conditions favours rusting?
(a) presence of air
(b) presence of moisture
(c) mixing of other metals
(d) both (a) and (b)

19. Select the set having correct statements.
(i) During electrolytic reduction of molten metals, they are obtained at the cathode.
(ii) Aluminium metal is extracted by the electrolysis of its chloride.
(iii) Zinc oxide can be reduced by carbon.
(iv) Iron oxide can be reduced by aluminium.
(a) (i), (ii) (b) (iii), (iv)
(c) (i), (ii), (iii) (d) (i), (iii), (iv)

20. Match the following.

Column I (Metal)	Column I (Metal)
A. Aluminium	(i) Cuprite
B. Iron	(ii) Cinnabar
C. Copper	(iii) Hematite
D. Mercury	(iv) Bauxite

(a) (A)—(iv), (B)—(iii), (C)—(ii), (D)—(i)
(b) (A)—(iv), (B)—(i), (C)—(iii), (D)—(ii)
(c) (A)—(iii), (B)—(ii), (C)—(i), (D)—(iv)
(d) (A)—(iv), (B)—(iii), (C)—(i), (D)—(ii)

Assertion Reason Questions

Directions: In the following questions two statements (Assertion) A and Reason (R) are given. Mark
(a) if A and R are both correct and R is the correct explanation of A;
(b) if A and R are both correct but R is not the correct explanation of A;
(c) A is true but R is false;
(d) A is false but R is true.

Assertion	Reason
1. Most of the metals are not present in the free state in nature.	**1.** Silver and gold are present in the free state in nature.
2. All the minerals are ores.	**2.** Metal is extracted from an ore.
3. Carbon can reduce magnesium oxide.	**3.** Copper and Hg undergo self reduction in air.
4. A sulfide ore is converted into metal oxide by roasting.	**4.** In roasting, non-metals like C, P are removed as their oxides.
5. Aluminium can reduce manganese oxide into manganese.	**5.** It is a highly exothermic reaction.
6. Copper can be purified by electrolytic refining.	**6.** In electrolytic refining, pure copper deposits at the anode.
7. Rusting is the reverse of metallurgy.	**7.** Red rust is $Fe_2O_3.xH_2O$.
8. An alloy is a homogenous mixture of two or more metals.	**8.** Stainless steel has iron, chromium and nickel.

Section B: Very Short Answer Questions (2 Marks)

1. What are minerals?
2. What is an ore?
3. Name two metals which are found in nature in the free state.
4. What is meant by metallurgy?
5. What do you understand by gangue or matrix?
6. Which metals do not corrode easily?
7. Name any two sulfide ores.
8. Name any two oxide ores.
9. What is the difference between roasting and calcination?
10. What is the process of depositing zinc on iron called?
11. Aluminium and zinc do not corrode easily even though they are reactive metals. Give reasons for your answer.
12. Name one ore each of zinc and mercury and give their chemical formulae.
13. Why do we apply paint on iron articles?
14. Why is carbon not used for reducing aluminium from aluminium oxide?
15. What happens when $ZnCO_3$ is heated in the absence of air? Give the relevant equation.
16. What is thermite reaction?
17. Name the constituent metals of bronze.
18. Name any two ores which are concentrated by electromagnetic separation.
19. What are the constituents of the alloy called German silver?
20. What is stainless steel?

Case Study or Passage Based Questions

An alloy is a homogenous mixture having two or more metals or a metal with a small amount of non-metals. An alloy can be prepared by mixing the various metals in their molten state in the required proportions followed by cooling the mixture. Every alloy has some useful properties and it also differs in many properties from the constituent metals.

21. Write any three properties of an alloy which make it better than the metal.
22. Solder is an alloy of which pair of metals?

23. Household utensils are mainly made of __________.

24. The electrical conductivity of an alloy is __________ than that of the pure metals.

25. Why are chromium and nickel added in iron to make steel?

Section C: Short Answer Questions (3 Marks)

1. Why is calcium found in the form of compounds?

2. Aluminium occurs in the combined state whereas gold is found in the free state in nature. Why?

3. What is the difference between a mineral and an ore?

4. Differentiate between roasting and calcination processes in metallurgy.

5. What are alloys? Give any three examples of alloys.

6. Explain the steps for the extraction of copper from its sulfide ore. Write the balanced equations involved in the process.

7. Name a metal/non-metal:
   (i) which makes iron hard and strong.
   (ii) which is alloyed with any other metal to make an amalgam.
   (iii) which is used to galvanise iron articles.
   (iv) whose articles when exposed to air form a black coating.

8. Name one metal each which is extracted by:
   (i) reduction with carbon
   (ii) electrolytic reduction
   (iii) reduction with aluminium
   (iv) reduction with heat alone

9. What are the constituents of solder? Which property of solder makes it suitable for welding electrical wires?

10. Corrosion is a serious problem. Every year, an enormous amount of money is spent to replace damaged iron. What steps can be taken to prevent this damage?

11. How are tarnished copper vessels are cleaned?

12. Zinc is a metal found in the middle of the activity series of metals. In nature, it is found as the carbonate ore, $ZnCO_3$. Mention the steps carried out for its extraction from the ore. Support your answer with equations.

13. (i) Carbon cannot be used as reducing agent to obtain Mg from MgO. Why?
    (ii) How is copper obtained from its sulfide ore? Give equations of the reactions.

14. What is cinnabar? How is metal extracted from cinnabar? **[CBSE 2015]**

15. Mention the names of the metals for the following:
    (i) Two metals which are alloyed with iron to make stainless steel.
    (ii) Two metals which are used to make jewellery. **[CBSE 2015]**

Section D: Long Answer Questions (5 Marks)

1. How is the method of extraction of metals high up in the reactivity series different from that for metals in the middle? Why can't the same process be applied for them? Explain the extraction of sodium, giving relevant equations.

2. What is meant by corrosion?
   (i) Why do aluminium sheets not corrode easily?
   (ii) Why are copper vessels covered with a green coating in the rainy season?

3. Define alloys. List the properties of alloys that make them useful over pure metals? Explain this fact with suitable examples.

4. (i) 'Sodium is a highly reactive metal and it cannot be obtained from its oxide by heating with carbon.' Give reasons.
   (ii) How can sodium be obtained from sodium chloride?

5. Which method will you use to reduce the following?
   (i) Oxides of less reactive metals.
   (ii) Oxides of moderately reactive metals.
   (iii) Oxides of highly reactive metals.
   Explain by giving a suitable example.

6. In a thermite reaction, a compound of iron reacts with a metal.
   (i) Name the metal used in this reaction.
   (ii) After completion of this reaction, a metal is obtained in the molten state. Identify the metal.
   (iii) Represent this reaction in the form of a balanced chemical equation.
   (iv) Mention the most common use of this reaction.

7. Zinc is the metal which lies in the middle of the activity series. This metal is extracted from its sulfide ore. Outline the steps involved in the process of extraction of zinc metal with the help of a balanced chemical equation for each step.

8. (a) Define the terms alloy and amalgam. Name the alloy used for welding electric wires together. What are its constituents?
   (b) Name the constituents of the following alloys:
   (i) brass
   (ii) stainless steel
   (iii) bronze
   State one property in each of these alloys, which is different from its main constituents.

9. (a) Write the chemical name of the coating that forms on silver and copper articles when these are left exposed to moist air.
   (b) Explain what galvanisation is. What purpose is served by it?
   (c) Define an alloy. How are alloys prepared? How do the properties of iron change when:
   (i) a small quantity of carbon is mixed in it?
   (ii) nickel and chromium are mixed with it?

10. A copper coin is kept immersed in silver nitrate solution for some time. What change will take place in the coin and the colour of the solution? Write a balanced chemical equation of the reaction involved.

11. Also write the chemical equations for the reactions involved in the process. **[CBSE 2019]**

12. List three differentiating features between the processes of galvanisation and alloying. **[CBSE 2020]**

Section E: Case Study or Passage Based Questions (4 Marks)

Extraction of a metal from its ore is known as metallurgy. It includes the following steps—concentration of ore, calcination, roasting, conversion of ore into metal by reduction and refining of the metal. Ores are mainly oxides, sulfides, carbonates and sulfates. When metals are intermixed, a number of alloys are possible which have a lot of significance and mostly have more advantages than an individual metal.

1. Which of the following is not a sulfide ore?
   (a) cinnabar (b) copper glance
   (c) zinc blende (d) hematite

2. Which of the following metals can be obtained by electrolytic reduction of their molten salts?
   (a) sodium (b) calcium
   (c) iron (d) both (a) and (b)

3. Thermite reaction is used to get which metal?
   (a) aluminium (b) iron
   (c) mercury (d) copper

4. Calcination is used mainly for _____.
   (a) sulfate ores (b) sulfide ores
   (c) carbonate ore (d) both (b) and (c)

5. Which of the following alloys contain copper?
   (a) brass (b) solder
   (c) bronze (d) both (a) and (c)

High Order Thinking Skills (HOTS) Questions

1. Two ores P and Q were taken. On heating, ore P gives CO_2 whereas ore Q gives SO_2. What steps will you take to convert them into metals?

2. An aluminium can is used to store ferrous sulfate solution. It is observed that in a few days, holes appeared in the can. Explain the observation and write the chemical equation to support your answer.

3. A metal X left in moist air for a long time, loses its shiny brown surface and gains a green coat. Why has this happened? Name and give the chemical formula of the given coloured compound and identify the metal. List two ways to prevent this process.

4. A metallic ore X reacts with dilute HCl to liberate a gas which turns lime water milky. Another ore Y gives off a gas with the smell of rotten eggs on treatment with the same acid. Which metallurgical processes are used for the extraction of the metals X and Y?

5. Giving one example of each, explain the method of obtaining the following metals from their compounds.
Metal A: Which is low in the activity series.
Metal B: Which is in the middle in the activity series.
Metal C: Which is towards the top of the activity series.

6. Explain how the following metals are obtained from their compounds by reduction.
   (i) Metal X which is low in the reactivity series.
   (ii) Metal Y which is in the middle of the reactivity series.
   (iii) Metal Z which is high in the reactivity series.

7. A metal M is found in nature as its carbonate. It is used in the galvanisation of iron. Identify M and name its ore. How will you convert this ore into the free metal? **[Board 2016]**

Answers

Section A: Multiple Choice Questions

1. (b)	2. (d)	3. (b)	4. (c)
5. (c)	6. (d)	7. (c)	8. (d)
9. (d)	10. (a)	11. (b)	12. (c)
13. (c)	14. (b)	15. (d)	16. (c)
17. (b)	18. (d)	19. (d)	20. (d)

Assertion Reason Questions

1. (b)	2. (d)	3. (d)	4. (b)
5. (b)	6. (c)	7. (b)	8. (b)

Section B: Very Short Answer Questions

1. The elementary state or the compounds in the form of which the metals occur in nature, associated with impurities of gangue, are called minerals.

2. The mineral from which a metal can be extracted conveniently and economically is called an ore.

3. Gold and platinum are two metals that are found in nature in the free state.

4. The extraction of metals from their ores and then refining them for use is known as metallurgy.

5. The earthy, sandy and rocky impurities associated with the mineral are called gangue or matrix.

6. Metals which are less reactive and lie at the bottom of reactivity series (silver, gold, platinum) do not react with atmospheric gases and hence do not corrode easily.

7. Iron pyrites (FeS_2) and zinc blende (ZnS).

8. Bauxite ($Al_2O_3.2H_2O$) and hematite (Fe_2O_3).

9. Roasting is heating the ore strongly in the presence of air. Calcination is heating the ore strongly in the absence of air.

10. Galvanisation.

11. Aluminium and zinc react with moist air to form a thin protective layer of oxide on their surfaces. This layer prevents further corrosion of the metals.

12. (i) Zinc blende (ZnS), (ii) Cinnabar (HgS).

13. Paint forms a protective coating on the surface of iron. Then, oxygen and moisture present in the air cannot have a direct contact with the metal. Therefore, the surface gets protected against rusting.

14. Because aluminium has greater affinity for oxygen than for carbon, carbon cannot reduce alumina (Al_2O_3) to aluminium.

15. ZnO(*s*) and CO_2(*g*) are formed.

$$ZnCO_3 \rightarrow ZnO + CO_2$$

16. Reduction of iron oxide to iron by aluminium is called thermite reaction.

$$Fe_2O_3 + 2Al \rightarrow 2Fe + Al_2O_3 + Heat$$

17. Constituents metals of bronze: copper, tin.

18. Chromite ore ($FeO.Cr_2O_3$) and magnetite ore (Fe_3O_4) which is a magnetic oxide of iron.

19. Cu (60%), zinc (20%), nickel (20%). Please note that it does not contain any silver.

20. Stainless steel is an alloy of 74% iron, 18% chromium and 8% nickel.

Case Study or Passage Based Questions

An alloy is a homogenous mixture having two or more metals or a metal with small amounts of non-metals. An alloy can be prepared by mixing the various metals in their molten state in required proportions followed by cooling the mixture. Every alloy has some useful properties and it also differs in many properties from the constituent metals.

21. Strength, hardness and resistance to corrosion.

22. Solder is an alloy of Cu and Sn.

23. Household utensils are mainly made of stainless steel.

24. Less.

25. Addition of chromium and nickel makes steel strong, tough and does not rust at all.

Section C: Short Answer Questions

1. Although calcium is the fifth most abundant element in the earth's crust, it is never found free in nature since it easily forms compounds by reacting with oxygen and water.

2. Gold is very unreactive whereas aluminium is a very reactive element. Therefore, aluminium reacts with other elements or compounds much more than gold. Thus, aluminium occurs in the combined state while gold occurs in the free state.

3. Difference between a mineral and an ore

	Mineral	Ore
(i)	Natural materials in which metals occur in the form of their compounds.	Minerals from which metals are extracted profitably.
(ii)	NaCl, mica.	Bauxite ($Al_2O_3.2H_2O$).

4. Difference between calcination and roasting

Calcination	Roasting
It is a process of heating an ore in the absence of air to remove moisture and volatile impurities and to convert carbonate ores into oxides. $ZnCO_3(s) \xrightarrow{\Delta} ZnO(s) + CO_2(g)$	It is a process in which the ore is heated in the presence of air to obtain metal oxides, which can be reduced easily to get the free metal. Sulfide ores are converted into oxides by roasting them in air. $2Zns(s) + 3O_2(g) \xrightarrow{\Delta} 2ZnO(s) + 2SO_2(g)$

5. Alloys are homogeneous mixtures of two or more metals or a metal and non-metals. Some examples of alloys are:

Brass	Cu + Zn	metal + metal
Steel	Fe + Ni + Cr + C	metals + non-metals
Bronze	Cu + Zn	metal + metal

6. (i) $2Cu_2S + 3O_2 \rightarrow 2Cu_2O(s) + 2SO_2(g)$
   (ii) $2Cu_2O + Cu_2S \rightarrow 6Cu(s) + SO_2(g)$

7. (i) Carbon (non-metal) makes iron hard and strong.
   (ii) Mercury is alloyed to make an amalgam.
   (iii) Zinc is used to galvanise iron articles.
   (iv) Silver articles form a black coating when exposed to air.

   Note: Pure iron is a very soft and flexible (on heating) metal. When mixed with a small amount of carbon (approximately 0.05%), it becomes hard.

8. (i) zinc (ii) aluminium
   (iii) manganese (iv) mercury

9. Constituents of solder are lead and tin. Solder has a low melting point to electrically connect components together. It is frequently used to join wires to leads of components such as switches or to join components of all kinds to a printed circuit board.

10. The rusting of iron can be prevented by painting, oiling, greasing, galvanising, chrome plating, anodising or making alloys.

11. Copper vessels get tarnished by reacting with air and water, due to which a layer is formed that mainly contains black copper oxide and

green copper hydroxide which is basic in nature. The citric acid in lemon dissolves this layer and washes off easily.

12. **Calcination:** Heating the ore strongly in a limited amount of oxygen.

$ZnCO_3 \xrightarrow{\Delta} ZnO + CO_2$

Reduction: Converting ZnO to Zn with the help of carbon.

$ZnO + C \rightarrow Zn + CO$

13. (i) Mg has more affinity for oxygen than carbon.

(ii) $2Cu_2S + 3O_2 \rightarrow 2Cu_2O + 2SO_2$

$2Cu_2O + Cu_2S \rightarrow 6Cu + SO_2$

14. Cinnabar is HgS. Mercury metal is obtained by roasting of cinnabar. HgO formed is thermally unstable and gives mercury.

$2HgS(s) + 3O_2(g) \rightarrow 2HgO(s) + 2SO_2(g)$

$2HgO(s) \xrightarrow{\Delta} 2Hg(l) + O_2(g)$

Or $HgS + O_2 \rightarrow Hg + SO_2$

Now, mercury can be purified by distillation.

15. (i) Nickel and chromium are alloyed with iron to make stainless steel.

(ii) Gold and platinum are used to make jewellery.

Section D: Long Answer Questions

1. Metals high up in the reactivity series are extracted by electrolytic reduction while those in the middle are extracted first by converting into the oxide and then reducing using carbon. The same method cannot be used because metals have more affinity for oxygen than carbon. Molten sodium chloride is taken for electrolytic reduction. The metals are deposited at the cathode and chlorine is liberated at the anode.

At cathode: $Na^+ + e^- \rightarrow Na$

At anode: $2Cl^- \rightarrow Cl_2 + 2e^-$

2. (i) Slow eating away of metals due to attack by atmospheric gases which converts the metal into its oxide, sulfide or carbonate is called corrosion.

(ii) A protective layer is formed on its surface to prevent it from corrosion (Al_2O_3).

(iii) Formation of copper oxide on it from direct contact with air and moisture due to corrosion of copper.

3. Alloys are a homogeneous mixture of two or more metals or a metal and a non-metal that cannot be separated into their components by physical methods.

(i) Electrical conductivity. For example, brass and bronze are not good conductors of electricity whereas copper is used in making electrical circuits.

(ii) Melting point of an alloy is less than that of the pure metal. For example, solder has a low melting point.

4. (i) Metallic compounds like sodium cannot be reduced by carbon or any other reducing agent due to their high affinity with oxygen.

(ii) Sodium is obtained from sodium chloride by electrolytic reduction. Sodium chloride in molten form is electrolysed. Sodium is deposited at the cathode and chlorine is liberated at the anode.

At cathode: $Na^+ + e^- \rightarrow Na$

At anode: $2Cl^- \rightarrow Cl_2 + 2e^-$

5. (i) Self reduction

$2HgO \rightarrow 2Hg + O_2$

$2Cu_2O + Cu_2S \rightarrow 6Cu + SO_2$

(ii) Reduction using carbon: $ZnO + C \rightarrow Zn + CO$

Sometimes, some highly reactive metals are used as reducing agents. For example,

$3MnO_2 + 4Al \rightarrow 3Mn + 2Al_2O_3 + heat$

or $Fe_2O_3 + 2Al \rightarrow 2Fe + Al_2O_3 + heat$

(iii) Electrolytic reduction. For example, Na, Mg and Ca are obtained by electrolysis of their molten chlorides.

6. (i) Aluminium (Al) (ii) Iron (Fe)

(iii) $Fe_2O_3 + 2Al \rightarrow 2Fe + Al_2O_3 + Heat$

(iv) Use: To join railway tracks or cracked machine parts.

7. **Steps**

(i) **Roasting:** Sulfide ores are converted into oxides by heating strongly in the presence of excess air.

$2ZnS + 3O_2 \xrightarrow{Heat} 2ZnO + 2SO_2$

(ii) **Reduction:** Metal oxides are then reduced to the corresponding metal by using a suitable reducing agent (carbon).

$ZnO + C \rightarrow Zn + CO$

8. (a) An alloy is a homogeneous mixture of two or more metals or a metal and a non-metal, mixed in the molten state. An amalgam is an alloy of a metal with mercury.
Solder is an alloy of lead and tin.

(b) (i) Brass: copper and zinc.
(ii) Stainless steel: iron, nickel and chromium.
(iii) Bronze: copper and tin.

Brass and bronze have lower electrical conductivity than their constituents. Stainless steel does not corrode as easily as iron does.

9. (a) Silver sulfide, copper carbonate coating is formed on Ag and Cu articles.

(b) Galvanisation is a method of protecting steel and iron from rusting by coating them with a thin layer of zinc. The galvanised article is protected against rusting even if the zinc coating is broken.

(c) An alloy is a homogeneous mixture of two or more metals and non-metals. It is prepared by first melting the primary metal and then dissolving the other elements in it in definite proportion. It is then cooled at room temperature.
(i) On mixing carbon, it becomes hard and strong,
(ii) On mixing Ni and Cr, it becomes hard and does not rust.

10. Copper is more reactive than silver so it will displace silver from silver nitrate solution. So, there will be a deposition of silver on the copper coin. The colour of the solution will turn blue.

$$Cu(s) + 2AgNO_3(aq) \rightarrow Cu(NO_3)_2(aq) + 2Ag(s)$$

11. This ore is a carbonate ore and metal can be obtained from the enriched ore using the following steps.

(a) **Calcination:** Carbonate ore is heated in a limited supply of air to obtain its oxide. For example,

$$ZnCO_3(s) \xrightarrow{\Delta} ZnO(s) + CO_2(g)$$

(b) **Reduction:** Oxide ore is heated with carbon.

$$ZnO(s) + C(s) \xrightarrow{\Delta} Zn(s) + CO(g)$$

12.

Galvanisation	Alloying
Galvanisation is the process of depositing a thin layer of zinc metal on iron objects to protect them for rusting.	A homogeneous mixture of two or more metals or a metal and a non-metal to change the properties of pure metals and protect them from rusting.
It can be done by spraying molten zinc on the iron surface and dipping iron objects into molten zinc.	It can be done by alloying an iron object with the chromium and nickel.
It does not affect the properties of iron (metal).	It changes the hardness, tensile strength and electrical resistance of a metal.

Case Study or Passage Based Questions

1. (d) **2.** (d) **3.** (b) **4.** (c)
5. (d)

High Order Thinking Skills (HOTS) Questions

1. Ore P gives CO_2 and another ore Q gives SO_2. Therefore, the ores are MCO_3 and MS.
P can be obtained by

$$MCO_3 \xrightarrow{\text{Calcination}} MO + CO_2$$

$$MO + C \xrightarrow{\text{Reduction}} M + CO$$

Q can be obtained by

$$2MS + 3O_2 \xrightarrow{\text{Roasting}} 2MO + 2SO_2$$

$$MO + C \rightarrow M + CO$$

2. It is a displacement reaction. Aluminium is more reactive than iron. So, Fe is displaced by Al and aluminium sulfate is formed and iron is precipitated.

$$3FeSO_4 + 2Al \rightarrow Al_2(SO_4)_3 + 3Fe$$

3. The metal has corroded because of being exposed to moist air. The green compound is basic copper carbonate, $[CuCO_3.Cu(OH)_2]$. The metal is copper.
Two ways to prevent this process: painting, greasing, oiling, galvanising **(any two)**.

4. X is a carbonate ore since it gives off carbon dioxide on treatment with HCl. So, X is extracted by the process of calcination. Y gives off hydrogen sulfide gas on treatment with dilute HCl. So, it is a sulfide ore which is extracted by roasting.

5. (i) By heating the ore/oxide:
$HgS + O_2 \xrightarrow{\Delta} Hg + SO_2$
(ii) By heating the oxide in the presence of a reducing agent like coke:
$ZnO + C \rightarrow Zn + CO$
(iii) By the electrolysis of the molten compound:
$2NaCl \rightarrow 2Na + Cl_2$ (molten)

6. (i) Metal X is obtained simply by heating its oxide with carbon. For example, mercury is obtained from cinnabar.
$2HgS(s) + 3O_2(g) \rightarrow 2HgO(s) + 2SO_2(g)$
$2HgO(s) \rightarrow 2Hg(l) + O_2$
(ii) Metals in the middle of the activity series can be obtained by heating with carbon. For example,
$ZnO(s) + C(s) \rightarrow Zn(s) + CO(g)$
(iii) Metals high in the reactivity series are obtained by electrolytic reduction of molten ores as in NaCl.
$NaCl \rightarrow Na+ + Cl-$
$Na + e- \rightarrow Na$ (at cathode)
$2Cl- \rightarrow Cl_2 + 2e-$ (at anode)

7. Zinc: The ore of zinc is zinc blend (ZnS). The sulfide ore is heated below its melting point in the presence of air to convert it into the metal oxide. This is called roasting.
$2ZnS(s) + 3O_2(s) \xrightarrow{\text{Roasting}} 2ZnO(s) + 2SO_2(g)$
Reduction of oxide ore: It is the process of conversion of the metal oxide ore into the metal. It can be done by heating the oxides.
$ZnO(s) + C(s) \rightarrow Zn(s) + CO(g)$

Hints

Section A: Multiple Choice Questions

19. Aluminium metal is extracted by the electrolysis of its oxide and not chloride.

20. A. Aluminium (iv) Bauxite ($Al_2O_3.2H_2O$)
B. Iron (iii) Hematite (Fe_2O_3)
C. Copper (i) Cuprite (Cu_2O)
D. Mercury (ii) Cinnabar (HgS)

Assertion Reason Questions

2. All ores are minerals but all minerals are not ores.

3. Carbon cannot reduce magnesium.

6. Pure copper deposits at the cathode.

Section E: Case Study or Passage Based Questions

1. Hematite is an oxide ore while the rest are sulfide ores.

5. Brass contains copper and zinc, bronze contains copper and tin while solder contains lead and tin.

Activities with Discussion and Conclusion

Activity 3.1: To check that metals have lustre, that is, a shining surface.
- Take samples of iron, copper, aluminium and magnesium and note the appearance of each sample.
- Clean the surface of each sample by rubbing them with sand paper and note their appearance again.

Discussion: The surfaces of the metals are dull as they are covered with a layer of oxide, hydroxide or carbonate, due to attack of the gases of the air on their surfaces. On rubbing the surface with sand paper, this layer is removed and a shiny surface appears.

Conclusion: Metals in the pure state, or freshly prepared, or just cut, have a shiny surface.

Activity 3.2: To test that metals are hard and the hardness varies from metal to metal.
- Take small pieces of iron, copper, aluminium and magnesium. Try to cut these metals with a sharp knife.
- Hold a piece of sodium metal with a pair of tongs.

Caution: Always handle sodium metal with care. Dry it by pressing between the folds of a filter paper.

- Put it on a watch glass and try to cut it with a knife.

Discussion and conclusion: All the four metals (Fe, Cu, Al and Mg) are found to be cut with difficulty. This shows that these metals are hard. The ease of cutting is found to be in the order Mg > Al > Cu > Fe. This shows that hardness varies from metal to metal as sodium can be cut very easily. Sodium is soft and it is an exception.

Activity 3.3: To test that metals are malleable, that is, can be hammered into sheets.

- Take pieces of iron, zinc, lead and copper.
- Place any one metal on the block of iron and strike it four or five times with a hammer.
- Repeat with other metals.
- Record the change in shape of these metals.

Discussion and conclusion: It is observed that metals can be beaten into thin sheets; this means they are malleable.

Activity 3.4: To justify that metals are ductile, that is, can be drawn into wires.

- Consider some metals such as iron, copper, aluminium, lead.
- Check which of these metals are available in the form of wires.

Discussion and conclusion: As wires of iron, copper and aluminium are easily available, this shows that metals can be drawn into wires, that is, they are ductile.

Activity 3.5: To test that metals are good conductors of heat and have high melting points.

- Take an aluminium or copper wire. Clamp the wire on a stand as shown in Fig. 3.3 in the text.
- Fix a pin to the free end of the wire using wax.
- Heat the wire with a spirit lamp, candle or a burner near the place where it is clamped.

Now answer:

- What do you observe after some time?
- Does the metal wire melt?

Discussion and conclusion: We observe that on heating the wire near the clamp, after some time the pin falls down. This shows that heat flows through the wire and melts the wax. Further, the wire does not melt even after heating for a long time. This shows that metals have high values of melting points.

Activity 3.6: To test that metals are good conductors of electricity.

- Set up an electric circuit as shown in Fig. 3.3 in the text.
- Place the metal to be tested in the circuit between terminals A and B.

Now answer:

- Does the bulb glow? What does this indicate?

Discussion: The bulb glows. This shows that electric current flows through the metal.

Conclusion: Metals are good conductors of electricity.

Activity 3.7: To study the physical properties of non-metals.

- Collect samples of carbon (coal or graphite), sulfur and iodine.
- Carry out the Activities 3.1 to 3.6 with these non-metals.

Discussion and conclusion: **(i)** Non-metals have no lustre (except iodine which has a lustre).

(ii) Non-metals are soft and brittle.

(iii) They are neither malleable nor ductile.

(iv) They are not good conductors of electricity except graphite.

Activity 3.8: To show that metal oxides are basic while non-metal oxides are acidic.

- Take a magnesium ribbon and some sulfur powder.
- Burn the magnesium ribbon. Collect the ashes formed and dissolve them in water.
- Test the resultant solution with both red and blue litmus paper.
- Now burn the sulfur powder. Place a test tube over the burning sulfur.
- Add some water to the above test tube and shake.
- Test this solution with blue and red litmus paper.

Now answer:

- Is the product formed on burning magnesium acidic or basic?
- Is the product formed on burning sulfur acidic or basic?
- Can you write equations for these reactions?

Discussion: The solution obtained on dissolving the ashes after burning magnesium turns red litmus blue while solution obtained on dissolving gases from burning sulfur turns blue litmus red. This shows that magnesium oxide (MgO) is basic while the oxides of sulfur (SO_2, SO_3) are acidic in nature. The reactions are:

$$\underset{\text{Magnesium}}{2Mg(s)} + O_2(g) \rightarrow \underset{\text{Magnesium oxide}}{2MgO(s)}$$

$$MgO(s) + H_2O(l) \rightarrow \underset{\text{Magnesium hydroxide}}{Mg(OH)_2(aq)}$$

$$\underset{\text{Sulfur}}{S(s)} + O_2(g) \rightarrow \underset{\text{Sulfur dioxide}}{SO_2(g)}$$

$$SO_2(g) + H_2O(l) \rightarrow \underset{\text{Sulphurous acid}}{H_2SO_3(aq)}$$

Conclusion: Metal oxides are basic while non-metal oxides are acidic in nature.

Activity 3.9: To study the burning of metals in air to form oxides.

Caution: This activity may be performed in the presence of the teacher; also wear eye protection.

- Take samples of sodium, potassium, magnesium, copper and aluminium.
- Hold the sample with a pair of tongs and try to burn it over the flame. Repeat with the other metal samples.
- Collect the product if formed.
- Let the products and the metal surface cool down.

Now answer:

- Which metals burn easily?
- What flame colour did you observe when the metal burned?
- How does the metal surface appear after burning?
- Arrange the metals in the decreasing order of their reactivity towards oxygen.
- Are the products soluble in water?

Discussion: Potassium, sodium and magnesium burn easily. Copper and aluminium do not burn but on heating in air form black copper(II) oxide and white aluminium oxide (Al_2O_3) respectively.

$$\underset{\text{Copper}}{2Cu(s)} + O_2(g) \rightarrow \underset{\text{Copper (II) oxide (Black)}}{2CuO(s)}$$

$$4Al(s) + 3O_2(g) \rightarrow \underset{\text{Aluminium oxide (white)}}{2Al_2O_3(s)}$$

These metal oxides are found to be insoluble in water. However, alkali oxides (Na_2O, K_2O) dissolve in water to form alkalies. The order of reactivity of these metals with oxygen is:

$$K > Na > Mg > Al > Cu$$

At ordinary temperature, the surfaces of metals such as magnesium, aluminium, zinc and lead are covered with a thin layer of the oxide. The protective layer of the oxide prevents the metal from further oxidation.

Conclusion: Almost all metals combine with oxygen to form metal oxides.

$$\text{Metal} + \text{Oxygen} \rightarrow \text{Metal oxide}$$

Activity 3.10: To study the reactivity of metals with water.

Caution: This activity needs a teacher's assistance.

- Collect samples of sodium, potassium, calcium, magnesium, aluminium, zinc and copper.
- Put small pieces of the samples separately in beakers half-filled with cold water.
- Put the metals that did not react with cold water in beakers half filled with hot water.
- For the metals that did not react with hot water, arrange the apparatus as shown in Fig. 3.7 in the text and observe their reaction with steam.

Now answer:

- Which metals reacted with cold water? Arrange them in increasing order of their reactivity with cold water.
- Did any metal produce fire on water?
- Does any metal start floating after some time?
- Which metals did not react even with steam?

Discussion: K, Na and Ca reacted with cold water. K and Na produced fire on water. Ca and Mg start floating after some time. Al, Fe and Zn reacted with steam. Cu did not react even with steam.

Conclusion: Decreasing order of reactivity of these metals with water is: K > Na > Ca > Mg > Al > Zn > Fe > Cu.

Activity 3.11: To study the reaction of metals with acids.

- Collect all the metal samples except of sodium and potassium. If the samples are tarnished, rub them clean with sand paper.

Caution: Do not take sodium and potassium as they react vigorously even with cold water.

- Put the samples separately in a test tube having dilute hydrochloric acid.
- Suspend thermometers in the test tubes so that their bulbs are dipped in the acid.
- Observe the rate of formation of bubbles with care.

Now answer:

- Which metals reacted vigorously with dilute hydrochloric acid?
- With which metal did you record the highest temperature?
- Arrange the metals in the decreasing order of reactivity with dilute acids.

Discussion: Mg reacts most vigorously followed by Al, Zn and then Fe, copper did not react at all. The temperature was found to rise in the case of all the metals that reacted with dilute acid showing that the reaction is exothermic. The rise of temperature is maximum in the case of magnesium.

Conclusion: The decreasing order of reactivity with dilute HCl is: Mg > Al > Zn > Fe > Cu.

Activity 3.12: To study the reaction of metals with solutions of other metal salts.

- Take a clean wire of copper and an iron nail.
- Put the copper wire in a solution of iron sulfate and the iron wire in a solution of copper sulfate taken in test tubes (Fig. 3.8 in the text).
- Record your observations after 20 minutes.

Now answer:

- In which test tube did you find that a reaction has occurred?
- On what basis can you say that a reaction has actually taken place?
- Can you correlate your observations for the activities 3.9, 3.10 and 3.11?
- Write a balanced chemical equation for the reaction that has taken place.
- Name the type of reaction.

Discussion: A reaction is found to occur in the tube having iron wire dipped in copper sulfate solution. This is because in this tube, the blue colour of the copper sulfate solution fades and a light green colour appears due to the formation of iron(II) sulfate. Moreover, a brown deposit of copper takes place on the iron nail. Hence, the following reaction takes place:

$$\underset{\text{Iron}}{Fe(s)} + \underset{\text{Copper sulfate (Blue)}}{CuSO_4(aq)} \rightarrow \underset{\text{Iron(II) sulfate (Green)}}{FeSO_4(aq)} + \underset{\text{Copper (Brown)}}{Cu(s)}$$

Conclusion: Iron is more reactive than copper and displaces copper from copper sulfate solution. In general, a more reactive metal displaces a less reactive metal from its salt in the solution.

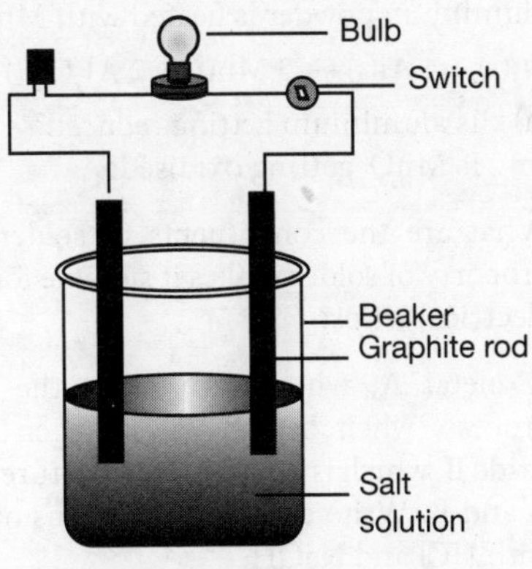

Activity 3.13: To study the properties of ionic compounds.

- Take samples of sodium chloride, potassium iodide, barium chloride or any other salt from the science laboratory.
- Take a small amount of a sample on a metal spatula and heat directly on the flame. Repeat with other samples.
- Try to dissolve the compounds in water and kerosene.
- Make a circuit as shown and insert the electrodes into a solution of one salt. Test the other salt samples too in this manner.

Now answer:

- What is the physical state of the salt taken?
- Did the samples impart any colour to the flame on heating?
- Did the compounds melt on heating?
- Are the compounds soluble in water or kerosene?
- Did the electric bulb glow on passing electric current?
- What is your inference about the nature of these compounds?

Discussion: All the salts taken are solids. Each salt imparts a specific colour to the flame. The compounds did not melt on heating. The compounds were soluble in water but not in kerosene. The electric bulb goes on passing electric current. All these properties show that the compounds are ionic in nature.

Conclusion: **(i)** Ionic compounds are generally solids.

(ii) They impart a characteristic colour to the flame.

(iii) They are soluble in a polar solvent like water and insoluble in non-polar solvents like kerosene, petrol.

(iv) Their melts or aqueous solutions conduct electricity.

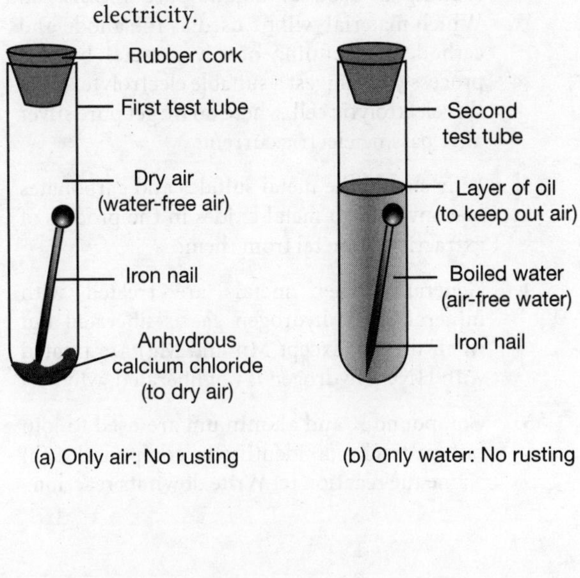

(a) Only air: No rusting (b) Only water: No rusting

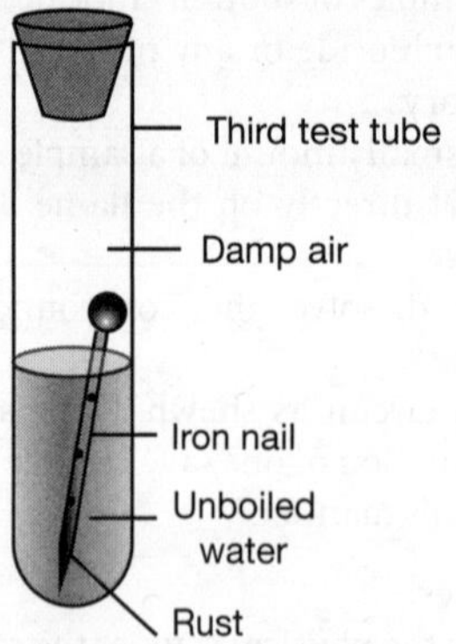

(c) Air and water together: rusting takes place

Activity 3.14: To prove that air and moisture are essential for rusting of iron.

- Take three test tubes and place clean iron nails in each of them.
- Label these test tubes as A, B and C. Pour some water in test tube C and cork it (Fig.).
- Pour boiled distilled water in test tube B and add about 1 mL of oil and cork it. The oil will float on water and prevent the air from dissolving in the water.
- Put some anhydrous calcium chloride in test tube A and cork it. Anhydrous calcium chloride will absorb the moisture, if any, from the air. Leave these test tubes for a few days and then observe.

Discussion: It is observed that iron nails rust in test tube C but they do not rust in test tubes B and A. In test tube C, the nails are exposed to both air and water. In test tube B, the nails are exposed to water only and the nails in test tube A are exposed to dry air free from water vapour.

Conclusion: Presence of both air and moisture is required for rusting to take place.

Exemplar Problems

Short Answer Questions

1. Iqbal treated a lustrous, divalent element M with sodium hydroxide. He observed the formation of bubbles in the reaction mixture. He made the same observations when this element was treated with hydrochloric acid. Suggest how he can identify the produced gas. Write chemical equations for both the reactions.

2. During the extraction of metals, electrolytic refining is used to obtain pure metals. (a) Which material will be used as the anode and cathode for refining of silver metal by this process? (b) Suggest a suitable electrolyte. (c) In this electrolytic cell, where do we get pure silver after passing electric current?

3. Why should the metal sulfides and carbonates be converted to metal oxides in the process of extraction of metal from them?

4. Generally, when metals are treated with mineral acids, hydrogen gas is liberated but when metals (except Mn and Mg) are treated with HNO_3, hydrogen is not liberated, why?

5. Compound X and aluminium are used to join railway tracks. (a) Identify the compound X (b) Name the reaction (c) Write down its reaction.

6. When a metal X is treated with cold water, it gives a basic salt Y with a molecular formula XOH (molecular mass = 40) and liberates a gas Z which easily catches fire. Identify X, Y and Z and also write the reaction involved.

7. A non-metal X exists in two different forms Y and Z. Y is the hardest natural substance, whereas Z is a good conductor of electricity. Identify X, Y and Z.

8. The following reaction takes place when aluminium powder is heated with MnO_2

   $$3MnO_2(s) + 4\,Al(s) \rightarrow 3\,Mn(l) + 2\,Al_2O_3(l) + \text{Heat}$$

   (a) Is aluminium getting reduced?
   (b) Is MnO_2 getting oxidised?

9. What are the constituents of solder? Which property of solder makes it suitable for welding electrical wires?

10. A metal A, which is used in the thermite process, when heated with oxygen gives an oxide B, which is amphoteric in nature. Identify A and B. Write down the reactions of oxide B with HCl and NaOH.

11. A metal that exists as a liquid at room temperature is obtained by heating its sulfide in the presence of air. Identify the metal and its ore and give the reaction involved.

12. Give the formulae of the stable binary compounds that would be formed by the combination of the following pairs of elements.
(a) Mg and N_2 (b) Li and O_2
(c) Al and Cl_2 (d) K and O_2

13. What happens when
(a) $ZnCO_3$ is heated in the absence of oxygen?
(b) a mixture of Cu_2O and Cu_2S is heated?

14. A non-metal A is an important constituent of our food and forms two oxides B and C. Oxide B is toxic whereas C causes global warming.
(a) Identify A, B and C.
(b) Which group of the periodic table does A belong to?

15. Give two examples each of the metals that are good conductors and poor conductors of heat respectively.

16. Name one metal and one non-metal that exist the in liquid state at room temperature. Also name two metals having their melting points less than 310 K (37°C).

17. An element A reacts with water to form a compound B which is used in white washing. The compound B on heating forms an oxide C which on treatment with water gives back B. Identify A, B and C and give the reactions involved.

18. An alkali metal A gives a compound B (molecular mass = 40) on reacting with water. The compound B gives a soluble compound C on treatment with aluminium oxide. Identify A, B and C and give the reaction involved.

19. Give the reaction involved during the extraction of zinc from its ore by
(a) roasting of zinc ore
(b) calcination of zinc ore

20. A metal M does not liberate hydrogen from acids but reacts with oxygen to give a black coloured product. Identify M and the black coloured product and also explain the reaction of M with oxygen.

21. An element forms an oxide A_2O_3 which is acidic in nature. Identify A as a metal or non-metal.

22. A solution of $CuSO_4$ was kept in an iron pot. After few days, the iron pot was found to have a number of holes in it. Explain the reason in terms of reactivity. Write the equation of the reaction involved.

Long Answer Questions

1. A non-metal A which is the largest constituent of air, when heated with H_2 in a 1:3 ratio in the presence of a catalyst (Fe) gives a gas B. On heating with O_2 it gives an oxide C. If this oxide is passed into water in the presence of air, it gives an acid D which acts as a strong oxidising agent.
(a) Identify A, B, C and D.
(b) Which group of the periodic table does this non-metal belong to?

2. Give the steps involved in the extraction of metals of low and medium reactivity from their respective sulfide ores.

3. Explain the following.
(a) Reactivity of Al decreases if it is dipped in HNO_3.
(b) Carbon cannot reduce the oxides of Na or Mg.
(c) NaCl is not a conductor of electricity in the solid state whereas it does conduct electricity in the aqueous solution as well as in the molten state.
(d) Iron articles are galvanised.
(e) Metals like Na, K, Ca and Mg are never found in their free state in nature.

4. (i) Given below are the steps for the extraction of copper from its ore. Write the reactions involved.
(a) Roasting of copper(1) sulfide.
(b) Reduction of copper(1) oxide with copper(1) sulfide.
(c) Electrolytic refining.
(ii) Draw a neat and well labelled diagram for the electrolytic refining of copper.

5. Of the three metals X, Y and Z, X reacts with cold water, Y with hot water and Z with steam only. Identify X, Y and Z and also arrange them in the order of increasing reactivity.

6. An element A burns with a golden flame in air. It reacts with another element B with atomic number 17 to give a product C. An aqueous solution of product C on electrolysis gives a compound D and liberates hydrogen. Identify A, B, C and D. Also write down the equations for the reactions involved.

7. Two ores A and B were taken. On heating, ore A gives CO_2 whereas ore B gives SO_2. What steps will you take to convert them into metals?

Answers and Solutions

Short Answer Questions

1. The produced gas can be identified by bringing a burning match stick near the reaction vessel, a pop sound is produced.
$M + 2NaOH \rightarrow Na_2MO_2 + H_2$
$M + 2HCl \rightarrow MCl_2 + H_2$
The element is a metal.

2. (a) Anode: Impure silver
Cathode: Pure silver
(b) Electrolyte: Silver salt, such as $AgNO_3$
(c) We get pure silver at the cathode.

3. It is easier to obtain a metal from its oxide, as compared to its sulfides and carbonates.

4. It is because HNO_3 is a strong oxidising agent. It oxidises the H_2 produced to H_2O.

5. (a) X: Fe_2O_3 (b) Thermite reaction
(c) $Fe_2O_3(s) + 2Al(s) \rightarrow 2Fe(l) + Al_2O_3(s)$ + Heat

6. X: Na, Y: NaOH, Z: H_2
$2Na + 2H_2O \rightarrow 2NaOH + H_2$ + Heat energy

7. X: carbon; Y: diamond and Z: graphite.

8. (a) No, because oxygen is added to aluminium, it is getting oxidised.
(b) No, since manganese has lost oxygen, it is getting reduced.

9. Solder is an alloy of lead and tin. The low melting point of solder makes it suitable for welding electrical wires.

10. A: Al; B: Al_2O_3
$Al_2O_3 + 6HCl \rightarrow 2AlCl_3 + 3H_2O$
$Al_2O_3 + 2NaOH \rightarrow 2NaAlO_2 + H_2O$

11. Metals low in the activity series can be obtained by reducing their sulfides or oxides by heating. Mercury is the only metal that exists as a liquid at room temperature. It can be obtained by heating cinnabar (HgS), the sulfide ore of mercury. The reactions are as follows:
$2HgS + 3O_2 \xrightarrow{Heat} 2HgO + 2SO_2$
$2HgO \xrightarrow{Heat} 2Hg + O_2$

12. (a) Mg_3N_2 (b) Li_2O (c) $AlCl_3$ (d) K_2O

13. (a) It undergoes calcination. The chemical reaction can be given as:
$ZnCO_3 \xrightarrow{Heat} ZnO + CO_2$
(b) It undergoes auto reduction forming copper and sulfur dioxide.
$2Cu_2O + Cu_2S \xrightarrow{Heat} 6Cu + SO_2$

14. (a) A is carbon, B is carbon monoxide and C is carbon dioxide.
(b) A belongs to group 14 of the periodic table.

15. (a) Good conductor: Ag and Cu
(b) Poor conductor: Pb and Hg

16. Metal: mercury (Hg); Non-metal: bromine (Br). Two metals with melting points less than 310 K are cesium (Cs) and gallium (Ga).

17. A: Ca; B: $Ca(OH)_2$; C: CaO
$Ca(s) + 2H_2O \rightarrow Ca(OH)_2(aq) + H_2(g)$
$Ca(OH)_2 \xrightarrow{Heat} CaO + H_2O$

18. A: Na; B: NaOH; C: $NaAlO_2$
$2Na + 2H_2O \rightarrow 2NaOH + H_2$
$Al_2O_3 + 2NaOH \rightarrow 2NaAlO_2 + H_2O$

19. (a) $2ZnS(s) + 3O_2 \xrightarrow{Heat} 2ZnO(s) + 2SO_2(g)$
(b) $ZnCO_3(s) \xrightarrow{Heat} ZnO(s) + CO_2(g)$

20. M = Cu; Black product: CuO
$2Cu + O_2 \rightarrow 2CuO$

21. Since the oxide of the element is acidic in nature, A will be a non-metal.

22. Fe is more reactive as compared to Cu. Therefore, Fe displaces Cu from $CuSO_4$ and forms $FeSO_4$.
Fe + CuSO4 $\rightarrow FeSO_4$ + Cu

Long Answer Questions

1. (a) A: N_2; B: NH_3; C: NO; D: HNO_3
(b) Element A belongs to group 15 of the periodic table.

2. Same as in answer 4.

3. ***Hint***: (a) Due to the formation of a layer of oxide, that is, Al_2O_3.
(b) Na or Mg are more reactive metals as compared to carbon.
(c) In solid NaCl, the movement of ions is not possible due to its rigid structure but in aqueous solution or molten state, the ions can move freely.
(d) To protect from corrosion.
(e) They are highly reactive.

4. (i) (a) Roasting of sulfide ore.

(i) $2Cu_2S(s) + 3O_2(s) \xrightarrow{\text{Heat}} 2Cu_2O(s) + 2SO_2(g)$

(b) $2Cu_2O + Cu_2S \xrightarrow{\text{Heat}} 6Cu(s) + SO_2(g)$
This reaction is known as auto-reduction.

(c) Reaction for electrolytic refining:
At cathode: $Cu^{2+}(aq) + 2e^- \rightarrow Cu(s)$
At anode: $Cu(s) \rightarrow Cu^{2+}(aq) + 2e^-$

(ii) Diagram for electrolytic refining of copper:

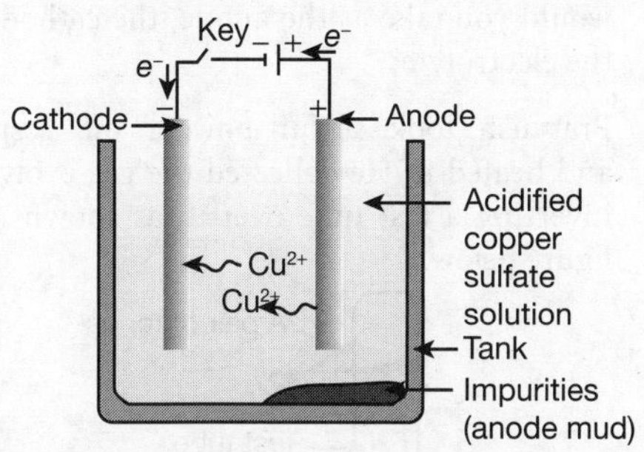

5. X is an alkali metal, Na or K.
Y is an alkaline earth metal, Mg or Ca.
Z is Fe.
Increasing reactivity series: Na > Mg> Fe

6. A = Na; B = Cl_2; C = NaCl; D = NaOH
$2Na + Cl_2 \rightarrow 2NaCl$
$2NaCl(aq) + 2H_2O(l) \rightarrow 2NaOH(aq) + Cl_2(g) + H_2(g)$

7. Since ore A gives CO_2 and ore B gives SO_2, the ores are MCO_3 and MS.
A can be obtained:

$MCO_3 \xrightarrow{\text{Calcination}} MO + CO_2$

$MO + C \xrightarrow{\text{Reduction}} M + CO$

B can be obtained:

$2MS + 3O_2 \xrightarrow{\text{Roasting}} 2MO + 2SO_2$

$MO + C \rightarrow M + CO$

NCERT CORNER

In-Text Questions

1. Give an example of a metal which
(i) is a liquid at room temperature,
(ii) can be easily cut with a knife,
(iii) is the best conductor of heat,
(iv) is a poor conductor of heat.

2. Explain the meanings of malleable and ductile.

3. Why is sodium kept immersed in kerosene?

4. Write equations for the reactions of
(i) iron with steam
(ii) calcium and potassium with water.

5. Samples of four metals A, B, C, and D were taken and added to the following solutions one by one. The results obtained have been tabulated as follows.

Metal	Iron (IIII) sulfate	Copper (IIII) sulfate	Zinc sulfate	Silver nitrate
A	No reaction	Displace-ment		
B	Displace-ment		No reaction	
C	No reaction	No reaction	No reaction	Displace-ment
D	No reaction	No reaction	No reaction	No reaction

Use the table above to answer the following questions about the metals A, B, C and D.
(i) Which is the most reactive metal?
(ii) What would you observe if B is added to a solution of copper(II) sulfate?
(iii) Arrange the metals A, B, C and D in order of decreasing **reactivity.**

6. Which gas is produced when dilute hydrochloric acid is added to a reactive metal? Write the chemical reaction when iron reacts with dilute H_2SO_4.

7. What would you observe when zinc is added to a solution of iron(II) sulfate? Write the chemical reaction that takes place.

8. (i) Write the electron dot structures for sodium, oxygen and magnesium.
(ii) Show the formation of Na_2O and MgO by the transfer of electrons.
(iii) What are the ions present in these compounds?

9. Why do ionic compounds have high melting points?

10. Define the terms (a) mineral, (b) ore, (c) gangue.

11. Name two metals which are found in nature in the free state.

12. What is the chemical process that is used for obtaining a metal from its oxide?

13. Metallic oxides of zinc, magnesium and copper were heated with the following metals.

Metal	Zinc	Magnesium	Copper
Zinc oxide			
Magnesium oxide			
Copper oxide			

In which cases will you find displacement reactions taking place?

14. Which metals do not corrode easily?

15. What are alloys?

Textbook Exercises

1. Which of the following pairs will give displacement reactions?
   (a) NaCl solution and copper metal
   (b) $MgCl_2$ solution and aluminium metal
   (c) $FeSO_4$ solution and silver metal
   (d) $AgNO_3$ solution and copper metal

2. Which of the following methods is suitable for preventing an iron frying pan from rusting?
   (a) applying grease
   (b) applying paint
   (c) applying a coating of zinc
   (d) all of the above

3. An element reacts with oxygen to give a compound with a high melting point. This compound is also soluble in water. The element is likely to be
   (a) calcium (b) carbon
   (c) silicon (d) iron

4. Food cans are coated with tin and not with zinc because
   (a) zinc is costlier than tin
   (b) zinc has a higher melting point than tin
   (c) zinc is more reactive than tin
   (d) zinc is less reactive than tin

5. You are given a hammer, a battery, a bulb, wires and a switch.
   (i) How could you use them to distinguish between samples of metals and non-metals?
   (ii) Assess the usefulness of these tests in distinguishing between metals and non-metals.

6. What are amphoteric oxides? Give two examples of amphoteric oxides.

7. Name two metals which will displace hydrogen from dilute acids, and two metals which will not.

8. In the electrolytic refining of a metal M, what would you take as the anode, the cathode and the electrolyte?

9. Pratyush took sulfur powder on a spatula and heated it. He collected the gas evolved by inverting a test tube over it, as shown in the figure below.

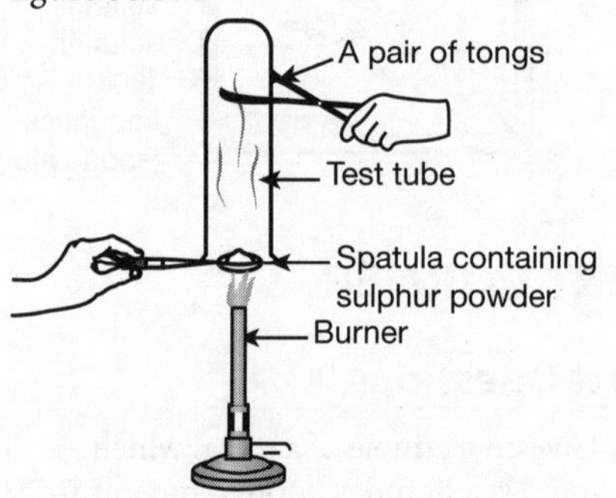

Collection of gas

   (a) What will be the action of the gas on
   (i) dry litmus paper?
   (ii) moist litmus paper?
   (b) Write a balanced chemical equation for the reaction taking place.

10. State two ways to prevent the rusting of iron.

11. What type of oxides are formed when non-metals combine with oxygen?

12. Give reasons.
   (i) Platinum, gold and silver are used to make jewellery.
   (ii) Sodium, potassium and lithium are stored under kerosene.
   (iii) Aluminium is a highly reactive metal, yet it is used to make utensils for cooking.
   (iv) Carbonate and sulfide ores are usually converted into oxides during the process of extraction.

13. You must have seen tarnished copper vessels being cleaned with lemon or tamarind juice. Explain why these sour substances are effective in cleaning the vessels.

14. Differentiate between a metal and non-metal based on their chemical properties.

15. A man went door to door posing as a goldsmith. He promised to bring back the glitter of old and dull gold ornaments. An unsuspecting lady gave a set of gold bangles to him which he dipped in a solution. The bangles sparkled like new, but their weight was reduced drastically. The lady was upset but after a futile argument the man beat a hasty retreat. Can you play the detective to find out the nature of the solution he had used?

16. Give reasons why copper is used to make hot water tanks and not steel (an alloy of iron).

Answers and Solutions

In-Text Questions

1. (i) Mercury (Hg) is a metal which is a liquid at room temperature.
(ii) Sodium and potassium are metals that can be easily cut with a knife.
(iii) Silver is the best conductor of heat.
(iv) Lead is a poor conductor of heat.

2. **Malleable:** A metal is said to be malleable if it can be beaten into thin sheets. For example, silver and gold are highly malleable metals.
Ductile: A metal is said to be ductile when it can be drawn into thin wires. For example, platinum and gold are highly ductile metals.

3. Sodium is a highly reactive metal. It reacts with the oxygen present in the air at room temperature and this reaction is highly exothermic. To prevent this, sodium is kept immersed in kerosene since it does not react with kerosene.

4. (i) The reaction of iron with steam is given as:
$3Fe(s) + 4H_2O(g)$ steam $\rightarrow Fe_2O_4(s) + 4H_2(g)$
(ii) The reaction of calcium with water is:
$Ca(s) + 2H_2O(l)$ Cold water $\rightarrow CaOH_2(aq) + H_2(g)$
The reaction of potassium with water is: $2K(s) + 2H_2O(l)$ Cold water $\rightarrow 2KOH(aq) + H_2(g)$

5. From the given table we can write the following observations:
(i) A is less reactive than Fe but more reactive than Cu.
(ii) B is more reactive than Fe but less reactive than Zn.
(iii) C is more reactive than Ag but less reactive than Fe, Cu and Zn.
(iv) D is less reactive than Fe, Cu, Zn, Ag.
Hence:
(i) B is the most reactive.
(ii) The blue colour of copper sulfate solution disappears, and reddish-brown copper metal is deposited on the metal B.
(iii) B > A > C > D is the order of reactivity.

6. When a reactive metal reacts with hydrochloric acid, hydrogen gas is produced. A reactive metal displaces the hydrogen from the acid and releases hydrogen gas. The chemical reaction when iron reacts with dilute H_2SO_4 is,
$Fe(s)$ + dil. $H_2SO_4(aq) \rightarrow FeSO_4(aq) + H_2(g)$

7. Zinc is more reactive than iron, hence when added to iron(II) sulfate, it displaces the iron metal to form zinc sulfate. The colour of the solution fades from green to colourless due to the formation of zinc sulfate and greyish black iron metal is deposited. The chemical reaction is given as:
$Zn(s) + FeSO_4(aq)$ Pale green $\rightarrow ZnSO_4(aq)$ Colourless + $Fe(s)$

8. (i) Sodium Na (2,8,1)

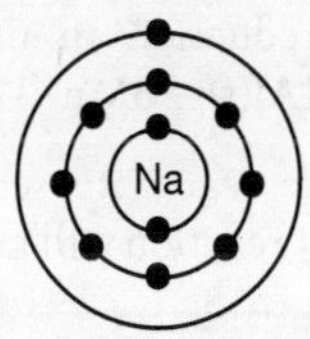

Oxygen O (2,8,6)

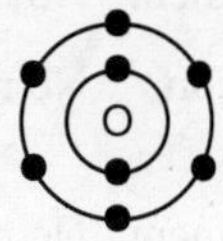

Magnesium Mg (2,8,2)

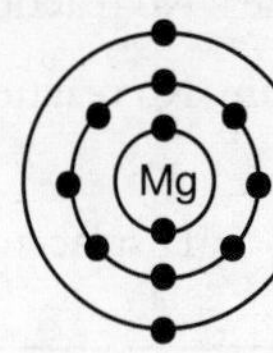

(ii) Na_2O

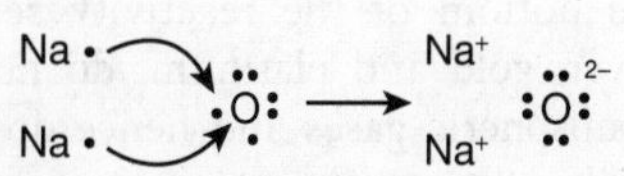

MgO

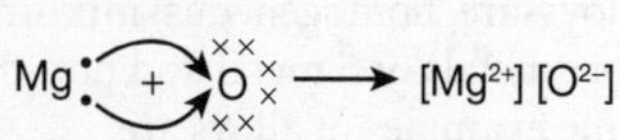

(iii) Ions in Na_2O are Na^+ and O^{2-}.
Ions in MgO are Mg^{2+} and O^{2-}.

9. In ionic compounds, the cations and anions are held together by strong electrostatic (or interionic) forces of attraction. To break or overcome these strong forces of attraction, a large amount of energy is required. As a result, the melting points of ionic compounds are quite high.

10. (a) Minerals: The naturally occurring chemical substances in the form of which metals occur in the earth along with the impurities are called minerals.
(b) Ore: The mineral from which a metal can be extracted conveniently and economically is called an ore.
(c) Gangue: The earth, sandy and rocky impurities associated with the mineral are called gangue or matrix.

11. Gold and platinum are two metals that are found in nature in the free state.

12. To obtain metals from their oxides, for metals of medium reactivity, one can use carbon as the reducing agent and the chemical process is called reduction. In the displacement reaction, the highly reactive metal acts as the reducing agent.
For metals of high reactivity, the method of electrolytic reduction in used. For example,
$3MnO_2(s) + 4Al(s) \rightarrow 3Mn(l) + 2Al_2O_3(s) + \text{heat}$

13. Based on the reactivity series of metals, the displacement reaction will take place as follows:

Metal	Zinc	Magnesium	Copper
Zinc oxide	No reaction	Displacement	No reaction
Magnesium oxide	No reaction	No reaction	No reaction
Copper oxide	Displacement	Displacement	No reaction

14. Metals which are less reactive and lie at the bottom of the reactivity series, that is, silver, gold and platinum, do not react with atmospheric gases and hence do not corrode easily.

15. Alloys are homogeneous mixtures of two or more metals or a metal and non-metal.
Some examples of alloys are:

Brass	Cu + Zn	metal + metal
Steel	Fe + Ni + Cr + C	metals + non-metals
Bronze	Cu + Zn	metal + metal

Textbook Exercises

1. (d)
Metals higher in the electrochemical series displace the metals lower in the series.
$NaCl + Cu \rightarrow$ no reaction
$MgCl_2 + Al \rightarrow$ no reaction
$FeSO_4 + Ag \rightarrow$ no reaction
$Cu + 2AgNO_3 \rightarrow 2Ag + Cu(NO_3)_2$
This reaction happens because copper is higher on the activity series than silver. If the free metal is below the metal bonded to the anion, the reaction would not proceed. When copper is added to silver nitrate (solution), copper dissolves to form copper nitrate.

2. (c)
By the galvanisation (applying a coating of zinc) we can protect the iron frying pan from rusting and it remains uncorroded for a long time. Grease and paint also prevent iron from rusting. However, we cannot use them on an iron frying pan because when the pan is heated and washed repeatedly, the coating of grease and paint would get destroyed.

3. (a)
$Ca + O_2 \rightarrow CaO$ high melting point $+ H_2O \rightarrow Ca(OH)_2$ Soluble
Thus, the element will be calcium.

4. (c)
Food cans are coated with tin and not with zinc because zinc is more reactive than tin.

5. (i) We can beat the sample with the hammer. If it can be beaten into thin sheets (that is, it is malleable), then it is a metal, if not, it is a non-metal.
(ii) We can use the battery, bulb, wires and a switch to set up a circuit with the sample. If the sample conducts electricity, then it is a metal, if not, it is a non-metal. Since metals have free electrons, they are good conductors of electricity while non-metals do not have free electrons to move. They are hence bad conductors of electricity.

The above tests are useful in distinguishing between metals and non-metals as these are based on the physical properties. No chemical reactions are involved in these tests.

6. Amphoteric oxides: The oxides that behave as both acidic and basic or that can react with acids as well as bases are called amphoteric oxides. Example: aluminium oxide Al_2O_3, zinc oxide ZnO.

(i) Amphoteric nature of zinc oxide.
Acidic nature: ZnO reacts with hydrochloric acid forming zinc chloride and water.
$ZnO+2HCl \rightarrow ZnCl_2+H_2O$
Basic nature: With sodium hydroxide (base), it forms sodium zincate and water.
$ZnO+2NaOH \rightarrow Na_2ZnO_2+H_2O$.

(ii) Amphoteric nature of aluminium oxide.
Acidic nature (reaction with alkalis)
$2Al(s) + 2NaOH(aq) + 6H_2O(l) \rightarrow 2Na^+[AlOH_4]^-(aq) + 3H_2(g)$
Sodium tetrahydroxoaluminate III
Basic nature (reaction with acids)
$2Al(s) + 6HCl(aq) \rightarrow 2Al^{3+}(aq) + 6Cl^-(aq) + 3H_2(g)$

7. Metals that are more reactive than hydrogen displace it from dilute acids. For example, sodium and potassium. Metals that are less reactive than hydrogen do not displace it. Example: copper and silver.

8. Electrolytic refining: it is a method to obtain the pure metal from the impure metal that has been obtained through a reduction process. In the electrolytic refining of a metal M such as Zn, Ni, Ag, Sn, Au,
Anode: Impure metal M
Cathode: Thin strip of pure metal M
Electrolyte: Solution of salt of the metal M

9. When sulfur powder is heated on a spatula, SO_2 will get formed.

(a) $S(s) + O_2(g) \rightarrow SO_2(g)$ (Sulfur dioxide)

(i) There will be no action on dry litmus paper.

(ii) Since the gas is sulfur dioxide, SO_2, it turns moist blue litmus paper to red because sulfur dioxide reacts with moisture to form sulfurous acid.

(b) $SO_2(g) + H_2O(l) \rightarrow H_2SO_3(aq)$ (Sulfurous acid)

It turns red litmus to blue.

10. Two ways to prevent the rusting of iron are given below.

(i) Oiling, greasing or painting: By applying oil, grease or paint, the surface becomes waterproof and the moisture and oxygen present in the air cannot come into direct contact with iron. Hence, rusting is prevented.

(ii) Galvanisation: An iron article is coated with a thin layer of zinc metal, both iron and zinc are in contact, zinc corrodes preferentially and releases electrons faster than iron. So, the chances of iron corrosion are reduced as zinc releases an electron quickly and that electron is accepted by iron. Whenever a metal accepts an electron it will not undergo corrosion rapidly. This prevents the iron from meeting oxygen and moisture. Hence, rusting is prevented.

11. Non-metals combine with oxygen to form neutral or acidic oxides.
$C(s) + O_2 \rightarrow CO_2$ (acidic oxide)
$N_2 + 1/2O_2 \rightarrow N_2O$ (neutral oxide)
$N_2 + O_2 \rightarrow NO$ (neutral oxide)
$N_2 + 2O_2 \rightarrow 2NO_2$ (acidic oxide)
$N_2 + 3/2O_2 \rightarrow N_2O_3$ (acidic oxide)
$N_2 + 5/2O_2 \rightarrow N_2O_5$ (acidic oxide)
$S(s) + O_2(g) \rightarrow SO_2(g)$ (acidic oxide)

12. (i) Platinum, gold and silver are used to make jewellery because they are highly malleable and ductile as well as lustrous. Also, they are less reactive and do not corrode easily.

(ii) Sodium, potassium and lithium are highly reactive metals and react very vigorously with air as well as water. Therefore, they are kept immersed in kerosene to prevent contact with air and moisture.
When sodium is heated in an excess of oxygen or air, sodium peroxide Na_2O_2 is the product.
$2Na + O_2 \rightarrow Na_2O_2$
When the reaction is carried out with a limited supply of oxygen or air, the major product will be sodium oxide.
$4Na+O_2 \rightarrow 2Na_2O$
Since the reaction is very exothermic, a part of the hydrogen reacts with the oxygen of the air to give water. Lithium, sodium and potassium metals react with water to form a colourless solution.

Sodium reacts with water to form sodium hydroxide (NaOH) with the evolution of hydrogen gas H_2.

The reaction is $Na + H_2O \rightarrow NaOH + H_2$

(iii) Though aluminium is a highly reactive metal, it is resistant to corrosion. This is because aluminium reacts with the oxygen present in air to form a thin layer of aluminium oxide. This oxide layer is very stable and prevents further reaction of aluminium with oxygen.

$4Al(s) + 3O_2(l) \rightarrow 2Al_2O_3(s)$

Also, Al is light in weight and a good conductor of heat. Hence, it is used for making cooking utensils.

(iv) Carbonate and sulfide ores are usually converted into oxides during the process of extraction because metals can be easily extracted from their oxides rather than from their carbonates and sulfides.

13. Copper reacts with moist carbon dioxide in the air to form copper carbonate and as a result, the copper vessel loses its shiny brown surface forming a green layer of copper carbonate which is basic in nature. The citric acid present in the lemon or tamarind neutralises the basic copper carbonate and dissolves the layer by a neutralisation reaction. This the reason why tarnished copper vessels are cleaned with lemon or tamarind juice to give the surface of the copper vessel its characteristic lustre.

14.

Metals	Non-metals
Metals are electropositive.	Non-metals are electronegative.
They react with oxygen to form basic oxides. $4Na + O_2 \rightarrow 2Na_2O$ $Na + O_2 \rightarrow Na_2O_2$ These have ionic bonds.	They react with oxygen to form acidic or neutral oxides. $C + O_2 \rightarrow CO_2$ (acidic oxide) $C + 1/2O_2 \rightarrow CO$ (neutral oxide) $N_2 + 1/2O_2 \rightarrow N_2O$ (neutral oxide) $N_2 + O_2 \rightarrow 2NO$ (neutral oxide) $N_2 + 2O_2 \rightarrow 2NO_2$ (acidic oxide) $N_2 + 3/2O_2 \rightarrow N_2O_3$ (acidic oxide) $N_2 + 5/2O_2 \rightarrow N_2O_5$ (acidic oxide) These have covalent bonds.
They react with water to form oxides and hydroxides. Some metals react with cold water, some with hot water and some with steam. $2Na + 2H_2O \rightarrow 2NaOH + H_2$	They do not react with water.
They react with dilute acid to form a salt and evolve hydrogen gas. However, Cu, Ag, Au, Pt, Hg do not react. $2Na + 2HCl \rightarrow 2NaCl + H_2$	They do not react with dilute acids. These are not capable of replacing hydrogen.
They react with the salt solutions of metals, depending on their reactivity. $CuSO_4 + Zn \rightarrow ZnSO_4 + Cu$	These react with the salt solutions of non-metals.
They act as reducing agents (as they can easily lose electrons). $Na \rightarrow Na^+ + e^-$	Generally, these act as oxidising agents (as they can gain electrons). $Cl_2 + 2e^- \rightarrow 2Cl^-$

15. He must have dipped the gold and silver metals in the solution of aqua regia (3:1 mixture of conc. HCl and conc. HNO_3). Aqua regia is a fuming, highly corrosive liquid which can even dissolve noble metals like gold and platinum. After dipping the gold ornaments in aqua regia, the outer layer of gold gets dissolved and the inner shiny layer appears. Therefore, the weight of the gold ornament is reduced.

16. Copper does not react with cold water, hot water, or steam. However, iron reacts with steam. If the hot water tanks are made of steel (an alloy of iron), then iron would react vigorously with the steam formed from hot water. 3Fe Iron + $4H_2O$ Steam $\rightarrow$ Fe_3O_4 Iron oxide + $4H_2$ Hydrogen

Fe_3O_4 is a mixture of ferrous oxide and ferric oxide. That is why copper is used to make hot water tanks and not steel.

CHAPTER 4

Carbon and Its Compounds

Learning Objectives

After studying this unit, you will be able to understand:

- the properties of carbon such as catenation and tetravalency
- all about the allotropes of carbon
- organic compounds and their classification
- hydrocarbons and their types
- isomers, isomerism
- functional groups
- homologous series
- nomenclature of organic compounds
- petroleum and other fuels
- various types of organic reactions
- important organic compounds
- soaps and detergents

4.1 Introduction

Carbon is an element with the symbol C and a non-metal. The name carbon is taken from the Latin word 'carbo' which means 'coal', as carbon is the main constituent of coal. The amount of carbon present in the earth's crust and the atmosphere is very small. For example, the earth's crust contains only 0.02% carbon in the form of minerals like carbonates, coal and petroleum, and the atmosphere has only 0.03% of carbon dioxide gas. In spite of the small amount of carbon available in nature, this element has a great importance in every sphere of life. The importance of carbon can be gauged from the fact that we are ourselves made of carbon compounds known as organic compounds. In fact, all living things – plants and animals – are made up of carbon-based organic compounds. That is, carbon is present in all living things.

A large number of things which we consume in our daily lives are made of carbon compounds. Food materials such as grains, pulses, sugar, tea, coffee, fruits and vegetables, are carbon compounds. Materials such as cotton, silk, wool, nylon and polyester which are used for making clothes, are carbon compounds. The fuels like wood, coal, kerosene, LPG (Liquefied Petroleum Gas), natural gas, CNG (Compressed Natural Gas), petrol and diesel which we use for cooking food and running vehicles, are carbon compounds. Paper, plastics, leather, rubber, drugs and dyes, are also made of carbon compounds (Fig. 4.1).

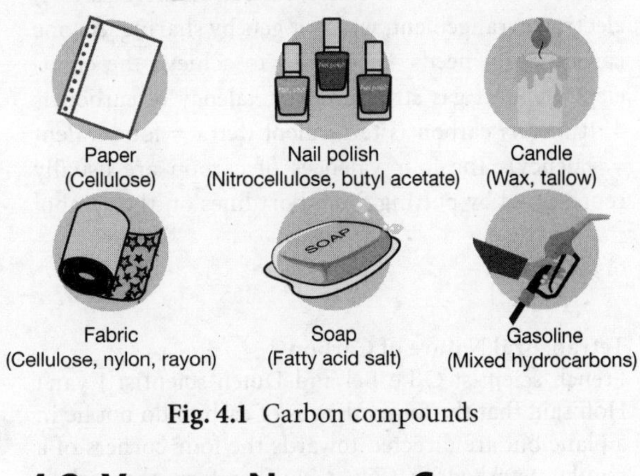

Fig. 4.1 Carbon compounds

4.2 Versatile Nature of Carbon

In our daily lives, most of the things we use and we ourselves are made of carbon. The number of carbon compounds known today are more than 3 million; this makes carbon a versatile element. Here are its unique properties.

4.2.1 Covalent Bonding in Carbon Compounds

Carbon always forms covalent bonds in its compounds. As the atomic number of carbon is 6, a neutral atom of carbon contains 6 electrons and its electronic configuration is 2 (K), 4 (L). That is, carbon has 4 electrons in the outermost shell (L orbit) of its atom. As a carbon atom has 4 electrons in its valence shell, it can either lose 4 electrons or gain 4 electrons to achieve the inert gas electron configuration and become stable. As the carbon atom is very small, its outermost electrons are strongly held by the nucleus, so it is not possible to remove 4 electrons from a carbon atom to give it the inert gas electron arrangement. It is also not possible to add as many as 4 electrons to a carbon atom to acquire the inert gas configuration due to energy considerations. It is obvious that the carbon atoms have to acquire the inert gas structure of 8 electrons in their outermost shell by the sharing of electrons. As the carbon atom can achieve the inert gas electron arrangement only by the sharing of electrons, carbon always forms only covalent bonds.

Carbon is Tetravalent

This characteristic of carbon atom by virtue of which it forms four covalent bonds is generally referred to as the tetra-covalency of carbon. As a carbon atom has 4 electrons in its valence shell (L), it needs 4 more electrons to achieve the stable, 8-electron inert gas (Ne) electron arrangement, which it gets by sharing. As one carbon atom needs 4 electrons to achieve the eight-electron inert gas structure, the valency of carbon is 4. It means carbon is tetravalent (tetra = four; valent = valency). The four valences of carbon are usually represented by putting four short lines on the symbol of carbon –C–.

Tetrahedral Nature of Carbon

French scientist C Le Bel and Dutch scientist J van't Hoff said that the four valences of carbon do not lie in a plane but are directed towards the four corners of a regular tetrahedron (Fig. 4.2). Therefore, the carbon atom is tetrahedral in nature and the angle between adjacent valences is 109.5° or 109°28' (109 degree and 28 minutes, 1° = 60 minutes).

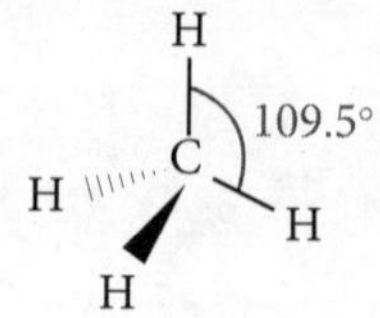

Fig. 4.2 Tetrahedral structure of carbon

4.2.2 Catenation or Self Combination

The most unique property of carbon is its ability to combine with itself, that is, atom to atom, to form long chains. This is called *catenation* (Fig. 4.3). For example, octane (C_8H_{18}), one of the constituents of petrol, has a chain of 8 carbon atoms, and some of the organic compounds like starch and cellulose contain chains of hundreds of carbon atoms. This property of self-linking or combination of carbon atoms to form long chains is quite useful as it gives rise to an extremely large number of carbon compounds or organic compounds. A long chain of carbon atoms acts as a backbone to which other atoms can attach in a number of ways to form a very large number of carbon compounds.

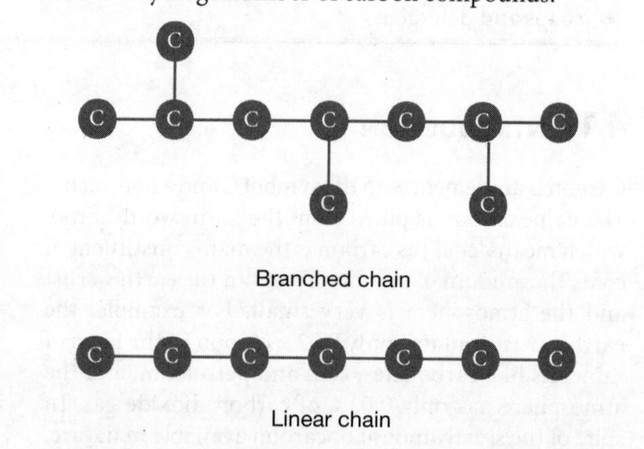

Fig. 4.3 Catenation of carbon

The covalent bonds formed between the various carbon atoms are very strong and cannot be not broken easily. Due to the small size of the carbon atom, their nuclei hold the shared pairs of electrons between the atoms strongly, leading to the formation of strong covalent bonds. It is important to note that carbon atoms can also form strong covalent bonds with the atoms of other elements like hydrogen, oxygen, nitrogen, sulfur, chlorine and phosphorus. The formation of strong bonds by carbon atoms among themselves and with other elements makes the carbon compounds exceptionally stable. Now, we can say that the strong catenation power of carbon atom is due to its small size, unique electronic configuration and the great strength of the carbon–carbon bonds.

Tendency to form Multiple Bonds

Due to its small size, the carbon atom can easily form multiple bonds (double and triple bonds) with another carbon atom or oxygen, nitrogen, and so on (Fig. 4.4). This multiplicity of carbon–carbon, carbon–oxygen, carbon–nitrogen bonds further increases the number of carbon compounds.

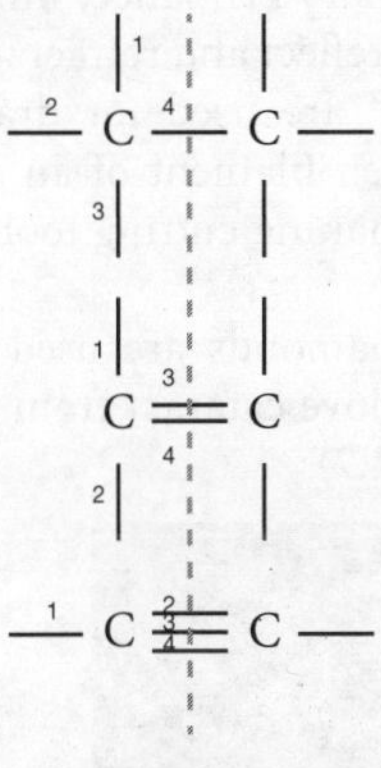

Fig. 4.4 Single and multiple bond formation by carbon

4.3 Occurrence of Carbon

Carbon occurs in nature in the free state as an element as well as in the combined state in the form of compounds with other elements. It exists in two allotropic forms—crystalline and amorphous.

In the free state, carbon occurs in nature mainly in two forms—diamond and graphite. One other form of carbon is fullerene (buckminsterfullerene). Only a small amount of carbon occurs as the free element in the earth's crust. In the combined state, carbon occurs in nature in the form of compounds such as (i) carbon dioxide gas in air, (ii) carbonates (limestone, marble, chalk), (iii) fossil fuels like coal, petroleum and natural gas, (iv) organic compounds like hydrocarbons, carbohydrates, fats and proteins, and (v) wood, cotton and wool.

4.3.1 Allotropes of Carbon

The phenomenon of existence of a chemical element in two or more forms differing in physical properties but having almost same chemical nature is known as *allotropy* and these forms are known as *allotropes.* Carbon shows crystalline (diamond, graphite and fullerene) and amorphous forms (coke, charcoal and so on) of allotropes (Fig. 4.5). Here we shall discuss only the crystalline allotropes of carbon.

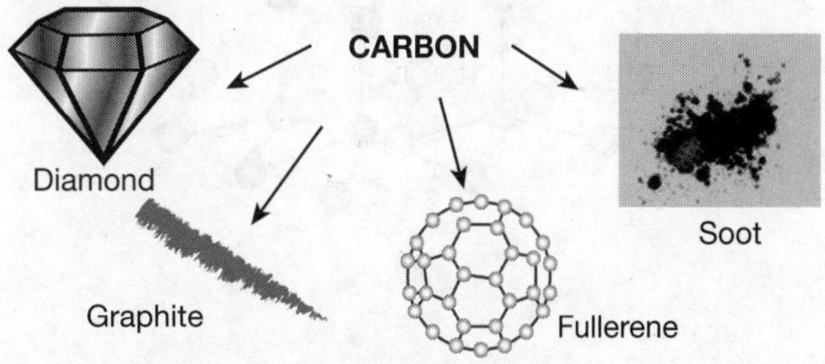

Fig. 4.5 Allotropes of carbon

Carbon occurs in three crystalline allotropic forms which are the purest form of carbon having similar chemical properties but differing in physical properties due to a difference in the arrangement of atoms. Let us discuss these allotropes separately.

(i) Diamond

It is a colourless, transparent, quite heavy substance having an amazing brilliance (lustre). It is the hardest known natural substance. It does not conduct electricity. It is the purest form of carbon with a high density (3.5 g/mL), melting point (7350°C) and refractive index (2.45). The size of diamonds is measured in carats (1 carat = 200 mg). For example, the world famous *Kohinoor diamond* is nearly 109 carats now.

Diamonds occur in nature in South Africa, Australia, and can also be synthesised from pure carbon at very high pressures (50000–60000 atm) and very high temperatures of around 1870 K. Such diamonds are known as synthetic diamonds. Such diamonds are small but are otherwise indistinguishable from natural diamonds.

Structure: A diamond crystal is a rigid and a giant molecule having carbon atoms only. Every carbon atom in the diamond crystal is linked to four other carbon atoms by strong covalent bonds. The four surrounding carbon atoms are at the four vertices (four corners) of a regular tetrahedron. In diamond, the carbon atoms are sp^3 hybridised with a tetrahedral structure. Here, one carbon atom is attached to four other carbon atoms by covalent bonds and this gives diamond a three dimensional polymeric structure (Fig. 4.6a). The C–C bond length is 1.54 Å and bond angle is 109°28′ (Fig. 4.6b).

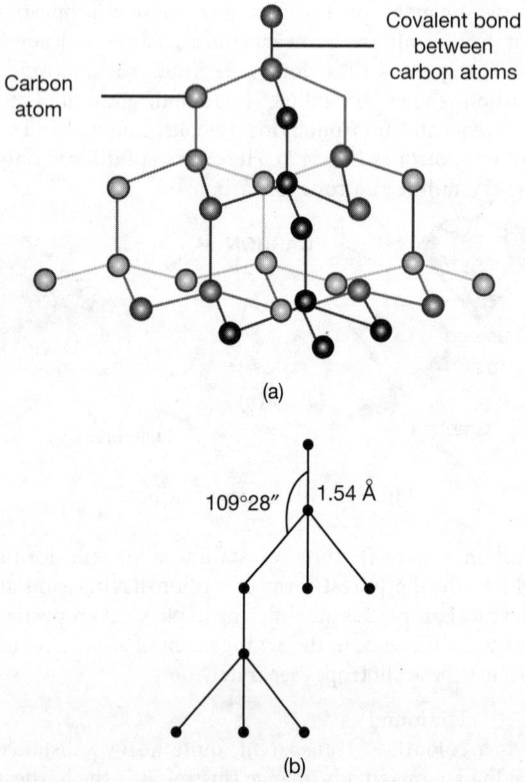

Fig. 4.6 (a) Structure of diamond; (b) Arrangement of carbon atoms in diamond

Properties

- It is a transparent solid with a specific gravity of 3.52.
- Diamond is very hard with very high density due to the compact and rigid 3-dimensional arrangement of carbon atoms.
- The high value of boiling point, melting point and density are due to its three dimensional structure which needs high energy for bond breaking. Light rays, X-rays (highest) can pass through it.
- Due to its high refractive index on proper cutting, it produces maximum total internal reflection.
- It is a bad conductor due to the absence of free electrons as each carbon atom in diamond shares all the four valence electrons. It is insoluble in all solvents.
- Pure diamonds are colourless while impure diamonds may be coloured.
- Diamond on heating in vacuum at 1800–2000°C gets converted into graphite.
- Being chemically inert, it does not react with acids, bases, and so on.
- On strong heating at 1173 K, diamond gives only carbon dioxide and nothing is left behind. It proves that diamond is composed of only carbon; that is why it is represented by C.

$$C + O_2 \rightarrow CO_2$$

Uses

- As diamond is very hard, it is the right material for cutting and grinding other hard materials, and for drilling holes in the earth's rocky layers. Hence diamonds can be used in cutting instruments like glass cutters, saw for cutting marble and in rock drilling equipment.
- Diamonds are used for making jewellery due to their extraordinary brilliance, which is due to their great ability to reflect and refract light.
- Diamond 'dies' are used for drawing thin wires like the tungsten filament of an electric bulb and often used in making cutting tools. It is used as an abrasive.
- Sharp-edged diamonds are used by eye-surgeons as a tool to remove cataract from eyes with a great precision (Fig. 4.7).

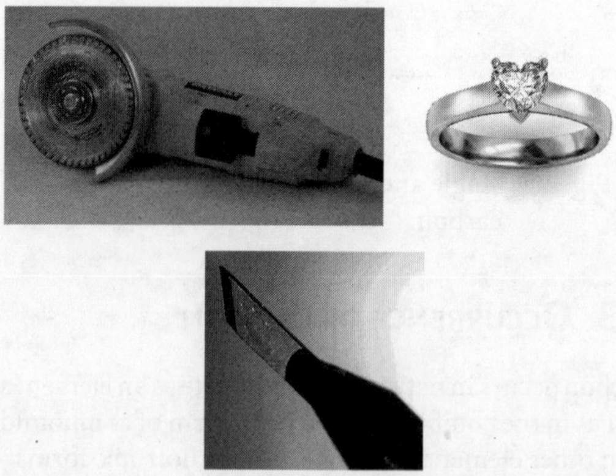

Fig. 4.7 Uses of diamond in cutting, jewellery and in eye surgery (keratoma)

(ii) Graphite

It is a gray-black opaque substance with a metallic lustre, soft and slippery in nature. It is more stable and reactive than diamond. It can conduct electricity. On strong heating, it gives only CO_2 leaving no other residue so it is composed of only carbon hence its symbol is C. It occurs in nature in China, South Korea and Russia, mixed with mica, quartz and silica. It can also be prepared from carbon in an electric furnace at a very high temperature of 2273–2773 K.

$$\underset{\text{Carbon}}{3C} + \underset{\text{Silica}}{SiO_2} \xrightarrow[-2CO]{2273\text{ K}} \underset{\text{Silicon carbide}}{SiC} \xrightarrow[\text{Onwards}]{2273\text{ K}} \underset{\text{Graphite}}{C} + \underset{\text{Silicon}}{Si}$$

Structure: Graphite differs a lot in its structure from diamond and has a crystal structure with layers or sheets of carbon atoms. It has a two dimensional sheet-like hexagonal structure in which each carbon atom is sp^2 hybridised and it is attached to three other carbon atoms in a hexagonal planar structure. Each carbon atom in a graphite layer is joined to other three other carbon atoms by strong covalent bonds to flat hexagonal rings. The various layers of carbon atoms in graphite are quite far apart so that no covalent bonds can exist between them. The various layers of carbon atoms in graphite are held together by weak van der Waals forces so they can slide over one another. Due to their sheet-like structure, graphite is a soft substance. Here, the C–C bond length is 1.42 Å and the distance between two successive layers is 3.35 Å. These layers are held together by weak van der Waals forces and is slippery in nature; so, graphite is soft and has low density. The C–C bond length is shorter than that in diamond because a pi bond is formed by the fourth valence electron present in each carbon atom (which is free). Due to these pi-electrons which are able to move in these slippery structures, graphite can conduct electricity (Fig. 4.8).

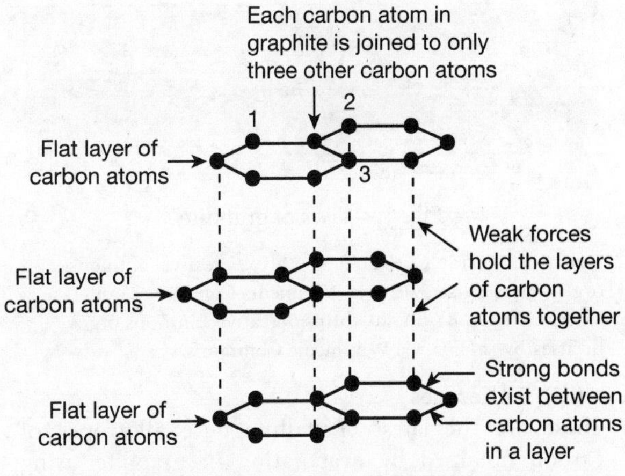

Fig. 4.8 Structure of graphite

Properties

- It is a soft, greyish, greasy crystalline solid with a density of 2.5 g/mL.
- Graphite has the lowest energy among carbon allotropes.
- It is a conductor of electricity due to the presence of free electrons in its slippery hexagonal sheets. Its conductivity increases with an increase in temperature.
- It is called plumbago or black lead as it leaves a black mark on paper.
- It is thermodynamically more stable than diamond and it has 1.9 kJ less free energy of formation than diamond.
- On heating at 1873 K and at an extremely high pressure (50,000–60,000 atm), it changes into diamond.
- It is a more reactive form than diamond; however, it does not react with dilute acids or alkalies.

Uses

- Due to its softness, powdered graphite is used as a lubricant for the fast-moving parts of machinery. It is non-volatile, so it can be used for lubricating those machine parts which operate at very high temperatures. It is a better lubricant on earth than on the moon.
- Graphite can be used as a dry lubricant in the form of graphite powder or mixed with petroleum jelly to form graphite grease.
- As graphite is a good conductor of electricity, it can be used for making carbon electrodes or graphite electrodes in dry cells and electric arcs. For example, the black coloured anode of a dry cell is made of graphite.
- The carbon brushes of electric motors are also made of graphite.
- As graphite is black in colour and quite soft, it marks black lines on paper. That's why graphite mixed with clay is used for making the black cores of our pencils—pencil leads.
- It is also used in electroplating, electrotyping, painting, stoves and in making electrodes and refractory crucibles.
- As graphite does not melt easily, tiles of graphite are used on the nose cone of space shuttles (Fig. 4.9).

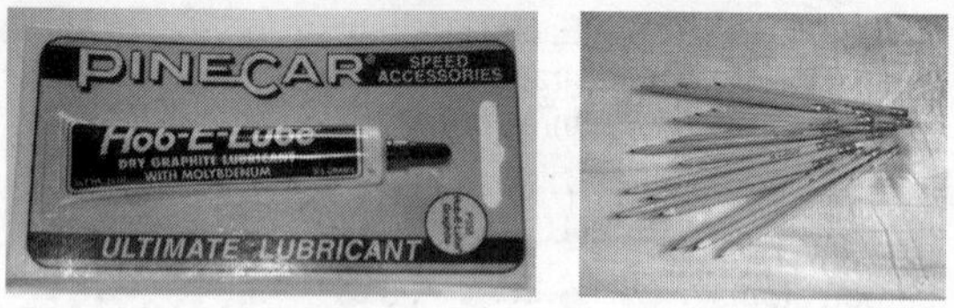

Fig. 4.9 Uses of graphite

Source: K. Murray, CC BY-SA 3.0 <http://creativecommons.org/licenses/by-sa/3.0/>, via Wikimedia Commons; Santeri Viinamäki, CC BY-SA 4.0 <https://creativecommons.org/licenses/by-sa/4.0>, via Wikimedia Commons

(iii) Fullerenes

These are the latest crystalline pure allotropes of carbon obtained by evaporation of graphite using a laser. C_{60} type of fullerene was discovered by H W Proto, R Smalley and Robert Curt (received the Nobel prize in Chemistry in 1996) by applying a laser beam on a sample of graphite. These are the only pure form of carbon as unlike diamond and graphite, they do not have any dangling edges or surface bonds which are instrumental in attracting other atoms. These are large spheroidal cage like molecules of carbon having the molecular formula C_{60}, C_{70} and so on.

C_{60} is called buckminster fullerene in honour of Robert Buckminster Fuller and its structure is similar to that of a soccer ball. It is a dark solid at room temperature which is neither very hard nor very soft. It has 12 five-membered rings and 20 six-membered rings. The five-membered rings are connected only to six-membered rings while six membered rings are connected both types of rings. The C–C and C=C bond lengths are 145.3 and 138.3 pm respectively. C_{70} has also been discovered and these are collectively referred to as buckyballs (Figs 4.10, 4.11).

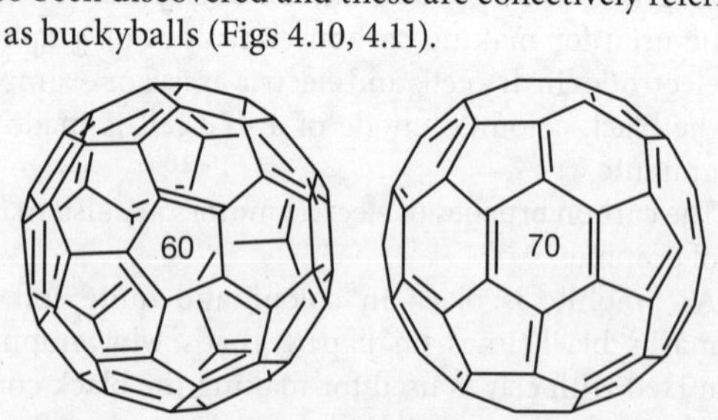

Fig. 4.10 Buckyballs

Fullerenes are soluble in organic solvents giving coloured solutions. For example, C_{60} gives a purple colour in toluene while C_{70} give an orange colour.

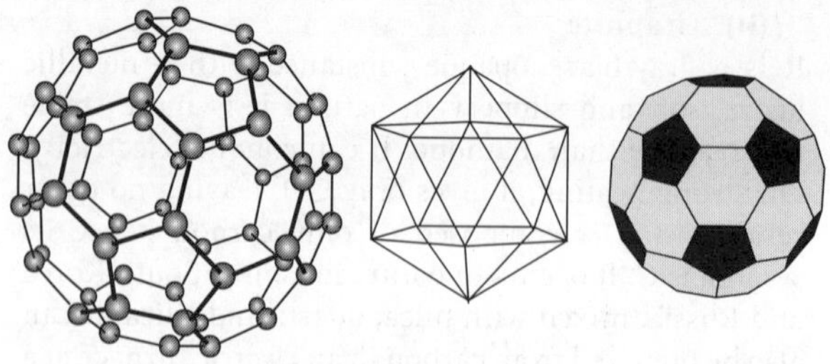

Fig. 4.11 Structure of buckminsterfullerene

TEST YOUR KNOWLEDGE

4.1 Why are covalent bonds formed between carbon atoms very strong?

Solution:

The covalent bonds formed between the various carbon atoms are very strong and cannot be broken easily. The reason for the formation of strong bonds by the carbon atom is its small size. Due to the small size of the carbon atom, their nuclei hold the shared pairs of electrons between the atoms strongly, leading to the formation of strong covalent bonds.

4.2 Write the decreasing order of catenation of the following atoms: carbon, sulfur, phosphorus, silicon.

Solution:

The decreasing order of catenation power is carbon > silicon > sulfur > phosphorus.

4.3 Explain why diamond has a high melting point.

Solution:

The high value of boiling point, melting point and density are due to its three dimensional structure which needs high energy for bond breaking.

4.4 If a boy sharpens a pencil at both the ends and then uses its back ends to complete an electric circuit, will the current flow through the electric circuit? Explain it and name the black substance of the pencil.

Solution:

The current will surely flow through the electric circuit as the black substance of a pencil is graphite and being a good conductor of electricity, the graphite core of pencil allows electric current to flow through it.

> **4.5** A piece of black electrode is used in the dry cell which on strong heating in air gives a colourless gas which turns limewater milky. What is the material of the electrode?
>
> **Solution:**
> Graphite is used for making the electrodes. So, the piece of black electrode used in the dry cell is made of graphite. This is confirmed by the fact that the piece of electrode, on strong heating in air, gives a colourless gas, carbon dioxide, which turns limewater milky. Hence, the material of the electrode is graphite.
>
> **4.6** How many hexagonal and pentagonal rings are present in C_{60} molecules?
>
> **Solution:**
> It contains 20 hexagonal and 12 pentagonal rings fused together.

4.4 Organic Compounds

The word, 'organic' means 'pertaining to life'. The compounds such as sugars, fats, oils, dyes, proteins and vitamins which were isolated directly or indirectly from living organisms like animals and plants are called *organic compounds*, and their study is called *organic chemistry*. The compounds such as common salt, alum, nitre, blue and green vitriol which were isolated from non-living sources, such as rocks and minerals are called *inorganic compounds*. CO, CO_2, CO_3^{-2}, HCO_3^-, HCN, and so on are not organic compounds.

Vital Force Theory (Berzelius Hypothesis)

According to this theory, organic compounds cannot be synthesised in laboratories, because they require the presence of a mysterious force, that is, vital force which exists only in the living organisms.

Fall of vital force theory (Wohler's synthesis)
According to Wohler, when ammonium cyanate (obtained by double decomposition of ammonium chloride and potassium cyanate) is heated, urea is formed. Urea was the first organic compound to be prepared in the laboratory.

$$NH_4Cl + KCNO \longrightarrow \underset{\text{Ammonium cyanate}}{NH_4CNO} + KCl$$

$$NH_4CNO \xrightarrow[\text{Rearrangement}]{\text{Heat}} \underset{\text{Urea}}{NH_2-\overset{\overset{\displaystyle O}{||}}{C}-NH_2}$$

Modern definition of organic compounds: Organic compounds may be defined as hydrocarbons and their derivatives. The study of hydrocarbons and their derivatives is called *organic chemistry*. For example, CH_4 (methane), CH_3OH (methyl alcohol), CH_3CN (methyl cyanide), CH_3CHO (acetaldehyde).

Reasons for large number of organic compounds: There are 5 million plus organic compounds today. The main reasons for such a large number or organic compounds are mainly (i) high catenation power of carbon, (ii) tetravalency of carbon, (iii) formation of multiple bonds, (iv) isomerism.

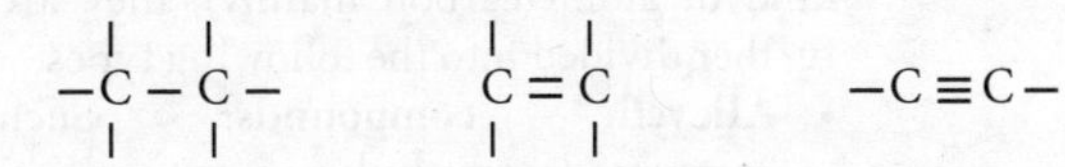

Single bond Double bond Triple bond

Carbon atoms may link to from:
(a) an open chain which may be straight, or branched;
(b) a closed chain.

(a) Open chain

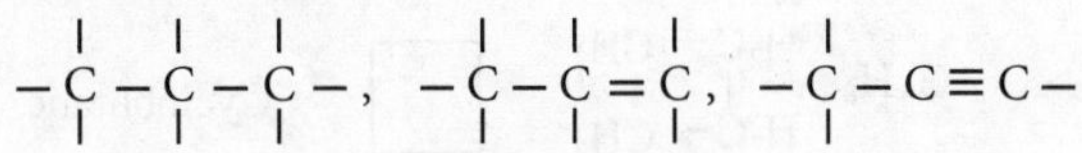

Straight chain

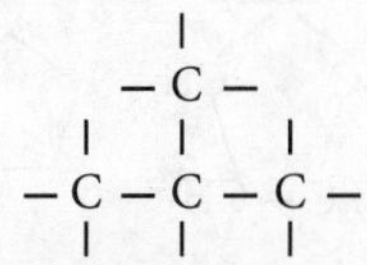

Branched chain

(b) Closed chain

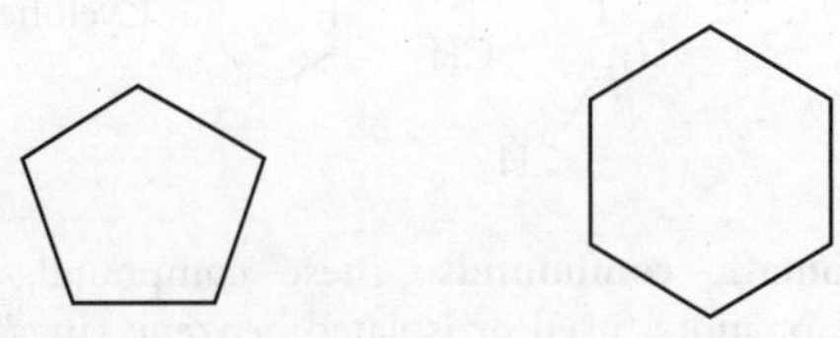

4.4.1 Classification of Organic Compounds

All known organic compounds have been divided into the following two classes.

(i) Acyclic compounds (aliphatic compounds).
(ii) Cyclic compounds (aromatic compounds).

(i) Acyclic compounds: The organic compounds in which all the carbon atoms are linked to one another to form open chains, either straight or branched, are called *acyclic* or *open chain* or *aliphatic compounds.* For example, hydrocarbons, haloalkanes, alcohols, aldehydes, ketones, acids having an open chain structure are aliphatic compounds. CH_4 (methane), C_2H_5OH (ethanol), HCHO (formaldehyde), CH_3COOH (acetic acid).

(ii) **Cyclic compounds:** Cyclic compounds are called *closed chain* or *ring compounds.* These compounds may be homocyclic or heterocyclic.

(a) **Homocyclic compounds:** In such cyclic compounds, the ring contains only one kind of atom (carbon mainly); they are further divided into the following types.

- **Alicyclic compounds:** Such compounds are cyclic in nature and resemble aliphatic compounds in their properties.

C_3H_6 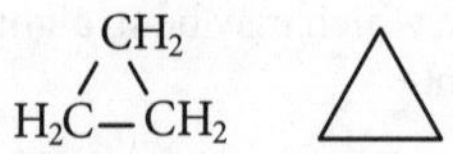 Cyclopropane

C_4H_8 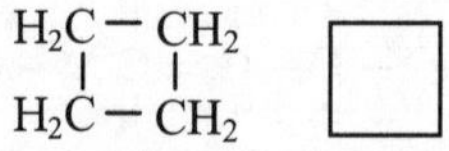 Cyclobutane

C_5H_{10} 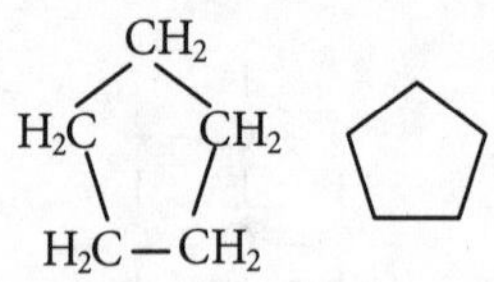 Cyclopentane

C_6H_{12} 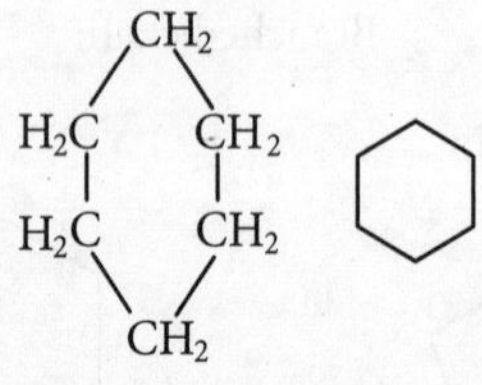 Cyclohexane

- **Aromatic compounds:** These compounds have one or more fused or isolated benzene rings. That is, benzene and its derivatives are called aromatic compounds. For example,

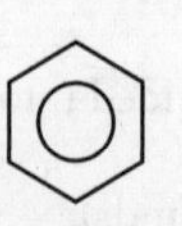
Benzene

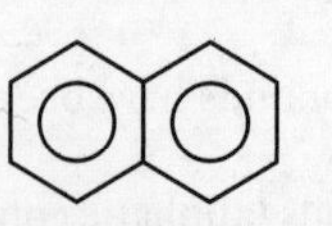
Naphthalene

Phenol

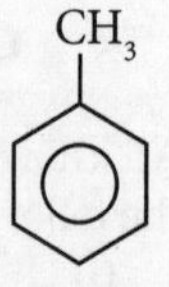
Toluene

(b) **Heterocyclic compounds:** The cyclic compounds having one or more heteroatoms (other than C atom, such as O, N, S) in the ring are called heterocyclic compounds.

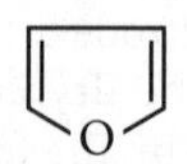
Furan

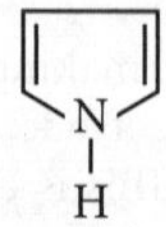
Pyrrole

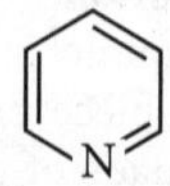
Pyridine

Competition Edge

Straight chain compounds: The compounds that contain straight carbon chains are called straight chain compounds. For example,

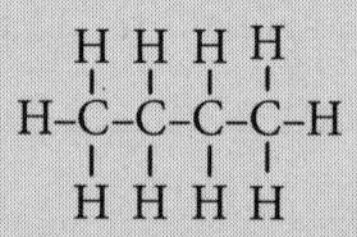

CH_3-CH_2-CH_2-CH_3
n-Butane

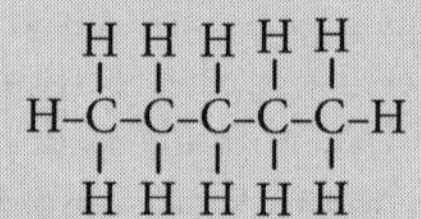

CH_3-CH_2-CH_2-CH_2-CH_3
n-Pentane

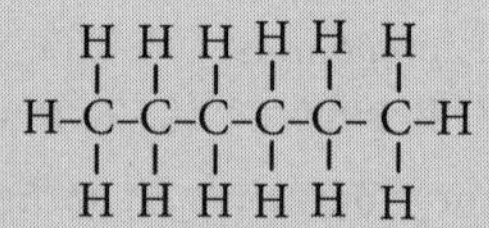

CH_3-CH_2-CH_2-CH_2-CH_2-CH_3
n-Hexane

Branched chain compounds: The compounds which are branched are called branched chain compounds. For example,

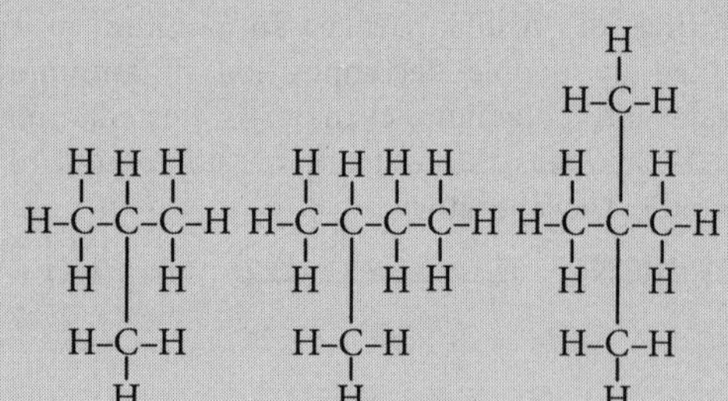

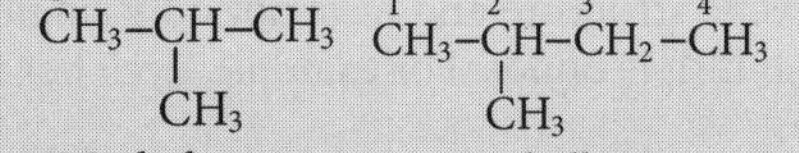

2-Methylpropane 2-Methylbutane

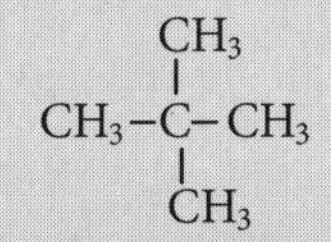

2,2-Dimethylpropane

Different types of formulae of organic compounds: Organic compounds can be represented by any of these formulas (a) molecular formula, (b) structural formula and (c) condensed formula.

Molecular formula: It indicates the kind of atoms and their numbers present in one molecule of the compound. Thus, the molecular formula of pentane is C_5H_{12} which indicates that pentane contains carbon and hydrogen as its constituent elements and the number of atoms of carbon and hydrogen are five and twelve respectively.

Structural formula: It indicates the arrangements of the atoms in a molecule in space. Bonds between atoms are shown by straight lines (shown below).

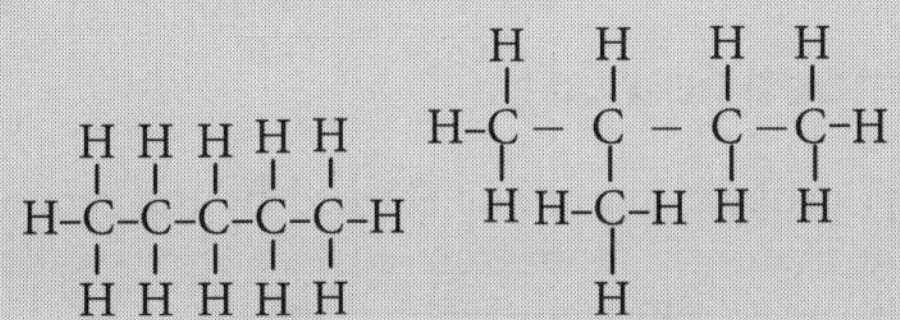

n-Pentane Isopentane

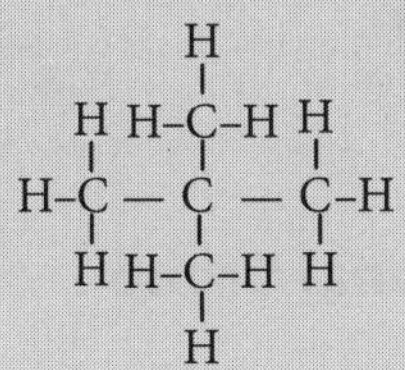

Neopentane

Condensed formula: It is a kind of structural formula which indicates the groups of atoms joined together to each of the carbon atoms in a line or branched line. The condensed formulae of the same compounds are shown below.

$CH_3-CH_2-CH_2-CH_2-CH_3$

n-Pentane

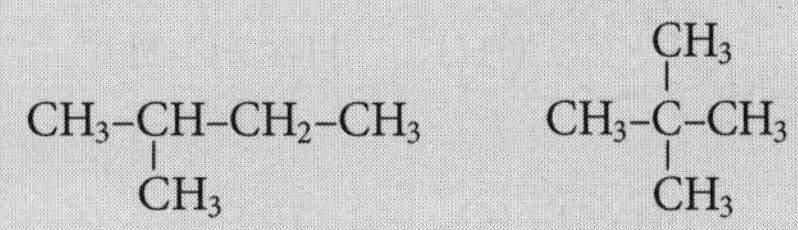

Isopentane Neopentane

4.5 Hydrocarbons

These are composed of carbon and hydrogen only (hydrogen + carbon = hydrocarbon) and their main source is petroleum or crude oil. Hydrocarbons can be further divided into aliphatic hydrocarbons and cyclic hydrocarbons.

4.5.1 Aliphatic Hydrocarbons

Aliphatic hydrocarbons have an open chain of carbon atoms and are of two types—saturated hydrocarbons (alkanes) and unsaturated hydrocarbons (alkenes and alkynes).

(i) **Saturated hydrocarbons (alkanes):** All compounds in which carbon and hydrogen are attached with single bonds are called *alkanes*. The general formula of an alkane from which all the members of the family can be derived is C_nH_{2n+2}. Here *n* is the number of carbon atoms. For example, CH_4 (methane), C_2H_6(ethane), C_3H_8(propene), C_4H_{10} (butane).

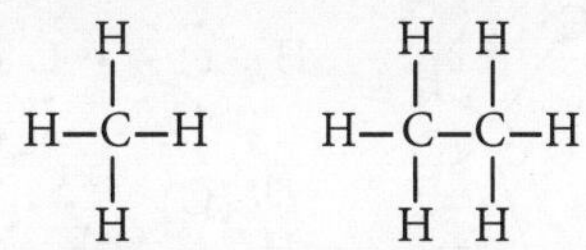

Methane Ethane

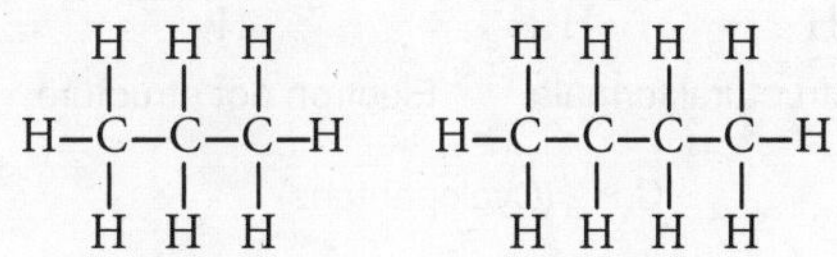

Propane Butane

(ii) **Unsaturated hydrocarbons:** Those hydrocarbons in which the valency of carbon is satisfied by double or triple bonds are called unsaturated hydrocarbons and they are further divided into two types—alkenes and alkynes.

Alkenes: Those unsaturated hydrocarbons which have one or more double bonds are called alkenes. Their general formula is C_nH_{2n}. For example, C_2H_4 (ethene), C_3H_6 (propene), C_4H_8 (butene), C_5H_{10} (pentene).

Ethene

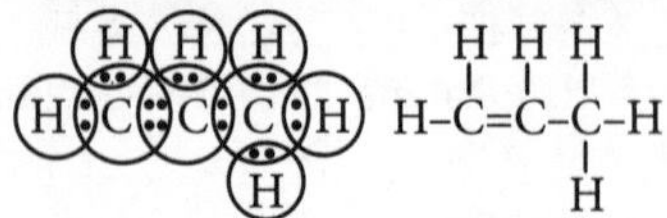

Propene

Alkynes: Those unsaturated hydrocarbons which contain one or more triple bonds are called *alkynes*. The general formula of alkynes is C_nH_{2n-2}. For example, C_2H_2 (ethyne), C_3H_4 (propyne), C_4H_6 (butyne), C_5H_8 (pentyne), C_6H_{10} (hexyne).

Ethyne

4.5.2 Cyclic Hydrocarbons

Hydrocarbons having a cyclic or ring structure are known as cyclo-hydrocarbons. They may be saturated cyclo-hydrocarbons and also unsaturated cyclo-hydrocarbons.

Saturated cyclo-hydrocarbons: Such compounds have a cyclic structure and have only single bonds. For example, cyclopentane and cyclohexane.

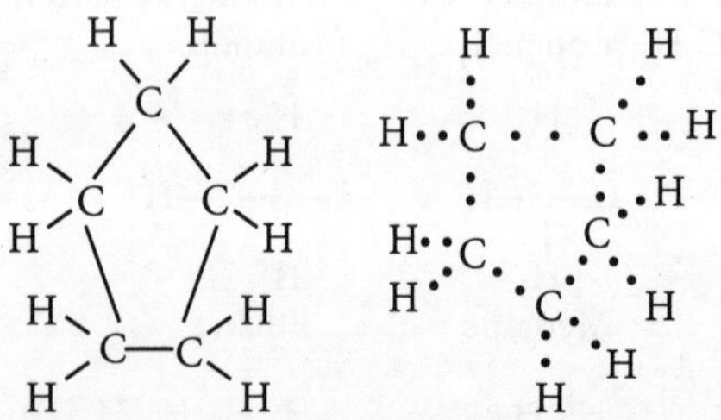

Structural formula Electron dot structure

C_5H_{10} (Cyclopentane)

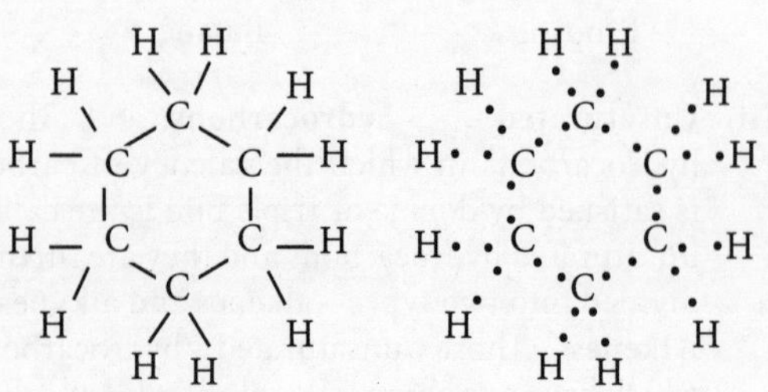

Structural formula Electron dot structure

C_6H_{12} (Cyclohexane)

Unsaturated cyclo-hydrocarbons: Such compounds have a cyclic structure and have not only single bonds but also multiple bonds. For example, cyclohexene and benzene.

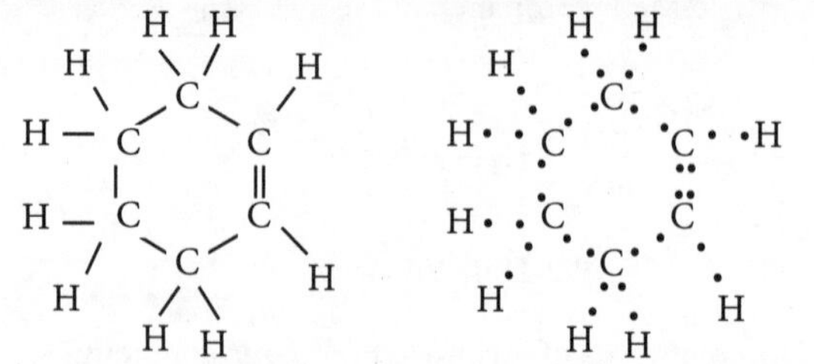

Structural formula Electron dot structure

C_6H_8 (Cyclohexene)

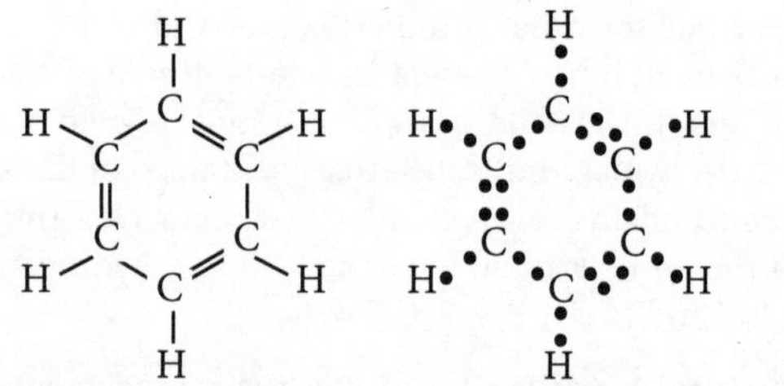

Structural formula Electron dot Structure

C_6H_6 (Benzene)

Competition Edge

Alkyl groups: The group formed by the removal of one hydrogen atom from an alkane molecule is known as an alkyl group. For example,

$$\underset{\text{Methane}}{CH_4} \xrightarrow{-H} \underset{\text{Methyl}}{CH_3},\ \underset{\text{Ethane}}{C_2H_6} \xrightarrow{-H} \underset{\text{Ethyl}}{C_2H_5},$$

$$\underset{\text{Propane}}{C_3H_8} \xrightarrow{-H} \underset{\text{Propyl}}{C_3H_7},\ \underset{\text{Butane}}{C_4H_{10}} \xrightarrow{-H} \underset{\text{Butyl}}{C_4H_9}$$

4.5.3 Functional Group

A functional group is an atom or group of atoms or reactive part of the compound which largely determines the chemical properties of the compound. That is, a functional group is the site of chemical activity in a molecule or compound. Here are some important functional groups.

–X (Halogen), –OH (Alcohol), –CHO (Aldehyde),

$-\overset{\overset{\large O}{\|}}{C}-$ (Ketone), $-\overset{\overset{\large O}{\|}}{C}-OH$ (Carboxylic acid), $-NH_2$ (Primary amine)

For example,

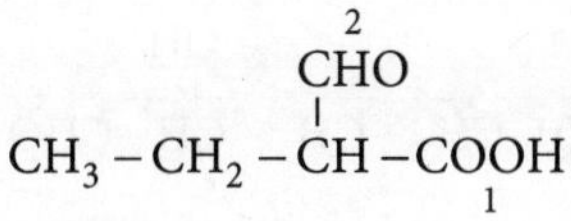

It has 2 functional groups.

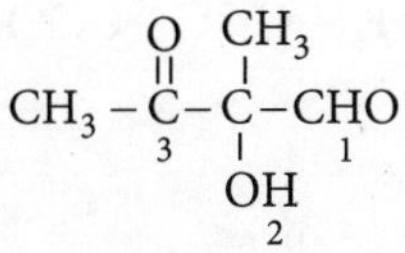

It has 3 functional groups.

4.6 Isomers and Isomerism

The compounds which differ in their properties but have same molecular formula are known as *isomers* and the phenomenon is known as *isomerism*. Isomerism is of three types—structural or constitutional isomerism, stereo or configurational isomerism and conformational isomerism. (*You will study these in higher classes.)

4.6.1 Structural or Constitutional Isomerism

In this type of isomerism, compounds possessing the same molecular formula differ in their properties due to the difference in the linkages of atoms inside the molecule (due to the difference in their structures). In this case, the isomers differ in the connectivity of carbon atoms, that is, differ in their structural formula. Structural isomers differ much in their physical as well as chemical properties. For example, CH_3CH_2OH and CH_3OCH_3. Both have the molecular formula C_2H_6O but they differ in their structures.

Types of Structural Isomerism

It is of the following types.

Chain or skeleton isomerism: In this type of isomerism, compounds possessing the same molecular formula differ in their properties due to the difference in the arrangement of carbon chain present in them. The chain of carbon atoms can be branched or unbranched. Here a chain of minimum 4 carbon atoms is necessary to exhibit this isomerism. Alkanes from butane onwards can show this type of isomerism.

For example,

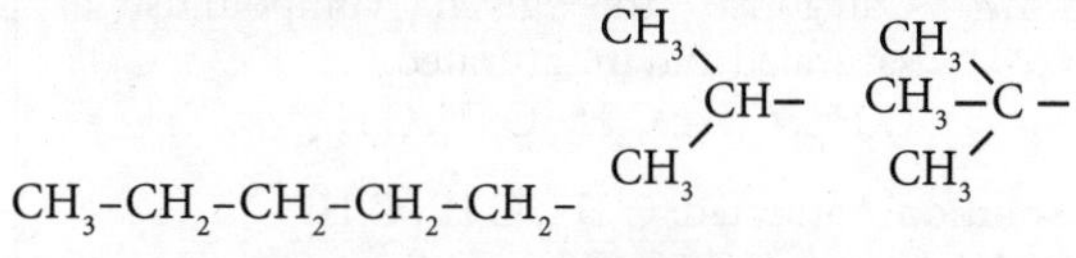

Normal alkane Iso alkane Neo alkane

A normal alkane has no side chain while an iso alkane has one side chain and a neo alkane has two side chains.

For example,

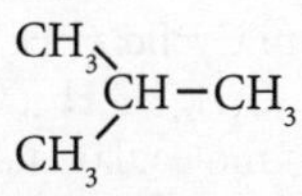

C_4H_{10}: $CH_3\text{-}CH_2\text{-}CH_2\text{-}CH_3$

n-butane Isobutane (2-methyl propane)

C_5H_{12}: $CH_3\text{-}CH_2\text{-}CH_2\text{-}CH_2\text{-}CH_3$

n-pentane

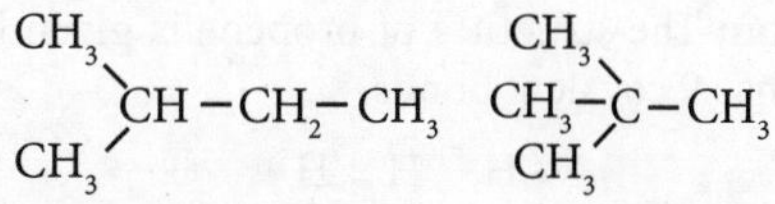

Isopentane Neopentane

Position isomerism: In position isomerism, the compounds possessing the same molecular formula differ in their properties due to the difference in the position of either the functional group or the multiple bond or the branched chain or the substituent attached to the main carbon chain.

For example, C_3H_8O: $CH_3\text{-}CH_2\text{-}CH_2\text{-}OH$ (n-propyl alcohol) and $CH_3\text{-}CH(OH)\text{-}CH_3$ (isopropyl alcohol) are position isomers as the position of the –OH group differs, that is, positions 1 and 2 respectively.

C_4H_8: $CH_2\text{=}CH\text{-}CH_2\text{-}CH_3$ (butene-1) and $CH_3\text{-}CH\text{=}CH\text{-}CH_3$ (but-2-ene) are position isomers as the position of the double bond differs, that is, positions 1 and 2 respectively.

Functional isomerism: In functional isomerism, the compounds possessing the same molecular formula differ in their properties due to the difference in their functional groups. For example, CH_3CH_2CHO (propanaldehyde) and CH_3COCH_3 (acetone) are functional isomers.

You will study other types of isomerism in class XI.

TEST YOUR KNOWLEDGE

4.7 Categorise the given compounds as saturated and unsaturated.
C_4H_{10}, C_6H_6, C_2H_6, C_2H_4, C_5H_{12}

Solution: Saturated: C_4H_{10}, C_2H_6, C_5H_{12}
Unsaturated: C_6H_6, C_2H_4

4.8 Categorise the given compounds as cyclic and acyclic compounds.
C_4H_{10}, C_6H_{12}, C_6H_6, C_2H_6, C_2H_4

Solution: Cyclic: C_6H_{12}, C_6H_6
Acyclic: C_4H_{10}, C_6H_{12}, C_2H_6, C_2H_4
As the molecular formula of cyclohexane and the alkene is the same, that is, C_nH_{2n}, it can be cyclohexane as well as hexene.

4.9 Write the structure of propene and find the total number of covalent bonds it has.

Solution: The structure of propene is given below and it has 9 covalent bonds.

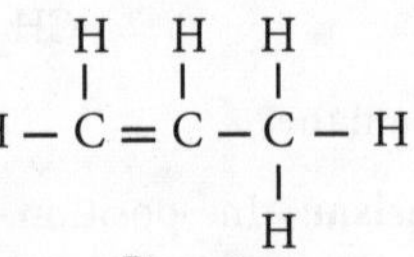

Propene

4.10 In the given compounds, how many functional groups are there?

(i) 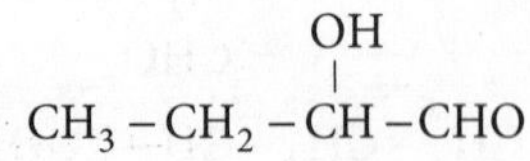

(ii) 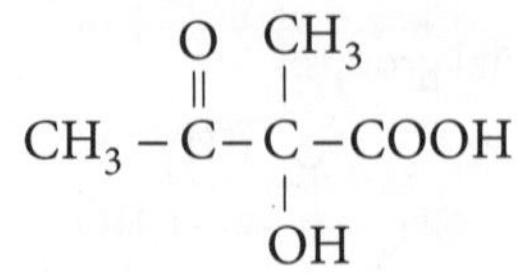

(iii) 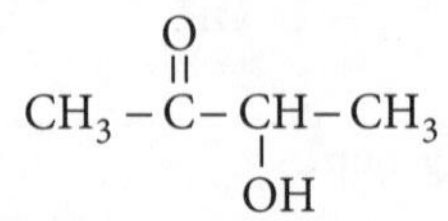

Solution: (i) 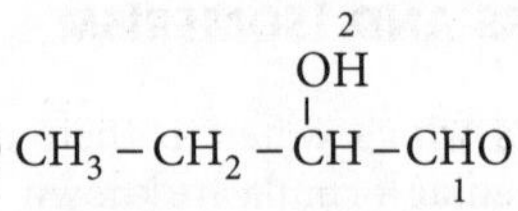

(ii) 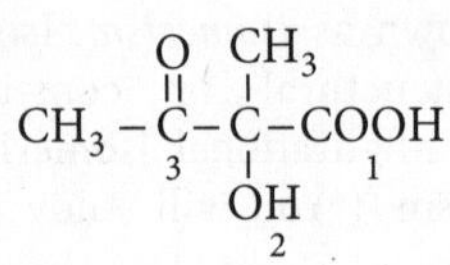

(iii) 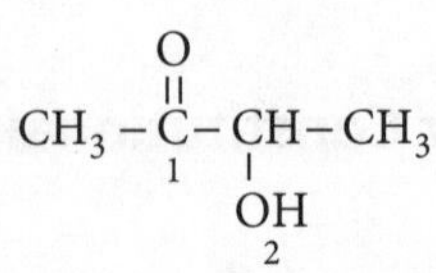

Compounds (i), (ii) and (iii) have 2, 3 and 2 functional groups respectively.

Quick Review

- This characteristic of the carbon atom by virtue of which it forms four covalent bonds is generally referred to as the **tetra-covalency** of carbon.
- **Catenation:** The most outstanding (or unique) property of carbon is its ability to combine with itself, atom to atom, to form long chains. For example, octane (C_8H_{18}).
- **Diamond** is a colourless, transparent, quite heavy substance possessing an amazing brilliance (lustre). It is the hardest natural known substance. It does not conduct electricity. It is the purest form of carbon.
- In diamond, the carbon atom is tetrahedral (*sp*³ hybridised), the C–C bond length is 1.54 Å and the bond angle is 109°28′.
- **Graphite** is a grey-black opaque coloured substance with a metallic lustre, and is soft and slippery. It is more stable and reactive than diamond.
- The various **hexagonal layers** of carbon atoms in graphite are held together by weak van der Waals forces so they can slide over one another. Due to their sheet-like structure, graphite is a soft substance. Here the C–C bond length is 1.42 Å and the distance between two successive layers is 3.35 Å.
- **Graphite** is a good conductor of electricity so it can be used for making carbon electrodes or graphite electrodes in dry cells and electric arcs. For example, the black coloured 'anode' of a dry cell is made of graphite.

- C_{60} is called **buckminster fullerene** in honour of Robert Buckminster Fuller, and its structure is similar to that of a soccer ball. It is a dark solid at room temperature which is neither very hard nor very soft. It has 12 five-membered rings and 20 six-membered rings.
- Organic compounds may be defined as hydrocarbons and their derivatives; the study of hydrocarbons and their derivatives is called **organic chemistry**. For example, CH_3OH, CH_3CN.
- **Catenation:** The property of self-linking of carbon atoms through covalent bonds to form long straight or branched chains and rings of different sizes is called catenation.
- **Tetravalency of carbon:** Carbon has four valence electrons and it can share all four electrons with other atoms of carbon as well as oxygen, hydrogen, nitrogen, sulfur and halogen.
- The organic compounds in which all the carbon atoms are linked to one another to form open chains – either straight or branched – are called acyclic or open chain or **aliphatic compounds**.
- **Cyclic compounds** are called closed chain or ring compounds.
- **Homocyclic compounds:** In such cyclic compounds, the ring has only one kind of atoms (carbon mainly).
- **Alicyclic compounds** are cyclic in nature and resemble aliphatic compounds in their properties.
- **Aromatic compounds** have one or more fused or isolated benzene rings. Benzene and its derivatives are called aromatic compounds.
- The compounds which resemble benzene and other aromatic compounds in most of their properties and possess a hetero atom are called **aromatic heterocyclic compounds**.
- A **functional group** is an atom or group of atoms or reactive part of compound which largely determines the chemical properties of compound (site of chemical activity). For example, –OH, $-NH_2$, –CHO.
- The compounds which differ in their properties but have the same molecular formula are called **isomers** and the phenomenon is known as **isomerism**.
- In **structural isomerism,** compounds possessing the same molecular formula differ in their properties due to the difference in the linkages of atoms inside the molecule, that is, due to the difference in their structures.
- **Chain or skeleton isomerism**: In this type of isomerism, compounds possessing the same molecular formula differ in their properties due to the difference in the arrangement of the carbon chain present in them.
- **Position isomerism**: In this type of isomerism, the compounds possessing the same molecular formula differ in their properties due to the difference in the position of either the functional group or the multiple bond or the branched chain or the substituent attached to the main carbon chain.
- **Functional isomerism**: It is the type of isomerism in which the compounds possessing the same molecular formula differ in their properties due to the difference in their functional groups.

Exercise 1

Section A: Multiple Choice Questions (1 Mark)

1. Which of the following is considered as a versatile non-metal element?
   (a) hydrogen (b) oxygen
   (c) nitrogen (d) carbon

2. The percentage of carbon in atmosphere is about
   (a) 0.02% (b) 0.03%
   (c) 0.3% (d) 1%

3. The electronic configuration of an element is found to be 2, 4. How many bonds can one carbon atom form in a compound?
   (a) 1 (b) 2
   (c) 4 (d) 6

4. Why does carbon form compounds mainly by covalent bonding?
   (a) There are four electrons in the outermost shell of carbon.
   (b) It requires a large amount of energy to form C^{4+} or C^{4-}.
   (c) It shares its valence electrons to complete its octet.
   (d) All of these.

5. Which of the following is not a characteristic of fullerenes?
   (a) Of all the fullerenes, the C_{60} allotrope is the most stable.
   (b) Its shape is similar to that of a soccer ball.
   (c) It contains only fused six-membered carbon–carbon rings.
   (d) Its hardness is lower than that of diamond.

6. In diamond, each carbon atom is bonded to four other carbon atoms to form
(a) a hexagonal array
(b) a rigid three-dimensional structure
(c) a structure in the shape of a football
(d) a structure of a ring

7. The image represents the structure of a few hydrocarbon compounds.

(A)

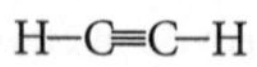

(B)

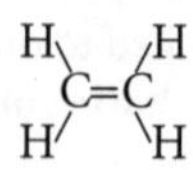

(C)

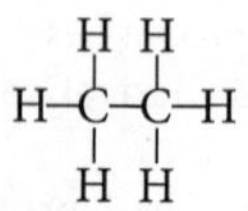

(D)

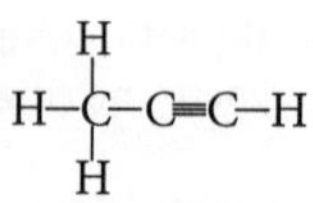

Which of these compounds can be classified as an alkyne?
(a) only (A) (b) only (B)
(c) both (A) and (D) (d) both (B) and

8. The image represents the structure of a carbon compound known as ethane.

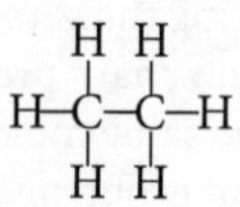

Which option explains the naming of ethane?
(a) the presence of a functional group connected with a single bond
(b) it contains two carbon atoms and a single bond connects the carbon atoms
(c) carbon compound with a total number of eight atoms are named as ethane
(d) it contains six hydrogen atoms and a single bond connects the carbon and hydrogen atom

9. A hydrocarbon has four carbon atoms. Give its molecular formula if it is an alkene.
(a) C_4H_{10} (b) C_4H_8
(c) C_4H_6 (d) C_4H_4

10. A carbon compound contains two atoms of carbon. Which name should the carbon compound bear?
(a) butane (b) ethene
(c) methane (d) propene

11. Which of the following is the molecular formula of cyclobutane?
(a) C_4H_{10} (b) C_4H_6
(c) C_4H_8 (d) C_4H_4

12. How many single bonds are present in methane?
(a) six (b) five
(c) four (d) three

13. A hydrocarbon should have a minimum of how many carbon atoms to show isomerism?
(a) three (b) four
(c) fived (d) six

14. The isomeric pair is
(a) ethane and propane
(b) propane and butane
(c) ethane and ethane
(d) butane and 2-methyl propane

15. The number of isomers of pentane is
(a) 2 (b) 3
(c) 4 (d) 5

16. Select the option having the set of correct statements.
(i) The ethene molecule is made up of 2 carbon atoms and 4 hydrogen atoms.
(ii) Each carbon atom shares three electrons with three hydrogen atoms to form three carbon–hydrogen single covalent bonds.
(iii) In ethane, the two carbon atoms share one pair of electrons among themselves to form one carbon–carbon single covalent bond.
(iv) All of these.
(a) (i), (ii) (b) (ii), (iii)
(c) (i), (ii), (iii) (d) (i), (ii), (iii), (iv)

17. Match the following.

Column I	Column II
1. Diamond	(i) Buckyball structure
2. Graphite	(ii) Contains methane
3. Fullerene	(iii) Hardest substance
4. Natural Gas	(iv) Conductor

(a) (1)—(ii), (2)—(iii), (3)—(i), (4)—(iv)
(b) (1)—(iii), (2)—(iv), (3)—(ii), (4)—(i)
(c) (1)—(ii), (2)—(iv), (3)—(i), (4)—(iii)
(d) (1)—(iii), (2)—(iv), (3)—(i), (4)—(ii)

Assertion Reason Questions

Directions: In the following questions two statements (Assertion) A and Reason (R) are given. Mark
(a) if A and R are both correct and R is the correct explanation of A;
(b) if A and R are both correct but R is not the correct explanation of A;
(c) A is true but R is false;
(d) A is false but R is true.

Assertion	Reason
1. Carbon always forms covalent bonds.	**1.** Due to energy considerations C^{4+} or C^{4-} are not possible so ionic bond is not possible.
2. Carbon has a tetrahedral structure.	**2.** Carbon has the highest catenation power.
3. Diamond can show conduction.	**3.** Graphite is soft.
4. Diamond and graphite are the purest forms of carbon.	**4.** The physical properties of allotropes are similar.
5. Sharp edged diamonds are used as a tool by eye surgeons.	**5.** Diamond has a high thermal conductivity.
6. Urea was the first organic compound to be prepared in the laboratory.	**6.** Organic compounds are hydrocarbons and their derivatives.
7. Benzene is a cyclic aromatic compound.	**7.** It is an isomer of cyclohexene.
8. The general formula of an alkyne is C_nH_{2n-2}.	**8.** The general formula of cycloalkenes is also C_nH_{2n-2}.

Section B: Very Short Answer Questions (2 Marks)

1. _______ is a versatile element that forms the basis for all living organisms and many of the things we use.
2. Mention the percentage of carbon in the earth's crust.
3. How does carbon attain noble gas configuration?
4. Name the element whose allotropic form is graphite.
5. Which of the following molecules is called buckminsterfullerene?
   C_{90}, C_{60}, C_{70}, C_{120}
6. What are the numbers of 5-membered and 6-membered rings present in buckminsterfullerene?
7. An allotrope of carbon has a molecular mass of 720, Identify it.
8. Which allotrope of carbon is used for making electrodes of dry cells?
9. What are the structural units of diamond and graphite?
10. Which element exhibits the property of catenation to the maximum extent and why? **[CBSE 2016]**
11. The unsaturated hydrocarbons which contain one or more double bonds are called _______.
12. The general formula for alkynes is _______.
13. Select saturated hydrocarbons from the following:
    C_3H_6, C_5H_{10}, C_4H_{10}, C_6H_{14}, C_2H_4. **[CBSE 2016]**
14. Write the number of covalent bonds in a molecule of ethane. **[CBSE 2015]**
15. Write the number of covalent bonds in a molecule of propane, C_3H_8. **[CBSE, 2015]**
16. The molecular formula of a hydrocarbon is C_3H_8. Draw its complete structure and write its name.
17. How many covalent bonds are present in pentane (C_5H_{12})?

Case Study or Passage Based Questions

Carbon is one of the most important non-metals. It can show allotropy and its allotropes differ in structure and many properties such as hardness, density and melting point. Carbon has the unique property of catenation so it can form a large number of compounds. Carbon can have all three types of covalent bonds. Organic compounds are present in many living organisms and can be further divided into many other types.

18. In the atmosphere, carbon is present in the form of its oxide. Name and write the formula of this oxide.

19. Why do carbon compounds have lower a value of melting and boiling points?

20. Which allotrope of carbon has the highest melting point?

21. Which allotrope of carbon is the softest?

22. What is the valency of carbon in CH_3–CH_3, CH_2=CH_2 and HC≡CH?

Section C: Short Answer Questions (3 Marks)

1. Why is carbon tetravalent?

2. Explain why carbon generally forms compounds by covalent bonds.

3. Why does carbon form a large number of compounds?

4. Why is carbon unique in nature?

5. Covalent compounds generally don't conduct electricity. Why?

6. Why is graphite slippery?

7. Why is graphite used as a lubricant?

8. What are hydrocarbons? Give at least two examples.

9. What is meant by isomers? Draw the structures of two isomers of butane (C_4H_{10}). Explain why we cannot have isomers of the first three members of the alkane series. **[CBSE 2015]**

10. Why is it not easy for carbon to take part in the formation of ionic compounds?

11. Write the number of covalent bonds in a molecule of butane, C_4H_{10}. **[Board, 2015]**

12. List two differences between saturated and unsaturated hydrocarbons.

13. Write the molecular, electronic and structural formulae of ethene.

14. What is meant by isomers? Draw the structure of two isomers of butane, C_4H_{10}. Explain why we cannot have isomers of the first three members of the alkane series. **[Board, 2015]**

15. What is structural isomerism? Draw the isomers of pentane (C_5H_{12}).

16. Name and draw the chain structure and dot structure of the first two alkanes.

Section D: Long Answer Questions (5 Marks)

1. (i) Define allotropy.
(ii) Write the names of any three allotropes of carbon.
(iii) Write any four differences between diamond and graphite.
(iv) Why does graphite show conductance?

2. (a) State any three physical properties of carbon compounds.
(b) Carbon is a versatile element. Justify this statement. **[Board, 2015]**

3. A solid element P has four electrons in the outermost shell of its atom. An allotrope Q of this element is used as a dry lubricant in machinery and also in making pencil leads.
(a) What is element P?
(b) Name the allotrope Q.
(c) State whether allotrope Q is a good conductor or a non-conductor of electricity.
(d) Name one use of allotrope Q (other than lubrication and pencil leads).
(e) Name two other allotropes of element P.

4. Carbon, a member of group 14, forms a large number of compounds, estimated to be about three million. Why is this property not exhibited by other elements of this group? Explain.

5. (a) State the reason why carbon can neither form C^{4+} cations nor C^{4-} anions, but forms covalent bonds. Also state the reasons to explain why covalent compounds
(i) are bad conductors of electricity
(ii) have low melting and boiling points
(b) Write the structure of benzene.

6. (i) Why are most carbon compounds poor conductors of electricity?
(ii) Write the name and structure of a saturated compound in which the carbon atoms are arranged in a ring. Give the number of single bonds present in this compound. **[CBSE 2018]**

7. What are hydrocarbons? Write the name and general formula of:
   (i) saturated hydrocarbons
   (ii) unsaturated hydrocarbons
   Draw the structure of one hydrocarbon of each type. How can an unsaturated hydrocarbon be made saturated?

8. Carbon, a member of group 14, forms a large number of carbon compounds estimated to be about three million. Why is this property not exhibited by other elements of this group? Explain. [CBSE 2020]

Section E: Case Study or Passage Based Questions (4 Marks)

Hydrocarbons are organic compounds having only carbon and hydrogen atoms. They are further classified into saturated and unsaturated hydrocarbons. Hydrocarbons may be aliphatic as well as cyclic. All other organic compounds are actually derivatives of hydrocarbons. Hydrocarbons can show substitution, addition, oxidation and combustion reactions mainly. Hydrocarbons contain covalent bonds only and can show isomerism also.

1. Which of the following cannot be a hydrocarbon?
   (a) C_2H_6 (b) C_2H_4
   (c) C_2H_2 (d) NH_2CONH_2

2. Which of the following pairs contains cyclic unsaturated hydrocarbons?
   (a) propene, cyclopropane
   (b) cyclohexene, cyclohexane
   (c) cyclopentene, cyclobutene
   (d) cyclohexane, benzene

3. Which of the following compounds can show combustion reaction?
   (a) ethane (b) ethene
   (c) ethyne (d) all of these

4. Which of the following hydrocarbons cannot show isomerism?
   (a) cyclohexane
   (b) propane
   (c) butane
   (d) pentane

5. The number of covalent bonds in ethene and ethyne are respectively.
   (a) 5, 5 (b) 6, 6
   (c) 5, 6 (d) 6, 5

High Order Thinking Skills (HOTS) Questions

1. The solid element X exhibits the property of catenation. It is also present in the form of a gas Y in the air which is utilised by plants in photosynthesis. An allotrope Z of this element is used in glass cutters.
   (a) What is element X?
   (b) What is the gas Y?
   (c) Name the allotrope Z.
   (d) State another use of allotrope Z (other than in glass cutters).
   (e) Name another allotrope of element X which exists as spherical molecules.
   (f) Name a yet another allotrope of element X which conducts electricity.

2. Carbon has the unique property to form bonds with other atoms of carbon.

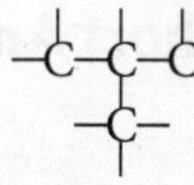

   (i) Name the characteristic property of carbon as depicted above.
   (ii) Give the reason for this unique property of carbon.
   (iii) Draw the structure of cyclohexane.
   (iv) What are the properties of carbon that lead to the formation of fullerenes. Name two other compounds which are allotropes of C. In which of these compounds are all the four valences satisfied?

3. Why are certain compounds called hydrocarbons? Write the general formula for the homologous series of alkanes, alkenes and alkynes and also draw the structure of the first member of each series. Write the name of the reaction that converts alkenes into alkanes and also write a chemical equation to show the necessary conditions for the reaction to occur.

4. *Shashwat read about the formation of covalent bonds and came to know the properties of carbon. Carbon has four valence electrons. It cannot lose four electrons since a very high amount of energy will be required to lose four electrons to form C^{4+}. Carbon cannot gain four electrons to form C^{4-} ion as six protons cannot hold 10 electrons easily and there will be strong interelectronic repulsion.

(i) What is the atomic number and electronic configuration of carbon?

(ii) Explain how covalent bonding is important between carbon compounds. Give any two suggestions.

Answers

Section A: Multiple Choice Questions

1. (d)	**2.** (b)	**3.** (c)	**4.** (d)
5. (c)	**6.** (b)	**7.** (c)	**8.** (b)
9. (b)	**10.** (b)	**11.** (c)	**12.** (c)
13. (b)	**14.** (d)	**15.** (b)	**16.** (c)
17. (d)			

Assertion Reason Questions

1. (a)	**2.** (b)	**3.** (d)	**4.** (c)
5. (b)	**6.** (b)	**7.** (c)	**8.** (b)

Section B: Very Short Answer Questions

1. Carbon.

2. 0.02%.

3. Carbon can form bonds by sharing of its electrons with the electrons of another carbon atom or with another element and attain the noble gas configuration.

4. Graphite is an allotropic form of carbon.

5. C_{60} molecule is called buckminsterfullerene.

6. The number of 5-membered and 6-membered rings in buckminsterfullerene is 12 and 20 respectively.

7. Atomic mass of C = 12 and molecular mass of an allotrope of carbon 720.
So, number of C atoms = 720/12 = 60.
Hence, the allotrope of carbon is C_{60}, that is, buckminsterfullerene.

8. Graphite.

9. The structural units of diamond and graphite are tetrahedral and hexagonal respectively.

10. Carbon exhibits this property as it forms strong covalent bonds (more energy for it).

11. Alkenes.

12. C_nH_{2n-2}.

13. Saturated hydrocarbons are C_6H_{14} and C_4H_{10}.

14. There are 7 covalent bonds in ethane.

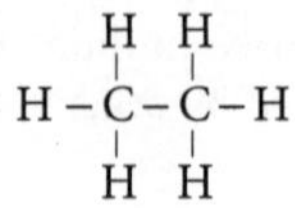

15. Ten covalent bonds.

16.

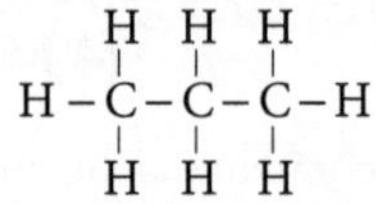

The name of the hydrocarbon is propane.

17. Sixteen (16) Covalent bonds.

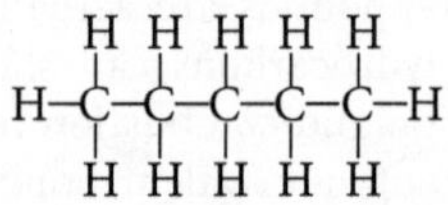

Case Study or Passage Based Questions

18. Carbon dioxide (CO_2).

19. Carbon compounds have lower values of melting and boiling points as the force of attraction between the molecules are weak.

20. Diamond has the highest melting point (4203 K).

21. Graphite.

22. The valency of carbon in all its compounds whether saturated or unsaturated is 4.

Section C: Short Answer Questions

1. The outer-most shell of the carbon atom contains four valence electrons so the valency of carbon is four (tetravalent). It is difficult to hold four extra electrons and would require a large amount of energy to remove four electrons, so, it forms bonds by sharing of electrons.

2. Carbon generally forms compounds through covalent bonds because carbon can neither donate nor accept four electrons for completing its octet due to very high energy considerations.

So, it shares its four electrons with other atoms forming covalent bonds.

3. Carbon is an element which shows the unique property of catenation, making it possible to form a large number of compounds. It is the property by which it can form bonds with carbon atoms, to form long chains.

4. Due to its four valence electrons and small size, carbon is able to form covalent bonds with four different atoms in its neutral form. Because of its highest catenation power, carbon can form large heavily branched compounds by stringing together with other carbon atoms and a few other non-metallic atoms in various arrangements.

5. A covalent bond is formed between non-metal atoms, which combine together by sharing electrons. Covalent compounds have no free electrons and no free ions so they don't conduct electricity.

6. In graphite, layers of the carbon atoms are quite far apart and are held together by weak van der Waals forces which enables them to slide over one another.

7. As the structure of graphite contains flat hexagonal parallel layers joined together by weak van der Waals forces, these layers can slide over one another easily making graphite a good lubricant.

8. Hydrocarbons are organic compounds that are made of only hydrogen and carbon atoms. For example, methane (CH_4), ethane (C_2H_6), ethene (C_2H_4).

9. Isomers are compounds having the same molecular formula but different structural formulae. $CH_3CH_2CH_2CH_3$ and

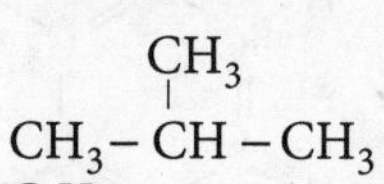

are the two isomers of C_4H_{10}.
Isomers are not possible for the first three members of the alkane series because there is no branching.

10. In order to form ionic bonds, the carbon atom either has to lose four electrons to form the C^{4+} ion or gain four electrons to form the C^{4-} ions. The loss or gain of four electrons is not easy for carbon since the energy needed is very high.

11. Butane contains a total of thirteen covalent bonds (3 C–C bonds and 10 C–H bonds).

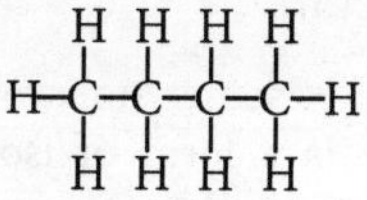

12.

S. No	Saturated hydrocarbons	Unsaturated hydrocarbons
(i)	Hydrocarbons having single bonds between carbon atoms are called saturated hydrocarbons.	Hydrocarbons having one or more multiple bonds between carbon atoms are called unsaturated hydrocarbons.
(ii)	They give a clean flame on burning.	They give a yellow flame with a lot of black smoke on burning.

13. Molecular formula: C_2H_4

Electron dot formula: 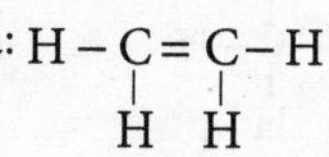

Structural formula: 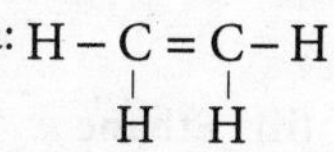

14. Isomers are the compounds which have the same molecular formula but different structural formula.
Isomers of butane:

(i) 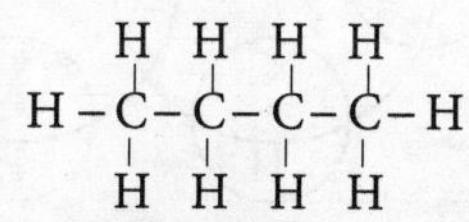

n-butane

(ii) 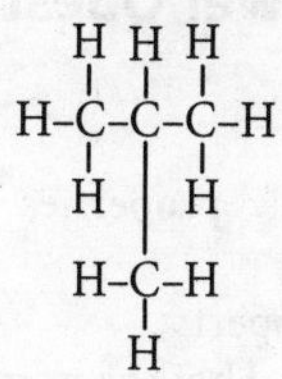

Iso-butane

We cannot have isomers of the first three members of the alkane series because of the following laws of isomers.

(i) The parent chain should have the most number of carbon atoms.

(ii) The branching cannot be done from the first or the last atom carbon atom of the structure.

15. Structural isomerism or constitutional isomerism is a form of isomerism in which molecules with the same molecular formula have different structure formulae. Pentane has the following three isomers.

$$CH_3-CH_2-CH_2-CH_2-CH_3$$

n-Pentane

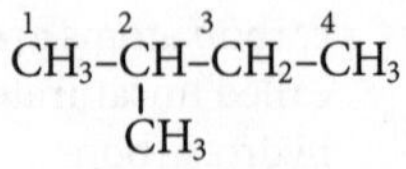

iso-pentane

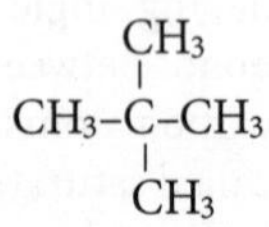

neopentane

16. (i) Methane

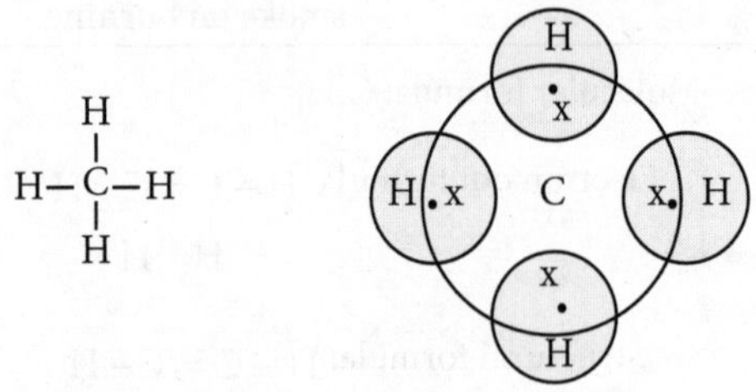

(ii) Ethane

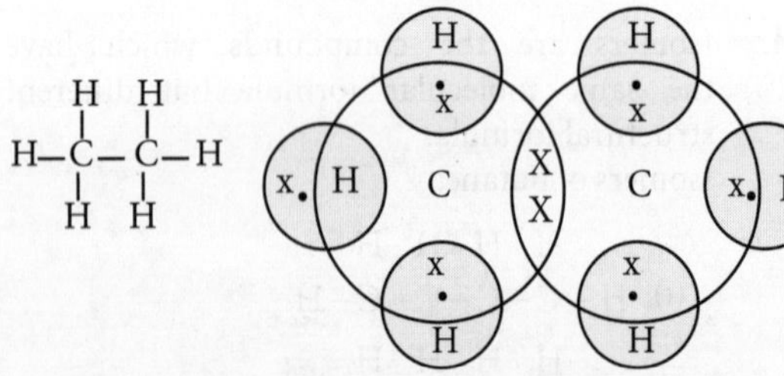

Section D: Long Answer Questions

1. See text (Section 4.2).

2. (a) Three physical properties of carbon compounds are:
   (i) catenation property,
   (ii) low melting and boiling points compared to ionic compounds, and
   (iii) poor conductors of electricity.

   (b) Carbon is a versatile element because it forms covalent bonds with a large number of elements and has the catenation capacity to form compounds by a chain of bonds with itself.

3. (a) Carbon, (b) graphite, (c) good conductor of electricity, (d) electrodes, (e) diamond and buckminsterfullerene.

4. (i) Tetravalency: Carbon can easily form four covalent bonds to other atoms.
   (ii) Catenation: Carbon can easily form covalent bonds with other carbon atoms and the carbon–carbon bonds are more stable than the bonds formed by the other members of group 14 (Si–Si, Ge–Ge).
   (iii) Formation of multiple bonds: Carbon can form not only single bonds but it can also form double or triple bonds with the same or different elements. Since these properties are not exhibited by other elements of this group, the number of carbon compounds is very large as compared to the other elements of group 14.

5. (a) Carbon cannot lose four electrons to form C^{4+} ions as very high energy is required to remove four electrons. Also, carbon cannot gain four electrons to form C^{4-} ions as 6 protons cannot hold 10 electrons and energy has to be supplied to overcome this repulsion. So, carbon shares four electrons and forms covalent compounds.
   (i) Covalent compounds are bad conductors of electricity as they are non-polar and they do not form ions, that is, there are no free ions or electrons to facilitate conduction.
   (ii) They have low melting and boiling points because of weak van der Waals forces of attraction, that is, due to weak forces of attraction between the molecules, less energy is required to break the bonds.

   (b)

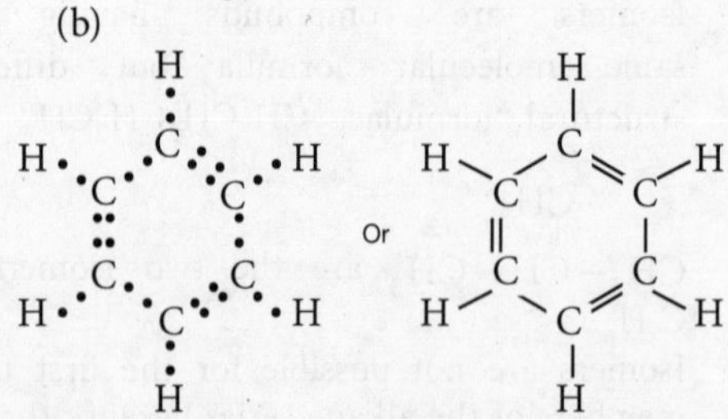

Benzene

6. (i) Most of the carbon compounds are poor conductors of electricity. Due to the presence of covalent bonds in them they do not have free ions or free electrons to move to show conduction.

(ii) **Name of compound:** Cyclohexane (C_6H_{12}).
Structure: It is a saturated compound since there are only single covalent bonds present in it. The number of single bonds in the structure of this compound is 18.

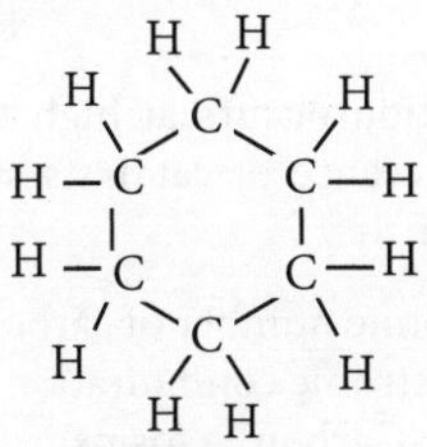

Cyclohexane

7. Hydrocarbons are compounds of carbon and hydrogen atoms only. They are of two types:
   (i) **Saturated hydrocarbons:** Alkanes such as methane.
   General formula: C_nH_{2n+2}
   Structure of methane (CH_4):

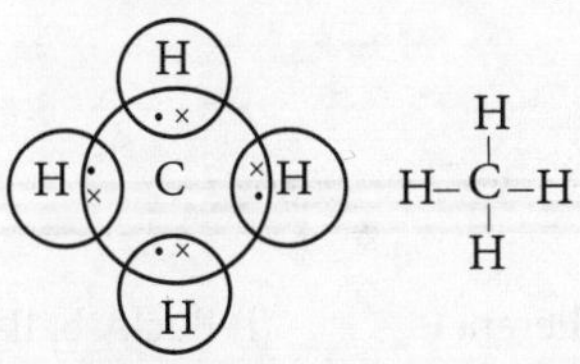

   (ii) **Unsaturated hydrocarbons:** Alkenes and alkynes.
   General formula of alkenes: C_nH_{2n}, for example, ethene.
   General formula of alkynes: C_nH_{2n-2}, for example, ethyne.
   Structure of ethene (C_2H_4):

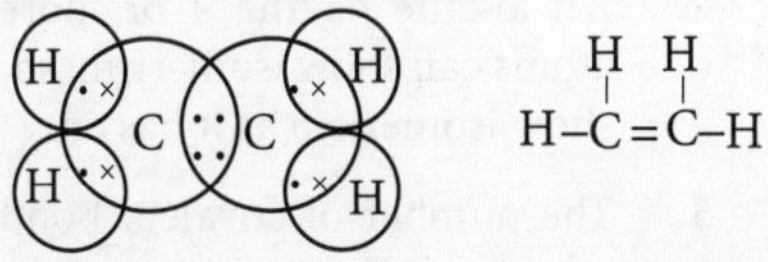

Structure of ethyne (C_2H_2):

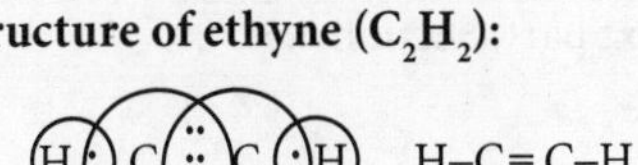

An unsaturated hydrocarbon can be made saturated by hydrogenation of unsaturated hydrocarbon in the presence of nickel or palladium as a catalyst. For example,

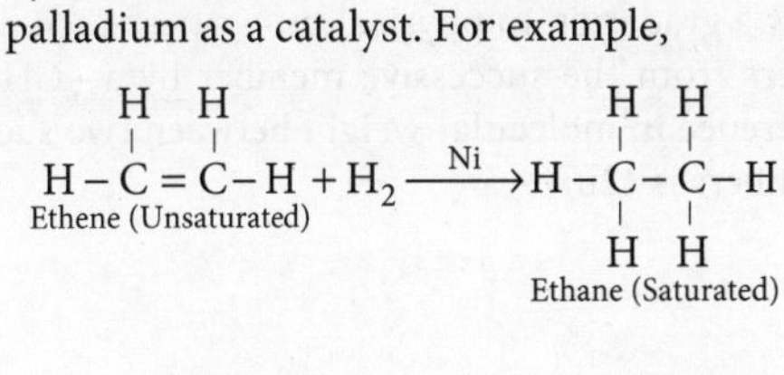

8. (i) **Tetravalency:** Carbon can easily form four covalent bonds with other atoms.
   (ii) **Catenation:** Carbon can easily form covalent bonds with other carbon atoms and the carbon–carbon bonds are more stable than 'other element to same element' bonds formed by other member of group 14.
   (iii) **Formation of multiple bonds:** Carbon can form not only single bonds but it can also form double or triple bonds with same or different elements. Since these properties are not exhibited by other elements of this group, the number of carbon compounds is very large as compared to the other elements of group 14.

Section E: Case Study or Passage Based Questions

1. (d) **2.** (c) **3.** (d) **4.** (b)
5. (d)

High Order Thinking Skills (HOTS) Questions

1. (a) Carbon, (b) carbon dioxide, (c) diamond, (d) for making jewellery, (e) buckminsterfullerene, (f) graphite.

2. (i) Catenation.
   (ii) Carbon forms a large number of compounds due to the following reasons.
   - **Catenation:** Carbon forms bonds with other atoms of carbon.
   - **Tetravalency:** Carbon shares four electrons with other atoms.

   (iii) Structure of cyclohexane:

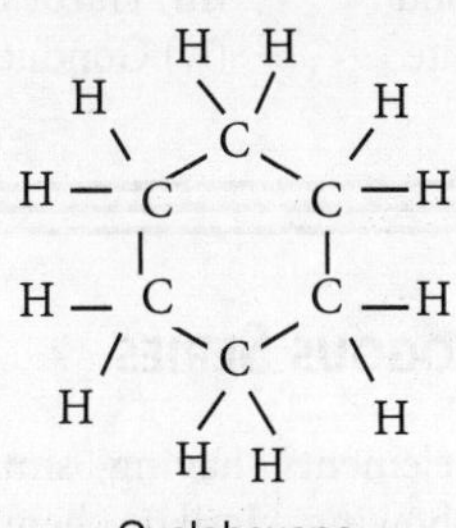

Cyclohexane

   (iv) Tetravalency and catenation are the properties of carbon that helps in the formation of fullerenes. Because of these properties, carbon forms two allotropes—diamond and graphite, out of which, all the four valencies of carbon are satisfied in diamond.

3. Organic compounds having carbon and hydrogen are known as hydrocarbons. Oxides, carbonates, hydrogen carbonates, hydrogen cyanide are not included in hydrocarbons and are considered as inorganic compounds.

	General formula		First member's structure	
Alkanes	C_nH_{2n+2}	where, $n = 1, 2, 3$	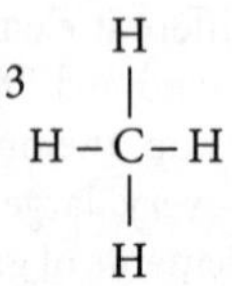	Methane
Alkenes	C_nH_{2n}	where, $n = 2, 3$	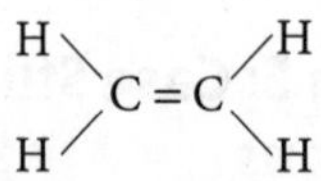	Ethene
Alkynes	C_nH_{2n-2}	where, $n = 2, 3$	$H-C\equiv C-H$	Ethyne

Addition reaction converts alkenes (unsaturated) to alkanes (saturated).

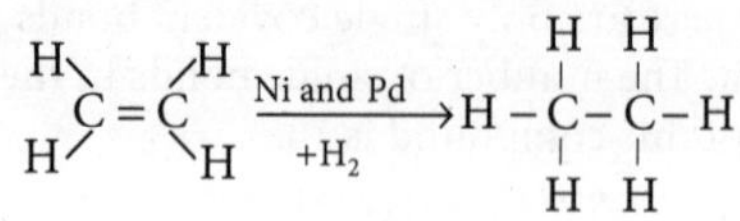

The reaction occurs at high temperature and in the presence of catalysts such as nickel or palladium.

4. (i) Atomic number of carbon = 6.
Electronic configuration of carbon = 2, 4.

(ii) (a) Carbon forms the maximum number of compounds due to covalent bonding.

(b) Carbon compounds have covalent bonds due to which they are poor conductors of electricity. They do not contain ions or free electrons to conduct electricity.

Hints

Section A: Multiple Choice Questions

3. As a carbon atom has 4 valence electrons, it can form 4 covalent bonds by sharing these 4 electrons.

5. It contains both fused 6-membered and 5-membered carbon–carbon rings.

9. The general formula of an alkene is C_nH_{2n}. So, if the number of carbon atoms is 4 (n), the number of hydrogen atoms will be 8 ($2n$). Hence, the alkene is C_4H_8.

17. 1. Diamond (iii) Hardest substance
2. Graphite (iv) Conductor
3. Fullerene (i) Bucky ball structure
4. Natural gas (ii) Contains methane

Section E: Case Study or Passage Based Questions

1. Urea is not a hydrocarbon.

4. An alkane having 4 or more than 4 carbon atoms can show isomerism; so, propane cannot show isomerism as it has only 3 carbon atoms.

5. The number of covalent bonds in ethene and ethyne are 6 and 5 respectively (see structures in text part, Section 4.5).

4.7 Homologous Series

Just as those elements having similar electronic structure and showing similar chemical properties are placed in the same group in the periodic table, all the organic compounds having similar structures and similar chemical properties can be put together in the same group or series—called a *homologous series.*

A homologous series is a series of compounds which are derived from the same general formula, having the same functional group, similar chemical properties and show a gradation in physical properties. Each member differs from the successive member by a $-CH_2-$. The difference in molecular weight between two successive members is 12u.

4.7.1 Features of a Homologous Series

- All members have the same general formula. For example, in the case of alkanes, it is C_nH_{2n+2}.
- All members have the same functional group. For example, in methyl alcohol (CH_3OH) and ethyl alcohol (C_2H_5OH), the functional group is –OH.
- All members have the same general methods of preparation.
- All members have similar chemical properties as the functional group is the same.
- Any two successive members differ in the molecular formula by a $-CH_2$ group and the molecular weight by 14. For example, methane (CH_4, molecular weight $= 12 + 4 \times 1 = 16$) and ethane (C_2H_6, molecular weight $= 12 \times 2 + 6 \times 1 = 30$).
- All members show a gradation in physical properties such as melting and boiling points with an increase in molecular weight. For example, the boiling points of alcohols increases with an increase in molecular weight.

Alkanes:

General formula molecular formula	C_nH_{2n+2} structural formula (where n is the number of carbon atoms)	Condensed structural formula	Name
When $n = 1$, CH_4	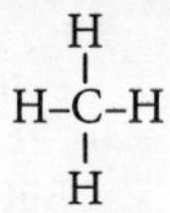	CH_4	Methane
When $n = 2$, C_2H_6	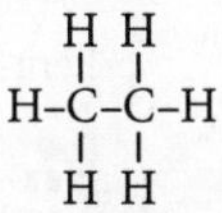	CH_3-CH_3	Ethane
When $n = 3$, C_3H_8	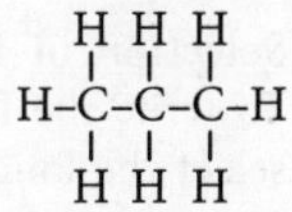	CH_3-CH_2-CH_3	Propane
For $n = 4$, C_4H_{10} has two isomers	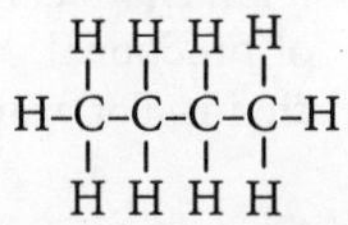	CH_3-CH_2-CH_2-CH_3	n-Butane
	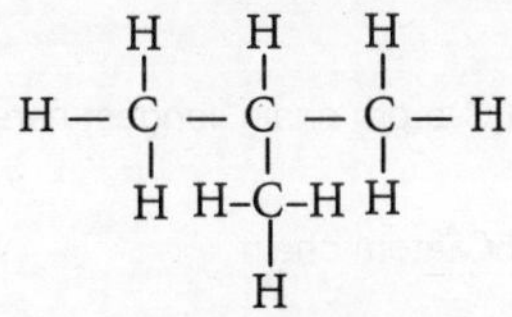	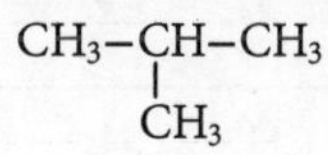	Isobutane
For $n = 5$, C_5H_{12} has three isomers	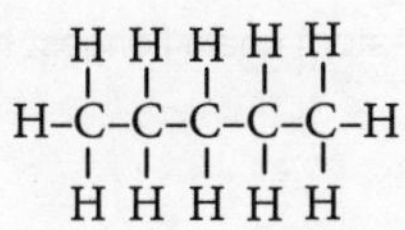	CH_3-CH_2-CH_2-CH_2-CH_3	n-Pentane
	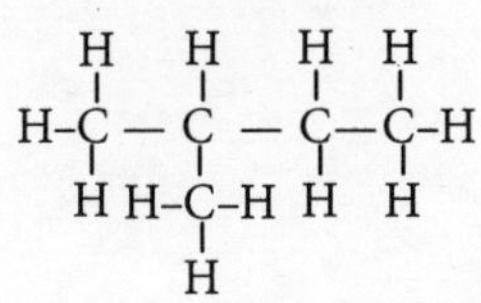	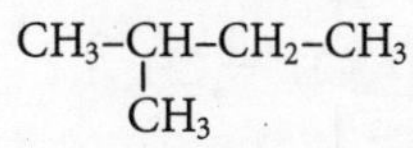	Isopentane
	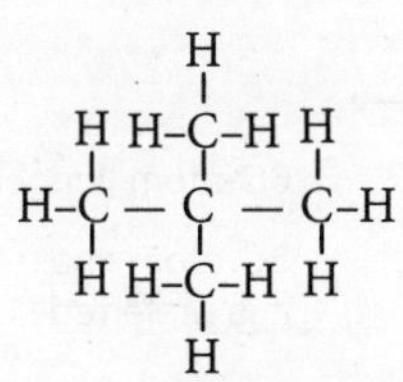	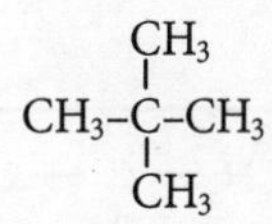	Neopentane

TEST YOUR KNOWLEDGE

4.11 Which of the following belong to the same homologous series?
C_3H_8, C_4H_8, C_4H_6, C_3H_6

Solution: C_4H_8, C_3H_6.

4.12 Write the name and formula of the 2nd member of the homologous series having the general formula C_nH_{2n+2}.

Solution: Name: Ethane; Formula: C_2H_6.

4.13 Write the name and molecular formula of the first member of the homologous series of alkynes. **[CBSE 2015]**

Solution: Ethyne (C_2H_2) is the first member of the homologous series of alkynes.

4.14 Write the name and formula of the second member of homologous series having the general formula C_nH_{2n+2}. **[CBSE 2015]**

Solution: Ethane (C_2H_6) is the second member of the homologous series having the general formula C_nH_{2n+2}.

4.15 Write the name and formula of the second member of the homologous series having the general formula C_nH_{2n}. **[CBSE 2015]**

Solution: Propene (C_3H_6 or $CH_2=CH-CH_3$) is the second member.

4.8 IUPAC System (International Union of Pure and Applied Chemists) Rules of Nomenclature

In order to name an organic compound, the following rules are used one by one.

Carbon chain length	Root word (Alk)	Carbon chain	Root word
C_1	Meth-	C_7	Hept-
C_2	Eth-	C_8	Oct-
C_3	Prop-	C_9	Non-
C_4	But-	C_{10}	Dec-
C_5	Pent-	C_{11}	Undec-
C_6	Hex-	C_{12}	Dodec-

(i) **Selection of the longest chain of C atoms:** To write the IUPAC name, the first step is to select the longest chain, that is, the parent or main chain containing the maximum number of functional groups, multiple bonds and side chains. For example,

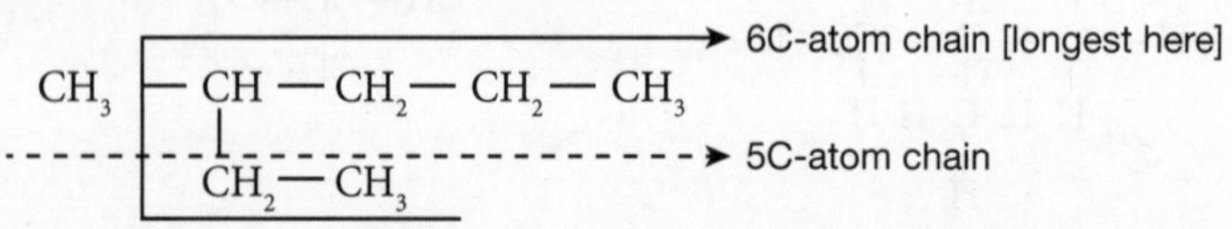

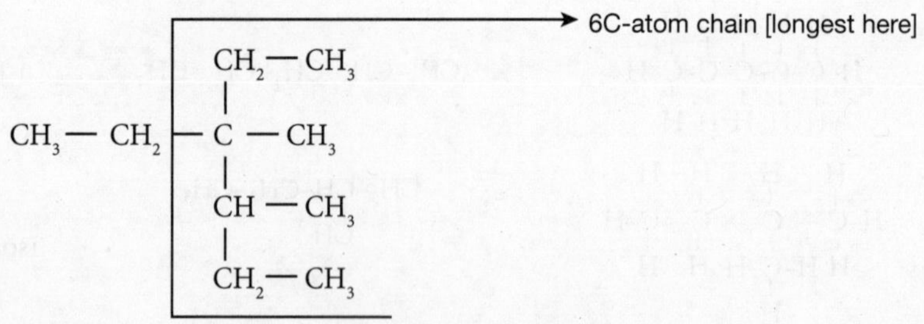

For example,

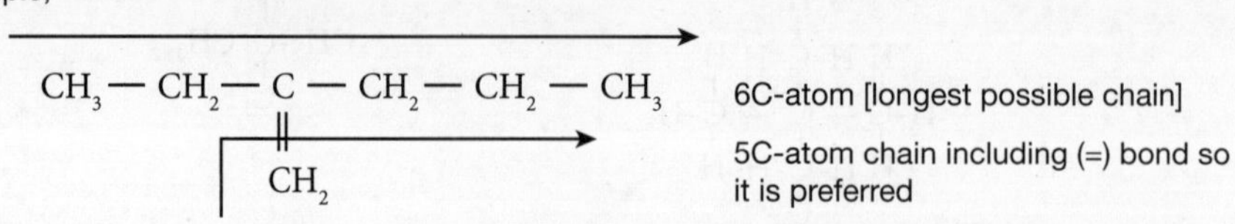

In a given compound, if there are two or more chains having a similar number of C atoms, then the chain with the maximum number of side chains is preferred. For example,

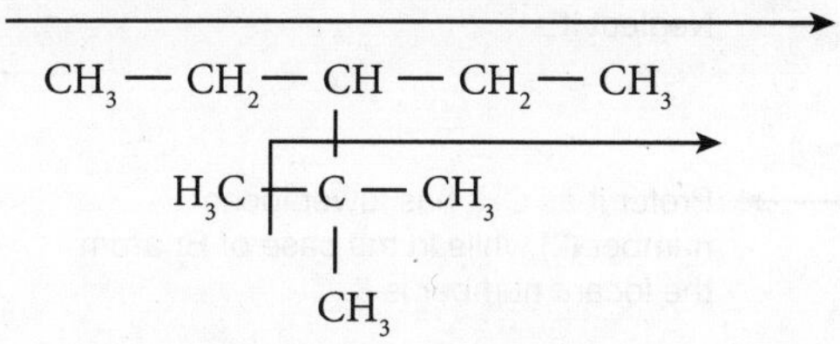

5C- atom chain with one-side chain

5C- atom chain with 3-side chains so it is preferred

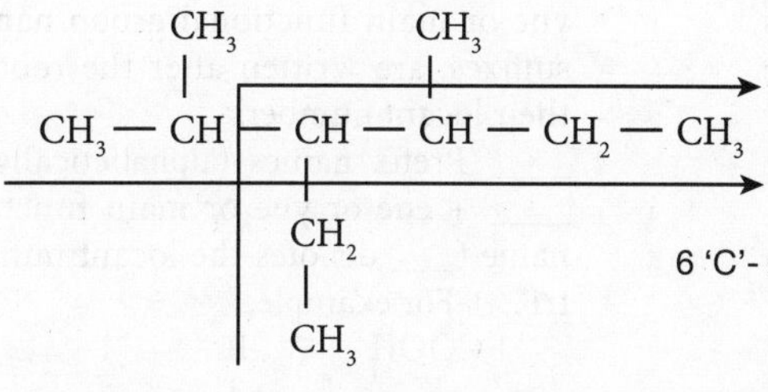

6 'C'- atom chain only two-side chains

6 'C'- atom chain with 3-side so it is preferred here.

(ii) Rules for numbering of C atoms of the chain: For numbering of C atoms of the chain, the priority order is given as follows.

- Functional group > (=) or (≡) > substituent or alkyl group. For example,

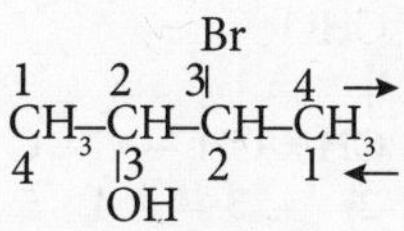

→ Prefer it (as functional group has a lower number)

← Neglect it

- In the case of (=) or (≡) bond, the preference for numbering the parent is given to the one which has a lower value of locant number. For example,

$$\overset{1}{CH_2}=\overset{2}{CH}-\overset{3}{C}\equiv\overset{4}{C}-\overset{5}{CH_2}-\overset{6}{CH_3} \longrightarrow \text{Prefer it}$$

6 5 4 3 2 1 ⟵ Neglect it

$$\overset{1}{CH}\equiv\overset{2}{C}-\overset{3}{CH_2}-\overset{4}{CH}=\overset{5}{CH}-\overset{6}{CH_2}-\overset{7}{CH_3} \longrightarrow \text{Prefer it}$$

7 6 5 4 3 2 1 ⟵ Neglect it

When both (=) and (≡) bond have the same locant numbers, priority is given to the (=) bond. For example,

$$\underset{1}{CH_2}=\underset{2}{CH}-\underset{3}{C}\equiv\underset{4}{CH}$$

Similarly, in the case of substituents and alkyl groups, the chain which has lower locant number values is preferred. For example,

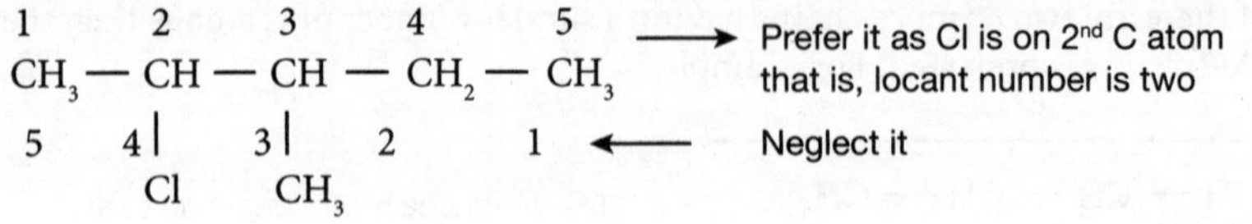

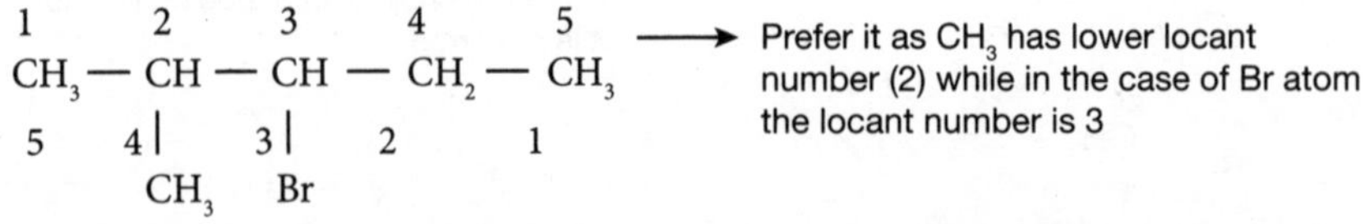

- In case any two substituents or alkyl groups have the same locant numbers, they are named alphabetically.
  For example,

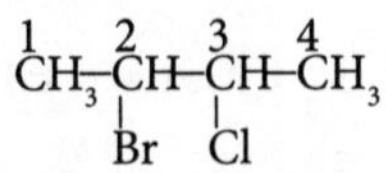

2-bromo-3-chlorobutane [correct]
3-bromo-2-chlorobutane [incorrect]

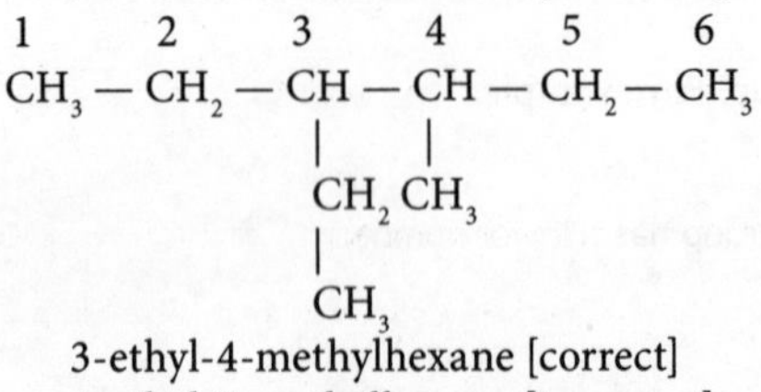

3-ethyl-4-methylhexane [correct]
4-ethyl-3-methylhexane [incorrect]

(iii) Naming of complex alkyl substituents: When a side chain further includes another side chain it is numbered and named as shown. For example,

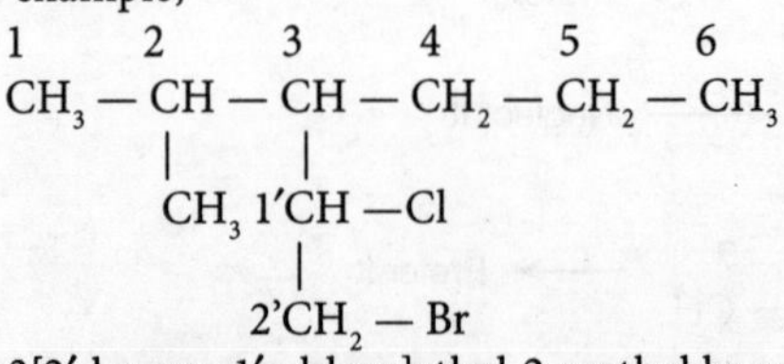

3[2′-bromo-1′-chloro]ethyl-2-methyl hexane

(iv) Prefixes: Use prefixes such as di, tri, tetra… in the case of simple substituents with the locant numbers written the same number of times. For example,

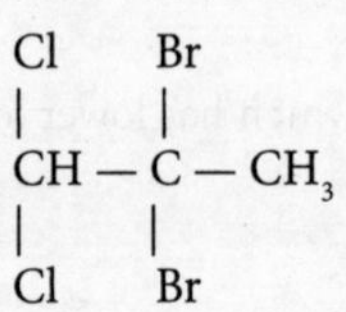

2,2-dibromo-1,1-dichloropropane

(v) All prefixes are written before alk (the root word) as in the above cases. However, ene, yne or main functional group names, that is, suffixes, are written after the root word with their locant numbers

____ Prefix names (alphabetically) + Alk. + ____ + ene or yne or main functional group name (___ denotes the locant number and di, tri …). For example,

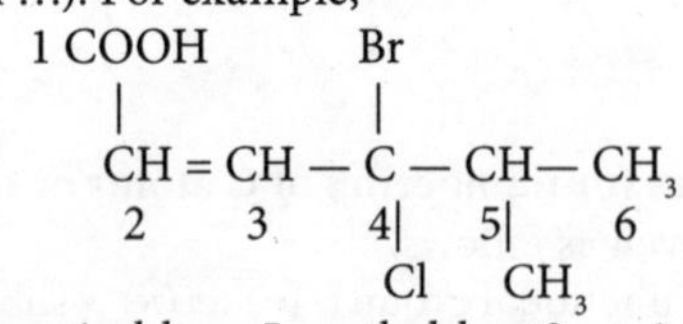

4-bromo-4-chloro-5-methyl-hex-2-en-1-oic acid

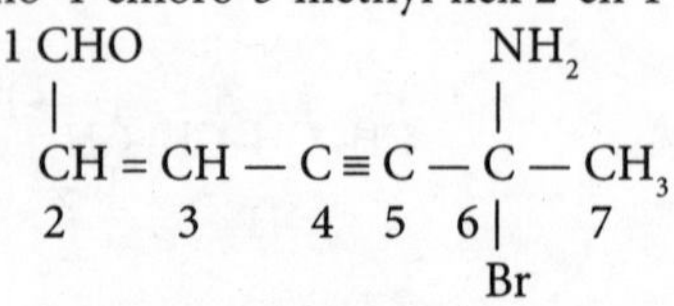

6-amino-6-bromohept-2-ene-4-yne-1-al

(vi) If a compound has many functional groups, the main functional group is named as the suffix while rest are given prefix names. The priority of the functional groups is given in Table 4.1.

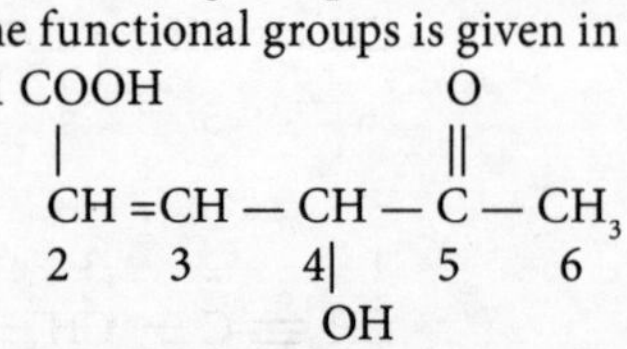

4-hydroxyl-5-keto-hex-2-en-1-oic acid

Table 4.1 Functional group preference table

Prefix name	Functional Group	Suffix name
Carboxy	—COOH	-oic acid
Aldo or formyl	—CHO	-al
Keto or oxo	>C=O	-one
Hydroxy	—OH	-ol
Amino	$—NH_2$	-amine

4.9 Examples of Nomenclature

4.9.1 Hydrocarbons

(i) **Alkanes:** They have general formula C_nH_{2n+2} where *n* is the number of carbon atoms and they are named as alkanes. Examples are given in Table 4.2.

Table 4.2 Examples of nomenclature of alkanes

General formula, molecular formula	C_nH_{2n+2} Structural formula (where *n* is the number of carbon atoms)	Condensed structural formula	IUPAC name
For *n* = 4, C_4H_{10} has two isomers	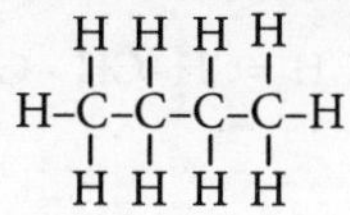	CH_3–CH_2–CH_2–CH_3	Butane
	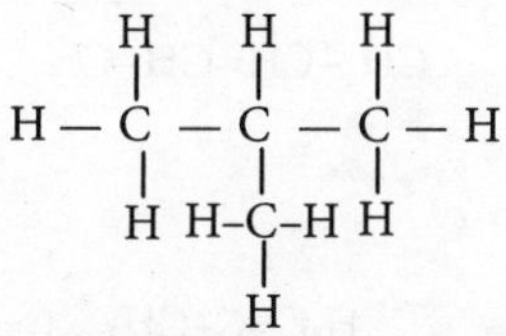	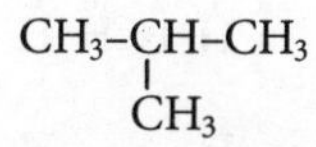	2-methylpropane
For *n* = 5, C_5H_{12} has three isomers	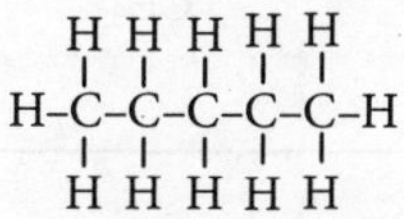	CH_3–CH_2–CH_2–CH_2–CH_3	Pentane
	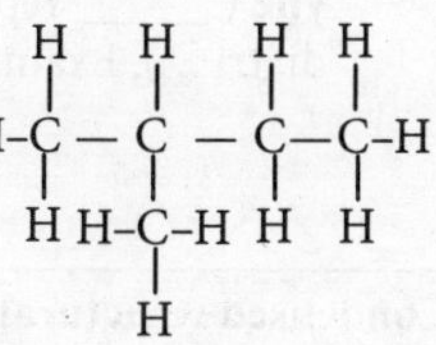	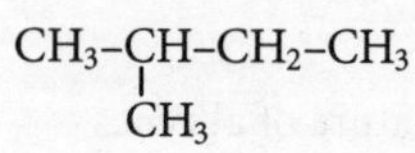	2-methylbutane
	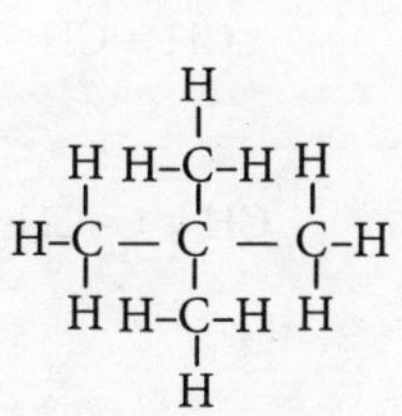	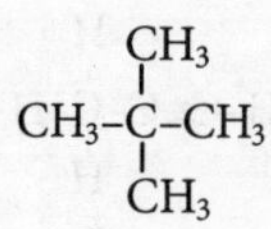	2,2-dimethylpropane

(ii) **Alkenes:** They have general formula C_nH_{2n} where *n* is the number of carbon atoms and they are named as _______ **prefix + alk + ____ ene** (____ represents locant number and di, tri…). Examples are given in Table 4.3.

Table 4.3 Examples of nomenclature of alkenes

Molecular formula	Structural formula	Condensed structural formula	IUPAC name
$n = 2$, C_2H_4	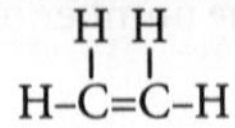	$CH_2=CH_2$	Ethene
$n = 3$, C_3H_6	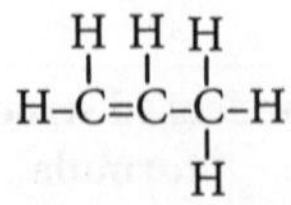	$CH_2=CH-CH_3$	Propene
$n = 4$, C_4H_8 has three isomers	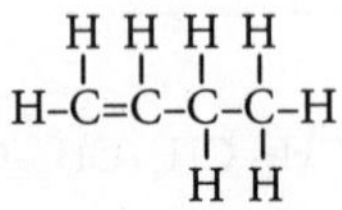	$CH_2=CH-CH_2-CH_3$	But-1-ene
	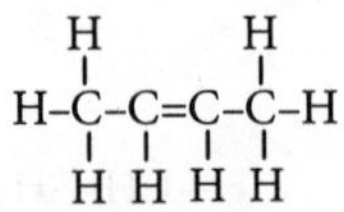	$CH_3-CH=CH-CH_3$	But-2-ene
	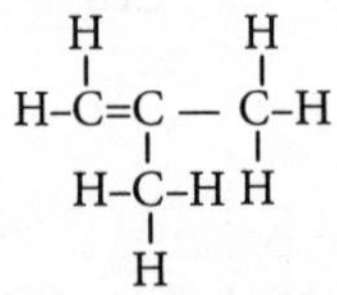	$H_2C=C(CH_3)-CH_3$	2-Methylpropene

(iii) **Alkynes:** Their general formula is C_nH_{2n-2} and they are named as ____ **prefix + alk +** ____ **yne** (_____ represents the locant number and di, tri ...). Examples are given in Table 4.4.

Table 4.4 Examples of nomenclature of alkynes

Molecular formula	Structural formula	Condensed structural formula	IUPAC name
$n = 2$, C_2H_2	$H-C\equiv C-H$	$CH\equiv CH$	Ethyne
$n = 3$, C_3H_4	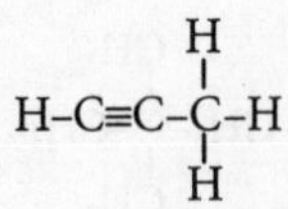	$CH\equiv C-CH_3$	Propyne
$n = 4$, C_4H_6 has two isomers	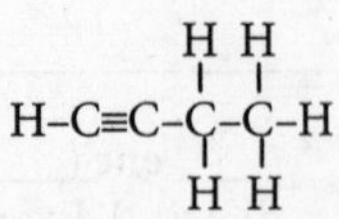	$CH\equiv C-CH_2-CH_3$	But-1-yne

(Continued)

Table 4.4 (Continued)

Molecular formula	Structural formula	Condensed structural formula	IUPAC name
	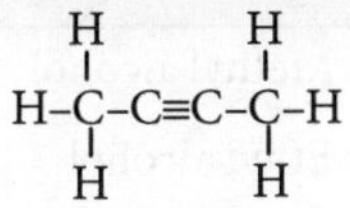	$CH_3-C \equiv C-CH_3$	But-2-yne
$n = 5$, C_5H_8 has three isomers	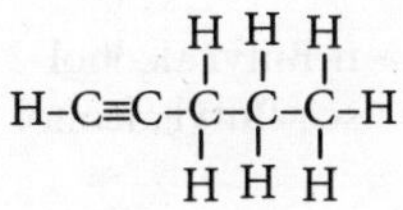	$CH \equiv C-CH_2-CH_2-CH_3$	Pent-1-yne
	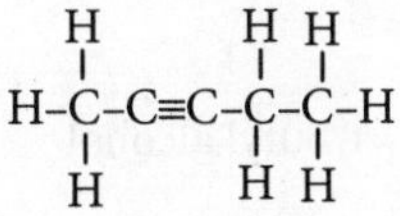	$CH_3-C \equiv C-CH_2-CH_3$	Pent-2-yne

4.9.2 Alkyl halides (Haloalkanes)

General formula is $C_nH_{2n+1}X$ (where X is Cl, Br, I, F) and they are named as in Table 4.5.

Table 4.5 Names of alkyl halides

Molecular formula	Structural formula	Common name	IUPAC name
$n = 1$ CH_3Cl	CH_3Cl	Methyl chloride	Chloromethane
$n = 2$ C_2H_5Cl	CH_3CH_2Cl	Ethyl chloride	Chloroethane
$n = 3$ C_3H_7Cl	$CH_3CH_2CH_2Cl$	*n*-Propyl chloride	1-Chloropropane
$n = 4$ C_4H_9Cl	$CH_3CH_2CH_2CH_2Cl$ $CH_3CHClCH_2CH_3$	n-Butyl chloride sec-Butyl chloride	1-Chlorobutane 2-Chlorobutane

Competition Edge

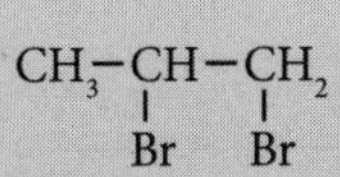

1,2-di-bromo propane

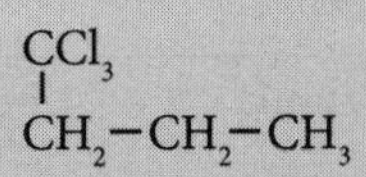

Trichlorobutane
or
1,1,1-trichlorobutane

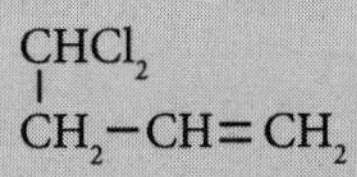

4,4-dichlorobutene-1

4.9.3 Alcohols (R-OH)

The general formula of alcohols is R–OH where R is an alkyl group and –OH is the functional group; these are named as alkanols.

Alkan + ol → Alkanol

The name of the alcohol is derived by replacing the –e in the name of the alkane from which it is derived by the suffix –ol. For example, methanol (CH_3OH), an alcohol, is derived by substituting the H of methane by –OH. Examples are given in Table 4.6.

Table 4.6 Examples of nomenclature of alcohols

Alkanes	Formula of alcohol	Common name	IUPAC name
CH_4 (Methane)	CH_3OH	Methyl alcohol	Methanol
C_2H_6 (Ethane)	C_2H_5OH	Ethyl alcohol	Ethanol
C_3H_8 (Propane)	C_3H_7OH	Propyl alcohol	Propanol
C_4H_{10} (Butane)	C_4H_9OH $CH_3CH_2CH_2CH_2OH$ $CH_3CHOHCH_2CH_3$	n-Butyl alcohol sec-Butyl alcohol	Butanol Butan-2-ol
	$CH_3-C(CH_3)_2-OH$	t-Butyl alcohol	2-methylpropan-2-ol

Competition Edge

$(CH_3)_2CH-OH$ — Propan-2-ol

$CH_2(OH)-CH(OH)-CH_2(OH)$ — Propan-1,2,3-triol

$CH_3-CH=CH-CH(Cl)-CH_2OH$ — 2-Chloro-pent-3-en-1-ol

4.9.4 Aldehydes (R-CHO)

Their general formula is $C_nH_{2n+1}-\overset{O}{\overset{\|}{C}}-H$. Aldehydes are named by replacing –e from the name of alkane by the suffix –al and ketones are named by replacing –e of alkane by the suffix –one.

Alkanals: Alkan + al → Alkanal

Examples are given in Table 4.7.

Table 4.7 Examples of nomenclature of aldehydes

Molecular formula	Structural formula	Common name	IUPAC name
$n = 0$ HCHO	$H-\overset{O}{\overset{\|}{C}}-H$	Formaldehyde	Methanal
$n = 1$ CH_3CHO	$CH_3-\overset{O}{\overset{\|}{C}}-H$	Acetaldehyde	Ethanal
$n = 2$ C_2H_5CHO	$CH_3CH_2-\overset{O}{\overset{\|}{C}}-H$	Propionaldehyde	Propanal
$n = 3$ C_3H_7CHO	$CH_3CH_2CH_2-\overset{O}{\overset{\|}{C}}-H$	Butyraldehyde	Butanal

4.9.5 Ketones ($R-\overset{O}{\overset{\|}{C}}-R$)

Their general formula is $C_nH_{2n+1}-\overset{O}{\overset{\|}{C}}-C_nH_{2n+1}$ and these are named as alkanones.

Alkane + one → Alkanone

Examples are given in Table 4.8.

Table 4.8 Examples of nomenclature of ketones

Molecular formula	Structural formula	Common name	IUPAC name
$n = 1$ CH_3COCH_3	$CH_3-\overset{O}{\overset{\|}{C}}-CH_3$	Acetone	Propan-2-one
$n = 1, 2$ $CH_3COC_2H_5$	$CH_3-\overset{O}{\overset{\|}{C}}-CH_2-CH_3$	Ethylmethylketone	Butan-2-one
$n = 1,3$ $CH_3COCH_2CH_2CH_3$	$CH_3-\overset{O}{\overset{\|}{C}}-CH_2-CH_2-CH_3$	Methylpropylketone	Pentan-2-one
$n = 1,4$ $CH_3COCH_2CH_2CH_2CH_3$	$CH_3-\overset{O}{\overset{\|}{C}}-CH_2-CH_2-CH_2-CH_3$	Butylmethylketone	Hexan-2-one

4.9.6 Carboxylic Acids (R-COOH)

The compounds containing the carboxyl (-COOH) group are known as carboxylic acids. Their general formula is $C_nH_{2n+1}-COOH$. They are named by substituting the –e of the corresponding alkane by –oic acid. These are named as alkanoic acids. Alkan + oic acid → Alkanoic acid. Examples are given in Table 4.9.

Table 4.9 Examples of nomenclature of carboxylic acids

Molecular formula	Structural formula	Common name	IUPAC name
$n = 0$ HCOOH	$H-\overset{O}{\overset{\|}{C}}-OH$	Formic acid	Methanoic acid
$n = 1$ CH_3COOH	$CH_3-\overset{O}{\overset{\|}{C}}-OH$	Acetic acid	Ethanoic acid
$n = 2$ C_2H_5COOH	$CH_3-CH_2-\overset{O}{\overset{\|}{C}}-OH$	Propionic acid	Propanoic acid
$n = 3$ C_3H_7COOH	$CH_3-CH_2CH_2-\overset{O}{\overset{\|}{C}}-OH$	Butyric acid	Butanoic acid

TEST YOUR KNOWLEDGE

4.16 Write the IUPAC names of the following compounds.

(i) $CH_3-CH=CH-CH=CH_2$

(ii) $CH_3-C\equiv C-CH_2-CH_3$

(iii) $CH_2=CH-C\equiv C-CH_3$

Solution: (i) $\underset{5}{CH_3}-\underset{4}{CH}=\underset{3}{CH}-\underset{2}{CH}=\underset{1}{CH_2}$

Penta-1,3-diene

(ii) $\underset{1}{CH_3}-\underset{2}{C}\equiv\underset{3}{C}-\underset{4}{CH_2}-\underset{5}{CH_3}$

Pent-2-yne

(iii) $\underset{1}{CH_2}=\underset{2}{CH}-\underset{3}{C}\equiv\underset{4}{C}-\underset{5}{CH_3}$

Pent-1-ene-3-yne

4.17 Write the IUPAC name and structure of an alcohol with four carbon atoms in its molecule. **[Board, 2016]**

Solution: 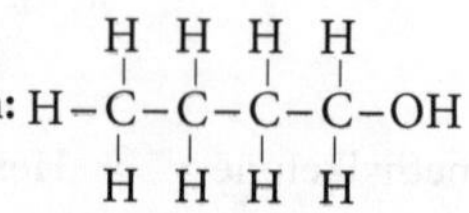

or $CH_3-CH_2-CH_2-CH_2-OH$

It is butanol-1 or butanol.

4.18 Write the IUPAC name and structure of an alcohol with five carbon atoms in its molecule.

Solution: 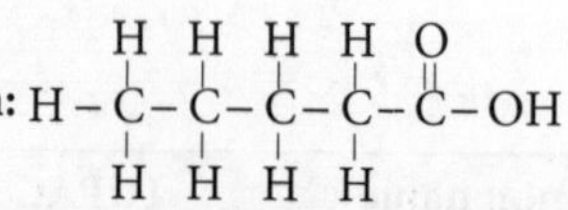

$CH_3-CH_2-CH_2-CH_2-COOH$

Pentan-1-oic acid or pentanoic acid

4.10 Coal and Petroleum

Fuel: A fuel is a material that has energy stored inside it and when it is burned, the energy is released mainly as heat along with some light. This heat energy can be used for various purposes such as cooking food, heating water and for generation of electricity in thermal power stations, machines in factories and engines of motor vehicles. Most of the common fuels are either free carbon (coke, coal and charcoal) or carbon compounds like petrol, diesel, LPG, CNG and natural gas.

Fig. 4.12 Coal and Petroleum

Source: CSIRO, CC BY 3.0 <https://creativecommons.org/licenses/by/3.0>, via Wikimedia Commons; Netojinn, CC BY-SA 4.0 <https://creativecommons.org/licenses/by-sa/4.0>, via Wikimedia Commons

When carbon in any form (coal, coke, charcoal) is burned in oxygen (air), it forms carbon dioxide gas and releases a large amount of heat and some light.

$$\underset{\text{Carbon (Coal, coke or charcoal)}}{C} + \underset{\text{Oxygen (From air)}}{O_2} \rightarrow \underset{\text{Carbon dioxide}}{CO_2} + \text{Heat} + \text{Light}$$

Most of the fuels that are used today are obtained from coal, petroleum and natural gas. Actually, coal, petroleum and natural gas are called fossil fuels. This is because they were formed by the decomposition of the remains of the pre-historic plants and animals (fossils buried under the earth long long ago).

4.10.1 Coal and its Formation

Coal is a complex mixture of compounds of carbon, hydrogen and oxygen and some free carbon. Fractional amounts of nitrogen and sulfur compounds are also present in coal.

Coal was formed due to the decomposition of large land plants and trees buried under the earth millions of years ago. It is supposed that millions of years ago, due to earthquakes and volcanoes, the forests were buried under the surface of the earth and got covered with sand, clay and water. As there is high temperature and high pressure inside the earth, in the absence of air, wood was converted into coal.

4.10.2 Petroleum and its Formation

Petroleum is a dark coloured, viscous and foul-smelling crude oil. The word petroleum means rock oil (petra

= rock; oleum = oil). It is known as petroleum as it is found under the crust of earth trapped in rocks. The crude petroleum is a complex mixture of several solid, liquid and gaseous hydrocarbons mixed with water, salts and earth particles. Hence, crude oil is not a single chemical compound but it is a mixture of compounds. The fuels such as petrol, kerosene, diesel and LPG are obtained from petroleum.

Petroleum and natural gas have been formed by the decomposition of the remains of extremely small plants and animals buried under the sea millions of years ago. It is supposed that millions of years ago, the microscopic plants and animals which lived in the seas died due to whatever reason. Their bodies sank to the bottom of the sea and were soon covered with mud and sand. The chemical effects of pressure, heat and bacteria converted the remains of these microscopic plants and animals into petroleum and natural gas, in a similar way as forest trees were converted into coal. This conversion took place in the absence of oxygen or air. The petroleum thus formed got trapped between two layers of impervious rocks, that is, non-porous rocks forming an oil trap. Natural gas lies above this petroleum oil.

As coal and petroleum contain some nitrogen and sulfur, when coal and petroleum fuels are burnt, they lead to the formation of oxides of nitrogen (NO_2) and sulfur (SO_2) which go into the air and cause air pollution.

4.10.3 Why do Substances Burn With a Flame or Without a Flame?

All of us have seen the burning of a candle flame and LPG flame. So, what exactly is a flame? A flame is the region where combustion or burning of gaseous substances takes place. So, a flame is produced only when gaseous substances burn. All the gaseous fuels burn with a flame but only those solid and liquid fuels which vaporise on heating to form a gas or vapour can burn with a flame. For example, cooking gas or LPG (Liquified Petroleum Gas) is a gaseous fuel which burns with a flame while wax (solid), kerosene (liquid) which vaporise or form gas on heating burn with a flame.

Flames are of two types—blue and yellow flame. When fuels burn, the type of flame produced depends on the proportion of oxygen or air which is available for the burning or combustion of fuel.

(i) **When the oxygen supply is sufficient:** When there is sufficient air or oxygen, the fuels can burn completely giving a blue flame. This blue flame cannot produce much light, so it is said to be a non-luminous flame, that is, a flame that gives no light. For example, in a gas stove, cooking gas or LPG burns with a blue, non-luminous flame because the gas stove has holes or inlets for air to mix property with cooking gas. As the cooking gas gets sufficient oxygen from this air, it burns completely giving a blue flame on complete combustion.

(ii) **When the oxygen supply is not sufficient:** When the air or oxygen is insufficient, the fuel burns incompletely, giving mainly a yellow flame. The yellow colour of flame is caused by the glow of hot, unburnt carbon particles produced due to the incomplete combustion of fuel. This yellow flame produces light, so it is known to be a luminous or light giving flame. For example, when wax is burned in the form of a candle, it burns with a yellow luminous flame because the candle wax melts and rises up the wick and gets converted into vapour. In a candle, the proper mixing of oxygen with the wax vapour does not occur. So, only incomplete combustion occurs giving small unburnt carbon particles. These solid carbon particles rise in the flame, get heated and glow to give a yellowish light along with soot and smoke.

(iii) **Solid and liquid fuels which cannot vaporise on heating:** In such a case these solids burn without producing a flame. For example, coal and charcoal burn in an *angithi* without giving a flame. Coal and charcoal contain some volatile substances and when these are ignited, the volatile substances present in them vaporise and they burn with a flame. Once all the volatile substances present in them get burnt, the remaining coal or charcoal just glows red and gives heat without giving any flame (Fig. 4.13).

Fig. 4.13 Burning of coal

TEST YOUR KNOWLEDGE

4.19 Give two examples of each of fuels (i) which burn with a flame and (ii) which burn without a flame.

Solution:

(i) LPG (cooking gas) and kerosene are fuels which burn with a flame.

(ii) Coal and charcoal are two fuels which do not burn with a flame.

4.20 Why are coal and petroleum categorised as natural resources? Give a reason as to why they should be used judiciously.

Solution: Coal and petroleum are known as natural resources as they were formed by the decomposition of the remains of the plants and animals buried under the earth millions of years ago. These should be used judiciously since (i) these resources are not renewable and will get exhausted one day and (ii) will not be available to future generations.

4.21 Does Compressed Natural Gas (CNG) (used these days in automobiles) undergo spontaneous or rapid combustion?

Solution: CNG undergoes rapid combustion in an automobile's internal combustion engine; it does not ignite by itself.

Quick Review

- A **homologous series** is a series of compounds which are derived from the same general formula, having the same functional group, with similar chemical properties and show a gradation in physical properties. Each member differs from the successive member by $-CH_2-$. The difference in molecular weight between two successive members is 12 u.
- **Alkanes** have a general formula C_nH_{2n+2} where *n* is the number of carbon atoms.
- **Alkenes** have a general formula C_nH_{2n} where *n* is the number of carbon atoms. They are named as ____ **prefix + alk +** ____ **ene** (____ represents the locant number and di, tri ...).
- **Alkynes** have a general formula C_nH_{2n-2} and they are named as ____ **prefix + alk +** ___ **yne.**
- **Alcohols** have a general formula R–OH where R is an alkyl group and –OH is the functional group. They are named as alkanols. Alkan + ol → Alkanol.
- **Ethers** are alkoxy derivatives of alkanes (R —O —R) having a general formula $C_nH_{2n+2}O$ and they are named as alkoxy alkanes.
- **Aldehydes [R —CHO]** have a general formula $C_nH_{2n+1}-CHO$. They are named by replacing –e from the name of alkane by the suffix –al.
- **Ketones** (R – CO – R) have a general formula $C_nH_{2n+1}-CO-C_nH_{2n+1}$ and these are named as alkanones. Alkane + one → Alkanone.
- **Carboxylic acids (R—COOH)** are compounds containing the carboxyl (-COOH) group and their general formula is C_nH_{2n+1}-COOH. They are named by substituting –e of the corresponding alkane by –oic acid. These are named as alkanoic acids. Alkan + oic acid → Alkanoic acid.
- A **fuel** is a material containing stored energy. When a fuel is burned, energy is released mainly as heat along with some light. This heat energy can be used for various purposes such as cooking food and heating water.
- **Petroleum** and **natural gas** were formed by the decomposition of the remains of extremely small plants and animals buried under the sea millions of years ago.
- The process of burning of a carbon compound in air to give CO_2, water, heat and light, is called **combustion**.

Exercise-2

Section A: Multiple Choice Questions (1 Mark)

1. In a homologous series the members have
(a) the same functional group
(b) the same general formula
(c) the same molecular weight
(d) both (a) and (b)

2. Two successive members in a homologous series differ in molecular weight by
(a) 12 (b) 14
(c) 16 (d) 18

3. C_3H_8 belongs to the homologous series of
(a) alkynes (b) alkenes
(c) alkanes (d) cycloalkanes

4. The first member of the alkyne homologous series is
(a) propyne (b) ethyne
(c) methane (d) ethene

5. Methane, ethane and propane are said to form a homologous series because all are
(a) hydrocarbons
(b) saturated compounds
(c) aliphatic compounds
(d) differ from each other by a CH_2 group

6. Which of the following belongs to the homologous series of alkynes?

C_6H_6, C_2H_6, C_2H_4, C_3H_4

(a) C_6H_6 (b) C_2H_4
(c) C_2H_6 (d) C_3H_4

7. Which of these functional groups can combine with carbon to produce alcohol?
(a) 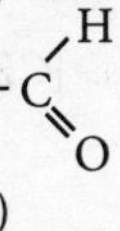
(c) 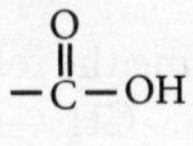
(b) – OH
(d) 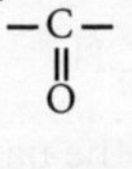

8. Name the functional group present in CH_3COCH_3.
(a) alcohol (b) carboxylic acid
(c) ketone (d) aldehyde

9. The IUPAC name of

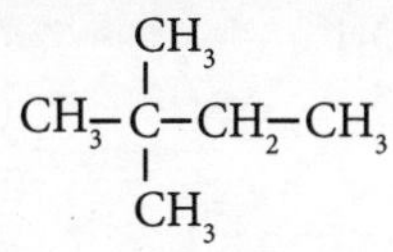

is
(a) 2-ethyl-2-methylpropane
(b) 2, 2-demethylbutane
(c) 1,1,1-trimethylpropane
(d) 2, 2-methylbutane

10. Which of the following is the formula of butanoic acid?
(a) $CH_3CH_2CH_2CH_2COOH$
(b) $COOH - CH_2 - CH_2 - CH_2 - CH_3$
(c)

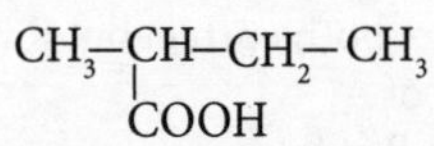

(d) $CH_2 - CH_2 - CH_2 - COOH$

11. Give the IUPAC name of $CH_3COOC_2H_5$.
(a) ethylethanoic acid
(b) butanoate
(c) ethylethanoate
(d) ethylmethylcarboxylic acid

12. Give the IUPAC name of $CH_3COCH_2CH_2CH_3$.
(a) pentanone (b) pentanone-2
(c) pentanone-4 (d) pentanal

13. The correct IUPAC name of $CH_3CHClCH_2CH_2CHO$ is
(a) 2-chloropentanal
(b) 2-chloropentanol
(c) 4-chloropentanal
(d) 4-chloropentanol

14. Select the option having the set of correct statements.
(i) Two successive members in a homologous series differ by a $-CH_2$ group.
(ii) Members of a homologous series may differ in physical properties.
(iii) The functional group decides the physical properties only.
(iv) During numbering, the longest chain of C atoms is selected.
(a) (i), (ii) (b) (ii), (iii)
(c) (i), (ii), (iii) (d) (i), (ii), (iv)

15. Match the following.

Column I (Functional group)	Column II (Name)
1. –COOH	(i) Alcohol
2. –CHO	(ii) Ketone
3. –CO–	(iii) Acid
4. –OH	(iv) Aldehyde

(a) (1)—(ii), (2)—(iii), (3)—(i), (4)—(iv)
(b) (1)—(iii), (2)—(iv), (3)—(ii), (4)—(i)
(c) (1)—(ii), (2)—(iv), (3)—(i), (4)—(iii)
(d) (1)—(iii), (2)—(iv), (3)—(i), (4)—(ii)

Assertion Reason Questions

Direction: In the following question, two statements (Assertion) A and Reason (R) are given. Mark

(a) if A and R are both correct and R is the correct explanation of A;
(b) if A and R are both correct but R is not the correct explanation of A;
(c) A is true but R is false;
(d) A is false but R is true.

Assertion	Reason
1. Members of a homologous series must have the same functional group.	**1.** Successive members differ in their molecular formula by a $-CH_2$ group.
2. C_2H_2 and C_3H_6 are successive homologues.	**2.** C_nH_{2n} is the general formula of alkenes.
3. During numbering, a functional group is given priority over a multiple bond.	**3.** Unsaturated hydrocarbons have multiple bonds.
4. CH_3CH_2CHO and CH_3COCH_3 have the same functional groups.	**4.** Compounds having the same functional group have the same chemical properties.
5. Coal and petroleum are fossil fuels.	**5.** They are formed by the decomposition of the remains of animals and plants buried under the earth.
6. LPG burns with a blue flame.	**6.** A gas stove has holes so that sufficient air is available for burning.
7. Unsaturated hydrocarbons burn with a yellow flame with lots of black smoke.	7. The sooty deposit is due to unburnt carbon produced due to incomplete combustion.

Section B: Very Short Answer Questions (2 Marks)

1. What is a homologous series of carbon compounds? **[CBSE 2016]**

2. Write the molecular formula of the first two members of the homologous series having the functional group –Cl. **[CBSE 2017]**

3. Write the molecular formula of the 2nd and 3rd members of the homologous series whose first member is methane. **[CBSE 2017]**

4. Write the molecular formula of the first two members of the homologous series having the functional group –Br.

5. Write the molecular formula of the first two members of the homologous series having the functional group –OH.

6. Write the molecular formula of the 2nd and 3rd members of the homologous series whose first member is ethene.

7. Write the molecular formula of the 2nd and 3rd members of the homologous series where the first member is ethyne.

8. Write the next homologue of each of the following.
(i) C_2H_4, (ii) C_4H_6 **[CBSE, 2016]**

9. Write the names of each of the following functional groups.
(i) –OH (ii) $-\underset{\displaystyle O}{\underset{\|}{C}}-$

10. Write the molecular formula of the alcohol that can be derived from butane. **[CBSE 2016]**

11. Write the name and structure of an alcohol with three carbon atoms in its molecule. **[CBSE 2016]**

12. Name the following compounds. **[CBSE 2016]**
(i) $CH_3—CH_2—OH$
(ii) $CH_3-\overset{\displaystyle H}{\overset{|}{C}}=O$

13. Write the names of the following compounds.
(a) HCOOH
(b) $CH_3COCH_2CH_3$.

14. Write the name and formula of the 2nd member of the series of carbon compounds whose general formula is $C_nH_{2n+1}OH$.

15. Name the following compounds. **[Board, 2016]**

(i) $CH_3–CH_2–OH$, (ii) $CH_3–\overset{\overset{H}{|}}{C}=O$

16. Write down structural formula of:
(a) propanone (b) hexanal

17. Name the simplest ketone.

18. What is the common name of CH_3COOH?

19. In an organic compound, which part largely determines its physical and chemical properties?

20. Is combustion an exothermic or endothermic process?

21. Why are carbon and its compounds very good fuels?

22. Name the main components present in natural gas and LPG.

23. When a candle burns, what colour of luminous flame is possible?

24. Why do gas stoves have inlets for air? **[Board, 2015]**

Case Study or Passage Questions

I. The chemical properties of organic compounds are decided on the basis of functional groups present in them. When organic compounds having the same functional group are placed in a series having members in the increasing order of number of carbon atoms, a homologous series is formed. Members of a homologous series show similar chemical properties but differ in their physical properties as their molecular weights differ. There is a general formula for a particular homologous series; for example, for alkanes, the general formula is C_nH_{2n}.

25. A group of organic compounds having similar structures and similar chemical properties in which the successive compounds differ by a CH_2 group is called a ________.

26. Write next two members of the homologous series: C_2H_6, C_3H_8.

27. Write the name and formula of the 2nd member of the homologous series having the general formula C_nH_{2n}.

28. Write the name and molecular formula of the first member of the homologous series of alkynes. **[Board, 2015]**

29. Write the name and formula of the 2nd member of the homologous series having the general formula C_nH_{2n-2}. **[CBSE 2015]**

II. A functional group is the site of chemical reactivcity in an organic compound and its chemical properties are decided on the basis of functional group present in it. There are many functional groups such as –OH (alcohol), –CHO (aldehyde), –CO– (ketone), –COOH (acid) and so on. There is a specific name for every functional group which is used for naming an organic compound. For example, for the –OH and –CHO groups, the suffixes used at the end of the IUPAC name are –ol and –al respectively.

30. Name the functional group present in
(i) CH_3CHO
(ii) C_2H_5COOH

31. Write the IUPAC name and structure of an alcohol with three carbon atoms in its molecule. **[Board, 2016]**

32. Write the IUPAC name and structure of an aldehyde with four carbon atoms in its molecule. **[Board, 2016]**

33. Name the functional group present in CH_3COCH_3 and give the IUPAC name of this compound.

34. Write the chemical names of $CH_3CH_2COCH_3$, C_2H_5COOH.

Section C: Short Answer Questions (3 Marks)

1. (i) What is a functional group?
(ii) State two properties of carbon which lead to the huge number of carbon compounds we see around us.

2. Identify the functional groups in the following compounds.
(a) HCHO (b) CH_3COOH
(c) CH_3CH_2OH (d) CH_3COCH_3

3. An acid as well as an ester can be represented by the same molecular formula, say $C_2H_4O_2$. Write their structures and name them. State the relation between the two in the language of science.

4. What is the difference between two consecutive members in a homologous series in alkanes in terms of:
(i) molecular mass
(ii) number of atoms of elements

5. The molecular formula of A is $C_{10}H_{18}$ and B is $C_{18}H_{36}$. Name the homologous series to which they belong.

6. Draw the structures of the following compounds and identify the functional group present in them:
(i) butanoic acid
(ii) bromopropane **[Board, 2015]**

7. State the meaning of functional group in a carbon compound. Write the functional group present in (i) ethanol and (ii) ethanoic acid. Draw their structures.

8. Write the name and general formula of a chain of hydrocarbons in which an addition reaction with hydrogen can take place. Stating the essential conditions required for an addition reaction to occur, write the chemical equation giving the name of the reactant and the product of such a reaction.

9. (i) Write the names of the following compounds.
CH_3CH_2COOH, C_6H_6
(ii) What is a homologous series? Write the formula of the functional group of a ketone and an aldehyde.

10. Write the IUPAC names of the following.
(i) $CH_3CH_2-C\equiv CH$ (ii) CH_3CH_2OH
(iii) CH_3COCH_3

11. An aldehyde as well as a ketone can be represented by the same molecular formula, say C_3H_6O. Write their structures and name them. State the relation between the two in the language of science. **[CBSE 2016]**

12. What is the relationship between 2-methylpropane and 2-methylbutane?

13. Write the structures and names of the isomers having the molecular formula, C_3H_7Cl.

14. Write the structures and IUPAC names of the two isomeric alcohols having the molecular formula, C_3H_8O.

15. List the products of combustion of fossil fuels. State the harmful effect of these products on the environment.

16. What is petrol and what is the composition of the natural gas used for cooking?

17. Explain why a fuel on burning gives mainly a yellow flame?

18. Why does cooking gas burn with a blue flame?

19. Under what conditions does the burning of LPG in a Bunsen burner produce a blue non-luminous flame and under what conditions does it produce a luminous yellow flame?

Section D: Long Answer Questions (5 Marks)

1. What is meant by a functional group in an organic compound? State in tabular form the structural formula and the functional groups present in:
(i) ethanol and
(ii) ethanoic acid **[CBSE 2015]**

2. Draw the electron dot structure for ethyne. A mixture of ethyne and oxygen is burnt for welding. In your opinion, why cannot we use a mixture of ethyne and air for this purpose? **[CBSE 2015]**

3. (a) Draw the structures for the following compounds:
(i) 2-bromopentane
(ii) 2-methylpropane
(iii) butanal
(iv) 1-hexyne
(b) Draw the electron dot structure for ethanoic acid.

4. What is meant by a homologous series of carbon compounds? Classify the following carbon compounds into two homologous series and name them.
C_3H_4, C_3H_6, C_4H_6, C_4H_8, C_5H_8, C_5H_{10} **[CBSE 2015]**

5. Why are homologous series of carbon compounds so called? Write the chemical formula of two consecutive members of a homologous series and state the part of these compounds that determines their

(i) physical properties and
(ii) chemical properties [CBSE 2015]

6. Write the name and general formula of a chain of hydrocarbons in which an addition reaction with hydrogen is possible. State the essential condition for an addition reaction. Stating this condition, write a chemical equation giving the name of the reactant and the product of the reaction. [CBSE 2015]

Section E: Case Study or Passage Based Questions (4 Marks)

Hydrocarbons are the parent organic compounds as other organic compounds are derived from these by replacing the hydrogen atom of a hydrocarbon with a functional group. A functional group decides the chemical properties of an organic compound. For example, the –COOH group is known as the carboxylic group. As there are millions of organic compounds, in order to name them systematically, an IUPAC system of nomenclature was introduced. By using certain rules, we can name any organic compound.

1. Which of the following is not a functional group?
   (a) –OH (b) –X
   (c) $-CH_3$ (d) –CHO

2. How many alcohols are possible for the formula C_3H_8O?
   (a) 1 (b) 2
   (c) 3 (d) 4

3. Which of the following compounds belong to the same homologous series?
   (i) C_3H_6O (ii) CH_4O
   (iii) $C_2H_4O_2$ (iv) C_3H_8O
   (a) I and II (b) II and IV
   (c) I and III (d) II and III

4. The correct IUPAC name of the given compound is:

$$CH_2=CH-\underset{\displaystyle CH_3}{\underset{|}{CH}}-CH_3$$

   (a) 3-methyl butane
   (b) 3-methyl butene
   (c) 2-methyl but-3-ene
   (d) 2-methyl butane

5. The correct IUPAC name of the given compound is:

$$CH_3-\overset{\displaystyle OH}{\overset{|}{CH}}-CH=CH_2$$

   (a) 3-hydroxybutene
   (b) 2-hydroxybut-3-ene
   (c) but-3-en-2-ol
   (d) but-1-en-3-ol

High Order Thinking Skills (HOTS) Questions

1. Write the IUPAC names of the given compounds.
   (i) $CH_3-\underset{\displaystyle CH_3}{\underset{|}{CH}}-CH_2-\underset{\displaystyle CH_3}{\underset{|}{CH}}-CH_2-CH_3$
   (ii) $CH_2=CH-\underset{\displaystyle CH_3}{\underset{|}{C}}=CH-CH_3$
   (iii) $CH_2=CH-\underset{\displaystyle CH_3}{\underset{|}{CH}}=CH_2$
   (iv) $CH_3-C\equiv C-\underset{\displaystyle CH_3}{\underset{|}{CH}}-CH_3$
   (v) $CH_3-C\equiv C-CH=CH_2$

2. Write the IUPAC names of the given compounds.

   (i) $\underset{\displaystyle CHO}{\overset{CHO}{|}}$ (ii) $\underset{\displaystyle HC(OH)-CHO}{\overset{HC(OH)-CHO}{|}}$ (iii) $CH_3-\underset{\displaystyle OH}{\underset{|}{CH}}-CHO$

   (iv) $C_6H_5-CH=CH-CHO$

3. Write the IUPAC names of the given compounds.

   (i) $\underset{\displaystyle COOH}{\overset{COOH}{|}}$ (ii) $\underset{\displaystyle HC(OH)-COOH}{\overset{HC(OH)-COOH}{|}}$ (iii) $CH_3-\underset{\displaystyle OH}{\underset{|}{CH}}-COOH$

   (iv) $C_6H_5-CH=CH-COOH$

Answers

Section A: Multiple Choice Questions

1. (d)	**2.** (b)	**3.** (c)	**4.** (b)
5. (d)	**6.** (d)	**7.** (b)	**8.** (c)
9. (b)	**10.** (d)	**11.** (c)	**12.** (b)
13. (c)	**14.** (d)	**15.** (b)	

Assertion Reason Questions

1. (b)	**2.** (d)	**3.** (b)	**4.** (d)
5. (a)	**6.** (a)	**7.** (a)	

Section B: Very Short Answer Questions

1. A homologous series is a series of organic compounds which has the same functional group and similar chemical properties arranged in increasing or decreasing order of C atoms.

2. These are CH_3Cl (chloromethane) and C_2H_5Cl (chloroethane).

3. These are ethane (C_2H_6) and propane (C_3H_8).

4. The molecular formula for two consecutive members of this series are CH_3Br (bromomethane) and C_2H_5Br (bromoethane).

5. The molecular formulae for two consecutive members of this series are CH_3OH (methanol) and C_2H_5OH (ethanol).

6. Propane (C_3H_6) and butane (C_4H_8).

7. These are propyne (C_3H_4) and butyne (C_4H_6).

8. (i) C_3H_6 (ii) C_5H_8

9. (i) –OH is an alcohol
(ii) $-\underset{\underset{O}{\|}}{C}-$ is a ketone

10. Butanol ($CH_3-CH_2-CH_2-CH_2-OH$)

11. IUPAC name: Propanol
Structure: $H_3C-CH_2-CH_2-OH$

12. (i) Ethanol (ii) Ethanal

13. (a) Methanoic acid (b) Butan-2-one

14. Ethanol (C_2H_5OH or CH_3CH_2OH) is the second member.

15. (i) Ethanol (ii) Ethanal

16. (a) Propanone: $CH_3-\overset{\overset{O}{\|}}{C}-CH_3$

(b) Hexanal:

$$CH_3-CH_2-CH_2-CH_2-CH_2-\overset{\overset{O}{\|}}{C}-H$$

17. The simplest ketone is acetone with a formula CH_3COCH_3.

18. Acetic acid or vinegar.

19. The alkyl part (carbon chain) of an organic compound determines its physical properties whereas the functional group determines its chemical properties.

20. Combustion is always an exothermic process as heat is evolved during combustion.

21. Carbon and its compounds are very good fuels. On burning in air, they provide a lot of heat energy.

22. Methane is the main component of natural gas and butane is the main component of LPG.

23. When wax of a candle burns, it creates a yellow luminous flame.

24. Sufficient supply of O_2 for complete combustion.

Case Study or Passage Based Questions

25. Homologous series

26. C_4H_{10}, C_5H_{12}

27. Name: Propene
Formula: C_3H_6

28. Ethyne, C_2H_2.

29. Propyne (C_3H_4) is the second member of the homologous series having the general formula C_nH_{2n-2}.

30. (i) Aldehyde (ii) Carboxylic acid

31. Propanol-1

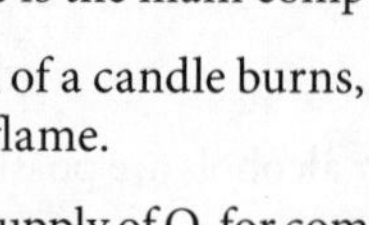

or $CH_3-CH_2-CH_2-OH$

32. Butanal, CH_3–CH_2–CH_2–CHO

or

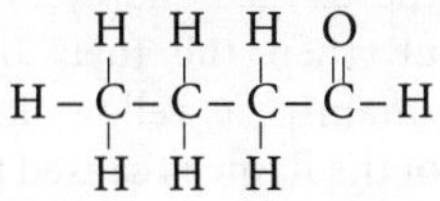

33. The functional group is ketone and name of the compound is propanone-2.

34. $CH_3CH_2COCH_3$: butanone-2 or butan-2-one. C_2H_5COOH: Propanoic acid

Section C: Short Answer Questions

1. (i) A functional group is an atom or a group of atoms in a carbon compound that gives the molecule its characteristic physical and chemical properties. It is the site of reactivity in an organic compound. For example, alcohol is –OH and carboxylic acid is –COOH.

(ii) Two properties of carbon which lead to the huge number of carbon compounds we see around us are catenation and tetravalency.

2. (a) Aldehyde (–CHO)
(b) Carboxylic acid (–COOH)
(c) Alcohol (–OH),(d) Ketone (–CO–)

3. (i) Propanal (aldehyde)
(ii) Propanone (ketone)
(iii) Isomers (possess the same molecular formula but different structural formula due to different functional group).

4. Difference in
(i) Molecular mass = 14 amu
(ii) Number of atoms of elements = CH_2

5. A belongs to alkynes as the general formula of an alkyne is $C_{2n}H_{2n-2}$.
B belongs to alkenes as the general formula of an alkene is C_nH_{2n}.

6. (i)

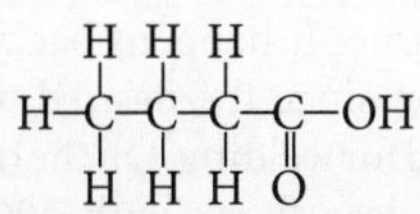

Carboxylic group - COOH

(ii)

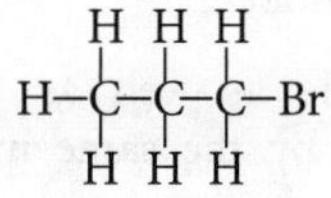

Halogen atom - Br

7. A functional group is an atom or a group of atoms that is bonded to a carbon chain. It defines the chemical property of the organic compound.

Compound	Functional group	Structure
Ethanol	Hydroxyl (–OH)	CH_3CH_2OH
Ethanoic acid	Carboxylic acid (–COOH)	CH_3COOH

8. The name and general formula of hydrocarbons undergoing addition reaction with hydrogen:

Name	General Formula
Alkene	C_nH_{2n}
Alkyne	C_nH_{2n-2}

Essential conditions required for the addition reaction to occur:

- Multiple bonds (double and triple bonds) must be present between carbon atoms in the hydrocarbon chain.
- Addition of hydrogen should be carried out in the presence of a catalyst such as nickel or platinum.

Chemical equation:

$$\underset{\text{Ethene}}{CH_2{=}CH_2} + H_2 \xrightarrow{(\text{Ni or Pt})} \underset{\text{Ethane}}{CH_3{-}CH_3}$$

$$\underset{\text{Ethyne}}{CH{\equiv}CH} + H_2 \xrightarrow{(\text{Ni or Pt})} \underset{\text{Ethene}}{CH_2{=}CH_2}$$

9. (i) Propanoic acid, benzene.
(ii) A series of compounds in which the compounds have similar chemical properties is called a homologous series.

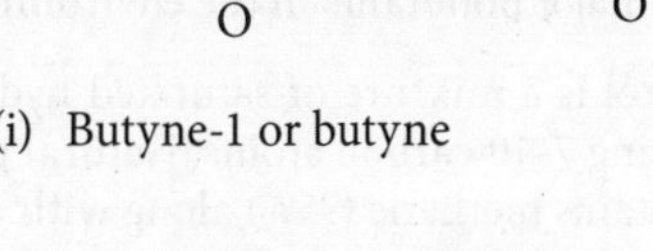

10. (i) Butyne-1 or butyne

(ii) Ethanol
(iii) Propanone-2

11. An aldehyde as well as a ketone can be represented by the same molecular formula, C_3H_6O.

Their structures are:

(i) Structure: $CH_3CH_2—CHO$
IUPAC name: Propanal

(ii) Structure: $CH_3-\overset{O}{\overset{\|}{C}}-CH_3$
IUPAC Name: Propanone

Such compounds with an identical molecular formula but different structures are called structural isomers. Therefore, in the language of science, they are called structural isomers.

12. The molecular formula of 2-methylpropane is C_4H_{10} and that of 2-methylbutane is C_5H_{12}. Since the difference between them is a CH_2 group, they are homologues.

13. The structures and the names of the two isomers are:

$CH_3-CH_2-CH_2-Cl$ (1-Chloropropane) $CH_3-\underset{Cl}{\underset{|}{CH}}-CH_3$ (2-Chloropropane)

14. The two isomeric alcohols are:

$CH_3-CH_2-CH_2-OH$ (1-Propanol or Propan-1-ol) $CH_3-\underset{OH}{\underset{|}{CH}}-CH_3$ (2-Propanol or Propan-2-ol)

15. Coal is a mixture of aromatic compounds. On combustion, it produces CO_2 and heat. Petroleum, on the other hand, is a mixture if open chain higher hydrocarbons. Hence, on combustion it produces CO_2, H_2O and a large amount of heat.

Harmful effects: Coal and petroleum also contain some amounts of nitrogen- and sulfur-containing compounds. Their combustion results in the formation of oxides of nitrogen, that is, nitrogen dioxide, nitric oxide and sulfur, that is, sulfur dioxide, sulfur trioxide which are the major pollutants in the environment.

16. Petrol is a mixture of saturated hydrocarbons having 7–10 carbon atoms. Natural gas mainly contains methane (95%) along with decreasing amounts of ethane, propane, butane and vapours of low boiling pentanes and hexanes.

17. When the oxygen supply or air supply is not sufficient, then the fuels burn incompletely giving mainly a yellow flame. The yellow colour of the flame is caused by the glow of hot, unburnt carbon particles produced due to the incomplete combustion of the fuel.

18. The gas stove has holes or inlets for air to mix properly with cooking gas. The cooking gas gets sufficient oxygen from this air and so burns completely producing a blue flame.

19. When the holes of the Bunsen burner are open, sufficient supply of air is available so LPG burns completely with a non-luminous blue flame. However, when the holes of the Bunsen burner are closed, a sufficient supply of air is not available so LPG burns with a luminous yellow sooty flame.

Section D: Long Answer Questions

1. An atom or a group of atoms which is the site of reactivity in the organic compound and which largely determines the properties of an organic compound is called the functional group.

Name the compound	Structural formula	Functional group
Ethanol	$H-\overset{H}{\underset{H}{C}}-\overset{H}{\underset{H}{C}}-OH$	Alcohol (–OH)
Ethanoic acid	$H-\overset{H}{\underset{H}{C}}-\overset{O}{\overset{\|}{C}}-OH$	Carboxylic acid (–COOH)

2. Electron dot structure for ethyne.

$\rightarrow H-C\equiv C-H$

Burning of ethyne in the presence of air gives a sooty flame. It happens because of incomplete oxidation since the desired temperature is not obtained for welding. On the other hand, oxygen gives a clear flame with 3000°C temperature and complete combustion takes place which is suitable for welding. So, a mixture of ethyne and oxygen is used for welding.

3.

(a) (i) 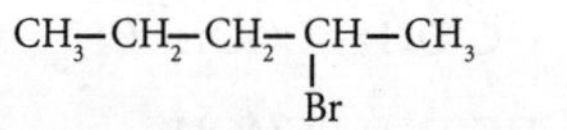

(ii) 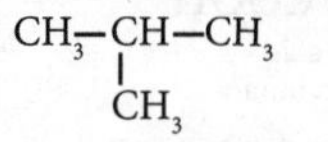

(iii) $CH_3CH_2CH_2CHO$

(iv) $HC\equiv C-CH_2-CH_2-CH_2-CH_3$

(b) Electrons dot structure of ethanoic acid

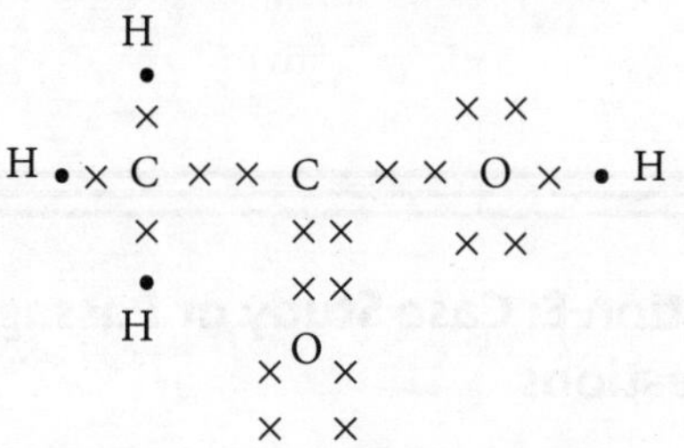

4. A homologous series is a series of organic compounds which have the same functional group, similar chemical properties but show a regular gradation in physical properties.

	Alkenes	
C_3H_6	Propene	$CH_3-CH=CH_2$
C_4H_8	1-Butene	$CH_2=CH-CH_2-CH_3$
C_5H_{10}	1-Pentene	$CH_2=CH-CH_2-CH_2-CH_3$
	Alkynes	
C_3H_4	Propyne	$CH\equiv C-CH_3$
C_4H_6	1-Butyne	$HC\equiv C-CH_2-CH_3$
C_5H_8	1-Pentyne	$HC\equiv C-CH_2-CH_2-CH_3$

5. A homologous series of carbon compounds is so called because the consecutive members of the series differ by a $-CH_2$ unit and their molecular mass differs by 14 u. Moreover, they have the same chemical properties and show a regular gradation in physical properties. For example, in the homologous series of alcohols, the first two members are methanol (CH_3OH) and ethanol (C_2H_5OH). The hydrocarbon (CH_3 and CH_3CH_2) determines the physical properties while the alcoholic part (–OH) determines the chemical properties.

6. C_nH_{2n} and C_nH_{2n-2} are the general formulae of alkenes and alkynes respectively in which addition reaction with hydrogen is possible. Hydrogen is added to unsaturated hydrocarbons (having double or triple bonds) in the presence of nickel as catalyst.

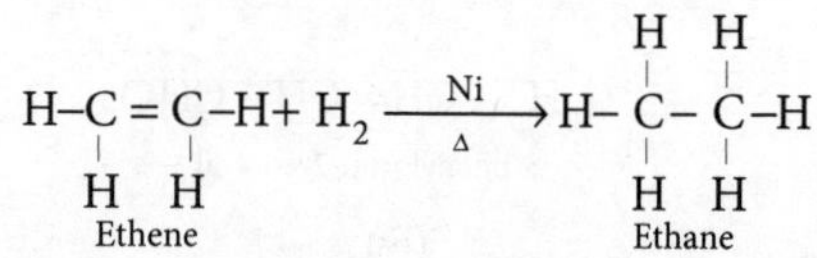

Section E: Case Study or Passage Based Questions

1. (c) **2.** (b) **3.** (b) **4.** (b)
5. (c)

High Order Thinking Skills (HOTS) Questions

1. (i)

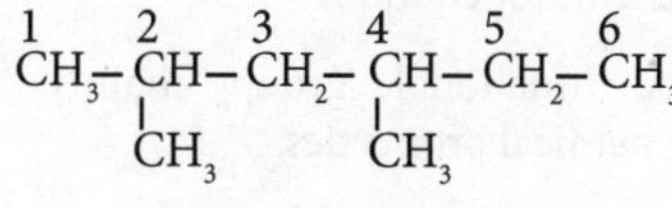

2,4-dimethylhexane

(ii)

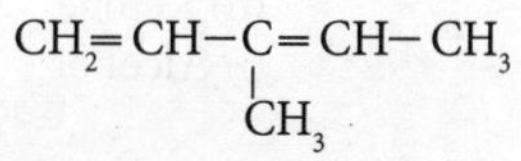

3-methylpenta-1,3-diene

(iii)

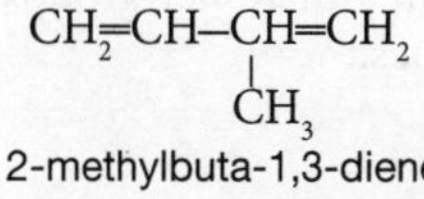

2-methylbuta-1,3-diene

(iv)

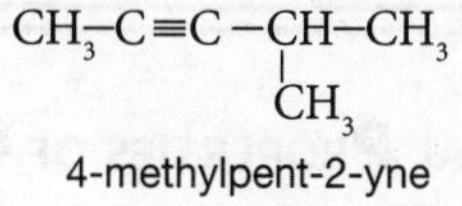

4-methylpent-2-yne

(v)

$CH_3-C\equiv C-CH=CH_2$
5 2 3 2 1

Pent-1-ene-3-yne

2.

CHO | CHO	HC(OH) – CHO | HC(OH) – CHO	CH_3 – CH – CHO | OH
Ethane-1,2-dial	2,3-Dihydroxybutan-1,4-dial	2-hydroxy propanal
(i)	(ii)	(iii)

$C_6H_5-CH=CH-CHO$
3-phenylprop-2-en-1-al
(iv)

3.

COOH | COOH	HC(OH) – COOH | HC(OH) – COOH	CH_3 – CH – COOH | OH
Oxalic acid Ethan1,2-dioic acid	Tartaric acid 2,3-Dihydroxybutan-1,4-dioic acid	Lactic acid 2-hydroxypropanoic acid
(i)	(ii)	(iii)

$C_6H_5-CH=CH-COOH$
3-phenylprop-2-en-1-oic acid
(iv)

Hints

Section A: Multiple Choice Questions

12. $\underset{1}{C}H_3\underset{2}{C}O\underset{3}{C}H_2\underset{4}{C}H_2\underset{5}{C}H_3$ it is pentan-2-one or pentanone-2.

13. $\underset{5}{C}H_3\underset{4}{C}H\ Cl\underset{3}{C}H_2\underset{2}{C}H_2\underset{1}{C}HO$ it is 4-chloropentanal.

14. A functional group mainly decides the chemical properties.

15.

1. –COOH	(iii) Acid
2. –CHO	(iv) Aldehyde
3. –CO–	(ii) Ketone
4. –OH	(i) Alcohol

Assertion Reason Questions

2. As C_2H_2 is ethyne and C_3H_6 is propene, they have different functional groups, and cannot be homologues.

Section E: Case Study or Passage Based Questions

1. $-CH_3$ is methyl, an alkyl group, and not a functional group.

2. C_3H_8O can be two alcohols, $CH_3CH_2CH_2OH$ (propanol-1) and $CH_3CHOHCH_3$ (propanol-2).

3. CH_4O (CH_3OH) and C_3H_8O ($CH_3CH_2CH_2OH$) are homologous as they have the same functional group.

4. $\overset{1}{C}H_2=\overset{2}{C}H-\overset{3}{C}H(CH_3)-\overset{4}{C}H_3$

   It is 3-methylbutene.

5. $\underset{1}{C}H_3-\underset{2}{C}H(OH)-\underset{3}{C}H=\underset{4}{C}H_2$

   It is but-3-en-2-ol.

4.11 Chemical Properties of Carbon Compounds

The most common carbon compounds are hydrocarbons. Let us study some of the chemical properties of hydrocarbons such as combustion reactions, substitution reactions and addition reactions. Combustion reactions are shown by all types of hydrocarbons, that is, saturated (alkanes) as well as unsaturated (alkenes and alkynes). Substitution reactions are shown by saturated hydrocarbons while addition reactions are shown by unsaturated hydrocarbons.

4.11.1 Combustion

The process of burning of a carbon compound in air to give carbon dioxide, water, heat and light, is called *combustion*. Combustion is also known as burning or complete oxidation. Most of the carbon compounds burn in air to produce a huge amount of heat. For example, alkanes burn in air to produce a huge amount of heat due to which they are excellent fuels. For example, when methane or natural gas burns in a sufficient supply of air, carbon dioxide and water vapour are formed and a lot of heat is also produced.

$$\underset{\substack{\text{Methane}\\\text{(Natural gas)}}}{CH_4} + \underset{\substack{\text{Oxygen}\\\text{(From air)}}}{2O_2} \xrightarrow{\text{Combustion}} \underset{\substack{\text{Carbon}\\\text{dioxide}}}{CO_2} + \underset{\text{Water}}{2H_2O} + \text{Heat} + \text{Light}$$

As natural gas or methane produces a lot of heat on burning, it is used as a fuel in homes, transport and in industry. The cooking gas (LPG) which we use in our homes is mainly an alkane known as butane (C_4H_{10}). When butane or LPG burns in air in the burner of a gas stove, it forms carbon dioxide and water vapour, with the evolution of a lot of heat and some light. This makes LPG an excellent fuel.

Competition Edge

Methane is the simplest hydrocarbon and is known as *marsh gas*. In a very limited supply of air, methane burns to give carbon black.

$$CH_4 + \underset{\text{Air (very limited)}}{O_2(g)} \xrightarrow{\Delta} \underset{\text{Carbon black}}{C} + 2H_2O$$

Difference in Flame of Saturated and Unsaturated Hydrocarbons

Saturated hydrocarbons burn in air with a blue non-sooty flame as the percentage of carbon in them is comparatively less. All of the carbon is oxidised into carbon dioxide. However, if oxygen or air is less, incomplete combustion occurs to give a sooty flame. Unsaturated hydrocarbons burn in air with a yellow sooty flame giving black smoke. However, if pure oxygen is used, a blue non-sooty flame is observed.

Ethyne or acetylene on burning in air gives very sooty blue flame which is known as an oxy-acetylene flame and used for welding of metals as it produces a very high temperature.

4.11.2 Substitution Reaction

In a substitution reaction, one atom or group is displaced by another atom or group. For example, B is substituted by C in the molecule below.

$$A–B + C \rightarrow A–C + B$$

Saturated hydrocarbons (alkanes) do not have any functional group or multiple bond so they are almost unreactive and do not react with many substances. Alkanes can undergo substitution reaction with halogens in the presence of sunlight. In this reaction, a hydrogen atom(s) of the alkane is(are) substituted one by one by chlorine atoms.

$$R–H + X_2 \xrightarrow{\text{Sunlight}} R–X + HX$$

$$\underset{\text{Methane}}{CH_4} + \underset{\text{Chlorine}}{Cl_2} \xrightarrow{\text{Sunlight}} \underset{\substack{\text{Chloromethane}\\\text{(or methyl chloride)}}}{CH_3Cl} + \underset{\text{Hydrogen chloride}}{HCl}$$

If the reaction of methane with chlorine in the presence of sunlight continues further, all the hydrogen atoms are successively substituted by chlorine atoms giving CH_3Cl (methyl chloride), CH_2Cl_2 (methylene dichloride), $CHCl_3$ (chloroform) and CCl_4 (carbon tetrachloride).

$$CH_3Cl \xrightarrow[-HCl]{Cl_2} CH_2Cl_2 \xrightarrow[-HCl]{Cl_2} CHCl_3 \xrightarrow[-HCl]{Cl_2} CCl_4$$

Other hydrocarbons such as ethane (C_2H_6), propane (C_3H_8), butane (C_4H_{10}), and so on, show this reaction with chlorine in the presence of sunlight.

4.11.3 Addition Reactions

The reaction in which an unsaturated compound (alkene or alkyne) reacts with another substance to give a single product is called an *addition reaction*. In an addition reaction, an alkene or alkyne react completely with molecule/s like hydrogen and chlorine.

(i) **Hydrogenation:** The addition of hydrogen to an unsaturated hydrocarbon (alkene or alkyne) to get a saturated hydrocarbon is called *hydrogenation*. It is done in the presence of a catalyst such as Ni, Pd or Pt. For example,

$$CH_2{=}CH_2 + H_2 \xrightarrow{\text{Ni, 120°C}} CH_3–CH_3$$

$$H–C{\equiv}C–H + H_2 \xrightarrow[\text{140°C}]{\text{Ni}}$$

$$CH_2{=}CH_2 \xrightarrow{Ni/H_2} CH_3–CH_3$$

(ii) **Hydrogenation of oils:** Vegetable oils such as mustard oil, groundnut oil and cotton seed oil are unsaturated compounds possessing double bonds. They are in the liquid state at room temperature. Due to the presence of double bonds, vegetable oils can undergo addition with hydrogen (hydrogenation) just like alkenes to form saturated products known as vegetable ghee or vanaspati ghee which are solid or semi-solid at room temperature. This process is called *hydrogenation of oils* and is used to prepare vanaspati ghee (dalda) from vegetable oils. For example,

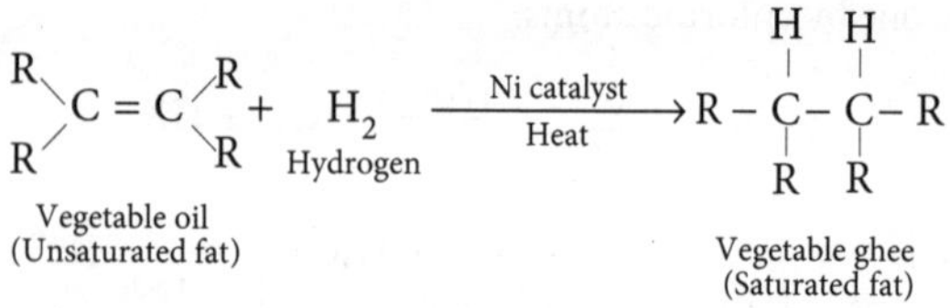

It is important to note that vegetable oils with having unsaturated fatty acids are good for our health and they must be used for cooking our food. Some good vegetable oils are mustard oil, peanut oil, sunflower oil and so on. Vegetable ghee having saturated fatty acids is not good for our health so their use must be avoided.

(iii) **Bromination:** The addition of bromine or bromine water to an alkene or alkyne is an example of an addition reaction. Unsaturated compounds can decolourise the red-brown colour of bromine water while saturated hydrocarbons do not give this reaction. Therefore, this reaction can be used to distinguish between saturated and unsaturated hydrocarbons. For example, cooking oil having unsaturated fatty acids can decolourise bromine water while butter or ghee having saturated fatty acids cannot do that so we can distinguish between cooking oil and ghee or butter.

$$CH_2{=}CH_2 + Br_2 \longrightarrow CH_2Br{-}CH_2Br$$

$$H{-}C{\equiv}C{-}H + Br_2 \longrightarrow$$

$$CHBr{=}CHBr \xrightarrow{Br_2} CHBr_2{-}CHBr_2$$

TEST YOUR KNOWLEDGE

4.22 What is a catalyst? Write the chemical equation to represent the hydrogenation of ethene.

Solution: A catalyst is a substance that causes a reaction to occur or proceed at a different rate without itself being affected.

$$CH_2{=}CH_2 \xrightarrow[H_2]{\text{Nickel catalyst}} CH_3{-}CH_3$$

4.23 Two carbon compounds X and Y have the molecular formulae C_4H_8 and C_5H_{12} respectively. Which one of these is most likely to show addition reactions? Justify your answer. Also give the chemical equation to explain the process of addition reaction in this case. **[CBSE 2017]**

Solution: Among these, C_4H_8 is most likely to show addition reactions as it is an unsaturated hydrocarbon.

Chemical equation: $C_4H_8 + H_2 \xrightarrow[\Delta]{Ni/Pd} C_4H_{10}$

4.24 What is meant by hydrogenation? With the help of a chemical equation, explain the role of this reaction in industry. **[Board Term II, Set-205, 2012]**

Solution: The process in which unsaturated compounds react with hydrogen in the presence of nickel (as a catalyst) to form saturated compounds is called hydrogenation. This reaction is commonly used in the hydrogenation of vegetable oils. Vegetable oils have long unsaturated carbon chains, which are converted into saturated fatty acids (vegetable ghee).

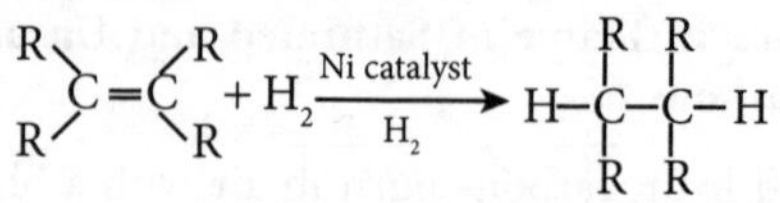

4.12 IMPORTANT CARBON COMPOUNDS

Many carbon compounds are invaluable to us. But here we shall study the properties of two commercially important compounds—ethanol and ethanoic acid.

4.12.1 Ethyl Alcohol (Ethanol, C_2H_5OH)

Ethanol is the second member of the homologous series of alcohols and the first member is methanol (CH_3OH). It is also called grain alcohol as it can be obtained from barley (used for beer manufacture). 100 per cent pure ethyl alcohol is called *absolute alcohol*. A mixture of 95% ethyl alcohol and 5% water is known as *rectified sprit*.

Preparation: Ethyl alcohol is produced on a large scale from sugarcane. Sugarcane juice is used to prepare sugar by crystallisation. During sugar manufacture, a dark brown liquid (molasses) is left behind having nearly 30% of uncrystallised sugar. Molasses on fermentation in the presence of many enzymes produces ethanol.

Properties of Ethanol

Physical properties

- Ethanol is a colourless liquid at room temperature and its melting and boiling points are 156 K and 351 K respectively.
- It has a burning taste and a distinct smell.
- It is soluble in water in any proportion and is lighter than water.
- It is a neutral compound and does not have any effect on litmus solution.

Chemical properties

(i) Reaction with sodium: Ethyl alcohol reacts with sodium leading to the evolution of hydrogen and sodium ethoxide (all alcohols show this reaction). This reaction is a test for ethyl alcohol and other alcohols.

$$2Na + 2CH_3CH_2OH \rightarrow \underset{\text{(Sodium ethoxide)}}{2CH_3CH_2O^-Na^+} + H_2$$

Experiment to detect alcohols

Take about 2 mL of absolute ethanol in a dry test tube (Fig. 4.14). Now take a small piece of freshly cut sodium metal. Dry it between the folds of a filter paper and add it to the ethanol. A brisk effervescence can be observed which is due to the evolution of H_2 gas. Collect this gas in another test tube inverted over the test tube containing the reaction mixture. Bring a burning splinter near the mouth of the inverted test tube. The H_2 gas burns with a pop sound.

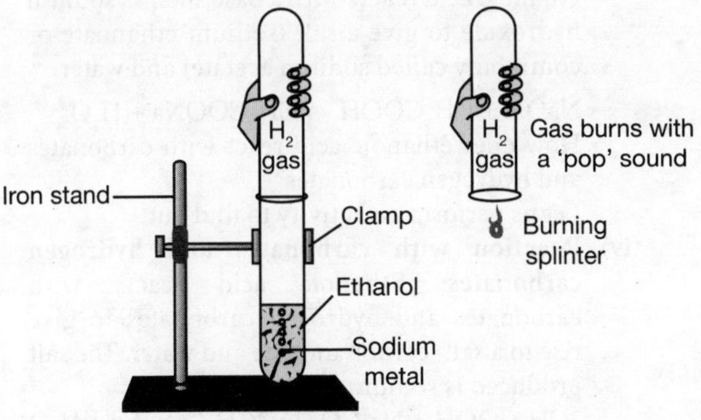

Fig. 4.14 Hydrogen gas produced by the action of sodium metal on absolute ethanol burns with a pop sound

(ii) Reaction to give unsaturated hydrocarbon: Heating ethanol at 443 K with an excess concentrated sulfuric acid results in the dehydration of ethanol to give ethene.

$$CH_3{-}CH_2OH \xrightarrow[H_2SO_4]{\text{Hot Conc.}} CH_2{=}CH_2 + H_2O$$

Concentrated sulfuric acid can be regarded as a dehydrating agent which removes water from ethanol.

(iii) Oxidation: When ethanol is heated with alkaline potassium permanganate solution or acidified potassium dichromate solution, it gets oxidised into ethanoic acid (acetic acid).

$$\underset{\text{Ethanol}}{CH_3CH_2OH} + \underset{\substack{\text{Nascent}\\\text{oxygen}\\\text{(From oxidizing}\\\text{agent)}}}{2[O]} \xrightarrow[\text{(or acidified } K_2Cr_2O_7)]{\text{Alkaline } KMnO_4\text{; Heat}} \underset{\text{Ethanoic acid}}{CH_3COOH} + H_2O$$

(iv) Combustion: As ethanol is a highly inflammable liquid, it burns readily in air with a blue flame producing CO_2, H_2O and lot of heat and some light.

$$\underset{\text{Ethanol (Ethyl alcohol)}}{CH_3CH_2OH} + \underset{\text{Oxygen (From air)}}{3O_2} \xrightarrow[\text{(Burning)}]{\text{Combustion}} \underset{\text{Carbon dioxide}}{2CO_2} + 3H_2O + \text{Heat} + \text{Light}$$

As ethanol burns with a clear flame and the evolution of a large amount of energy, it can be used as a fuel. In fact, now ethanol is an additive in petrol (Power alcohol having 80% petrol and 20% ethanol) which is a very good fuel for vehicles as no harmless product is formed.

Uses:

(i) Ethanol is commonly called alcohol and is the active ingredient of all alcoholic drinks. Consumption of small quantities of dilute ethanol causes drunkenness. Even though this practice is condemned, it is a socially widespread practice. However, intake of even a small quantity of pure ethanol (called absolute alcohol) can be lethal. Also, long-term consumption of alcohol leads to many health problems.

(ii) It is a good solvent for many other organic compounds.

(iii) Being a good solvent, it is also used in medicines such as tincture iodine, cough syrups and many tonics.

(iv) On mixing with petrol, it can be used as a clean fuel in vehicles.

(v) It is also used as an antiseptic to sterilise wounds and syringes.

How do alcohols affect living beings?

When large quantities of ethanol are consumed, it tends to slow the metabolic processes and to depress the central nervous system. This results in lack of coordination, mental confusion, drowsiness, lowering of normal inhibitions and finally stupor. The individual may feel relaxed without realising that his sense of judgement, sense of timing and muscular coordination have been seriously impaired.

Unlike ethanol, the intake of methanol even in very small quantities can cause death. Methanol is oxidised to methanal in the liver. Methanal reacts rapidly with the components of cells. It coagulates the protoplasm, in much the same way an egg is coagulated by cooking. Methanol also affects the optic nerve, causing blindness.

As ethanol is an important industrial solvent, to prevent the misuse of ethanol produced for industrial use, it is made unfit for drinking by adding poisonous substances like methanol to it. Dyes are also added to colour the alcohol blue so that it can be identified easily. This is known as denatured alcohol or denatured ethanol.

4.12.2 Acetic Acid (Ethanoic Acid, CH_3COOH)

Ethanoic acid is the second member of the homologous series of carboxylic acids as the first member is formic acid or methanoic acid (HCOOH). A 5%–8% solution of acetic acid in water is called vinegar and is used widely as a preservative in pickles.

Preparation

When ethanol is heated with alkaline potassium permanganate solution or acidified potassium dichromate solution, it gets oxidised into ethanoic acid (acetic acid).

$$\underset{\text{Ethanol}}{CH_3CH_2OH} + \underset{\substack{\text{Nascent oxygen}\\ \text{(From oxidising agent)}}}{2[O]} \xrightarrow[\text{(or acidified } K_2Cr_2O_7)]{\text{Alkaline } KMnO_4\text{; Heat}} \underset{\text{Ethanoic acid}}{CH_3COOH} + H_2O$$

Physical properties

- It is a colourless pungent smelling liquid.
- The melting point of pure ethanoic acid is 290 K and hence it often freezes during winter in cold climates. This gave rise to its name glacial acetic acid (100% pure acetic acid). Its boiling point is 391 K.
- It is soluble in water in any proportion.

Chemical properties

(i) **Acidic nature:** The group of organic compounds called carboxylic acids are obviously characterised by their acidic nature. However, unlike mineral acids such as HCl which are completely ionised, carboxylic acids are weak acids. Acetic acid can turn blue litmus red.

$$HCl(aq) \rightarrow H^+(aq) + Cl^-(aq) \quad \text{Completely ionised}$$

$$CH_3COOH(aq) \rightleftharpoons H^+(aq) + CH_3COO^-(aq) \quad \text{Partly ionised}$$

When universal indicator paper is used with dil. ethanoic acid, it turns orange showing that the pH is close to 4, which means ethanoic acid is a weak acid. When dil. hydrochloric acid is taken, it turns the universal indicator paper to red showing a pH of close to 1 which means hydrochloric acid is a much stronger acid.

(ii) **Reaction with active metals:** Active metals like sodium, potassium, calcium, magnesium, aluminium, iron and zinc which lie above hydrogen in the activity series react with ethanoic acid to form metal salts with the evolution of hydrogen gas. For example,

$$\underset{\substack{\text{Ethanoic acid}\\ \text{(Acetic acid)}}}{2CH_3COOH} + \underset{\text{Sodium}}{2Na} \rightarrow \underset{\substack{\text{Sodium ethanoate}\\ \text{(Sodium acetate)}}}{2CH_3COONa} + H_2$$

$$\underset{\substack{\text{Ethanoic acid}\\ \text{(Acetic acid)}}}{2CH_3COOH} + \underset{\text{Calcium}}{Ca} \rightarrow \underset{\substack{\text{Calcium ethanoate}\\ \text{(Calcium acetate)}}}{(CH_3COO)_2Ca} + H_2$$

(iii) **Reaction with a base:** Like mineral acids, ethanoic acid reacts with a base such as sodium hydroxide to give a salt (sodium ethanoate or commonly called sodium acetate) and water.

$$NaOH + CH_3COOH \rightarrow CH_3COONa + H_2O$$

How does ethanoic acid react with carbonates and hydrogen carbonates?

Let us perform an activity to find out.

(iv) **Reaction with carbonates and hydrogen carbonates:** Ethanoic acid reacts with carbonates and hydrogen carbonates to give rise to a salt, carbon dioxide and water. The salt produced is sodium acetate.

$$2CH_3COOH + Na_2CO_3 \rightarrow 2CH_3COONa + H_2O + CO_2$$

$$CH_3COOH + NaHCO_3 \rightarrow CH_3COONa + H_2O + CO_2$$

Experiment to show that carboxylic acids react with sodium carbonate or sodium bicarbonate to form carbon dioxide gas.

Set up the apparatus as shown in Fig. 4.15. Take a spatula full of sodium carbonate in a test tube and add 2 mL of dilute ethanoic acid to it. You will soon observe brisk effervescence due to the evolution of CO_2 gas. When this gas is passed through freshly prepared lime water, it turns the lime water milky (CO_2 confirmation test).

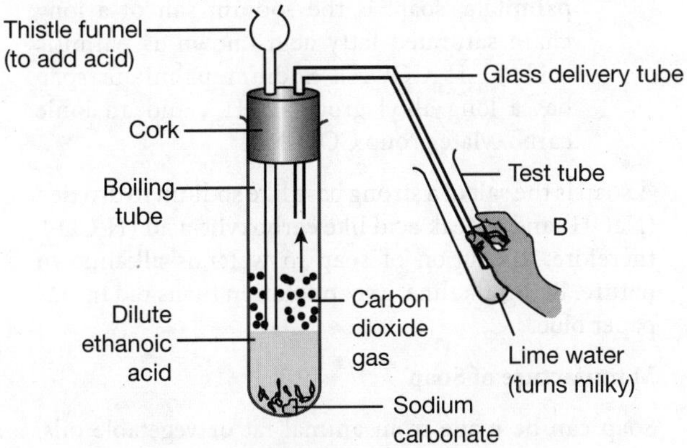

Fig. 4.15 Carbon dioxide formed by the action of sodium carbonate on ethanoic acid turns lime water milky

(v) Esterification reaction: Esters are most commonly formed by the reaction of an acid and an alcohol and this reaction is known as *esterification*. Ethanoic acid reacts with absolute ethanol in the presence of an acid catalyst to give an ester ethyl acetate or ethyl ethanoate which has a fruity smell.

$$\underset{\text{(Ethanoic acid)}}{CH_3{-}COOH} + \underset{\text{(Ethanol)}}{CH_3{-}CH_2OH} \xrightleftharpoons{\text{Acid}} \underset{\text{(Ester ethyl acetate)}}{CH_3{-}\overset{\displaystyle O}{\overset{\|}{C}}{-}O{-}CH_2{-}CH_3} + H_2O$$

Generally, esters are sweet-smelling substances and are used in making perfumes and as flavouring agents.

Hydrolysis of esters: On treating with sodium hydroxide, which is an alkali, an ester is converted back to an alcohol and the sodium salt of the carboxylic acid. This reaction is known as *saponification* because it is used in the preparation of soap.

$$CH_3COOC_2H_5 \xrightarrow{\text{NaOH}} C_2H_5OH + CH_3COONa$$

Experiment to show the formation of esters

Esters can be easily prepared in the laboratory as follows. Mix equal volumes of absolute ethanol (100% alcohol) and pure ethanoic acid (glacial acetic acid, 100% acetic acid) in a dry test tube. Add to it a few drops of conc. H_2SO_4 and keep the test tube in warm water (Fig. 4.16) for about five minutes. Thereafter, pour the contents of the test tube in about 20–30 mL of water and smell the reaction mixture. You will detect a fruity smell due to the formation of an ester, that is, ethyl ethanoate (ethyl acetate).

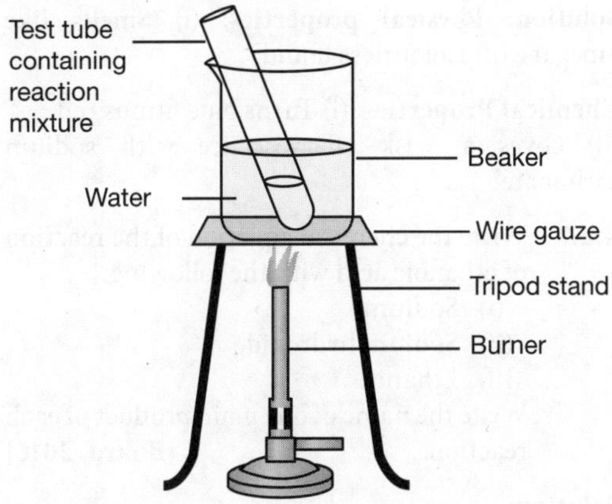

Fig. 4.16 Formation and detection of an ester

Uses of Acetic Acid

- Dilute acetic acid as vinegar is used as a food preservative in the making of pickles, sauces.
- It is a good dressing for salads.
- It is used in the manufacture of esters (on reaction with alcohols).
- It is used in making cellulose acetate (an important synthetic fibre used in fabrics).
- It is used in the preparation of plastics, dyes and pharmaceuticals.
- It is used to coagulate rubber from latex.

TEST YOUR KNOWLEDGE

4.25 How are the following products obtained from ethanol?
(i) Ethyl ethanoate
(ii) Sodium ethoxide

Solution: (i) Ethanol reacts with ethanoic acid in the presence of acid catalyst to give ethyl ethanoate.

$$CH_3COOH + CH_3CH_2OH \xrightarrow[\text{Catalyst}]{\text{Acid}} CH_3COOCH_2CH_3 + H_2O$$

(ii) When ethanol reacts with sodium, sodium ethoxide is produced along with the liberation of hydrogen.

$$2CH_3CH_2OH + 2Na \rightarrow 2CH_3CH_2ONa + H_2$$

4.26 A student is studying the properties of acetic acid in his school laboratory. List two physical and two chemical properties which he must observe and note in his record book. **[Board Term II, Foreign Set-I, 2016]**

Solution: Physical properties: (i) Smells like vinegar (ii) Colourless liquid

Chemical Properties: (i) Turns blue litmus red (ii) Gives a brisk effervescence with sodium carbonate

4.27 Write the chemical equation of the reaction of ethanoic acid with the following.

(i) Sodium
(ii) Sodium hydroxide
(iii) Ethanol

Write the name of one main product of each reaction. **[Board, 2016]**

Solution:

(i) $2CH_3COOH + 2Na \rightarrow \underset{\text{Sodium ethanoate/Sodium acetate}}{2CH_3COONa} + H_2$

(ii) $CH_3COOH + NaOH \rightarrow \underset{\text{Sodium ethanoate/Sodium acetate}}{CH_3COONa} + H_2O$

(iii) $CH_3COOH + C_2H_5OH \rightarrow \underset{\text{Ethyl ethanoate/Ester}}{CH_3COOC_2H_5} + H_2O$

4.13 Soaps and Detergents

All of us are aware of the fact that in order to clean our hands, body and clothes, we use some substances which are known as soaps and detergents. A substance which has a cleaning action in water or a substance which is used to remove dirt is known as a detergent. Detergents are of two types—soapy and non-soapy. In general language, soapy detergents are known as soaps while non-soapy detergents are known as detergents or synthetic detergents.

4.13.1 Soaps

A soap is the sodium salt or potassium salt of a long chain carboxylic acid, that is, a fatty acid having cleansing properties in water. A soap contains a large non-ionic hydrocarbon group and an ionic group, COO^-Na^+. For example, sodium stearate and sodium palmitate.

(i) Sodium stearate, $C_{17}H_{35}COO^-Na^+$: Sodium stearate 'soap' is the sodium salt of a long chain saturated fatty acid known as stearic acid ($C_{17}H_{35}COOH$). Sodium stearate soap has a long alkyl group, $C_{17}H_{35}$, and an ionic carboxylate group, COO^-Na^+.

(ii) Sodium palmitate, $C_{15}H_{31}COO^-Na^+$: Sodium palmitate 'soap' is the sodium salt of a long chain saturated fatty acid known as palmitic acid ($C_{15}H_{31}COOH$). Sodium palmitate soap has a long alkyl group, $C_{15}H_{31}$, and an ionic carboxylate group COO^-Na^+.

A soap is the salt of a strong base like sodium hydroxide (NaOH) and a weak acid like carboxylic acid (H_2CO_3), therefore, a solution of soap in water is alkaline in nature. Being alkaline, a soap solution turns red litmus paper blue.

Manufacture of Soap

Soap can be made from animal fat or vegetable oils. Fats and vegetable oils are naturally occurring esters of higher fatty acids (long chain carboxylic acids) and an alcohol like glycerol. When oils and fats (obtained from plants and animals) are heated with sodium hydroxide solution, they split to form the sodium salt of a higher fatty acid, that is soap, and glycerol. The process of making soap by the hydrolysis of fats and oils with alkalis is known as *saponification*. The reaction given here is a case of saponification and oils like castor oil, cotton seed oil, soyabean oil, linseed oil, coconut oil, palm oil and olive oil, which are also used for preparing soaps.

$$\underset{\text{(An ester)}}{\text{Fat or Oil}} + \underset{\text{(An alkali)}}{\text{Sodium hydroxide}} \xrightarrow{\text{Heat}} \underset{\text{(Sodium salt of fatty acid)}}{\text{Soap}} + \underset{\text{(An alcohol)}}{\text{Glycerol}}$$

Use of sodium chloride in soap making: Common salt (NaCl) is added to the mixture to make the soap come out of solution. Although most part of the soap separates out on its own, some of it remains in solution. Common salt is added to precipitate out all the soap from the aqueous solution. Actually, on adding common salt to the solution, the solubility of soap present in it decreases, as a result all the soap separates out from the solution in the form of a solid.

Preparation of soap in the laboratory or at home

The raw materials needed for the preparation of soap in the laboratory or at home are:

- Vegetable oil such as castor oil, cotton seed oil, linseed oil or soyabean oil.
- 20% solution of caustic soda (sodium hydroxide, NaOH).
- Sodium chloride (common salt, NaCl).

Procedure: Soap can be prepared in the laboratory as follows.

- Add about 20 mL of castor oil or linseed oil or any other oil in a beaker.
- Add 30 mL of 20% NaOH solution into the oil.
- Heat the mixture with constant stirring till a thick paste of soap is obtained.
-

$$\text{Vegetable oil} + \text{NaOH} \xrightarrow{\text{Heat}} \underset{\text{(Sodium salt or fatty acid)}}{\text{Soap}} + \underset{\text{(An alcohol)}}{\text{Glycerol}}$$

- Now add about 5–10 g of common salt to the soap paste.
- Stir the mixture well and allow it to cool. On cooling, almost all the soap separates out as NaCl decreases the solubility of soap in H_2O.
- The soap is filtered out and some antiseptic like Dettol or antibacterial like bithional and perfumes are added to the separated soap.
- The soap is now allowed to stand when it sets into a solid mass. The solid mass is finally cut into soap cakes or bars having fancy shapes.

Structure of a soap molecule: A soap molecule is composed of two parts—a long hydrocarbon part and a short ionic part having the $-COO^-Na^+$ group. The soap molecule has a tadpole structure.

As the long hydrocarbon chain is water-repelling (hydrophobic), it is insoluble in water but soluble in oil and grease. The ionic portion of the soap molecule is water-attracting (hydrophilic) due to the polar nature of the water molecule; so, the ionic portion of the soap molecule is soluble in water but insoluble in oil and grease. Now it is clear that the hydrocarbon part of the soap molecule is soluble in oil or grease, so it can attach to the oil and grease particles present on dirty clothes. The short ionic part of the soap molecule having a negative charge is soluble in water, so it can attach to the water particle in which the soap is dissolved and dirty clothes are dipped.

Cleansing Action of Soap or Detergent

Soaps and detergents consist of a large hydrocarbon tail with a negatively charged head (Fig. 4.17). The hydrocarbon tail is hydrophobic (water-hating or water repelling) and negatively charged head is hydrophilic (water-loving). In an aqueous solution, as the water molecules are polar in nature, they surround the ions and not the hydrocarbon part of the molecule.

When a soap or detergent is dissolved in water, the molecules associate into clusters called *micelles* [Fig. 4.17(C)]. Micelles are spherical and ionic and they remain suspended in water as a colloid and do not come closer to precipitate out, due to repulsion between the similar negative charges.

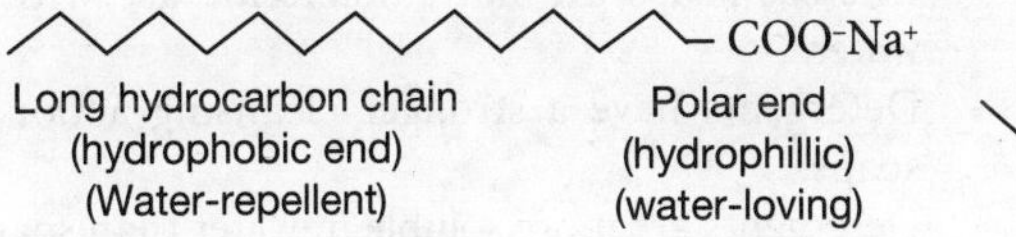

(a) Hydrophilic (water-loving) and hydrophobic (water-repellent) ends of a soap molecule

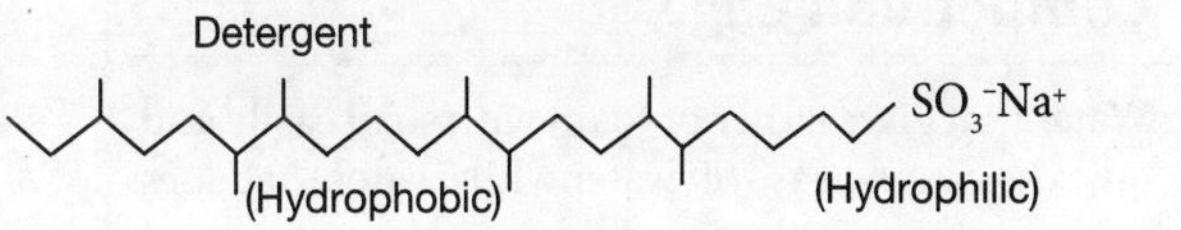

(b) Hydrophilic and hydrophobic ends of a detergent

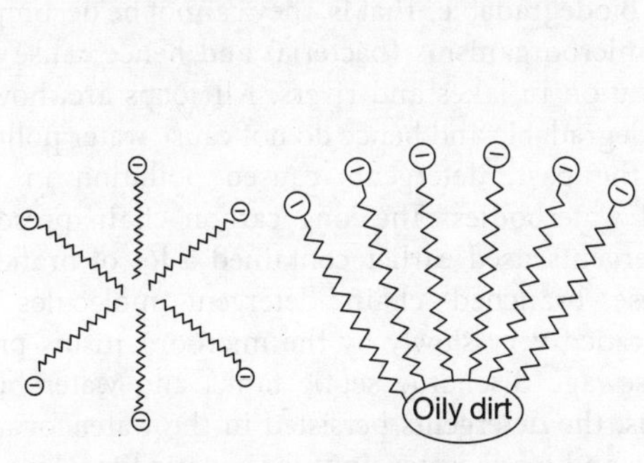

(c) Micelle formed by detergent molecules in water The hydrocarbon tails stick the oily dirt

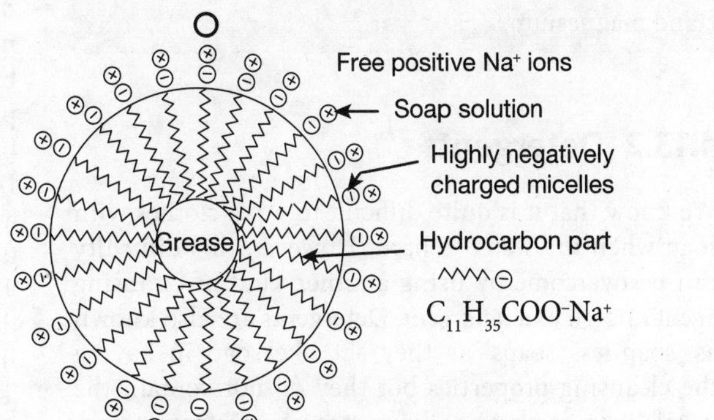

Cleansing action of soap. Soap micelle entraps the oily dirt particle

Fig. 4.17 Micelle formation

The tails stick inwards and the heads outwards.

In cleansing, the hydrocarbon tail attaches itself to the oily dirt. When water is agitated (shaken vigorously), the oily dirt tends to lift off from the dirty surface and dissociate into fragments. This gives an opportunity to the other tails to stick to the oil. The solution now contains small globules of oil surrounded by detergent molecules. The negatively charged heads present in water prevent the small globules from coming together and form aggregates. Thus, the oily dirt is removed.

Limitations of soap: Soaps are not suitable for washing clothes with hard water (water containing salts of calcium and magnesium) because of the following reasons.

(i) A soap does not give lather in good amount with hard water as it reacts with the calcium and magnesium ions present in hard water to form insoluble precipitates of calcium and magnesium salts of fatty acids.

(ii) When soap is used for washing clothes with hard water, more soap is needed as a large amount of soap is wasted in reacting with calcium and magnesium ions of hard water to give an insoluble precipitate (scum).

(iii) The scum or curdy precipitate formed here sticks to the clothes that are being washed and interferes with the cleaning ability of the soap, that is, it makes cleaning or washing difficult.

> **Competition Edge**
>
> Water that gives lather or foam with soap easily and quickly is known as soft water while water that does not give lather with soap readily is known as hard water. The hardness of water is due to the presence of bicarbonates, chlorides and sulfates of calcium and magnesium.

4.13.2 Detergents

We know that it is quite difficult to wash clothes with soap when the water is hard. However, this difficulty can be overcome by using another kind of cleansing agent known as a *detergent*. Detergents are also known as 'soap-less soaps' as they act like soap in having the cleansing properties but they do not contain the usual 'soap' such as sodium stearate. Detergents are better cleansing agents than soaps as they do not form insoluble calcium and magnesium salts with hard water, so they can be used for washing even with hard water. Unlike soap, a detergent can lather well even with hard water.

Structure of a detergent molecule: The structure of a detergent is quite similar to that of a soap. A detergent is the sodium salt of a long chain benzene sulfonic acid or the sodium salt of a long chain alkyl hydrogen sulfate which has cleansing properties in water. A detergent has a large non-ionic hydrocarbon group and an ionic group like the sulfonate group, $SO_3^-Na^+$ or the sulfate group $SO_4^-Na^+$. Examples of detergents are sodium *n*-dodecyl benzene sulfonate and sodium *n*-dodecyl sulfate.

$$CH_3{-}(CH_2)_{11}{-}C_6H_4{-}SO_3^-Na^+$$

Sodium n-dodecyl benzene sulphonate
(A common detergent)

$$CH_3{-}(CH_2)_{10}{-}CH_2{-}SO_4^-Na^+$$

Sodium n-dodecyl benzene sulphate
(Another detergent)

Advantages and Disadvantages of Detergents over Soap

Advantages: Detergents have many advantages over soaps due to which they are replacing soaps as washing agents. Detergents are better than soaps due to these reasons.

- Detergents can be used even with hard water whereas soaps are not suitable for use with hard water.
- Detergents have a stronger cleansing action than soaps.
- Detergents are more soluble in water than soaps.

Disadvantages: An important disadvantage of detergents over soaps is that some of the detergents are not biodegradable. That is, they cannot be decomposed by microorganisms (bacteria) and hence cause water pollution in lakes and rivers. All soaps are, however, biodegradable and hence do not cause water pollution. In the past, detergents caused pollution in rivers and waterbodies. The long carbon chain present in detergents used earlier contained a lot of branching. These branched chain detergent molecules were degraded very slowly by the microorganisms present in sewage discharge septic tanks and water bodies. Thus, the detergents persisted in the water for a long time and made water unfit for aquatic life. Nowadays, the detergents are made up of molecules in which branching is kept at a minimum. These are degraded more easily than branched chain detergents.

Experiment 1: Take about 10 mL of distilled water (or rain water) and 10 mL of hard water (from a tube well or hand pump) in separate test tubes. Add a few drops of soap solution to each test tube and shake vigorously for an equal period of time.

Observation: You will observe that the test tube containing soft water (distilled water) produces foam readily. On the other hand, the test tube having hard water produces curdy white precipitates due to the formation of calcium and magnesium soaps.

Experiment 2: Now take 10 mL of hard water in two separate test tubes. Add five drops of soap solution to one and five drops of a detergent solution to the other. Shake both the test tubes for the same period.

Observation: You will observe that the test tube having soap solution produced a curdy white precipitate with little or no foam, while the test tube having the detergent produced a large amount of foam.

Difference between soaps and detergents

Soaps	Detergents
1. Soaps are the sodium salts or potassium salts of long chain carboxylic acids (fatty acids). The ionic group in soaps is $-COO^-Na^+$.	1. Detergents are the sodium salts of long chain benzene sulfonic acids or long chain alkyl hydrogen sulfates. The ionic group in a detergent $—SO_3^-Na^+$ or $—SO_4^-Na^+$.
2. Soaps are not suitable for washing purposes when the water is hard.	2. Detergents can be used for washing even when the water is hard.
3. Soaps are biodegradable.	3. Some of the detergents are not biodegradable.
4. Soaps have a relatively weak cleansing action.	4. Detergents have a strong cleansing action.

TEST YOUR KNOWLEDGE

4.28 Name any two chemicals used for the preparation of toilet soap.

Solution: Fat and potassium hydroxide are used to prepare toilet soap.

4.29 We need 20% aqueous solution of sodium hydroxide for the study of the saponification reaction. When we open the lid of the bottle containing solid sodium hydroxide, which form do we observe it in?

Solution: Solid sodium hydroxide is stored in the form of white pellets/flakes.

4.30 When red litmus paper and blue litmus paper are dipped into the soap solution, what colour changes will you observe?

Solution: Red litmus paper turns blue and blue litmus paper remains unchanged, as the soap solution is basic in nature.

4.31 Why is common salt added during soap manufacturing?

Solution: Common salt is added to precipitate out all the soap from the aqueous solution. On adding common salt to the solution, the solubility of the soap present in it decreases; as a result, all the soap separates out from the solution in the form of a solid.

4.32 Give an important disadvantage of detergents over soaps.

Solution: An important disadvantage of detergents over soaps is that some detergents are not biodegradable, that is, they cannot be decomposed by microorganisms and hence cause water pollution in lakes and rivers.

Quick Review

- In substitution reactions, one atom is substituted by another atom or group.
- In **halogenation** of alkanes, hydrogen atoms are substituted by halogen free radicals in the presence of sunlight.
- Unsaturated compounds show addition reactions; in addition reactions, one molecule completely reacts to form an adduct.
- All organic compounds undergo combustion reaction in the presence of air to give carbon dioxide and water along with heat (exothermic).

- Alkanes on **combustion** give CO_2 and water along with heat.
- Alcohols on oxidation by acidic $K_2Cr_2O_7$ or alkaline $KMnO_4$ give acids.
- Ethanol (C_2H_5OH) is the most common alcohol used for drinking and also in petrol as power alcohol.
- Alcohol is made undrinkable by adding methyl alcohol (denatured sprit).
- Alcohols can react with sodium metal to liberate hydrogen gas.
- Acetic acid (CH_3COOH) is known as vinegar and is used in pickles.
- Reaction of acid and alcohol to form an ester is known as esterification. Esters have a sweet smell and are therefore used in perfumes.
- Acids can react with sodium bicarbonate to give carbon dioxide.
- A **soap** is the sodium salt (or potassium salt) of a long chain carboxylic acid (fatty acid) having cleansing properties in water. A soap contains a large non-ionic hydrocarbon group and an ionic group, COO^-Na^+. For example, sodium stearate and sodium palmitate.
- A **detergent** is the sodium salt of a long chain benzene sulfonic acid or the sodium salt of a long chain alkyl hydrogen sulfate which has cleansing properties in water.
- When a soap or detergent is dissolved in water, the molecules associate together as clusters called **micelles.**

Exercise 3

Section A: Multiple Choice Questions (1 Mark)

1. The chemical reaction shows the addition of chlorine gas to hydrocarbon in the presence of sunlight.
   $CHCl_3 + Cl_2 \rightarrow CCl_4 + HCl$
   How does chlorine react to a hydrocarbon compound in the presence of sunlight?
   (a) it adds hydrogen into the compound
   (b) it adds an oxygen atom into the compound
   (c) it substitutes the hydrogen atom from the compound
   (d) it breaks the double and triple bonds into single bonds

2. The chemical reaction shows the addition of chlorine to methane in the presence of sunlight.
   $$CH_4 + Cl_4 \rightarrow X$$
   What is likely to be the product of the reaction represented by X?
   (a) $CH_4 + H_2SO_4$ (b) $CH_3Cl + HCl$
   (c) $CHCl_3 + HCl$ (d) $CH_3Cl + H_2SO_4$

3. Addition reactions are undergone by
   (a) saturated hydrocarbons (alkanes)
   (b) only alkenes
   (c) only alkynes
   (d) both (b) and (c)

4. Which of the following will undergo addition reactions?
   (a) CH_4 (b) C_3H_8
   (C) C_2H_6 (d) C_2H_4

5. Which is denatured spirit?
   (a) ethanol only
   (b) ethanol and methanol (50%)
   (c) ethanol and methanol (5%)
   (d) methanol only

6. Tertiary butane gets oxidised with oxidising agents like alkaline $KMnO_4$ to
   (a) isobutane
   (b) tert-butyl alcohol
   (c) secondary-propyl alcohol
   (d) all of the above

7. Which of the following is used to oxidise ethanol to ethanoic acid?
   (a) alkaline $KMnO_4$ (b) conc. H_2SO_4
   (c) acidified $K_2Cr_2O_7$ (d) all of above

8. Vinegar is a solution of
   (a) 30%–40% acetic acid in alcohol
   (b) 5%–8% acetic acid in alcohol
   (c) 5%–8% acetic acid in water
   (d) 15%–20% acetic acid in water

9. Which of the following can be used for the denaturation of ethyl alcohol?
   (a) methyl alcohol (b) pyridines
   (c) copper sulfate (d) all of above

10. When ethanoic acid is treated with $NaHCO_3$ the gas evolved is _______
    (a) H_2 (b) CO_2
    (c) CH_4 (d) CO

11. Ethanol on complete oxidation gives _______
    (a) acetic acid/ethanoic acid

(b) CO_2 and water
(c) ethanal
(d) acetone/ethanone

12. Which of the following will give the pleasant smell of ester when heated with ethanol and a small quantity of sulfuric acid?
(a) CH_3COOH (b) CH_3CH_2OH
(c) CH_3OH (d) CH_3CHO

13. Identify A in the following reaction.

$$CH_3COOH + Na_2CO_3 \rightarrow A + CO_2 + H_2O$$

(a) CH_3COONa
(b) $CH_2(Na)COOH$
(c) NaOH
(d) $NaHCO_3$

14. Identify A and B.

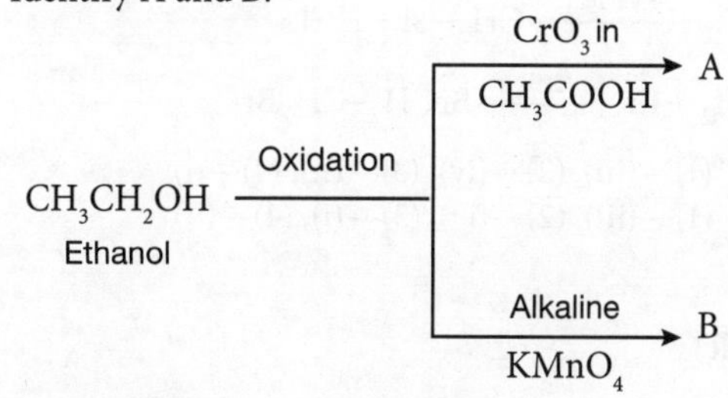

(a) CH_3CHO, Ethanal, CH_3COOH Ethanoic acid
(b) CH_3COOH, Ethanoic acid CH_3CHO Ethanal
(c) CH_3CHO, Ethanal $CO_2 + H_2O$
(d) CH_3COOH, Ethanoic acid $CO_2 + H_2O$

15. A soap molecule has a
(a) hydrophobic head and hydrophobic tail
(b) hydrophobic head and hydrophilic tail
(c) hydrophilic head and hydrophilic tail
(d) hydrophilic head and hydrophobic tail

16. Soaps are formed by the saponification of
(a) alcohols (b) glycosides
(c) simple esters (d) carboxylic acids

17. The substance not responsible for the hardness of water is
(a) sodium nitrate
(b) calcium hydrogen carbonate
(c) calcium carbonate
(d) magnesium carbonate

18. The byproduct of soap is
(a) isoprene (b) glycerol
(c) butene (d) ethylene glycol

19. The equation represents a chemical reaction where ethanol is oxidised using potassium dichromate and sulfuric acid.

$$\underset{\text{Ethanol}}{CH_3CH_2OH} \xrightarrow{K_2Cr_2O_7/H_2SO_4} x$$

Which option represents the product X?
(a) CH_2O (b) CH_3CHO
(c) CH_3OH (d) CH_3COOH

20. A student studies that vinegar, which is a diluted form of ethanoic acid, freezes during winter. What does this suggest about the physical properties of pure ethanoic acid?
(a) it has a low boiling point
(b) it has a low melting point
(c) it has a very high boiling point
(d) it has a very high melting point

21. A student studies that a soap molecule has two ends, one of which is an ionic end and the other is the carbonic chain. Which option explains the interaction of a soap molecule with oil?
(a) the ionic end of the soap interacts with the oil
(b) the closest end of the soap interacts with the oil
(c) the carbonic chain end of the soap interacts with the oil
(d) the ends of the soap randomly interact with the oil

22. A student studies that soap solution results in micelle formation which helps to remove dirt. It has a unique orientation which helps in keeping the dirt out of the water as shown.

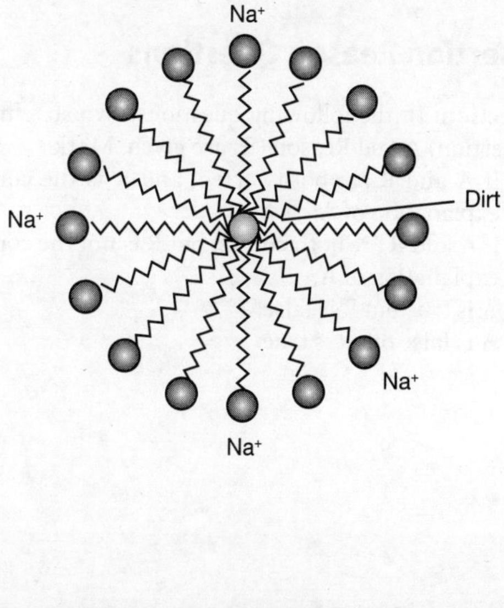

What helps the dirt to rise away?

(a) suspension of the dirt in the micelles
(b) a collection of water molecules in the centre of the micelle
(c) attraction between the ionic end and the dirt to remove it
(d) mixing of the soap molecules along with the dirt to make it heavier

23. Select the set having correct statements.

(i) Vinegar decomposes baking soda to evolve carbon dioxide.
(ii) Ethanol on oxidation with acidic $K_2Cr_2O_7$ gives ethanal.
(iii) Ethanol reacts with sodium metal to evolve hydrogen gas.
(iv) Hydrolysis of an ester with dil. H_2SO_4 is known as saponification.

(a) (i), (ii) (b) (ii), (iii)
(c) (i), (iii) (d) (i), (iii), (iv)

24. Match the following.

Column I (Type of reaction)	Column II (Example)
1. Substitution	(i) $CH_3CH_2OH + O_2 \xrightarrow[\text{Heat}]{KMnO_4/KOH} CH_3COOH + H_2O$
2. Addition	(ii) $CH_2{=}CH_2 + 3O_2 \xrightarrow{\Delta} 2CO_2 + 2H_2O$
3. Oxidation	(iii) $CH_4 + Cl_2 \xrightarrow{\text{hv, Heat}} CH_3{-}Cl + HCl$
4. Combustion	(iv) $CH_2{=}CH_2 + Br_2 \xrightarrow{CCl_4} BrCH_2{-}CH_2Br$

(a) (1)—(ii), (2)—(iii), (3)—(i), (4)—(iv) (b) (1)—(iii), (2)—(iv), (3)—(ii), (4)—(i)
(c) (1)—(ii), (2)—(iv), (3)—(i), (4)—(iii) (d) (1)—(iii), (2)—(iv), (3)—(i), (4)—(ii)

25. Match the following.

Column I (Type of reaction)	Column II (Example)
1. Hydrogenation	(i) $CH_3COOC_2H_5 + NaOH \xrightarrow{\text{Heat}} CH_3COONa + C_2H_5OH$
2. Dehydration	(ii) $CH_3COOH + C_2H_5OH \xrightarrow[\text{Heat}]{\text{Conc. } H_2SO_4} CH_3COOC_2H_5 + H_2O$
3. Esterification	(iii) Vegetable oil + $H_2 \xrightarrow{\text{Ni, Heat}}$ Fat
4. Saponification	(iv) $CH_3CH_2OH \xrightarrow[\text{433–443 K}]{\text{Conc. } H_2SO_4} CH_2{=}CH_2 + H_2O$

(a) (1)—(ii), (2)—(iii), (3)—(i), (4)—(iv)
(b) (1)—(iii), (2)—(iv), (3)—(ii), (4)—(i)
(c) (1)—(ii), (2)—(iv), (3)—(i), (4)—(iii)
(d) (1)—(iii), (2)—(iv), (3)—(i), (4)—(ii)

Assertion Reason Questions

Direction: In the following questions, two statements (Assertion) A and Reason (R) are given. Mark

(a) if A and R are both correct and R is the correct explanation of A;
(b) if A and R are both correct but R is not the correct explanation of A;
(c) A is true but R is false;
(d) A is false but R is true.

Assertion	Reason
1. Addition reactions are shown by unsaturated compounds.	**1.** During addition reactions, unsaturated compounds change into saturated.
2. Both ethene and ethyne decolourise bromine solution.	**2.** It is a case of substitution reaction.

Assertion	Reason
3. Methane can react with chlorine even in the dark.	**3.** It is a case of substitution reaction.
4. Drinking of methyl alcohol can cause blindness.	**4.** Ethyl alcohol is drinkable.
5. Higher fatty acids are used for soap manufacture.	**5.** Soap can give lather easily with hard water also.
6. Soaps are biodegradable while detergents may or may not be.	**6.** Detergents having highly branched chains are not biodegradable.
7. Esterification is a process in which a sweet-smelling compound is formed.	**7.** In esterification, an acid and an alcohol molecule react together.
8. Ethanoic acid decomposes sodium bicarbonate to evolve CO_2 gas.	**8.** It is due to the acidic nature of ethanoic acid.

Section B: Very Short Answer Questions (2 Marks)

1. Identify, from the following, the hydrocarbons that can undergo addition reactions: C_3H_4, C_2H_6, CH_4, C_2H_4. Justify your answer.

2. Explain why propene undergoes addition reaction.

3. Name the process by which unsaturated fats are changed to saturated fats.

4. ______ are sweet-smelling substances which are used in making perfumes.

5. Complete the following reactions.
(a) $CH_4 + O_2 \rightarrow$ (b) $CH_4 + Cl_2 \rightarrow$

6. What happens when methane is burnt in air?

7. What happens when bromine water is added to ethene gas?

8. Name the gas evolved when ethanoic acid is added to sodium carbonate.

9. Write a balanced chemical equation for the dehydration of ethanol by hot conc. H_2SO_4.

10. What is vinegar? Give its uses.

11. Complete the following reaction.
$CH_3COOH + NaHCO_3 \rightarrow$

12. Write the chemical equations for the conversion of ethanol to ethanoic acid in the presence of $KMnO_4$.

13. What type of reaction takes place between CH_3COOH and NaOH solution? Where do you find the use of ethanoic acid in food products?

14. A student puts a drop of the reaction mixture of a saponification reaction first on a blue litmus paper and then on a red litmus paper. Write down what he will observe.

15. During saponification, Rupal takes refined oil in a test tube and adds 20% NaOH solution. She then stirs the reacting mixture. Write down what she will observe?

16. Rupal took a liquid A in a test tube and added sodium hydroxide solution to it. The mixture was stirred for some time. Glyceride and solid B are formed after the reaction. Identify A and B.

17. For determining the saponification reaction for the preparation of soap, write down the names of any five apparatus that we require.

18. Do soaps create water pollution? Give reasons in support of your answer.

19. Name any two fatty acid salts that are are used in the making of soaps.

Case Study or Passage Based Questions

Organic compounds are hydrocarbons and their derivatives. Hydrocarbons mainly show substitution, addition and oxidation reactions. Other organic compounds like alcohols, aldehydes, ketones and acids can show chemical properties on the basis of the functional group present in them. For example, alcohols on oxidation give acids. Acids and alcohols combine together to form an ester. There are many reagents which are used to distinguish organic compounds.

20. When an alkane reacts with chlorine in the presence sunlight, what is the reaction known as?

21. By which reaction can alkanes and alkenes be distinguished?

22. What is the conversion of vegetable oil into ghee known as?

23. Between substitution, addition and combustion, which one is exothermic in nature?

24. What compound is obtained by the oxidation of propanol by alkaline potassium permanganate?

Section C: Short Answer Questions (3 Marks)

1. Differentiate between addition reactions and substitution reactions shown by hydrocarbons.

2. Why is the conversion of ethanol to ethanoic acid an oxidation reaction?

3. Which is better for health, butter or vegetable oil? Why?

4. List any four differences between soaps and detergents.

5. Describe the addition reaction of carbon compounds with its applications. State the function of a catalyst in this reaction. How is this reaction different from a substitution reaction? **[Board, 2015]**

6. (a) Complete the following reactions.
   (i) $CH_4 + Cl_2 \xrightarrow{\text{Sunlight}}$ .... + HCl
   (ii) $CH_3CH_2OH \xrightarrow{\text{Hot.conc.}}$ .... + H_2O
   (b) How is scum formed?

7. (i) Describe the mechanism of cleansing action of soaps.
   (ii) Why do soaps not work with hard water?

8. What is meant by denatured alcohol? What is the need to denature alcohol?

9. Why is soap not suitable for washing clothes when the water is hard?

10. The structural formula of an ester is:

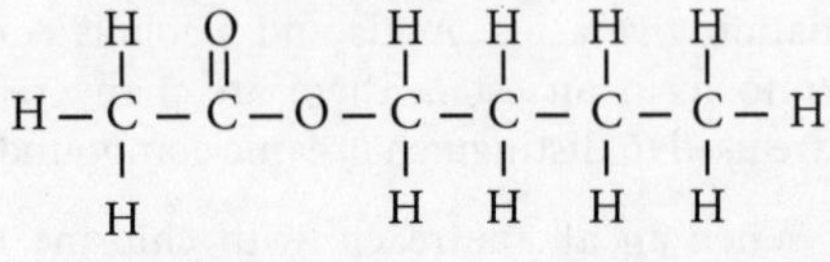

Write the structural formula of the acid and the alcohol from which it might be prepared. Name the process of formation of an ester. **[Board, 2015]**

11. Write balanced chemical equation for the following.
    (i) Methane is burned in sufficient air.
    (ii) Ethanol is treated with sodium.
    (iii) Ethanoic acid is reacted with sodium hydroxide.

12. What is the cause of hardness of water? Why does soap not form lather with hard water? Mention the disadvantage of cleaning clothes with soap using hard water.

13. Explain why it is difficult to wash clothes with soap when the water is hard. How do detergents help in overcoming this problem?

14. While studying the saponification reaction for the preparation of soap, a teacher suggested to a student to add a small quantity of common salt to the reaction mixture. Write down the function of common salt in this reaction.

15. While studying the saponification reaction, a student measures the temperature of the reaction mixture and also finds its nature using blue/red litmus paper. What conclusion can be drawn from his observations?

16. Take a test tube and put soap and water and shake it 10 times by placing your thumb on its mouth. What will you observe?

17. Write the name and structural formula of the compound formed when ethanol is heated at 443 K temperature with excess of conc. H_2SO_4. What is the role of conc. H_2SO_4 in this reaction? Also give the chemical equation for the reaction. **[Board, 2015]**

Or

A compound X on heating with excess conc. sulfuric acid at 443 K gives an unsaturated compound Y. X also reacts with sodium metal to evolve a colourless gas Z. Identify X, Y and Z. Write the equation for the chemical reaction of formation of Y and also write the role of sulfuric acid in the reaction. **[CBSE 2018]**

18. What is an oxidising agent? What happens when an oxidising agent is added to propanol? Explain with the help of a chemical equation. **[CBSE 2016]**

19. Write two tests you would perform to detect whether the given colourless liquid is acetic acid or not. **[CBSE 2013]**

20. (i) Why is ethanol used in making of tincture, iodine, cough syrup, tonic?
(ii) What is the role of conc. H_2SO_4 in making ethane from ethanol?

21. Mention the essential material (chemicals) to prepare soap in the laboratory. Describe in brief the test for determining the nature (acidic/alkaline) of the reaction mixture of the saponification reaction.

22. Consider the following oils:
(i) Mobil oil (ii) castor oil
(iii) turpentine (iv) kerosene
(v) mustard oil (vi) coconut oil
Which of these can be used for the preparation of soap?

23. A student takes about 6 mL of distilled water in each of the four test tubes P, Q, R and S. He then dissolves an equal amount of four different salts namely, sodium chloride in P, potassium chloride in Q, calcium chloride in R and magnesium chloride in S. Next, he adds 10 drops of soap solution to each test tube and shakes its contents. In which of these test tubes is scum (insoluble substance) formed with soap?

24. What is an oxidising agent? What happens when an oxidising agent is added to propanol? Explain with the help of a chemical equation. **[Board, 2016]**

Section D: Long Answer Questions (5 Marks)

1. (i) Give a chemical test to distinguish between saturated and unsaturated hydrocarbons.
(ii) Name the products formed when ethane burns in air. Write the balanced chemical equation for the reaction showing the types of energies liberated.
(iii) Why is reaction between methane and chlorine in the presence of sunlight considered a substitution reaction? **[CBSE 2015, 16]**

2. (i) With the help of an example, explain the process of hydrogenation. Mention the essential conditions for the reaction and state the change in physical property with the formation of the product. **[CBSE 2015]**
(ii) How does a saturated hydrocarbon react with chlorine? Write a chemical equation for it. What type of reaction is it called and why?

3. (i) Define the term 'isomer'.
(ii) Two compounds have the same molecular formula C_3H_6O. Write the name of these compounds and their structural formulae.
(iii) How would you bring about the following conversions?
(a) Ethanol to ethene
(b) Propanol to propanoic acid
[Board 2019]

4. Distinguish between esterification and saponification reaction of organic compounds with the help of a chemical equation for each. What is the use of: (i) esters and (ii) the saponification process? **[CBSE 2017]**

5. Write the chemical equations to show what happens when:
(i) an ester reacts with a base.
(ii) methane is treated with chlorine in the presence of sunlight.
(iii) ethanol reacts with ethanoic acid in the presence of sulfuric acid. **[CBSE 2014]**

6. A carboxylic acid ($C_2H_4O_2$) reacts with an alcohol in the presence of H_2SO_4 to form a compound X. The alcohol on oxidation with alkaline $KMnO_4$ followed by acidification gives the same carboxylic acid ($C_2H_4O_2$). Write the name and structure of the:
(i) carboxylic acid,
(ii) alcohol,
(iii) compound X. **[CBSE 2014]**

7. Write the structural formula of ethanol. What happens when it is heated with excess of conc. H_2SO_4 at 443 K? Write the chemical equation for the reaction stating the role of conc. H_2SO_4 in this reaction. **[CBSE 2017]**

OR

Write the name and structural formula of the compound obtained when ethanol is heated at 443 K with excess of H_2SO_4. Also write the chemical equation for the reaction stating the role of conc. H_2SO_4. **[CBSE 2015]**

8. When ethanoic acid reacts with sodium hydrogen carbonate, a salt X is formed along with a gas Y. Name X and Y. Describe an activity and draw the diagram of the apparatus used to prove that the gas Y is one which you have named. Also write the chemical equation for the reaction involved.

9. An organic compound X on heating with conc. H_2SO_4 forms a compound Y which on the addition of one molecule of hydrogen in the presence of nickel forms a compound Z. One molecule of compound Z on combustion forms two molecules of CO_2 and three molecules of H_2O. Identify giving reasons, the compounds X, Y and Z. Write the chemical equations for all the chemical reactions involved.

10. Both soap and detergent are same type of salts. What is the difference between them? Describe in brief the cleansing action of soap. Why do soaps not form lather in hard water? List two problems that arise due to the use of detergents instead of soaps. **[CBSE 2017]**

11. Explain the following reactions with one example for each giving relevant chemical equations.
- (i) Hydrogenation reaction
- (ii) Oxidation reaction
- (iii) Substitution reaction
- (iv) Combustion reaction
- (v) Saponification reaction

12. (i) Differentiate between alkanes and alkenes. Name and draw the structure of one member of each.
(ii) Alkanes generally burn with a clean flame. Why?

13. What are esters? How they are prepared? List two uses of esters.

Or

What are esters? How are esters prepared? Write the chemical equation for the reaction involved. What happens when an ester reacts with sodium hydroxide? Write the chemical equation for the reaction and also state the name and use of this reaction.

14. Name the oxidising agents used for the conversion of ethanol to ethanoic acid. Distinguish between ethanol and ethanoic acid on the basis of (i) litmus test, (ii) reaction with sodium carbonate.

15. List two tests for experimentally distinguishing between an alcohol and a carboxylic acid and describe how these tests are performed. **[CBSE 2015]**

16. Give reasons.
- (i) Why is pure ethanoic acid known as glacial acetic acid?
- (ii) Why is ethanol used as an antifreeze?
- (iii) Why is ethanol mixed with methanol known as denatured alcohol?
- (iv) Why is an oxyacetylene flame used for welding purposes?
- (v) What is gasol?

17. *Ethanol, commonly called alcohol, is an excellent solvent, it is used in medicines and is an important chemical compound involved in the synthesis of many chemical compounds. However, in spite of its benefits to man, its impact on social behaviour has always been questioned. Media has often shown abnormal behaviour of people while drunk. It is considered as a curse in the lives of those who are addicted to alcohol. Alcoholic people are not only lowering their metabolism and affecting their central nervous system, they are also in danger of liver damage. Anger and rude behaviour are some of its ill effects.
- (i) Comment on the statement: 'Should production of alcohol be banned'. Give three valid reasons to justify.
- (ii) As a student, what initiative would you take in the common concern of 'Save life, do not drink'. Give two suggestions.

18. On the basis of the study of saponification reactions, a group of students noted down the following comments in their notebooks.
- (i) The nature of the reaction mixture is basic.
- (ii) In saponification reactions, heat is evolved.
- (iii) Saponification reaction is not a neutralisation reaction.
- (iv) Soap is nothing but a salt of fatty acids.

Which of these comments is correct according to you?

19. What are detergents chemically? List two merits and two demerits of using detergents for cleaning. State the reason for the suitability of detergents for washing, even in the case of water having calcium and magnesium ions.

20. Both soaps and detergents are the same type of salts. What is the difference between them? Describe in brief the cleansing action of soap. Why do soaps not form lather in hard water? List two problems that arise due to the use of detergents instead of soaps.

21. What are micelles? Why do they form when soap is added to water? Will a micelle be formed in other solvents such as ethanol also? State briefly how the formation of micelles help to clean the clothes having oily spots. **[Board, 2016]**

Section E: Case Study or Passage Based Questions (4 Marks)

There are many types of reactions shown by organic compounds such as hydrocarbons and their derivatives. For example, hydrocarbons can mainly show substitution, addition and oxidation reactions. Other organic compounds like alcohols, aldehydes, ketones and acids can show chemical properties on the basis of functional groups present in them. For example, alcohols on oxidation give acids. An acid and an alcohol combine together to form an ester. There are many reagents which are used to distinguish between organic compounds. These reactions have many advantages commercially and in our daily lives.

1. Which of the following can show an addition reaction with bromine water solution?
(a) C_2H_6 (b) C_2H_4
(c) C_3H_6 (d) both (b) and (c)

2. Which of the following reagents can be used to convert an alcohol into an acid?
(a) alkaline $KMnO_4$ (b) bromine water
(c) acidic $K_2Cr_2O_7$ (d) both (a) and (c)

3. Which of the following cannot show substitution reaction with chlorine in sunlight?
(a) ethane (b) cyclohexane
(c) hexene (d) butane

4. Which of the following compounds can be used to prepare an ester?
(a) ethanoic acid (b) ethyl alcohol
(c) acetone (d) both (a) and (b)

5. Combustion of an organic compound gives mainly:
(a) CO_2 (b) H_2O
(c) lots of heat (d) all of these

High Order Thinking Skills (HOTS) Questions

1. (i) An organic compound burns with a sooty flame. Is it a saturated or an unsaturated compound? Justify.
(ii) Why is pure ethanoic acid called glacial ethanoic acid (or glacial acetic acid)?

2. *What do you observe when you add a few drops of acetic acid to a test-tube containing:
(i) phenolphthalein
(ii) distilled water
(iii) universal indicator
(iv) sodium hydrogen carbonate powder **[Board, 2016]**

3. An organic compound A of molecular formula C_2H_4 on reduction gives another compound B of molecular formula C_2H_6. B on reaction with chlorine in the presence of sunlight gives C of molecular formula C_2H_5Cl.
(i) Name the compounds A, B and C.
(ii) Write the chemical equation for the conversion of A to B and name the type of reactions.

4. (i) What would be observed on adding a 5% alkaline potassium permanganate drop by drop to some warm ethanol taken in a test tube?
(ii) Write the name of the compound formed during the chemical reaction.

5. (i) An organic compound X reacts with sodium metal to form sodium ethoxide and a gas Y. Identify X and Y.
(ii) What happens when ethanol is heated at 443 K with conc. H_2SO_4?

6. An organic acid X is a liquid which often freezes during winter time in cold countries.
It has a molecular formula, $C_2H_4O_2$. On warming with ethanol in the presence of a few drops of conc. H_2SO_4, a compound Y with a sweet smell is formed.
(i) Identify X and Y.
(ii) Write chemical equations for the reactions involved.

7. Two compounds A and B have the molecular formula C_3H_8 and C_3H_6 respectively. Which one of the two is most likely to show addition reactions? Justify your answer. Explain with the help of a chemical equation, how an addition reaction is useful in the vegetable ghee industry.

8. You have three unlabelled test tubes containing ethanol, ethanoic acid and soap solution. Explain the method you would use to identify the compounds in different test tubes by chemical tests using litmus paper and sodium metal.

9. An organic compound A is widely used as a preservative in pickles and has a molecular formula $C_2H_4O_2$. This compound reacts with ethanol to form a sweet-smelling compound B.
   (i) Identify the compound A.
   (ii) Write the chemical equation for its reaction with ethanol.
   (iii) Name the products formed.
   (iv) Name the process involved in the reaction.
   (v) How can we get back the compound A from B?

10. Two compounds A and B have the same molecular formula $C_3H_6O_2$. One of them reacts with sodium metal and $NaHCO_3$ to liberate H_2 gas and a gas C, respectively. C turns lime water milky. The other compound does not react with either Na metal or $NaHCO_3$ but undergoes saponification to give the sodium salt of a carboxylic acid and wood spirit, D. Identify A, B, C and D.

Answers

Section A: Multiple Choice Questions

1. (c)	2. (b)	3. (d)	4. (d)
5. (c)	6. (b)	7. (d)	8. (c)
9. (d)	10. (b)	11. (b)	12. (a)
13. (a)	14. (a)	15. (d)	16. (c)
17. (a)	18. (b)	19. (d)	20. (b)
21. (c)	22. (a)	23. (c)	24. (d)
25. (b)			

Assertion Reason Questions

1. (b)	2. (c)	3. (d)	4. (b)
5. (c)	6. (a)	7. (a)	8. (a)

Section B: Very Short Answer Questions

1. C_3H_4, C_2H_4 will undergo addition reactions because they are unsaturated compounds.

2. CH_3–CH=CH_2 undergoes an addition reaction because of the presence of the double bond.

3. Hydrogenation.

4. Esters.

5. (a) $CH_4 + O_2 \xrightarrow{\text{Sunlight}} CO_2 + 2H_2O$ + Heat + Light
   (b) $CH_4 + Cl_2 \xrightarrow{\text{Sunlight}} CH_3Cl + HCl$

6. Methane burns in air with the formation of carbon dioxide and water.
   $CH_4 + 2O_2 \rightarrow CO_2 + 2H_2O$+ Heat energy

7. Ethene gas decolourises the brown colour of bromine water.

8. $2CH_3COOH + Na_2CO_3 \rightarrow 2CH_3COONa + H_2O + CO_2\uparrow$

9. $CH_3CH_2OH \xrightarrow[H_2SO_4,\ 443\ K]{\text{Hot conc.}} CH_2{=}CH_2 + H_2O$

10. 5% to 8% solution of acetic acid in water is called vinegar. It is used as a preservative in pickles.

11. $CH_3COOH + NaHCO_3 \rightarrow CH_3COONa + H_2O + CO_2$

12. $\underset{\text{Ethanol}}{CH_3CH_2OH} \xrightarrow{KMnO_4,[O]} \underset{\text{Ethanoic acid}}{CH_3COOH}$

13. Neutralisation reaction. It is used in the form of vinegar, which is a dilute solution of ethanoic acid in water.

14. There is no change in the colour of blue litmus paper as soap solution is alkaline.

15. She will observe that soap settles down in the test tube.

16. Olive oil (A) and sodium hydroxide forms soap (B).

17. Beaker, measuring cylinder, thermometer, test tube, glass rod.

18. No, because soaps are biodegradable.

19. Sodium stearate and sodium palmitate are used in the making of soaps.

Case Study or Passage Based Questions

20. It is known as substitution reaction.

21. On reaction with bromine water, the alkene will decolourise it while the alkane does not react with it.

22. The conversion of vegetable oil into ghee is known as hydrogenation.

23. Combustion is always exothermic in nature.

24. The oxidation of propanol by alkaline potassium permanganate gives propanoic acid.

$$CH_3-CH_2-CH_2OH + O_2 \xrightarrow{\text{Potassium permanganate}} CH_3-CH_2-COOH + H_2O$$

Section C: Short Answer Questions

1. (i) Reactions between unsaturated hydrocarbons with simple substances to form a single saturated product are called addition reactions.

(ii) Those reactions in which one or more hydrogens of a saturated hydrocarbon are replaced by an atom or a group of atoms are known as substitution reactions.

2. $$\underset{\text{Ethanol}}{CH_3CH_2OH} \xrightarrow[-H_2O]{2[O]} \underset{\text{Ethanoic acid}}{CH_3COOH}$$

The addition of oxygen to any compound or element is called oxidation. As the conversion of ethanol to ethanoic acid shows the addition of oxygen to ethanol, it is an oxidation reaction.

3. Butter is mostly made up of heart-healthy saturated fat and is low in polysaturated fat. Saturated fats are good for health. A vegetable oil is a triglyceride extracted from a plant, that are liquids at room temperature. These oil contain very large amounts of biologically active fats called omega-6 polyunsaturated fatty acids, which are harmful in excess.

4.

S. No	Soaps	Detergents
(i)	They are sodium salts of long chain fatty acids.	These are sodium or potassium salts of sulfonic acids of hydrocarbons.
(ii)	Soaps cannot be used with hard water.	Detergents work well with both hard and soft water.
(iii)	They are fully biodegradable.	They are non-biodegradable.
(iv)	They take time to dissolve in water.	They dissolve faster in water.

5. When any molecule like H_2 adds on to an unsaturated hydrocarbon due to the presence of a double or triple bond, it is called an addition reaction.

$CH_2=CH_2 + H_2 \rightarrow C_2H_6$

Use: Hydrogenation of vegetable oils.

Catalyst: Increases the rate of reaction.

Substitution reaction: $CH_4 + Cl_2 \rightarrow CH_3Cl + HCl$ one atom is replaced by another; two products only, one product in addition reaction.

6. (a) Complete reactions are:

(i) $CH_4+Cl_2 \xrightarrow{\text{Sunlight}} CH_3Cl+HCl$

(ii) $CH_3CH_2OH \xrightarrow{\text{Hot.conc.}} H_2C=CH_2+H_2O$

(b) When soap reacts with hard water, the minerals present in water react with soap and form a white curdy substance known as scum.

It reduces the cleansing ability of soap.

7. (i) Soap molecules form micelles where the ionic end of soap dissolves in water while the carbon chain dissolves in oil. The emulsion in water thus formed helps in dissolving the dirt in water and we can wash our clothes clean.

(ii) Soaps react with calcium or magnesium ions to form an insoluble substance which results in wastage of soap.

8. Ethyl alcohol mixed with methyl alcohol is called denatured alcohol.

Methyl alcohol is added to ethyl alcohol in order to make it unfit for drinking.

9. The magnesium and calcium salts present in hard water react with the soap molecules to form an insoluble product called scum. This scum creates difficulty in the cleansing action.

10. $CH_3COOH + C_4H_9OH \rightarrow CH_3COOC_4H_9 + H_2O$
Acid Alcohol Ester

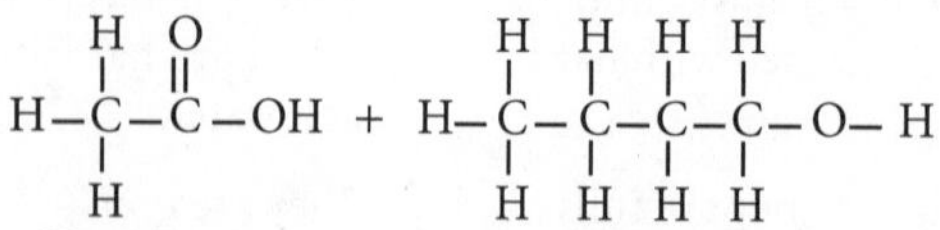

Ethanoic acid Butanol

The process is esterification.

11. (i) $CH_4 + 5O_2 \rightarrow CO_2 + H_2O$ + heat + light
(ii) $2CH_3CH_2OH + 2Na \rightarrow 2CH_3CH_2ONa + H_2$
(iii) $CH_3COOH + NaOH \rightarrow CH_3COONa + H_2O$

12. The presence of Ca^{2+} ions and Mg^{2+} ions.
Formation of insoluble calcium and magnesium salts with soap.
Soap gets wasted as it does not form lather with soap.

13. In hard water, soap reacts with calcium and magnesium salts which are present in hard water and forms insoluble substances called scum.
Detergents are ammonium or sulfonate salts of long chain carboxylic acids. The charged ends of these compounds do not form insoluble precipitates with calcium and magnesium ions in hard water.

14. Common salt is added in saponification reaction, as it helps in the precipitation of soap by decreasing the solubility of soap.

15. Saponification reaction is a highly exothermic reaction. The reaction mixture of this reaction is basic in nature and hence, this will turn red litmus blue.

16. When you put soap and water in a test tube and shake it well for 10 times by placing a thumb on its mouth, it will form foam.

17. X: Ethanol (C_2H_5OH)
Y: Ethene (C_2H_4)
Z: Hydrogen gas ($H_2 \uparrow$)

The chemical equation involved is:

$$\underset{\text{Ethanol}}{C_2H_5OH} \xrightarrow[-H_2O]{\text{Conc. } H_2SO_4} \underset{\text{Ethene}}{H_2C{=}CH_2}$$

The role of sulfuric acid (conc. H_2SO_4) is to act as a dehydrating agent and remove water molecules from ethanol. It also acts as a catalyst here.

18. Oxidising agents are the substances which add oxygen to other compounds/elements. Examples are alkaline $KMnO_4$ and acidified $K_2Cr_2O_7$. When an oxidising agent is added to propanol, propanoic acid is formed.
Chemical equation:

$$\underset{\text{Propanol}}{CH_3CH_2CH_2OH} + 2[O] \xrightarrow{KMnO_4/H_2SO_4} \underset{\text{Propanoic acid}}{CH_3CH_2COOH} + H_2O$$

19. Two tests: (i) When we put a drop of the given colourless liquid on blue litmus paper, if the blue litmus paper changes to red, the given acid is acetic acid. (ii) On smelling the given liquid, if it smells like vinegar, the given acid is acetic acid.

20. (i) Ethanol is a solvent used to make tincture of iodine because ethanol can dissolve the cell membranes of microorganisms which is made up of lipids and kill the microorganisms (which may be pathogenic).
(ii) Concentrated sulfuric acid acts as a dehydrating agent, that is, it removes water molecules from ethanol and also catalyses the reaction and converts ethanol to ethene.

21. The essential materials/chemicals used to prepare soap in the laboratory are vegetable oil/fat and sodium hydroxide. The test to determine the nature of the reaction mixture of the saponification reaction is the red litmus test. In this test, the red litmus paper turns blue.

22. Castor oil, turpentine and coconut oil can be used in the preparation of soap.

23. Scum is formed by the reaction of soap with calcium and magnesium salts in water. Test tubes R and S contain calcium chloride and

magnesium chloride, respectively. Therefore, scum will be formed in these two test tubes.

24. (i) It is a substance which can give oxygen to other substances.

(ii)

$$\underset{\text{Propanol}}{CH_3-CH_2-CH_2-OH} \xrightarrow[\text{Acidified } K_2Cr_2O_7 + \text{Heat}]{\text{Alkaline } KMnO_4 + \text{Heat}} \underset{\text{Propanoic acid}}{CH_3-CH_2-COOH}$$

(iii) Propanol is oxidised to propanoic acid.

Section D: Long Answer Questions

1. (i) Saturated hydrocarbons burn with a blue and non-smoky flame as there is complete combustion and unsaturated hydrocarbons generally burn with a sooty flame as there is incomplete combustion. Also, unsaturated compounds decolourise bromine and their colour, therefore, disappears.

$$\underset{\text{Ethene}}{CH_2=CH_2} + \underset{\text{Bromine (Brown)}}{Br_2(aq)} \rightarrow \underset{\text{1, 2-dibromoethane (Colourless)}}{\underset{Br\qquad Br}{CH_2-CH_2}}$$

Saturated compounds do not decolourise the bromine water.

(ii) Carbon dioxide and water are formed when ethane burns in air.

$$C_2H_6 + \frac{7}{2}O_2 \rightarrow 2CO_2 + 3H_2O + \text{Light} + \text{Heat}$$

(iii) The reaction in which a reagent substitutes (replaces) an atom or a group of atoms of the reactant (substrate) to form the product are called substitution reactions. These are generally shown by saturated compounds and benzene. When chlorine is added to hydrocarbons at a rapid rate in the presence of sunlight, Cl replaces the H atoms one by one to give carbon tetrachloride as the final product.

$$CH_4 + Cl_2 \xrightarrow{\text{Sunlight}} CH_3Cl + HCl$$

$$CH_3Cl + Cl_2 \rightarrow \xrightarrow{\text{Sunlight}} CH_2Cl_2 + HCl$$

$$CH_2Cl_2 + Cl_2 \xrightarrow{\text{Sunlight}} CHCl_3 + HCl$$

$$CHCl_3 + Cl_2 \xrightarrow{\text{Sunlight}} CCl_4 + HCl$$

2. (i) Hydrogenation reaction involves the addition of hydrogen to an unsaturated hydrocarbon in the presence of nickel as catalyst to form saturated hydrocarbons (alkene to alkyne). This reaction is used for the conversion of vegetable oil into fats (vanaspati ghee).

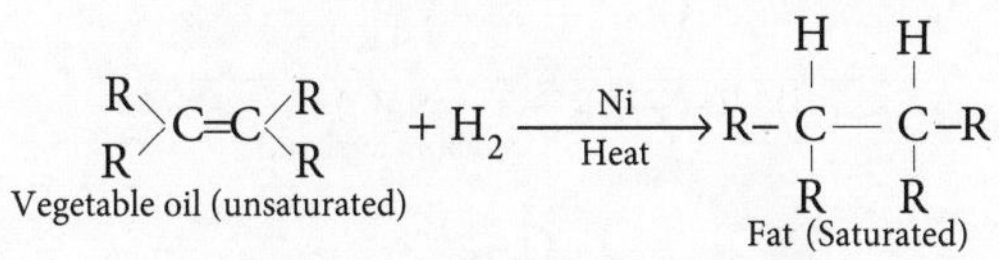

Physical properties: Vegetable oil, a liquid, gets converted into fats which is solid.

(ii) Saturated hydrocarbons, that is, alkanes react with chlorine in the presence of sunlight to form the chloro-derivatives of alkanes. For example,

$$\underset{\text{Methane}}{CH_4} + Cl_2 \xrightarrow{\text{Sunlight}} \underset{\text{Chloromethane}}{CH_3Cl} + HCl$$

Type of reaction: Substitution reaction. It is called so, as in the above reaction, hydrogen atoms in methane are replaced by chlorine atoms.

3. (a) Isomers are those compounds which have the same molecular formula but different structural formula.

(b) Propanal: CH_3CH_2CHO
Propanone: CH_3COCH_3

(c) (i) $CH_3CH_2OH \xrightarrow[\text{Conc. } H_2SO_4]{443\text{ K}} H_2C=CH_2 + H_2O$

(ii) $CH_3CH_2OH \xrightarrow[\text{Heat}]{\text{Alkaline } KMnO_4} CH_3CH_2COOH + H_2O$

4.

Esterification reaction	Saponification reaction
This reaction is used for the preparation of esters in which an alcohol is made to react with a carboxylic acid in the presence of conc. H_2SO_4. For example, $\underset{\text{Ethanol}}{C_2H_5OH} + \underset{\text{Ethanoic acid}}{CH_3COOH} \xrightarrow{\text{Conc. } H_2SO_4} \underset{\text{Ester (Ethyl ethanoate)}}{CH_3COOC_2H_5} + H_2O$ It is used in making perfumes.	The alkaline hydrolysis of esters (that is, the breakdown of esters) to form the sodium salt of the esters and alcohol is called saponification reaction. Example, $CH_3COOC_2H_5 + NaOH \xrightarrow{\text{Heat}} \underset{\text{Sodium ethanoate}}{CH_3COONa} + C_2H_5OH$ It is used in the preparation of soaps.

5. (i) $\underset{\text{Ester}}{R-\overset{\overset{\displaystyle O}{||}}{C}-OR'} + \underset{\text{Sodium hydroxide}}{NaOH} \rightarrow$

$$\underset{\text{Sodium carboxylate}}{R-\overset{\overset{\displaystyle O}{||}}{C}-ONa} + \underset{\text{Alcohol}}{R'-OH}$$

$$CH_3COOC_2H_5 \xrightarrow{NaOH} C_2H_5OH + CH_3COONa$$

(ii) $CH_4 + Cl_2 \xrightarrow{\text{Sunlight}} CH_3Cl + HCl$

The reaction does not stop here and all the hydrogen in the methane can in turn be replaced by chlorine atoms.

$$CH_3Cl + Cl_2 \xrightarrow{\text{Sunlight}} CH_2Cl_2 + HCl$$

$$CH_2Cl_2 + Cl_2 \xrightarrow{\text{Sunlight}} CHCl_3 + HCl$$

$$CHCl_3 + Cl_2 \xrightarrow{\text{Sunlight}} CCl_4 + HCl$$

(iii)

$$\underset{\text{Ethanol}}{C_2H_5OH} + \underset{\text{Ethanoic acid}}{CH_3COOH} \xrightarrow{H_2SO_4} \underset{\text{Ester}}{CH_3COOC_2H_5} + H_2O$$

6.

$$\underset{\text{Carboxylic acid}}{CH_3COOH} + \underset{\text{Alcohol}}{CH_3CH_2OH} \underset{KMnO_4}{\overset{H_2SO_4}{\rightleftharpoons}} \underset{\text{Compound X}}{CH_3COOCH_2CH_3} + H_2O$$

(i) **Carboxylic acid:** CH_3COOH (Ethanoic acid)

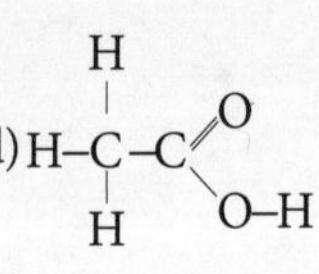

(ii) **Alcohol:** CH_3CH_2OH (Ethanol)

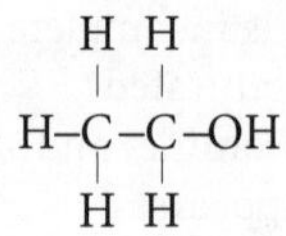

(iii) X = $CH_3COOCH_2CH_3$ (Ethyl ethanoate)

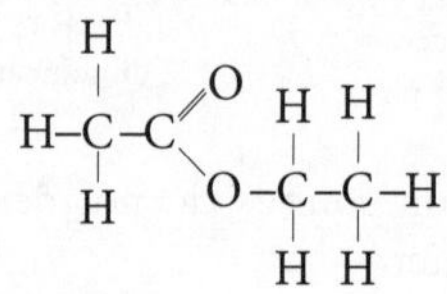

7. The structural formula of ethanol is

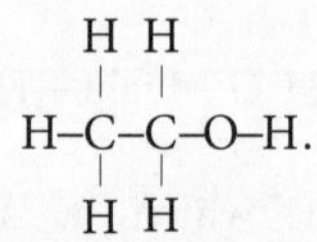

When ethanol is heated at 443 K in excess of conc. H_2SO_4, it gets dehydrated and results in the formation of ethane.

The structural formula of ethane is

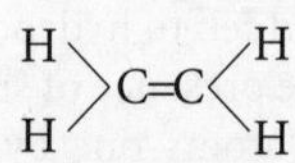

Chemical equation:

$$\underset{\text{Ethanol}}{C_2H_2OH} \xrightarrow[-H_2O]{\text{Conc. } H_2SO_4 (443K)} \underset{\text{Ethene}}{H_2C{=}CH_2}$$

In the above reaction, conc. H_2SO_4 acts as a dehydrating agent and removes water molecules from ethanol.

8. $CH_3COOH + NaHCO_3 \rightarrow CH_3COONa + H_2O + CO_2\uparrow$

The salt X is sodium ethanoate. The gas Y evolved is carbon dioxide.

Activity: Set up the apparatus as shown. Take some ethanoic acid and sodium hydrogen carbonate in the test tube as well as some limewater in a curved tube.

Observation: Limewater will turn milky, which is a characteristic property of CO_2 gas.

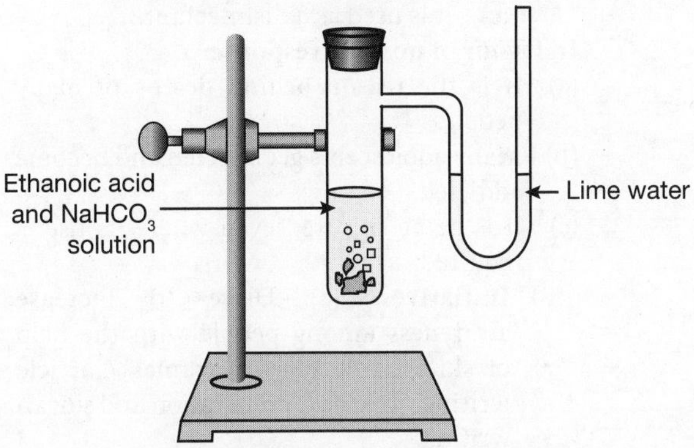

Reaction involved:

$$CH_3COOH + NaHCO_3 \rightarrow CH_3COONa + H_2O + CO_2$$

$$Ca(OH)_2 + CO_2 \rightarrow CaCO_3 + H_2O$$

9. Compound X is ethanol, Y is ethene and Z is ethane.

$$\underset{\text{Ethanol X}}{C_2H_5OH} \xrightarrow[-H_2O]{\text{Conc. } H_2SO_4} \underset{\text{Ethene Y}}{C_2H_4}$$

$$\underset{\text{Ethene}}{H_2C{=}CH_2} + H_2 \xrightarrow[\Delta]{Ni} \underset{\text{Ethane Z}}{CH_3{-}CH_3}$$

$$\underset{\text{Ethane}}{C_2H_6} + \frac{7}{2}O_2 \xrightarrow{\text{Combustion}} 2CO_2 + 3H_2O$$

10. **Difference between soap and detergent:** Soaps are sodium or potassium salts of higher fatty acids. Detergents are sodium or potassium salts of sulfonic acids of benzene or alkene type hydrocarbons. Soaps have the –COONa group, whereas detergents have the $-SO_3Na$ or $-SO_4Na$ group.

Cleansing action of soap: Soaps have a hydrophobic part which attracts dirt and grease whereas hydrophilic part attracts water. Dirt and grease are washed away by water.

Action of soap in hard water: Soaps react with the Ca^{2+} and Mg^{2+} ions present in hard water to form scum (calcium and magnesium salts of fatty acids) and soap gets wasted.

Problems: The two problems are:

- detergents are more expensive than soap,
- some detergents are non-biodegradable and create water pollution.

11. See text (Section 4.11).

12. (i) Difference between alkanes and alkenes.

S. No	Alkanes	Alkenes
(i)	In all the compounds, carbon and hydrogen are attached with single bonds.	These are unsaturated hydrocarbons which have one or more double bonds.
(ii)	Member: Methane (CH_4) 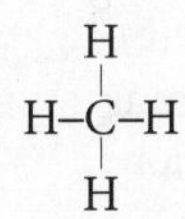	Member: Ethene ($CH_2{=}CH_2$) 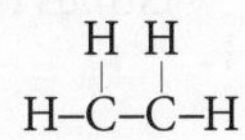

(ii) Alkanes are saturated hydrocarbons which always burn with a clean blue flame because complete combustion takes place in sufficient oxygen to give CO_2 and H_2O with the liberation of a large amount of heat and light.

13. Esters are the pleasant fruity smelling organic compounds derived from carboxylic acids which contain the -COOR group. The -OH group in an acid is replaced by -O-alkyl or aryl group to form an ester. Esters are represented by the formula R_1COOR_2 where R_1 and R_2 represent an alkyl or aryl group.

Preparation of esters: When carboxylic acids are heated with alcohols in the presence of an acid catalyst, esters are produced.

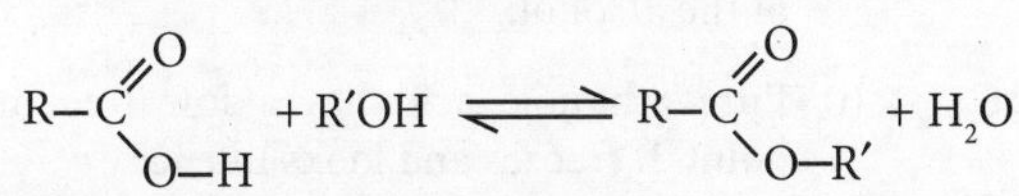

$$CH_3COOH + C_2H_5OH \xrightarrow{\text{Acid}} \underset{\text{Ester}}{CH_3COOC_2H_5} + H_2O$$

When an ester reacts with sodium hydroxide, sodium ethanoate (sodium salt of ester) is formed. The name of the reaction is saponification and this reaction is used in the preparation of soap.

$$CH_3COOC_2H_5 + NaOH \rightarrow CH_3COONa + C_2H_5OH$$

Uses of esters: (i) In food as flavouring agents and fragrances.
(ii) In making of soaps.
(iii) As solvents.
(iv) In medicines.
(v) As emulsifying agents. **(Any two)**

14. The oxidising agents used for the conversion of ethanol to ethanoic acid are alkaline potassium permanganate ($KMnO_4$) and acidified potassium dichromate ($K_2Cr_2O_7$).

S. No	Test	Ethanol	Ethanoic acid
(i)	Litmus test	No change in colour of litmus solution	Blue litmus solution turns red
(ii)	Reaction with sodium carbonate	No brisk effervescence	Brisk effervescence due to evolution of CO_2

15. (i) **Litmus test:** Add a small amount of blue litmus solution to the given compound. If the blue litmus solution turns red, it indicates that the organic compound is a carboxylic acid. But alcohol does not give this test.
(ii) **Sodium bicarbonate test:** To a small portion of the organic compound in a test tube, add a pinch of solid sodium bicarbonate. Evolution of carbon dioxide with a brisk effervescence shows the presence of carboxylic acid. But there is no release of carbon dioxide in the case of the alcohol.

16. (i) Pure ethanoic acid has a low freezing point. It freezes and looks like ice.
(ii) Ethanol mixed with water freezes at a temperature lower than the freezing point of water. Thus, a mixture of water and ethanol is used in radiators of cars in cold countries.
(iii) When methanol is mixed with ethanol, the latter becomes undrinkable.
(iv) Acetylene mixed with oxygen burns giving a large amount of heat which is used for welding purposes.
(v) Gasol is a mixture of petrol and alcohol.

17. (i) **In favour of negative response:**
(a) It is an excellent industrial solvent.
(b) It is used in so many ways such as in medicines and ornamentation.
(c) It is used as a disinfectant.

In favour of positive response:
(a) It is the reason behind deaths of many people.
(b) Many adolescents get affected and become addicted.
(c) It is being misused even where its use is required.

(ii) **Initiatives:** (a) Drive to increase awareness among people with the help of skits / role plays / dramas / article writing. (b) Chart preparation and slogan writing.

18. The correct comments are:
(i) In saponification reactions, heat is evolved.
(ii) Saponification reaction is not a neutralisation reaction.
(iii) Soap is nothing but a salt of a fatty acid.

19. Detergents are chemically ammonium or sulfonate salts of long chain carboxylic acids.
Merits of using detergents for cleansing:
(i) They can work effectively even in hard water.
(ii) They have better cleansing action than soaps.

Demerits of using detergents of cleansing:
(i) They are non-biodegradable and thus harm aquatic life.
(ii) They are prepared from petroleum hydrocarbons.

Detergents can work effectively even if calcium and magnesium ions are present in water. Since the charged ends of detergents do not form an insoluble precipitate called scum with hard water, they clean the clothes more effectively than soaps.

20. **Differences between soap and detergent:** Soaps are sodium or potassium salts of higher fatty acids. Detergents are sodium or potassium salts of sulfonic acids of benzene or alkene type hydrocarbons. Soaps have the $-COONa$ group, whereas detergents have the $-SO_3Na$ or $-SO_4Na$ group.

Cleansing action of soap: Soaps have a hydrophobic part which attracts the dirt and grease whereas the hydrophilic part attracts water. Dirt and grease are washed away by water.

Action of soap in hard water: Soaps react with Ca^{2+} and Mg^{2+} ions present in hard water to form scum (calcium and magnesium salts of fatty acids) and soap gets wasted.

Problems: The two problems are:
(i) detergents are more expensive than soap,
(ii) some detergents are non-biodegradable and create water pollution.

21. (i) Soap molecules have two ends—one end is the hydrocarbon chain which is water repellent (hydrophobic), whereas the other end is the ionic part which is the water-soluble end (hydrophilic). When soap is dissolved in water it forms a group of many molecules, known as a micelle.
(ii) These micelles are formed as their hydrocarbon chains come together and the polar ends are projected outwards.
(iii) Micelle formation in ethanol will not occur as the hydrocarbon chain end of the soap will dissolve in ethanol.
(iv) Soaps in the form of micelles are able to clean dirty clothes having oily spots, as the oily dirt is collected in the centre of the micelle, which forms an emulsion in water and on rinsing, the water washes away the micelles with dirt attached to them.

Section E: Case Study or Passage Based Questions

1. (d) 2. (d) 3. (c) 4. (d)
5. (d)

High Order Thinking Skills (HOTS) Questions

1. (i) It is an unsaturated compound because it has more carbon. It burns with a sooty or smoky flame.
(ii) When pure ethanoic acid is cooled, it freezes to form a colourless, ice-like liquid. This pure ethanoic acid is also called acetic acid (or glacial acetic acid).

2. (i) No change (remains colourless)
(ii) No change
(iii) Turns pink/orange
(iv) Evolution of colourless gas

3. (i) A: $CH_2=CH_2$ Ethene
B: CH_3-CH_3 Ethane
C: CH_3-CH_2-Cl Chloroethane

(ii) $\underset{(A)}{CH_2=CH_2} \xrightarrow[\text{Nickel}]{\text{Hydrogen}} \underset{(B)}{CH_3-CH_3}$

Type of reaction: Addition reaction.

4. (i) Colour of $KMnO_4$ disappears because it takes part in the oxidation of ethanol.
(ii) Ethanol is oxidised to produce ethanoic acid.

5. (i) $2C_2H_5OH + 2Na \rightarrow 2C_2H_5ONa + H_2$
X is ethanol and Y is hydrogen.
(ii) $CH_3CH_2OH \xrightarrow{\text{Conc. } H_2SO_4} CH_2=CH_2 + H_2O$

6. (i) X is CH_3COOH and Y is $CH_3COOCH_2CH_3$.
(ii) $CH_3COOH + C_2H_5OH \longrightarrow CH_3COOC_2H_5 + H_2O$

7. A is a saturated hydrocarbon, propane.
B is an unsaturated hydrocarbon $CH_2=CH-CH_3$ which will undergo an addition reaction.

$CH_2=CH-CH_3 + Br_2 \rightarrow \underset{\text{(1, 2 - dibromo-propane)}}{CH_2Br-CHBr-CH_3}$

$\underset{\text{(Unsaturated)}}{\text{Vegetable oil}} + H_2 \rightarrow \underset{\text{(Saturated)}}{\text{Vegetable ghee}}$

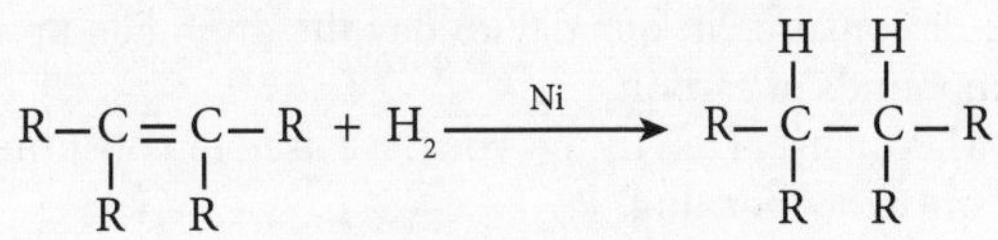

8.

Solution	Blue litmus paper	Red litmus paper	Sodium metal
Ethanol	No change	No change	Hydrogen gas
Ethanoic acid	Turns red	No change	Hydrogen gas
Soap	No change	Turns blue	Hydrogen gas

9. (i) CH_3COOH (Ethanoic acid).
(ii) $CH_3-COOH + C_2H_5OH \rightarrow CH_3-COOC_2H_5 + H_2O$
(iii) Ethyl ethanoate.
(iv) Esterification.
(v) Through saponification.

10. Two possible compounds with the molecular formula $C_3H_6O_2$ are propanoic acid (CH_3CH_2COOH) and methyl acetate (CH_3COOCH_3). Propanoic acid reacts with Na metal to give sodium propanoate and H_2 gas. With $NaHCO_3$, it gives CO_2 gas which turns limewater milky. Methyl acetate on saponification gives sodium acetate and wood spirit—methanol. Hence, A is propanoic acid, B is methyl acetate, C is carbon dioxide and D is methanol.

Hints

Section A: Multiple Choice Questions

24. 1. Substitution (iii) $CH_4 + Cl_2 \xrightarrow{hv, Heat} CH_3-Cl + HCl$
2. Addition (iv) $CH_2=CH_2 + Br_2 \xrightarrow{CCl_4} BrCH_2-CH_2Br$
3. Oxidation (i) $CH_3CH_2OH + O_2 \xrightarrow[Heat]{KMnO_4/KOH} CH_3COOH + H_2O$
4. Combustion (ii) $CH_2=CH_2 + 3O_2 \xrightarrow{\Delta} 2CO_2 + 2H_2O$

25. 1. Hydrogenation (iii) Vegetable oil + $H_2 \xrightarrow{Ni, Heat}$ Fat
2. Dehydration (iv) $CH_3CH_2OH \xrightarrow[433-443 K]{Conc. H_2SO_4} CH_2=CH_2 + H_2O$
3. Esterification (ii) $CH_3COOH + C_2H_5OH \xrightarrow[Heat]{Conc. H_2SO_4} CH_3COOC_2H_5 + H_2O$
4. Saponification (i) $CH_3COOC_2H_5 + NaOH \xrightarrow{Heat} CH_3COONa + C_2H_5OH$

Section E: Case Study or Passage Based Question

3. As hexene is an alkene, it does not show substitution reaction.

Activities with Discussion and Conclusion

Activity 4.1: To illustrate that most of the things we use or consume in our day to day life are made up of compounds of carbon.

- Make a list of ten things you have used or consumed since the morning.
- Compare the list with the lists made by your classmates and then sort the items into the following table.
- If there are items which are made up of more than one material, put them into both the relevant columns.

Things made of metals	Things made of glass/ clay (ceramics)	Others
—	—	Water
Bucket	—	Bucket (plastic)
Mug	Mug	Mug (plastic)
Cooking pans	Cooking pans	—
—	—	Furniture (Wood/plastic) Soaps and detergents
—	—	Newspaper, books, magazines

Things made of metals	Things made of glass/ clay (ceramics)	Others
—	—	Food items (bread, butter, sugar, milk, tea leaves, eggs, juice)
—	—	Medicines and vitamins

Discussion: Except water and cooking utensils (made up of metals, glass/clay or ceramics), a large number of things like soaps, detergents, newspaper, books, furniture (made up of wood and plastic) and various food items are made up of only compounds of carbon. However, some of the things such as mugs, cups/plates are usually made up of glass or ceramics while buckets are made up of metals or plastics.

Conclusion: Most of the things we use or consume in our day to day life are made up of compounds of carbon.

Activity 4.2: To illustrate the concept of homologous series.

- Calculate the difference in the formulae and molecular masses for (a) CH_3OH and C_2H_5OH (b) C_2H_5OH and C_3H_7OH and (c) C_3H_7OH and C_4H_9OH.
- Is there any similarity in these three?

Now answer:

- Arrange these alcohols in the order of increasing carbon atoms to get a family. Can we call this family a homologous series?
- Generate the homologous series for compounds containing up to four carbons for the other functional groups given in the table.

Discussion:

Difference in formulae	Difference in molecular masses
(a) C_2H_5OH — CH_3OH=CH_2	[(2×12+5 × 1+1×16+1×1) – [(1×12+3×1+1×16+1×1) = 46 – 32 = 14 u
(b) C_3H_7OH — C_2H_5OH=CH_2	[(3×12+7×1+1×16+1×1) – [(2×12+5×1+1×16+1×1) = 60 – 46 = 14 u
(c) C_4H_9OH — C_3H_7OH=CH_2	[(4×12+9×1+1×16+1×1) – [(3×12+7×1+1×16+1×1) = 74 – 60 = 14 u

The difference in formulae in each of these three cases is the same, that is, CH_2.

Similarly, the difference in molecular masses of in each of these cases is also the same, that is, 14 u.

- The order of increasing carbons of these four alcohols is:
  CH_3OH C_2H_5OH C_3H_7OH C_4H_9OH
- Yes, these four alcohols constitute a homologous series.

For homologous series of compound containing up to four carbons for the other functional groups.

Conclusion: The common difference in the molecular formulae and molecular masses of any two successive (adjacent members) of a homologous series is a CH_2 unit and 14 u (mass units) respectively.

Activity 4.3: To illustrate that saturated hydrocarbons burn with a non-luminous blue flame while unsaturated hydrocarbons give a luminous yellow sooty flame.

- Take some carbon compounds (naphthalene, camphor, alcohol) one by one on a spatula and burn them.
- Observe the nature of the flame and note whether smoke is produced.
- Place the metal plate above the flame. Is there a deposition on the plate in the case of any of the compounds?

Discussion: Camphor and alcohol burn with a blue non-luminous flame and there is no sooty deposit on the metal plate. Hence, camphor and alcohol are saturated compounds.

In contrast, naphthalene burns with a luminous yellow flame with lots of black smoke. Hence, naphthalene is an unsaturated compound.

Conclusion: Saturated compounds burn with a non-luminous blue flame while unsaturated compounds burn with a luminous yellow sooty flame.

Activity 4.4: To illustrate that complete combustion of a fuel gives a blue flame while incomplete combustion produces a yellow sooty flame.

- Light the Bunsen burner and adjust the air hole at the base to get different types of flames/presence of smoke.

Now answer:

- When do you get a yellow sooty flame?
- When do you get a blue flame?

Discussion and conclusion: When the hole of the Bunsen burner is closed, enough supply of air is not available and so the gas burns incompletely to produce a yellow sooty flame. However, when the hole is open, complete combustion of gas occurs and a blue flame is obtained.

Activity 4.5: To study oxidation of alcohols with alkaline potassium permanganate.

- Take about 3 mL of ethanol in a test tube and warm it gently in a water bath.
- Add a 5% solution of alkaline potassium permanganate drop by drop to this solution.

Now answer:

- Does the colour of potassium permanganate persist when it is added initially?
- Why does the colour of potassium permanganate not disappear when excess is added?

Discussion: When a 5% solution of alkaline $KMnO_4$ is added to warm alcohol drop by drop, the purple colour of $KMnO_4$ disappears initially because it oxidises ethanol to ethanoic acid and is itself reduced to manganese dioxide (MnO_2). However, if the $KMnO_4$ solution added is more than what is required to oxidise ethanol to ethanoic acid, then some of the $KMnO_4$ solution remains unused and hence its colour will not disappear.

Conclusion: Alkaline $KMnO_4$ solution oxidises ethanol to ethanoic acid while is itself reduced to MnO_2.

Activity 4.6: To identify the gas evolved when sodium reacts with ethanol.

- Drop a small piece of sodium, about the size of a couple of grains of rice, into ethanol (absolute alcohol).

Now answer:

- What do you observe?
- How will you test the gas evolved?

Discussion and conclusion: When a small piece of sodium is added to absolute alcohol (100% alcohol), a brisk effervescence due to the evolution of H_2 gas takes place. Hydrogen is a combustible gas, so when a burning splinter is brought near it, it burns with a pop sound.

Activity 4.7: To compare the strength of dilute acetic acid and dilute hydrochloric acid.

- Compare the pH of dilute acetic acid and dilute hydrochloric acid using both litmus paper and universal indicator.

Now answer:

- Are both acids indicated by the litmus test?
- Does the universal indicator show them as equally strong acids?

Discussion: Both dilute hydrochloric and dilute acetic acid turn blue litmus red but this action does not tell whether hydrochloric acid is a stronger acid than acetic acid. But, when we compare the pH of the aqueous solution of these two acids using the universal indicator, we find that the pH of the dilute acetic acid is around 4 (yellowish pink colour) while that of hydrochloric acid is around 2 (light pink colour). It means acetic acid is a weaker acid than hydrochloric acid.

Conclusion: As HCl is completely ionised while acetic acid is only partly ionised, hydrochloric acid is a stronger acid than acetic acid.

Activity 4.8: To study the formation and identification of an ester.

- Take 1 mL ethanol (absolute alcohol) and 1 mL of glacial acetic acid along with a few drops of concentrated sulfuric acid in a test tube.
- Warm in a water bath for a least five minutes as shown in Fig. 4.16 (of the textbook).
- Pour into a beaker containing 20–50 mL of water and smell the resulting mixture.

Discussion and conclusion: Absolute ethanol (100% alcohol) reacts with glacial (100%) acetic acid in the presence of a few drops of conc. H_2SO_4 to give a sweet-smelling ester—ethyl ethanoate. Conc. H_2SO_4 removes the H_2O formed during the reaction and thus shifts the equilibrium in the forward direction.

Activity 4.9: To demonstrate that carboxylic acids react with sodium carbonate or bicarbonate to evolve carbon dioxide gas.

- Set up the apparatus as shown in Fig. 4.15 (in the textbook).
- Take a spatula full of sodium carbonate in a test tube and add 2 mL of dilute ethanoic acid.
- What do you observe?
- Pass the gas produced through freshly prepared limewater. What do you observe?
- Can the gas produced by the reaction between ethanoic acid and sodium carbonate be identified by this test?
- Repeat this activity with sodium hydrogen carbonate.

Discussion and conclusion: When sodium carbonate is added to a dilute solution of ethanoic acid, a brisk effervescence due to the evolution of carbon dioxide (CO_2) gas is observd. When CO_2 gas is passed through freshly prepared limewater, it turns the limewater milky due to the formation of insoluble calcium carbonate. The same observations are made when sodium bicarbonate is used instead of sodium carbonate.

Activity 4.10: To demonstrate that oil dissolves in soap solution.

- Take about 10 mL of water each in two test tubes.
- Add a drop of oil (cooking oil) to both the test tubes and label them as A and B.

- To test tube B, add a few drops of soap solution.
- Now shake both the test tubes vigorously for the same period of time.
- Can you see the oil and water layers separately in both the test tubes immediately after you stop shaking them?
- Leave the test tubes undisturbed for some time and observe. Does the oil layer separate out? In which test tube does this happen first?

Discussion and conclusion: When test tube B containing cooking oil, water and soap solution was allowed to stand after vigorous shaking, only one layer was observed. This shows that oil dissolves in soap and hence helps in cleaning. When test tube A containing water and cooking oil was allowed to stand after vigorous shaking, two separate layers—one of water and the other of oil were seen. This shows that oil does not dissolve in water and hence water alone does not clean clothes.

Activity 4.11: To demonstrate that soaps produce foam with soft water but a curdy white precipitate with hard water.

- Take about 10 mL of distilled water or rain water and 10 mL of hard water (from a tube well or hand pump) in separate test tubes.
- Add a couple of drops of soap solution to both.
- Shake the test tubes vigorously for an equal period of time and observe the amount of foam formed.

Now answer:

- In which test tube do you get more foam?
- In which test tube do you observe a white curdy precipitate?

Discussion: The test tube that contains soft water (distilled water or rain water), produces foam readily. On the other hand, the test tube containing hard water produces a curdy white precipitate due to the formation of insoluble calcium or magnesium soaps.

Conclusion: Soft water readily produces foam with soaps while hard water produces a curdy white precipitate.

Activity 4.12: To demonstrate that hard water produces foam with a detergent but only a curdy white precipitate with soap.

- Take two test tubes with about 10 mL of hard water in each.
- Add five drops of soap solution to one and five drops of detergent solution to the other.
- Shake both the test tubes for the same period.

Now answer:

- Do both test tubes have the same amount of foam?
- In which test tube is a curdy solid formed?

Discussion: Both the test tubes do not have the same amount of foam. The test tube to which soap solution was added produced a curdy white precipitate with little or no foam at all, while the test tube containing the detergent produced a large amount of foam with little or no white precipitate.

Conclusion: Detergent can be used for washing both in soft and hard water while soaps can be used only in soft water.

Exemplar Problems

Short Answer Questions

1. Draw the electron dot structure of ethyne and also draw its structural formula.

2. Write the names of the following compounds

(a) 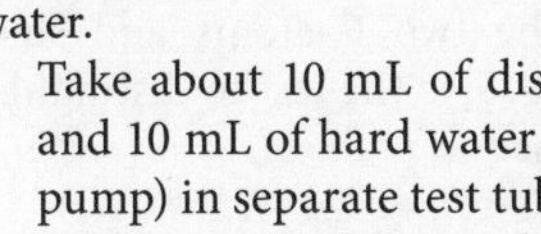

(b) 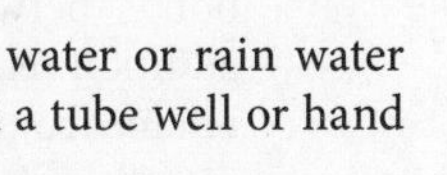

(c) 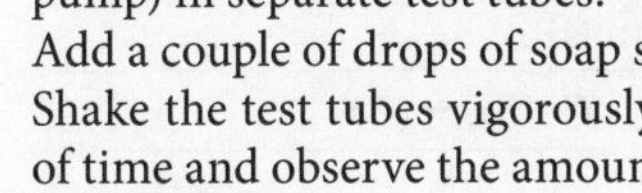

(d) 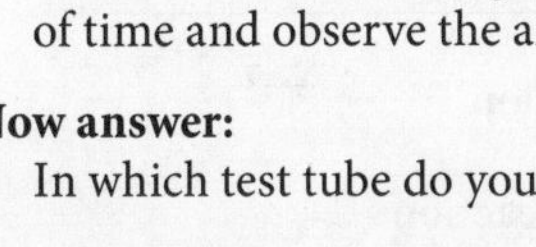

3. Identify and name the functional groups present in the following compounds.

(a) 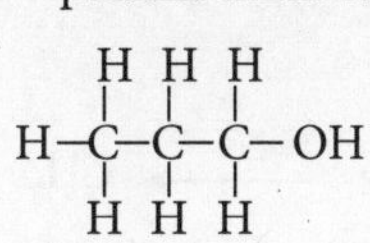

(b) 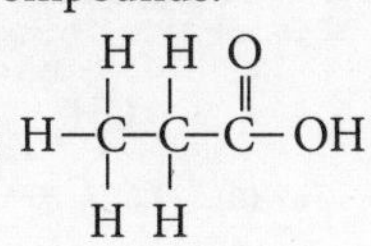

(c) 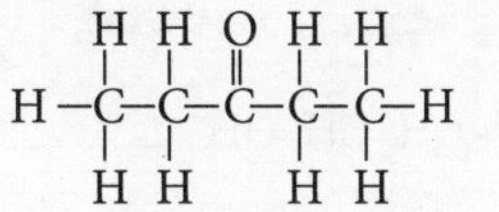

(d) 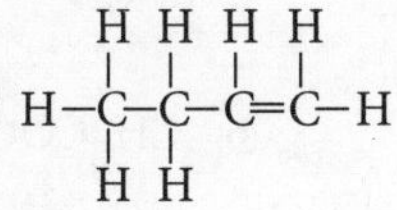

4. A compound X is formed by the reaction of a carboxylic acid $C_2H_4O_2$ and an alcohol in the presence of a few drops of H_2SO_4. The alcohol on oxidation with alkaline $KMnO_4$ followed by acidification gives the same carboxylic acid as used in this reaction. Give the names and

structures of (a) carboxylic acid, (b) alcohol and (c) the compound X. Also write the reaction.

5. Why are detergents better cleansing agents than soaps? Explain.

6. Name the functional groups present in the following compounds
(a) $CH_3COCH_2CH_2CH_2CH_3$
(b) $CH_3CH_2CH_2COOH$
(c) $CH_3CH_2CH_2CH_2CHO$
(d) CH_3CH_2OH

7. How is ethene prepared from ethanol? Give the reaction involved in it.

8. Intake of small quantity of methanol can be lethal. Comment.

9. A gas is evolved when ethanol reacts with sodium. Name the gas evolved and also write the balanced chemical equation of the reaction involved.

10. Ethene is formed when ethanol at 443 K is heated with excess of concentrated sulfuric acid. What is the role of sulfuric acid in this reaction? Write the balanced chemical equation of this reaction.

11. Carbon, a group 14 element in the periodic table, is known to form compounds with many elements. Write an example of a compound formed with
(a) chlorine (group 17 of periodic table)
(b) oxygen (group 16 of periodic table)

12. In the electron dot structure, the valence shell electrons are represented by crosses or dots.
(a) The atomic number of chlorine is 17. Write its electronic configuration.
(b) Draw the electron dot structure of the chlorine molecule.

13. Catenation is the ability of an atom to form bonds with other atoms of the same element. It is exhibited by both carbon and silicon. Compare the ability of catenation of the two elements. Give reasons.

14. Unsaturated hydrocarbons contain multiple bonds between the two C-atoms and show addition reactions. Give the test to distinguish ethane from ethene.

15. Match the reactions given in Column (A) with the names given in column (B).

Column (A)	Column (B)
(a) $CH_3OH + CH_3COOH \xrightarrow{H^+} CH_3COOCH_3 + H_2O$	(i) Addition reaction
(b) $CH_2{=}CH_2 + H_2 \xrightarrow{Ni} CH_3 - CH_3$	(ii) Substitution reaction
(c) $CH_4 + Cl_2 \xrightarrow{Sunlight} CH_3Cl + HCl$	(iii) Neutralisation reaction
(d) $CH_3COOH + NaOH \rightarrow CH_3COONa + H_2O$	(iv) Esterification reaction

16. Write the structural formulae of all the isomers of hexane.

17. What is the role of metal or reagents written on arrows in the given chemical reactions?

(a) 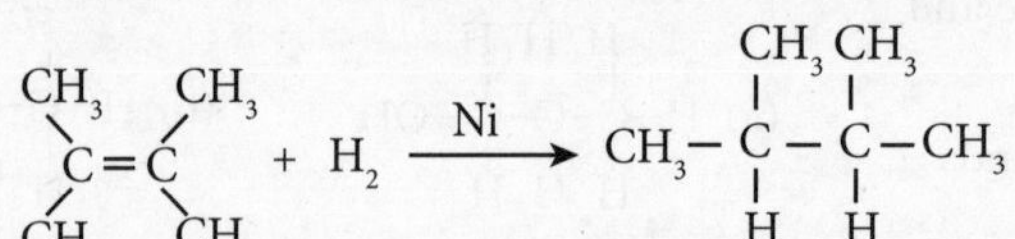

(b) $CH_3COOH + CH_3CH_2OH \xrightarrow{Conc.\ H_2SO_4} CH_3COOC_2H_5 + H_2O$

(c) $CH_3CH_2OH \xrightarrow[Heat]{Alk.\ KMnO_4} CH_3COOH$

Long Answer Questions

1. A salt X is formed and a gas is evolved when ethanoic acid reacts with sodium hydrogen carbonate. Name the salt X and the gas evolved. Describe an activity and draw the diagram of the apparatus to prove that the evolved gas is the one which you have named. Also, write the chemical equation of the reaction involved.

2. (a) What are hydrocarbons? Give examples.
   (b) Give the structural differences between saturated and unsaturated hydrocarbons with two examples each.
   (c) What is a functional group? Give examples of four different functional groups.

3. Name the reaction which is commonly used in the conversion of vegetable oils to fats. Explain the reaction involved in detail.

4. (a) Write the formula and draw the electron dot structure of carbon tetrachloride.
   (b) What is saponification? Write the reaction involved in this process.

5. Esters are sweet-smelling substances and are used in making perfumes. Suggest some activity and the reaction involved for the preparation of an ester with a well-labelled diagram.

6. A compound C (molecular formula, $C_2H_4O_2$) reacts with Na metal to form a compound R and evolves a gas which burns with a pop sound. Compound C on treatment with an alcohol A in the presence of an acid forms a sweet-smelling compound S (molecular formula, $C_3H_6O_2$). On addition of NaOH to C, it also gives R and water. S on treatment with NaOH solution gives back R and A. Identify C, R, A, S and write down the reactions involved.

7. Look at the figure and answer the following questions.
   (a) What change would you observe in the calcium hydroxide solution taken in tube B?

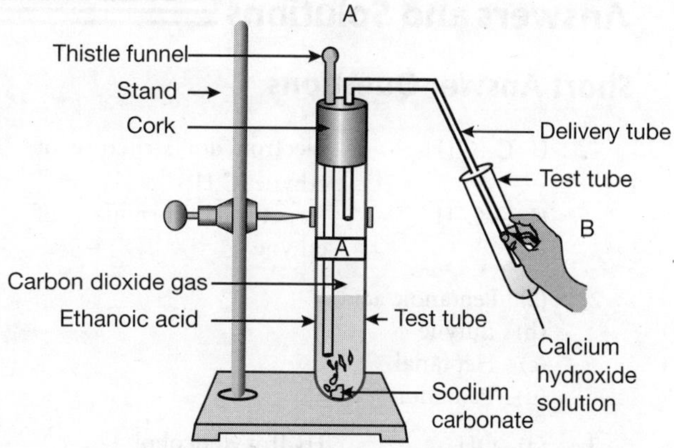

   (b) Write the reaction involved in test tubes A and B respectively.
   (c) If ethanol is given instead of ethanoic acid, would you expect the same change?
   (d) How can a solution of limewater be prepared in the laboratory?

8. How would you bring about the following conversions? Name the process and write the reaction involved.
   (a) Ethanol to ethene.
   (b) Propanol to propanoic acid.
   Write the reactions.

9. Draw the possible isomers of the compound with molecular formula C_3H_6O and also give their electron dot structures.

10. Explain the given reactions with examples.
   (a) Hydrogenation (b) Oxidation
   (c) Substitution (d) Saponification
   (e) Combustion

11. An organic compound A on heating with concentrated H_2SO_4 forms a compound B which on addition of one mole of hydrogen in the presence of Ni forms a compound C. One mole of compound C on combustion forms two moles of CO_2 and 3 moles of H_2O. Identify the compounds A, B and C and write the chemical equations of the reactions involved.

Answers and Solutions

Short Answer Questions

1. H:C⁚⁚C:H — Electron dot structure of ethyne (C_2H_2)

   H–C≡C–H — Structural formula of ethyne

2. (a) Pentanoic acid
   (b) Butyne
   (c) Heptanal
   (d) Pentanol

3. (a) –OH — Hydroxyl/alcohol

   (b) $-\overset{}{\underset{\underset{O}{\|}}{C}}-OH$ — Carboxylic acid

   (c) $-\overset{\overset{O}{\|}}{C}-$ — Ketone

   (d) $-\overset{|}{C}=\overset{|}{C}-$ — Alkene

4. (a) Carboxylic acid is ethanoic acid.
   (b) Alcohol is ethanol.
   (c) X is ethyl ethanoate.

$$\underset{\text{Ethanoic acid}}{CH_3\text{–}COOH} + \underset{\text{Ethanol}}{C_2H_5OH} \xrightarrow{H_2SO_4} \underset{\text{Ethyl ethanoate}}{CH_3\text{–}COOC_2H_5} + H_2O$$

5. As detergents work as cleansing agents both in hard and soft water, and the charged ends of detergents do not form insoluble precipitates with calcium and magnesium ions in hard water, they are better cleansing agents than soaps.

6. (a) Ketone
   (b) Carboxylic acid
   (c) Aldehyde
   (d) Alcohol

7. Ethanol on heating with excess concentrated sulfuric acid at 443 K results in the dehydration of ethanol to give ethene.

$$CH_3CH_2OH \xrightarrow[443\text{ K}]{\text{Hot conc. } H_2SO_4} CH_2{=}CH_2 + H_2O$$

8. Methanol is oxidised to methanal in the liver. Methanal reacts rapidly with the components of cells. It causes the protoplasm to coagulate. It also affects the optic nerve, causing blindness.

9. The gas evolved is hydrogen.

   $2CH_3CH_2OH + 2Na \rightarrow 2CH_3CH_2O^-Na^+ + H_2$

10. Sulfuric acid acts as a dehydrating agent.

$$CH_3CH_2OH \xrightarrow[443\text{ K}]{\text{Hot conc. } H_2SO_4} CH_2{=}CH_2 + H_2O$$

11. (a) Carbon tetrachloride (CCl_4)

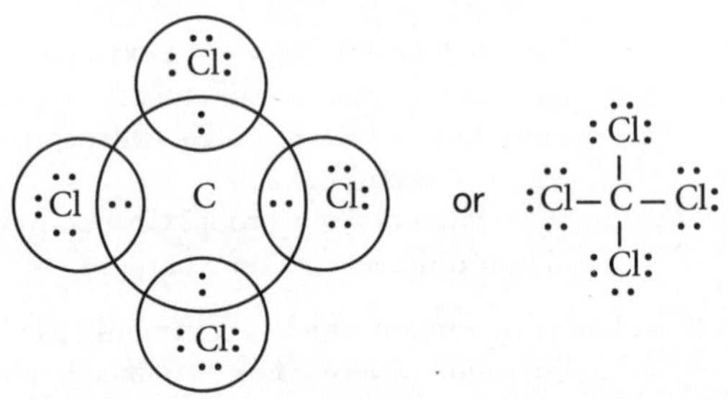

   (b) Carbon dioxide (CO_2)

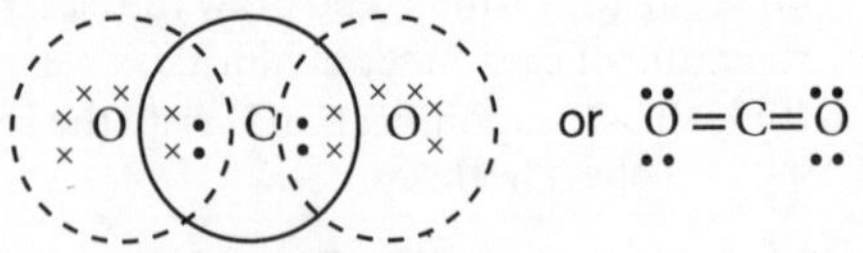

Carbon dioxide

12. (a) The electronic configuration of chlorine with atomic number 17 is given as

K	L	M
2	8	7

   (b) The electron dot structure of the chlorine molecule (Cl_2) is:

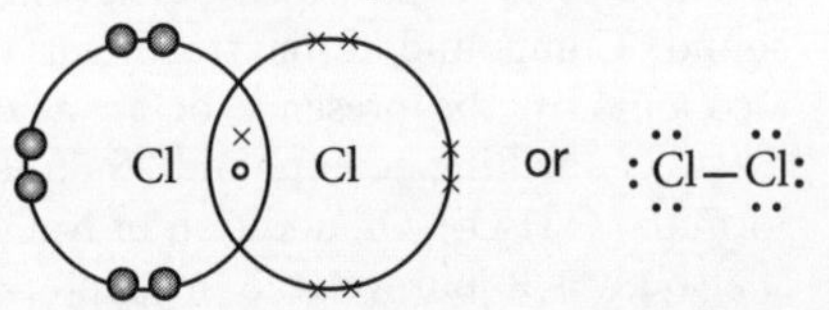

13. Carbon exhibits catenation much more than silicon or any other element due to its smaller size which makes the C–C bonds strong while the Si–Si bonds are comparatively weaker due to its large size.

14. The two can be distinguished by subjecting them to the flame. Saturated hydrocarbons generally give a clear flame while unsaturated hydrocarbons give a yellow flame with lots of black smoke. Ethane and ethene can also be distinguished by using Baeyer's reagent and addition of bromine. Ethene gives these tests while ethane does not.

15. (a)—(iv) (b)—(i) (c)—(ii) (d)—(iii)

16.

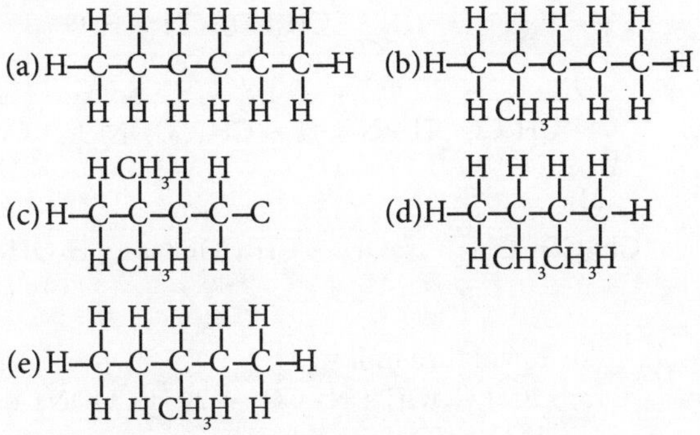

17. (a) Ni acts as a catalyst.

(b) Concentrated H_2SO_4 acts as a catalyst.

(c) Alkaline $KMnO_4$ acts as an oxidising agent.

Long Answer Questions

1. $CH_3COOH + NaHCO_3 \rightarrow CH_3COONa + H_2O + CO_2$

X is sodium ethanoate

Gas evolved is carbon dioxide

Hint: Activity

Limewater will turn milky, a characteristic property of CO_2 gas.

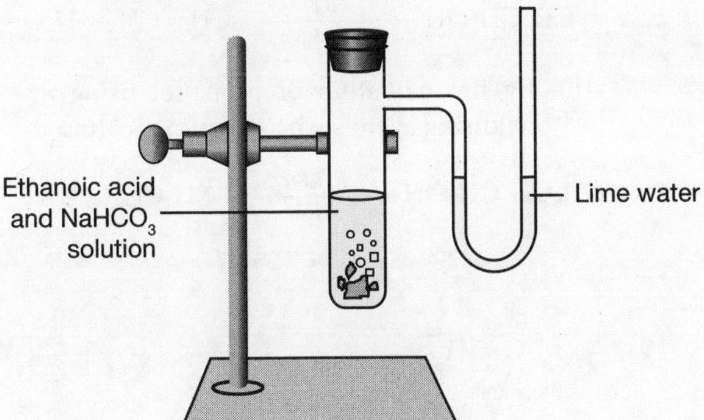

2. (a) Compounds of carbon and hydrogen are called hydrocarbons. Example, methane, ethane.

(b) Saturated hydrocarbons contain carbon–carbon single bonds. Unsaturated hydrocarbons contain at least one carbon–carbon double or triple bond.

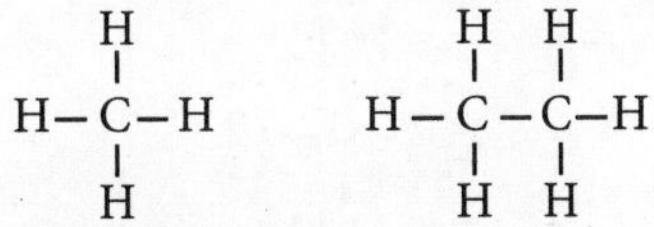

Methane Ethane

Saturated hydrocarbons

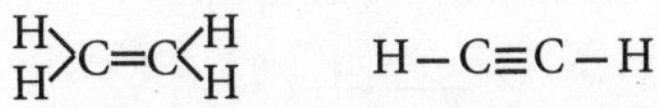

Ethene Ethyne

Unsaturated hydrocarbons

(c) Functional group: An atom/group of atoms joined in a specific manner which is responsible for the characteristic chemical properties of the organic compounds. Examples are hydroxyl group (-OH), aldehyde group (-CHO), carboxylic group (-COOH).

3. Addition of hydrogen to unsaturated hydrocarbons in the presence of a catalyst like nickel or palladium to form saturated hydrocarbons is known as hydrogenation. For example,

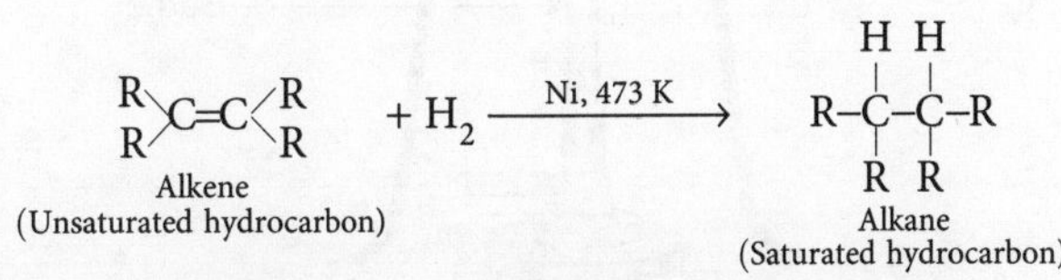

Industrial application: The process of hydrogenation is used in industry to convert vegetable oils to vanaspati ghee.

$$\text{Vegetable oil} + H_2 \xrightarrow{\text{Ni, 473 K}} \text{Vanaspati ghee}$$

(Cotton seed oil or groundnut oil) (Dalda, Rath, Gagan)

Vegetable oils contain a number of unsaturated carbon chains having double bonds between carbon atoms. When H_2O is bubbled through these vegetable oils in the presence of nickel as catalyst at 473 K, some of these double bonds add H_2 to form saturated carbon chains. As a result of this partial hydrogenation, oils are converted into solid fats (ghee).

4. (a) The formula of carbon tetrachloride is CCl_4 and its structure is:

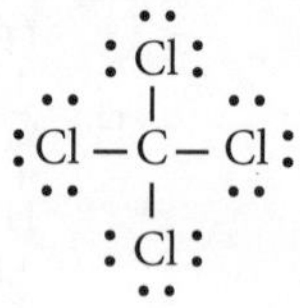

(b) Saponification is the process of converting esters into salts of carboxylic acids and ethanol by treating them with a base.

$$CH_3COOC_2H_5 \xrightarrow{NaOH} CH_3COONa + C_2H_5OH$$

5. **Activity**

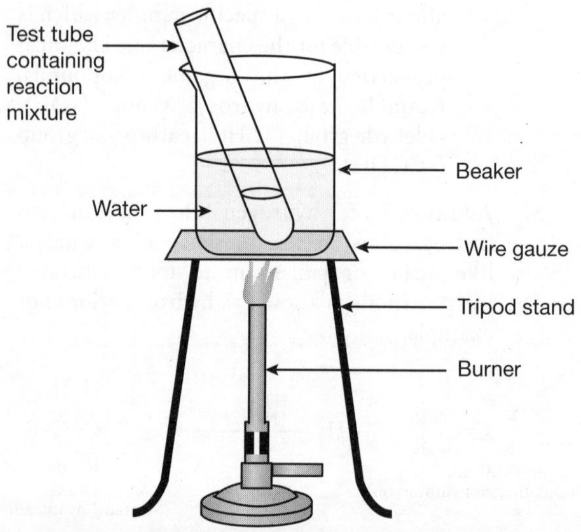

- Take 1 mL ethanol (absolute alcohol) and 1 mL glacial acetic acid along with a few drops of concentrated sulfuric acid in a test tube.
- Warm in a water bath at about 60°C for at least 15 minutes as shown (it should not be heated directly on flame as the vapours of ethanol catch fire).
- Pour into a beaker containing 20–50 mL of water and smell the resulting mixture.

$$\underset{\text{Ethanoic acid}}{CH_3COOH} + \underset{\text{Ethanal}}{CH_3CH_2OH} \xrightarrow{\text{Conc. } H_2SO_4} \underset{\text{Ester}}{CH_3COOCH_2CH_3} + H_2O$$

6. C: Ethanoic acid
R: Sodium salt of ethanoic acid (sodium acetate) and the gas evolved is hydrogen
A: Methanol
S: Ester (methyl acetate)
(a) $\underset{(C)}{2CH_3COOH} + 2Na \rightarrow \underset{(R)}{2CH_3COONa} + H_2$

(b) $\underset{(C)}{CH_3COOH} + \underset{(A)}{CH_3OH} \xrightarrow{\text{Conc. } H_2SO_4} \underset{(S)}{CH_3COOCH_3} + H_2O$

(c) $CH_3COOH + NaOH \rightarrow \underset{(R)}{CH_3COONa} + H_2O$

(d) $CH_3COOCH_3 + NaOH \rightarrow \underset{(R)}{CH_3COONa} + \underset{(A)}{CH_3OH}$

7. (a) It will turn milky.
(b) $2CH_3COOH + Na_2CO_3 \rightarrow 2CH_3COONa + H_2O + CO_2$ (Test tube A)
$Ca(OH)_2 + CO_2 \rightarrow CaCO_3 + H_2O$ (Test tube B)
With excess CO_2, the milkyness disappears.
$CaCO_3 + H_2O + CO_2 \rightarrow Ca(HCO_3)_2$
(c) As C_2H_5OH and Na_2CO_3 do not react, a similar change is not expected.
$C_2H_5OH + Na_2CO_3 \rightarrow$ No change
(d) The limewater is prepared by dissolving calcium oxide in water and decanting the supernatant liquid.

8. (a) By the dehydration of ethanol in the presence of concentrated H_2SO_4.

$$CH_3CH_2OH \xrightarrow[\text{Conc. } H_2SO_4]{\text{Hot}} CH_2{=}CH_2 + H_2O$$

(b) By the oxidation of propanol using an oxidising agent such as alkaline $KMnO_4$.

$$CH_3CH_2CH_2OH \xrightarrow[\text{Heat}]{\text{Alk. } KMnO_4} CH_3CH_2COOH$$

9. $CH_3{-}\overset{O}{\overset{\|}{C}}{-}CH_3$
Propanone

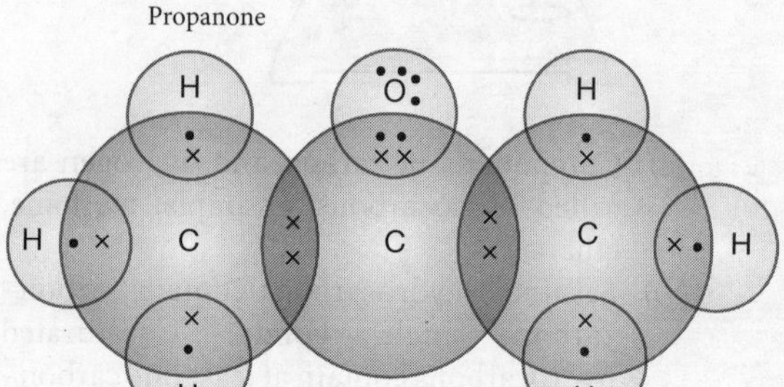

Electron dot structure of propanone

$CH_3–CH_2–CHO$
Propanal

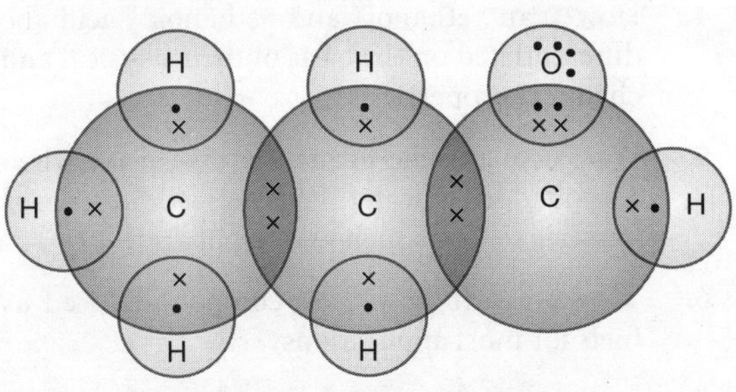

Electron dot structure of propanal

10. (a) In hydrogenation reactions, unsaturated hydrocarbons add on hydrogen in the presence of nickel catalyst to give saturated hydrocarbons.

$$R_2C{=}CR_2 + H_2 \xrightarrow{Ni} R_2\underset{H}{C}–\underset{H}{C}R_2$$

(b) Ethanol is oxidised to ethanoic acid in the presence of alkaline $KMnO_4$ on heating.

$$CH_3CH_2OH \xrightarrow[\text{Heat}]{\text{Alk. } KMnO_4} CH_3COOH$$

(c) In the presence of sunlight, chlorine is added to hydrocarbons by replacing the hydrogen atom. For example,

$$CH_4 + Cl_2 \xrightarrow{h\nu} CH_3Cl + HCl$$

(d)

$$\underset{\text{Ester}}{CH_3COOC_2H_5} + NaOH \rightarrow CH_3COONa + C_2H_5OH$$

It is saponification used in the preparation of soap.

(e) In combustion, carbon compounds release a large amount of heat and light on burning and give carbon dioxide and water. For example,

$$CH_4 + 2O_2 \rightarrow CO_2 + 2H_2O + \text{Heat and light}$$

11. Since the compound C gives 2 mol of CO_2 and 3 mol of H_2O, we can infer that it has the molecular formula C_2H_6 (ethane). C is obtained by the addition of one mol of hydrogen to compound B so the molecular formula of B should be C_2H_4 (ethene). Compound B is obtained by heating compound A with concentrated H_2SO_4 which shows it to be an alcohol. So compound A could be C_2H_5OH (ethanol).

$$\underset{A}{C_2H_5OH} \xrightarrow{\text{Hot conc. } H_2SO_4} \underset{B}{C_2H_4} + H_2O$$

$$\underset{B}{C_2H_4} + H_2 \xrightarrow{Ni} \underset{C}{C_2H_6}$$

$$\underset{C}{2C_2H_6} + 7O_2 \rightarrow 4CO_2 + 6H_2O + \text{Heat and light}$$

NCERT CORNER

In-Text Questions

1. What would be the electron dot structure of carbon dioxide which has the formula CO_2?

2. What would be the electron dot structure of a molecule of sulfur which is made up of eight atoms of sulfur?

3. How many structural isomers can you draw for pentane?

4. What are the two properties of carbon which lead to the huge number of carbon compounds we see around us?

5. What will be the formula and electron dot structure of cyclopentane?

6. Draw the structures for the following compounds.
(i) Ethanoic acid
(ii) Bromopentane
(iii) Butanone
(iv) Hexanal

7. How would you name the following compounds?
1. $CH_3–CH_2–Br$
2.

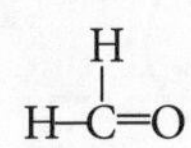

3.

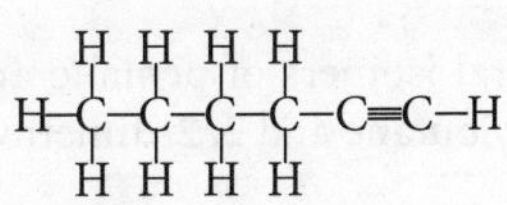

8. Why is the conversion of ethanol to ethanoic acid an oxidation reaction?

9. A mixture of oxygen and ethyne is burnt for welding. Can you tell why a mixture of ethyne and air is not used?

10. How would you distinguish experimentally between an alcohol and a carboxylic acid?

11. What are oxidising agents?

12. Would you be able to check if water is hard by using a detergent?

13. People use a variety of methods to wash clothes. Usually after adding the soap, they 'beat' the clothes on a stone, or beat it with a paddle, scrub with a brush or the mixture is agitated in a washing machine. Why is agitation necessary to get clean clothes?

Exercises

1. Explain the nature of the covalent bond using the bond formation in CH_3Cl.

2. Draw the electron dot structures for:
(a) ethanoic acid (b) H_2S
(c) propanone (d) F_2

3. What is a homologous series? Explain with an example.

4. How can ethanol and ethanoic acid be differentiated on the basis of their physical and chemical properties?

5. Why does micelle formation take place when soap is added to water? Will a micelle be formed in other solvents such as ethanol also?

6. Why are carbon and its compounds used as fuels for most applications?

7. Explain the formation of scum when hard water is treated with soap.

8. What change will you observe if you test soap with litmus paper (red and blue)?

9. Which of the following hydrocarbons undergo addition reactions: C_2H_6, C_3H_8, C_3H_6, C_2H_2 and CH_4?

10. Give a test that can be used to differentiate between butter and cooking oil.

11. Explain the mechanism of the cleaning action of soaps.

Answers and Solutions

In-Text Questions

1. 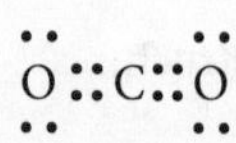

2. The eight atoms of sulfur are joined together in the form of a ring.

or

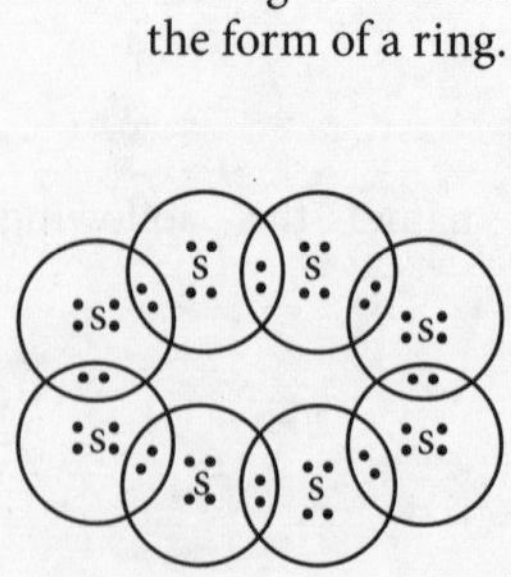

or

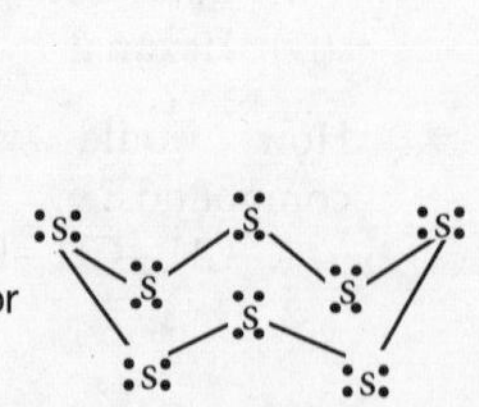

3. Structural isomers of pentane are: n-pentane, 2-methylbutane and 2, 2-dimethylpropane.

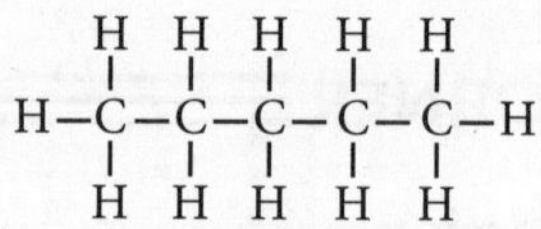

n-pentane

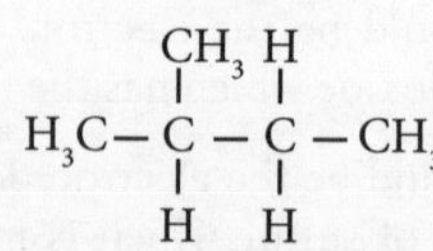

2-methylbutane

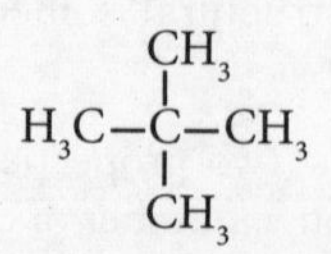

2, 2 dimethylpropane

4. Two properties of carbon which lead to the huge number of carbon compounds we see around us are:

- Carbon has six valence electrons which are actually a high valency.
- Covalent bonding happens easily with carbon atoms and numerous others such as oxygen, chlorine, nitrogen, sulfur, hydrogen.

5.

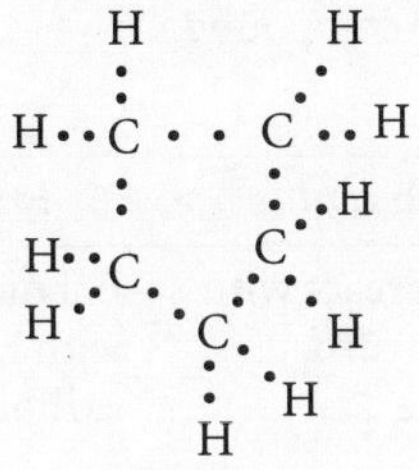

C_5H_{10}

6. (i)

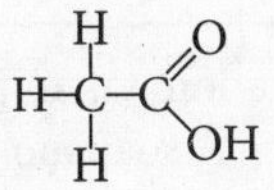

(ii)

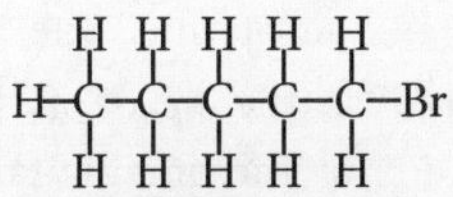

Bromopentane

(iii)

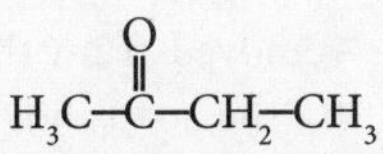

Butanone

(iv)

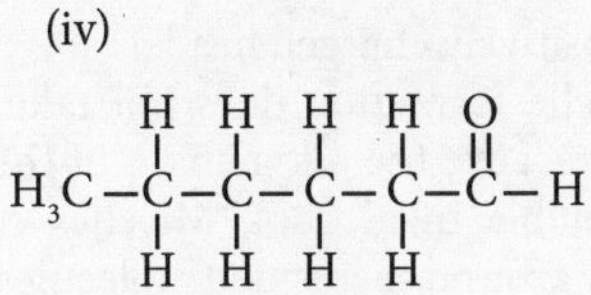

Hexanal

7. 1. Bromoethane

2. Methanal or formaldehyde

3. Hexyne

8.

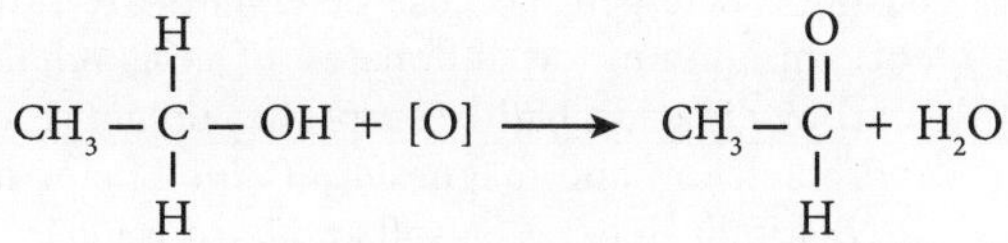

Ethanol Ethanal

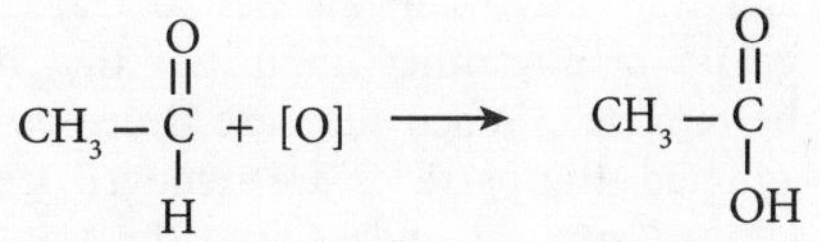

Ethanal Ethanoic acid

Conversion of ethanol to ethanoic acid involves the removal of hydrogen and the addition of oxygen. It is an oxidation reaction. In the first step, a H_2 molecule is removed from ethanol to form ethanal. As the loss of hydrogen is oxidation, the reaction is an oxidation reaction. Similarly, an oxygen atom is added to form ethanoic acid from ethanal. As gain of oxygen is called oxidation, the reaction is an oxidation reaction.

9. A mixture of oxygen and ethyne is burnt for welding instead of a mixture of ethyne and air because the production of heat is very important for welding metals. When oxygen and ethyne are burnt, it burns completely and produces a higher temperature than air and ethyne. Oxygen and ethyne produce a very hot blue flame but the mixture of air and ethyne gives out a sooty flame which means that there are unburnt particles, resulting in lesser heat.

10. On reaction with sodium carbonate, carboxylic acids produce carbon dioxide gas which turns limewater milky whereas alcohols do not give this reaction. This experiment can be used to distinguish an alcohol and a carboxylic acid. Reaction of carboxylic acid with sodium carbonate:

$$2CH_3COOH + Na_2CO_3 \rightarrow 2CH_3COONa + H_2O + CO_2$$

11. Oxidising agents are the substances that can oxidise other substances by adding oxygen to them. For example, alkaline or acidic $KMnO_4$ and acidic $K_2Cr_2O_7$ are oxidants.

$$CH_3{-}CH_2OH + 2[O] \xrightarrow{\text{Oxidant}} CH_3COOH + H_2O$$

12. It is not possible to check if water is hard by using a detergent because detergents are salts of ammonium or sulfonates of long chain carboxylic acids. Unlike soaps they do not react with calcium and magnesium (which helps us to distinguish between soft and hard water).

13. Agitation is necessary to get clean clothes as agitation aids soap micelles to trap the oil, grease or any other impurities that have to be removed. When they are being beaten or agitated, the particles are removed from the clothes' surfaces and go into the water, thus cleaning the clothes.

Exercises

1. Carbon can neither lose 4 electrons nor gain four electrons as these processes make the system unstable due to requirement of extra energy. Therefore, carbon completes its octet configuration by sharing its 4 electrons with hydrogen atoms and with a chlorine atom. Hence the bonding that exists in CH_3Cl is covalent bonding. Here, carbon requires 4 electrons to complete its octet, while each hydrogen atom requires one electron to complete its duplet. Also, chlorine requires an electron to complete the octet. Therefore, all of these share the electrons and as a result, carbon forms 3 bonds with hydrogen and one with chlorine.

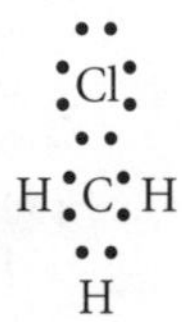

2. (a)

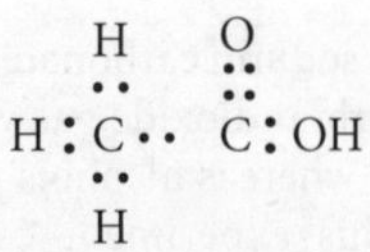

Ethanoic acid

(b)

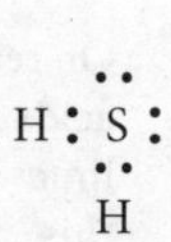

H_2S

(c)

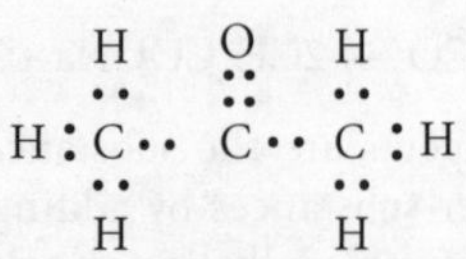

Propanone

(d)

F_2

3. A homologous series is a series of compounds which contains the same functional group. This has a similar general formula and chemical properties. Since there is a change in the physical properties, we can say that there would be an increase in the molecular size and mass. For example, methane, ethane, propane, butane, and so on, are all part of the alkane homologous series. The general formula of this series is C_nH_{2n+2}. Methane CH_4, ethane CH_3CH_3, propane $CH_3CH_2CH_3$, butane $CH_3CH_2CH_2CH_3$. It can be noticed that there is a difference of a $-CH_2$ unit between each successive compound.

4.

Ethanol	Ethanoic acid
Does not react with sodium hydrogen carbonate	Bubbles and fizzes with hydrogen carbonate
A good smell	Smells like vinegar
No action in litmus paper	Blue litmus paper to red
Burning taste	Sour taste

5. Micelle formation takes place when soap is added to water. A soap molecule is a long chain molecule with one end having an alkyl group which is hydrophobic, and the other end has $SO_4^-Na^+$, which is hydrophilic. The $SO_4^-Na^+$ can be considered as the head of the soap molecule and is ionic. The organic tail mixes and dissolves with the dirt whereas the oil or grease and ionic head dissolves and mixes with the water. Therefore, when the material to be cleaned is removed from the water, the dirt is taken off by the soap molecules in the water. Hence, the soap cleans by forming closed structures by the mutual repulsion of the micelles (positively charged heads).

No, micelle formation does not take place in ethanol because the alkyl chain of the soap becomes soluble in alcohol. Micelles can be formed only around suspended molecules of oil in water solution. Ethanol acts as a very good solvent and is able to dissolve even oil, thus forming a clear solution.

6. Carbon and its compounds are used as fuels for most applications for they have high calorific values and give out a lot of energy. Most of the carbon compounds give a lot of heat and light when burnt in air.

7. Scum is produced from reaction of hard water with soap. Calcium and magnesium present in the hard water form an insoluble white precipitate which is called scum. Soap molecules are long carbon chains with sodium or potassium at the other end. When soap is added to hard water, the calcium and magnesium ions present in the hard water displace the sodium or potassium ions from the soap molecules forming scum. A lot of soap gets wasted in the process.

8. When we test soap with litmus paper, the colour of litmus paper turns from red to blue. As soap is basic in nature, it will turn red litmus blue; however, the colour of blue litmus will remain blue because only an acidic substance can turn blue litmus into red.

9. Unsaturated hydrocarbons undergo addition reactions. C_3H_6 and C_2H_2 are unsaturated hydrocarbons which undergo addition reactions.

10. Butter has saturated compounds while cooking oil contains unsaturated compounds. Unsaturated compounds are oxidised by alkaline potassium permanganate (Baeyer's reagent) with the disappearance of its pink colour; so, when cooking oil is treated with a few drops of alkaline potassium permanganate solution, its pink colour is lost. In the case of butter, the pink colour of alkaline potassium permanganate does not disappear.

11. There are so many impurities and dirt mixed in water, and the dirt does not dissolve in the water as the dirt present on clothes is organic (non-polar) in nature. Therefore, it cannot be removed by only washing with water. Soap molecules are a combination of salts such as sodium or potassium. A soap molecule is a long chain molecule with one end having an alkyl group which is hydrophobic, and the other end has $SO_4^-Na^+$ which is hydrophilic. The molecules are from a long chain carboxylic acid. So, when the carbon chain is dissolved in oil and the ionic end is dissolved in the water, the soap starts cleansing and trapping the dirt. The soap molecules have different properties at different ends. The first end being the hydrophilic end which dissolves in the water and is attracted towards the water and the second one being the hydrophobic end is dissolved in the oil. The hydrophobic tail aligns itself along the surface of the water because it is not soluble in the water. When this happens, the soap molecules form structures that are called micelles which are used for capturing the oil droplets; the other end being ionic faces the water. This will then form an emulsion in water and help in dissolving the dirt or impurities when the clothes are washed.

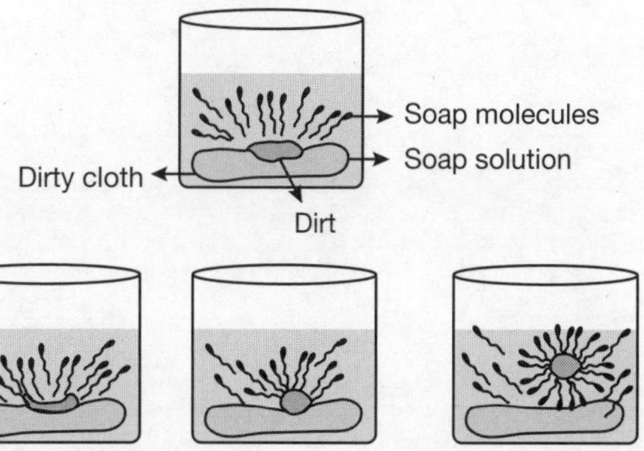

CHAPTER 5

Periodic Table and Periodic Properties

Learning Objectives

After studying this chapter, you will be able to:

- understand all periodic laws, need for periodic table, classification of elements
- compare the different periodic tables
- understand the various periodic properties and their trends
- compare the periodic properties for various elements

5.1 INTRODUCTION

We have studied that matter is present in the form of elements, compounds and mixtures and each element is composed of one type of atom. Elements like silver, gold, copper, tin, mercury and lead have been known since a long time. From the discovery of the first element, phosphorus, in 1649 we presently have 118 elements. All these elements have seemingly different properties. Scientists started recognising similarities and dissimilarities among these elements and tried various types of arrangements to make their study easy and systematic. This helped in the development of the periodic table. The periodic table is an arrangement of elements with similar properties placed together. It helps not only in the classification of elements but also in the easier study of elements.

5.1.1 Genesis of Periodic Classification

The earliest attempt to classify the elements was made by Lavosier who divided the elements known at that time into metals and non-metals in order to study their acidic and basic properties. In 1808, John Dalton proposed the atomic theory according to which, 'atoms of an element can be distinguished from the others on the basis of their atomic weight or atomic mass'. All the early attempts to classify the elements were based upon the atomic masses of the elements. Here are some important ones.

Dobereiner—The Law of Triads: In 1829, the German chemist Dobereiner (Fig. 5.1), found that a few elements having similar properties can be placed together in groups of three elements each. Such groups of three elements were called triads and in a particular triad all the three elements would possess similar chemical properties. Dobereiner made the first attempt to show a relationship between the properties of elements and their atomic masses and according to this principle, 'elements could be arranged in groups of three, in which the atomic weight of the middle element was the arithmetic mean of the first and last element and also the difference in the atomic weight of consecutive elements was atom constant'. Some examples are discussed.

Fig. 5.1 Johann Dobereiner (1780–1849)

Source: Carl August Schwerdgeburth, 1785-1878 (engraver), and Fritz Ries, 1826-1857 (painter), Public domain, via Wikimedia Commons

(i) The alkali metal group triad: It includes three elements lithium (Li), sodium (Na) and potassium (K) having similar physicochemical properties such as (i) metallic nature, (ii) reaction with water to form alkali and hydrogen gas, and (iii) a valency of one (monovalent).

$$_{3}Li^{9}\ ,\ _{11}Na^{23},\ _{19}K^{39}$$

The mean of the atomic masses of the first (Li) and the third (K) elements is (7 + 39)/2 u. The atomic mass of the middle element, sodium, Na is equal to 23 u.

(ii) **The alkaline earth metal group triad:** It includes three elements beryllium (Be), magnesium (Mg) and calcium (Ca) having similar physicochemical properties such as (i) metallic nature, (ii) basic or alkaline oxide formation, and (iii) a valency of two (bivalent). The mean of the atomic masses of the first (Be) and the third (Ca) elements is (9 + 40)/2 u. The atomic mass of the middle element, magnesium Mg, is equal to 24.5 u which is nearly equal to its exact atomic mass 24 u.

$$_{4}Be^{9},\ _{12}Mg^{24},\ _{20}Ca^{40}$$

One more example of Dobereneir's triads is $_{20}Ca^{40}$, $_{38}Sr^{88}$, $_{56}Ba^{137}$.

Mean of the atomic masses of the first and third elements = (40 + 137)/2 = 88.5 u.

(iii) **The halogen group triad:** It includes three elements chlorine (Cl), bromine (Br) and iodine (I) (Fig. 5.2) which have similar physicochemical properties such as (i) non-metallic nature, (ii) reaction with water to form acid, and (iii) a valency of one (monovalent).

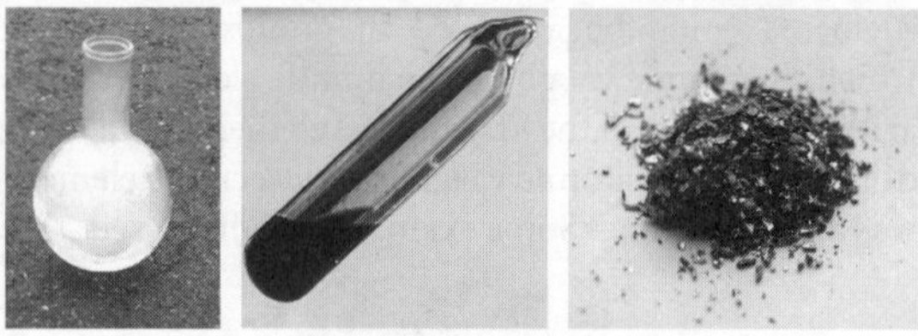

Fig. 5.2 (a) Chlorine (Cl) (b) Bromine (Br) (c) Iodine (I)

Source: Larenmclane, CC BY-SA 4.0 <https://creativecommons.org/licenses/by-sa/4.0>, via Wikimedia Commons; Hi-Res Images ofChemical Elements, CC BY 3.0 <https://creativecommons.org/licenses/by/3.0>, via Wikimedia Commons; Benjah-bmm27, Public domain, via Wikimedia Commons

$$_{17}Cl^{35.5},\ _{35}Br^{80},\ _{53}I^{127}$$

Mean of the atomic masses of the first and third elements = (35.5 + 127)/2 = 81.25 u

However, the actual atomic mass of the second element (Br) = 80 u which is very close to the mean value.

(iv) **The oxygen or chalcogen group triad:** It includes three elements sulfur (S), selenium (Se) and tellurium (Te). The mean of the atomic masses of the first (S) and the third (Te) elements is (32 + 127.6)/2 u. The atomic mass of the middle element, selenium Se, is equal to 79.8 u which is nearly equal to its exact atomic mass 78.96 u.

Limitation and failure: Dobereiner's idea of classification of elements into triads did not receive wide acceptance since he could arrange only a few elements in this manner. It is also found that some elements having similar properties don't follow the triad rule. Nitrogen (N), phosphorus (P) and arsenic (As) have similar physicochemical properties but the atomic mass of phosphorus is different from the mean value of the atomic masses of N, As.

Newlands (1864)—Law of Octaves: In 1866, J A Newlands (Fig. 5.3) arranged the elements in the increasing order of their atomic masses starting with hydrogen with the least atomic mass (1) and ended with thorium (Th). He observed that the properties of every eighth element was similar to the properties of the first element; so he introduced the *law of octaves*. According to this, 'if the elements are arranged in the order of increasing atomic weight, the eighth element starting from a given one, is a kind of repetition of the first like the eighth node in an octave of music' (Fig. 5.4).

Fig. 5.3 John Newlands

Source: The original uploader was Conget at Dutch Wikipedia., Public domain, via Wikimedia Commons.

$Li^{6.9}$	Be^{9}	$B^{10.8}$	C^{12}	N^{14}
O^{16}	F^{19}	Na^{23}	$Mg^{24.3}$	Al^{27}
$Si^{28.1}$	P^{31}	$S^{32.1}$	$Cl^{35.5}$	$K^{39.1}$
$Ca^{40.1}$				

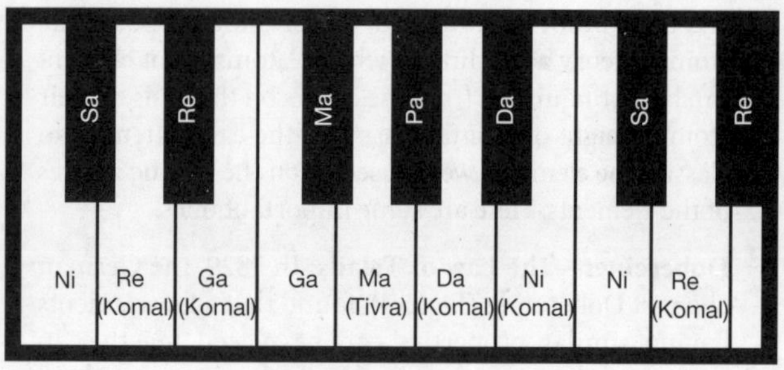

Fig. 5.4 Music notes

For example, starting from lithium (Li) the eighth element is sodium (Na); so the properties of lithium and sodium should be similar. Similarly, the properties of Na and K (potassium) should be similar.

Starting from Beryllium (Be) the eighth element is magnesium (Mg) so the properties of beryllium and magnesium should be similar. Similarly, the properties of Mg and Ca (calcium) should be similar.

Table 5.1 Newlands octaves

Serial No	1	2	3	4	5	6	7
Notes of musical scale	sa	re	ga	ma	pa	dha	ni
Elements with atomic mass up to 40 u	H F Cl	Li Na K	Be Mg Ca	B Al	C Si	N P	O S
Elements with atomic mass > 40 u	Co Ni Br	Cu Rb	Zn Sr	Cr Y Ce La	Ti In Zr	Mn As —	Fe Se —

Limitations and failure: (i) Newlands could arrange elements up to calcium (atomic weight 40) by using this law. At that time, more than 60 elements were known, so this law was not appreciated.

(ii) After calcium, every eighth element does not possess properties similar to that of the first one. For example, chromium (Cr) as the first element and the eighth element yttrium (Y) do not have similar chemical properties. Similarly, Ti, In and Mn, As do not have similar properties.

(iii) There are many elements which are placed in the same slot but do not have the same properties. For example, Co and Ni are placed with Br and they differ in properties from Br.

(iv) After the discovery of the noble gases, the properties of the eighth element was not similar to the first element. Actually, now it was first and the ninth element that had similar properties.

Competition Edge

Lothar Meyer (1869): He plotted various physical properties of the elements against their atomic weights and pointed out that the similar elements occupy similar positions on the curve. For example, alkali metals are present at the peaks of the curves.

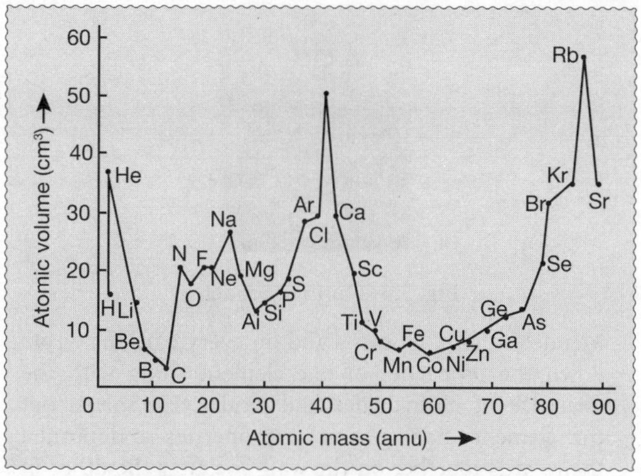

TEST YOUR KNOWLEDGE

5.1 If the atomic weight of phosphorus (P) is 31 and that of antimony (Sb) is 120, find the atomic weight of arsenic using the triad system.

Solution:

$^{31}P \quad ^{?}As \quad ^{126}Sb$

$$\text{Atomic weight of As} = \frac{31+120}{2} = 75.5$$

5.2 Mendeleev's Periodic Table

The Russian chemist, Dmitri Ivanovich Mendeleev, made the first real breakthrough in the classification of elements. In order to understand Mendeleev's (Fig. 5.5) periodic law and table, let us first understand the meaning of periodic table. A periodic table is defined as an arrangement or a chart which can classify all the known elements on the basis of their properties in such a manner that elements with similar properties are placed in the same vertical column and the dissimilar elements are separated from one another. At that time only 63 elements were known, so Mendeleev tried to correlate the atomic masses of the elements with their physical and chemical properties. In the case of chemical properties, he mainly focused on the compounds formed by the elements with hydrogen and oxygen. The reason for selecting hydrogen and oxygen was their highly reactive nature and ability to form compounds with almost all the elements. That is, the atomic masses and similarity in the formulae of the hydrides and the oxides formed by these elements were selected as the basis for their classification.

Fig. 5.5 Dmitri Mendeleev

Mendeleev used 63 cards and on every card, he wrote down the properties of one element along with the formulae of its hydrides and oxides. He sorted out the elements having similar properties and pinned their cards together on the wall one after the other in a vertical column. While doing that, he observed that most of the elements automatically got arranged in the increasing order of their atomic masses. He also found the occurrence of elements having similar properties after certain regular intervals. On the basis of these observation, he introduced Mendeleev's periodic law.

Mendeleev's Periodic Law

According to this law, 'the physical and chemical properties of elements are the periodic functions of their atomic weights'. Mendeleev considered atomic weight as the fundamental property of an element and he arranged different elements in the order of their increasing atomic weights and formed Mendeleev's periodic table. In this table, elements are arranged in increasing order of their atomic masses in horizontal rows (periods) in such a manner that elements with similar properties are placed in the same vertical columns (groups).

5.2.1 Features of Mendeleev's Periodic Table

Mendeleev arranged elements in horizontal rows (7 periods) and vertical columns (8) of a table in order of their increasing atomic weights in such a way that the elements with similar properties occupied the same vertical column or group. Mendeleev's system of classifying elements was more elaborate than that of Lothar Meyer's. He fully recognised the significance of periodicity and used a broader range of physical and chemical properties to classify the elements. In particular, Mendeleev relied on the similarities in the empirical formulae and properties of the compounds formed by the elements. He realised that some of the elements did not fit in with his scheme of classification if the order of atomic weight was strictly followed. He ignored the order of atomic weights, thinking that the atomic measurements might be incorrect, and placed the elements with similar properties together. For example, iodine with lower atomic weight than that of tellurium (group VI) was placed in group VII along with fluorine, chlorine, bromine because of similarities in properties (Fig. 5.8). At the same time, keeping his primary aim of arranging the elements of similar properties in the same group, he proposed that some of the elements were still undiscovered and, therefore, left several gaps in the table. For example, both gallium and germanium were unknown at the time Mendeleev published his periodic table. He left a gap under aluminium and a gap under silicon, and called these elements *eka-aluminium* and *eka-silicon*. Mendeleev predicted not only the existence of gallium and germanium, but also described some of their general physical properties. These elements were discovered later. Some of the properties predicted by Mendeleev for these elements and those found experimentally are listed in Table 5.2.

5.2.2 Merits of Mendeleev's Periodic Table

The following are some merits and uses of Mendeleev's periodic table.

- **Systematic and easy study of elements:** Mendeleev classified elements in the form of a table having periods and groups so their study became easier as elements with similar properties are placed within the same vertical column or group.
- **Prediction of new elements:** Mendeleev's periodic table helped in the discovery of new elements as he left a few gaps for those elements which were not known at that time and he also predicted that elements will be discovered to fill in these gaps. Mendeleev proposed a term *eka* (meaning first) to the name of the preceding element in the same group. For example, eka-aluminium for gallium (Ga), eka-silicon for germanium (Ge) and eka-boron for scandium (Sc). Later on, when these elements were discovered, Mendeleev's predictions were found to be correct.
- **Deciding and correction of atomic weights:** Mendeleev's periodic table helped in deciding atomic weights of new elements and in the correction of atomic weights of existing elements. For example, (i) the atomic weight of eka-aluminum predicted by Mendeleev is 68 which is very close to the actual value of its atomic mass 69.7. (ii) The atomic weight of indium is corrected from 76 to 113 and also for other elements like Be and U (Table 5.3).

Table 5.2 Mendeleev's periodic table

Group: Oxide: Hydride:	I R_2O A RH B	II RO A RH_2 B	III R_2O_3 A RH_4 B	IV R_2H_5 A RO_4 B	V R_2H_5 A RO_4 B	VI RO_3 A RH_2 B	VII R_2O_3 A RH B	VIII RO_4 Transition traids	Zero RO_4 Noble gases
1	H 1(At. No.) 1.008(AL.Wt.)								He 2 4.0026
2	Li 3 6.939	Be 4 9.012	B 5 10.811	C 6 12.011	N 7 14.007	O 8 15.999	F 9 18.998		Ne 10 20.183
3	Na 11 22.99	Mg 12 24.312	Al 13 26.981	Si 14 28.086	P 15 30.974	S 16 32.06	Cl 17 35.453		Ar 18 39.948
4 First series second series	K 19 39.102 Cu 29 63.54	Ca 20 40.08 Zn 30 65.37	Sc 21 44.96 Ga 31 69.72	Ti 22 47.90 Ge 32 72.59	V 23 50.94 As 33 74.92	Cr 24 51.99 Se 34 78.96	Mn 25 54.939 Br 35 79.909	Fe 26 C 27 Ni 28 55.85 58.93 58.71	Kr 36 83.80
5 First series second series	Rb 37 85.47 Ag 47 107.87	Sr 38 87.62 Cd 48 112.40	Y 39 88.905 In 49 114.82	Zr 40 91.22 Sn 50 118.69	Nb 41 92906 Sb 51 121.75	Mg 42 95.94 Te 52 127.60	Te 43 (99) 1 53 124.9014	Ru 44 Rh 45 Pd 46 101.07 102.91 106.4	Xe 54 131.30
6 First series second series	Ca 55 132.90 Au 79 196.97	Ba 56 137.34 Hg 80 200.59	*Rare earths 57-71 Tl 81 204.37	Hf 72 178.49	Ta 73 180.948 Bi 83 208.98	W 74 183.85 Po 84 (210)	Re 75 186.2 At 85 (210)	Os 76 Ir 77 Pt 78 190.2 192.2 195.09	Rn 86 (222)
7	Fr 87 (223)	Ra 88 (226)	Actinided elements 89-103	Pb 82 207.19	Ha 105				

Table 5.3 Predicted and discovered elements

Property	Eka-aluminium (R) (predicted)	Gallium (found)	Eka-silicon (R) (predicted)	Germanium (found)
Atomic mass	68	69.7	72	72.6
Density (g cm^{-3})	5.9	5.94	5.5	5.36
Melting point (K)	Low	303.2	High	1231
Formula of oxide	R_2O_3	Ga_2O_3	RO_2	GeO_2
Formula of chloride	RCl_3	$GaCl_3$	RCl_4	$GeCl_4$

- **Position of noble gases:** At the time of Mendeleev, these gases were not present but when these gases (helium, neon and so on) were discovered, they were placed in a new separate group without disturbing the original arrangement of Mendeleev's periodic table. These gases are chemically unreactive so were placed separately. These gases were discovered late as they are very unreactive (almost inert) and present in only in traces in the atmosphere. This group is known as the zero group.

5.2.3 Demerits of Mendeleev's Periodic Table

Although Mendeleev's periodic table played a great role in the study of elements, it had a few anomalies or limitations.

- **Anomalous position of hydrogen:** The position of hydrogen is not justified as it has been placed in group I and also in group VII as it resembles both the alkali metals (group I) and the halogens (group VII) in properties.

As hydrogen has one electron (like alkali metals) in its outermost shell, it is placed with them. Hydrogen resembles the alkali metals in that, it reacts with halogens, oxygen and sulfur to give compounds having similar formulae. For example, HCl, NaCl, H_2O, Na_2O, H_2S, Na_2S. Due to similar valency and reactivity, its position with the alkali metals is justified.

Hydrogen needs one electron to attain the stable nearest noble gas configuration, that is, of helium (2 electrons). Similarly, the halogens also need one electron to attain the nearest noble gas configuration. Thus, hydrogen resembles the halogens in electronic configuration. Both hydrogen and halogens exist as gaseous diatomic molecules, that is, H_2, F_2, Cl_2. Both hydrogen and halogens combine with alkali metals to form similar types of compounds—NaH, NaCl.

Both hydrogen and halogens combine with non-metals (such as carbon) to form the same type of compound—CH_4, CCl_4, NH_3, NCl_3. Thus, we can say that hydrogen resembles the halogens too, in its properties.

From this discussion we can say that hydrogen can be placed at both the places. This means that Mendeleev's periodic law could not assign a correct position to hydrogen. In fact, the position of hydrogen with the alkali metals is more suitable as it has an atomic mass of 1, so, it is the first element of the periodic table.

- **Position of isotopes:** Isotopes are atoms of the same element having the same atomic number but different atomic masses. In Mendeleev's periodic table, elements are arranged in the increasing order of their atomic masses and isotopes have different atomic masses so they must have different positions also. However, there is no position or place for isotopes. Moreover, isotopes have the same chemical properties so they must have same position also. It means the position of isotopes is anomalous.
- **Anomalous position of pairs:** There are many anomalous pairs in which the element with a higher atomic weight is placed before an element having a lower atomic weight. For example, argon (Ar), tellurium (Te) and cobalt (Co) are placed before potassium (K), iodine (I) and nickel (Ni) respectively.

  Ar (39.94) K (39.10)
  Te (127.6) I (126.6)
  Co (58.9) Ni (58.6)
- **Anomalous placement:** Many elements with different properties are placed together. For example, less reactive coinage metals (Cu, Ag, Au) were placed with highly reactive alkali metals (Li, Na, K). In some cases, elements having similar properties are placed in different columns or groups. For example, copper and mercury are placed separately.
- **Position of lanthanides and actinides:** Lanthanides and actinides have been put along with lanthanum and actinium respectively, and there is no suitable separate position or place for these elements.
- **Uncertainty in the prediction of new elements:** In the table, the elements are arranged in increasing order of their atomic masses. As atomic masses do not increase in a regular manner in going from one element to the next element and the difference in atomic masses between two successive elements is small in the case of lighter elements and large in case of heavier elements, it is not possible to predict correctly how many new elements can be discovered between any two known elements.

Competition Edge

Today, Mendeleev's periodic table has no practical significance. However, its significance lies in the fact that the concept of periodicity in properties and the resultant arrangement of elements in the form of a table was originated by Mendeleev.

5.3 Modern Periodic Table

Considering the major limitations of Mendeleev's periodic table, scientists observed that the atomic mass is not the correct basis for the classification of elements and they thought that there must be some other fundamental property of elements which can be used to classify elements in a better manner. In 1913 Henry Mosely showed that the atomic number of an element is a more fundamental property than its atomic mass and considering that, he introduced the modern periodic law and the modern periodic table.

Modern Periodic Law: H G J Moseley (Fig. 5.6) in 1913 introduced the modern periodic law based on his work on X-ray spectra of elements. He introduced the atomic number (*Z*) and said that it is equal to the total nuclear charge or number of protons ($Z = p$). He established atomic number as a fundamental property and introduced the modern periodic law. According to it, 'the physical and chemical properties of elements are the periodic function of their atomic numbers'.

Modern Periodic Table

It is based on arranging different elements in the increasing order of their atomic numbers in a tabular form in which elements having similar properties will occur after fixed intervals or periods. As the elements are arranged in the increasing order of atomic numbers, there is a periodicity in the electronic configuration of these elements which leads to the periodicity in their chemical properties. That is, elements which have similar electronic configurations can show similar properties.

Fig. 5.6 Henry Moseley

Source: Nature magazine, Public domain, via Wikimedia Commons

5.3.1 Periodicity and its Cause

Periodicity in properties refers to the repetition of properties of elements after certain regular intervals. The cause of periodicity in the properties of the elements is the repetition of similar outer electronic configuration after certain regular intervals. For example, (i) in the case of alkali metals (Li, Na, K, Rb, Cs) the number of valence electrons is one, so they show similarity in properties (Fig. 5.7).

Element	Li	Na	K	Rb	Cs
Atomic number	3	11	19	37	55
Configuration	2,1	2,8,1	2,8,8,1	2,8,18,8,1	2,8,18,18,8,1

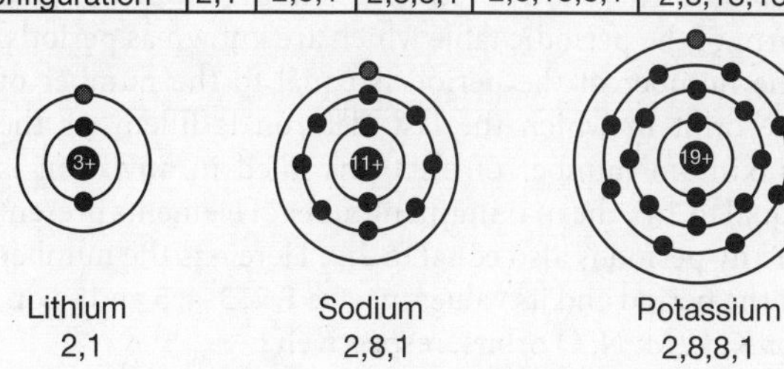

Fig. 5.7 Alkali metals

(ii) In the case of halogens (F, Cl, Br, I) the number of valence electron is seven so they show similarity in properties (Fig. 5.8).

K L   K L M   K L M N
F (9) = 2, 7   Cl (17) = 2, 8, 7   Br (35) = 2, 8, 18, 7

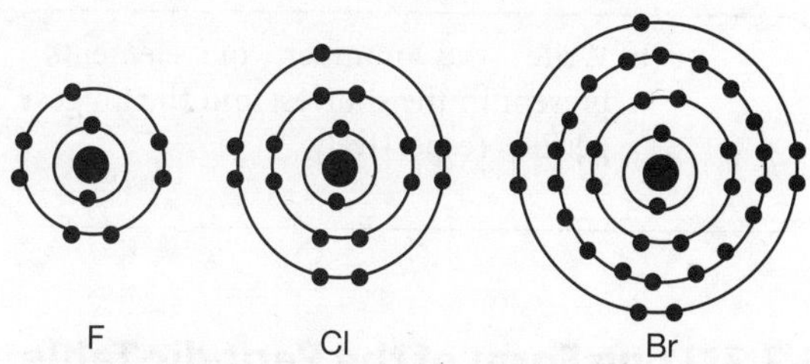

Fig. 5.8 Electronic configuration of halogens

Now it is clear that elements with similar electronic configurations (same number of valence electrons) show similar properties and exhibit periodicity of properties.

5.3.2 Features of the Modern Periodic Table

The main features of the modern periodic table are discussed below.

- Elements have been arranged in order of increasing atomic number.
- It is an ideal arrangement as elements with similar properties are placed together in the same column or group.
- It has seven horizontal rows known as periods.
- There are sixteen vertical columns which are called groups or families.
- The transition elements are placed in the middle of the long periods (Fig. 5.9).
- Isotopes of an element are assigned the same place, as they have the same atomic number.
- Dissimilar metals are placed in different groups. For example, noble metals and alkali metals are separated by placing them in group I and group II respectively.

TEST YOUR KNOWLEDGE

5.2 (a) Which group was not present in Mendeleev's periodic table?
(b) Give any two examples of anomalous pairs in Mendeleev's periodic table.

Solution: (a) Zero group was not present in Mendeleev's periodic table.
(b) Anomalous pairs are: K, Ar and Co, Ni.

5.3 (a) The number of horizontal and vertical columns in the modern periodic table are respectively ________ and ________.

(b) Write the number of elements present in the shortest and the longest periods respectively.

Solution: (a) The number of horizontal and vertical columns in the modern periodic table are 7 and 16 respectively.

(b) The shortest period is I with 2 elements while the longest period is VI with 32 elements.

5.3.3 Long Form of the Periodic Table

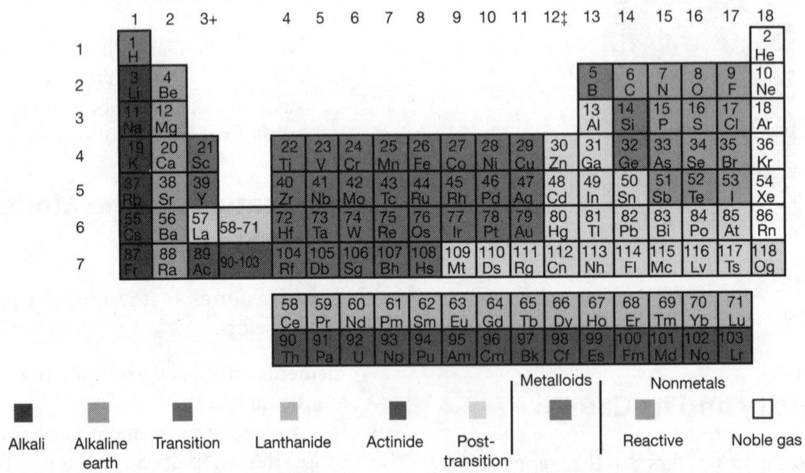

Fig. 5.9 Long form of the periodic table

The modern periodic table was prepared by Niels Bohr (Fig. 5.10) and developed by Range, Werner and Bury on the basis of the electronic configuration of elements. It is also known as the extended form of the periodic table or Moseley's periodic table. As it is based on the electronic configuration of elements those elements with similar properties (similar number of valence electrons) are placed in the same column or group.

Fig. 5.10 Niels Bohr

Source: The American Institute of Physics credits the photo [1] to AB Lagrelius & Westphal, which is the Swedish company used by the Nobel Foundation for most photos of its book series Les Prix Nobel., Public domain, via Wikimedia Commons

Features of the long form of the periodic table: The periodic table is divided into four blocks, *s*, *p*, *d* and *f*, on the basis of the electronic configuration. The *s*-block is on the left hand side of the table, the *p*-block is on the right hand side of the table and the *d*-block is in between the *s* and the *p*-blocks. The *f*-block is placed separately down the table. It has seven horizontal rows (known as periods) and 18 vertical columns (groups).

Periods: There are seven horizontal rows in the long form of the periodic table which are known as periods. The number of the period is equal to the number of the orbit in which the last electron is filled. As the maximum number of electrons filled in any orbit is equal to $2n^2$, the maximum number of elements present in any period is also equal to $2n^2$. Here n is the number of the period and its values may be 1, 2, 3, 4, 5 and so on, for K, L, M, N, O orbits, respectively.

- The first period (K shell) is the shortest period having only two elements (H, He) as $n = 1$ (so, maximum number of elements is equal to $2 \times 1^2 = 2$).
- The second (L shell) and third periods (M shell) are the short periods having eight elements each as the valence orbit cannot have more than eight electrons. Therefore, in the third period also, the maximum number of elements is eight and not eighteen.

- The fourth and the fifth periods are the long periods with eighteen elements each.
- The sixth period is the longest period with 32 elements.
- The seventh period also has 32 elements.
- The first element of each period is an alkali metal while last element is an inert gas. For example, in the second period the first element is Li (alkali metal) and the last element is Ne (inert gas).

Period	Name	Total number of elements
I	Shortest	2
II, III	Short	8, 8
IV, V	Long	18, 18
VI	Longest	32
VII	Incomplete	25*

Groups:

- There are 18 vertical columns and these are known as *groups*. Elements having similar outer electronic configurations in their atoms are arranged in vertical columns, referred to as *groups* or *families*.
- According to the recommendation of the International Union of Pure and Applied Chemistry (IUPAC), the groups are numbered from 1 to 18 replacing the older notation of groups IA … VIIA, VIII, IB … VIIB and 0.
- The IIIB group (third vertical row) is the largest group having 32 elements which includes the lanthanides and actinides.
- Groups IA, IIA (columns 1 and 2), IIIA, IVA, VA, VIA, VIIA, zero group (columns 13 to 18) have metals, non-metals and metalloids. These elements are known as representative elements because their valence electrons represent their group numbers. Groups IA (alkali metals) and IIA (alkaline earth metals) are in the *s*-block while group IIIA (boron family), IVA (carbon family), VA [nitrogen family or pnicogens (producing suffocation)], VIA [oxygen family or chalcogens (ore forming)], VIIA [halogens (sea salt forming)] and VIII are in the *p*-block. The 0 group (18th vertical row) consists of noble gases or inert gases.
- Groups IB, IIB, IIIB (only Sc, Y, La and Ac), IVB, VB, VIB, VIIB and VIII (column 3 to 12) have transition metals only and are in the *d*-block. There are 40 transition elements now. The word transition is used for these metals as their properties lie in between the properties of the elements of the *s* and *p*-block.
- In group IIIB, except Sc, Y, La and Ac, the other elements are known as the inner transition metals and are placed separately in two rows under the periodic table in the *f*-block. They belong to the VI period ($_{58}$Ce, to $_{71}$Lu) and the VII period ($_{90}$Th to $_{103}$Lr). These elements are further divided into two categories lanthanoids or 4*f*-block elements ($_{58}$Ce, to $_{71}$Lu) and actinoids or 5*f*-block elements ($_{90}$Th to $_{103}$Lr). All elements after uranium (atomic number 92) are known as the trans-uranium elements or trans-uranics. They are synthetic or man-made elements and show the phenomenon of radioactivity (emission of alpha, beta and gamma radiations).

Merits of Modern/Long Form of Periodic Table

This form of the periodic table has the following advantages and merits which also overcomes the drawbacks of Mendeleev's periodic table.

Easy to remember: The elements are arranged in the increasing order of atomic numbers in a serial manner so its very easy to remember and reproduce the modern periodic table.

Prediction of properties: Since the elements are arranged on the basis of their electronic configuration, by looking at the position of an element, its properties can be easily predicted.

Position of isotopes: All the isotopes of an element have the same atomic number so they are placed within the same column or group. For example, all the isotopes of hydrogen ($^{1}_{1}$H, $^{2}_{1}$H, $^{3}_{1}$H) are in group 1 and isotopes of carbon (C-12, C-13, C-14) are in group 4.

Justification of anomalous position of some pairs of elements: In Mendeleev's periodic table, cobalt with an atomic mass of 58.93 u was placed before nickel with an atomic mass of 58.71 u. The atomic number of cobalt is 27 and that of nickel is 28, so, with the help of the atomic number, the placement of cobalt before nickel is justified. Similarly, in the case of Te (atomic number 52, atomic mass 127.6) and I (atomic number 53, atomic mass 126.9), their positions are justified.

Uncertainty in prediction of new elements: Atomic masses do not increase in a regular pattern from one element to the next, so it was not possible to predict as to how many new elements could be discovered between two existing elements in Mendeleev's periodic table. This limitation is removed in the modern periodic table as the atomic number increases by 1 unit in going from one element to the next, that is, the number of new elements to be discovered is in between two existing elements is equal to the difference in their atomic numbers.

Defects in the Long Form of Periodic Table

The long form of periodic table has the following defects.

Position of hydrogen: The position of hydrogen is not fixed in this table also. It is placed either in IA or in the VIIA group. As the electronic configuration of hydrogen is similar to that of the alkali metals (1 valence electron) hydrogen can be placed in group I with the alkali metals. However, hydrogen is a non-metal with a very small size while alkali metals are large in size. The oxide of hydrogen (H_2O) is neutral while those of the alkali metals (M_2O, M is alkali metal) are basic in nature. Since there is a difference in properties, placing hydrogen with the alkali metals is still controversial.

Hydrogen needs one electron (like the halogens) to attain a stable configuration of the nearest inert gas and hydrogen is also diatomic and gaseous like the halogens. The oxide of hydrogen (H_2O) is neutral while those of the halogens (like Cl_2O_7) are acidic, it is still controversial to place hydrogen with the halogens in column 17. As the long form is based on the increasing order of atomic number and hydrogen has an atomic number of 1, it is the first element to be placed in the table and its position in group I is justified in this manner.

Position of helium: As helium has an atomic number of 2, it must be placed in group 2 according to electronic configuration. However, it has been placed in column 18 along with the inert gases as its properties resemble theirs.

Position of lanthanides and actinides: No individual places have been assigned to these 28 elements in the periodic table.

This table does not reflect the exact distribution of electrons among all the orbitals of the atoms of all the elements.

Competition Edge

In the *s*-block, except hydrogen, the rest of the elements are soft metals. The *d* and *f*-block include hard metals. The *p*-block consists of non-metals like C, N, P, O, S, F, Cl, Br, I and inert gases. The *p*-block also includes metalloids like B, Si, Ge, As, Sb, Te, Po (Fig. 5.11) which have properties in between those of metals and non-metals. The lower elements of the *p*-block such as Sn, Pb, Bi, At, are metals. There are a few liquid elements as well (Fig. 5.12).

Fig. 5.11 Metalloids

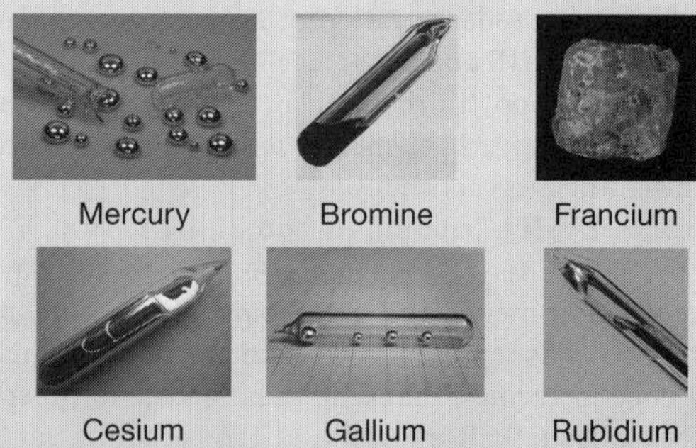

Fig. 5.12 Elements in the liquid state

TEST YOUR KNOWLEDGE

5.4 (a) What is the basis of arranging elements in the long form of the periodic table?
(b) What is the position of the lanthanides and actinides in the long form of the periodic table?

Solution:
(a) In the long form of the periodic table, elements are arranged on the basis of their electronic configurations.
(b) Both lanthanides and actinides are placed separately below the main body of the long form of the periodic table.

Nomenclature and Symbols of Elements with Atomic Numbers > 100

The elements with atomic number greater than 100 are collectively known as heavy elements. In order to name these elements, IUPAC has decided upon the following rules.

- The names are derived directly from the atomic numbers using numerical roots for 0 and the numbers 1–9 (given below) and adding the suffix '-ium'.
- In certain cases, the names have been shortened. For example, bi-ium tri-ium have been shortened to bium and trium and enn-nill has been shortened to ennil.

The IUPAC names and symbols for the element with atomic number 120 are given in Table 5.4. From the table, the roots for 1, 2 and 0 are un, bi and nil, respectively. Hence, the symbol and the name respectively are Ubn and unbinilium.

Table 5.4 Nomenclature of elements with atomic numbers above 100

Digit	Name	Abbreviation
0	nil	n
1	un	u
2	bi	b
3	tri	t
4	quad	q
5	pent	p
6	hex	h
7	sept	s
8	oct	0
9	enn	e

Atomic Number	Name	Symbol
101	Unnilunium	Unu
102	Unnilbium	Unb
103	Unniltrium	Unt
104	Unnilquadium	Unq
105	Unnilpentium	Unp
106	Unnilhexium	Unh
107	Unnilseptium	Uns
108	Unniloctium	Uno
109	Unnilennium	Une
110	Unnnillium	Uun
111	Unununnium	Uuu
112	Ununbium	Uub
113	Ununtrium	Uut
114	Ununquadium	Uuq
115	Ununpentium	Uup
116	Ununhexium	Uuh
117	Ununseptium	Uus
118	Ununoctium	Uuo

Quick Review

- **The periodic table** is an arrangement of elements with similar properties placed together. It helps in easier study of elements.
- **The law of triads:** elements could be arranged in groups of three, in which the atomic weight of the middle element was the arithmetic mean of the first and last element.
- **Law of octaves:** If the elements are arranged in the order of increasing atomic weight, the eighth element starting from a given one, is a kind of repetition of the first, like the eighth node in an octave of music.
- **Proust (1915)**: According to him, all elements are made up of hydrogen atoms.
- **Lothar Meyer (1869)**: He plotted various physical properties of elements against their atomic weights and pointed out that the similar elements occupy similar positions on the curve.
- **Mendeleev's periodic law**: The physical and chemical properties of elements are the periodic function of their atomic weights.
- Mendeleev considered **atomic weight** as the fundamental property of an element and he arranged different elements in the order of their **increasing** atomic weights and formed the periodic table.
- **Modern periodic law:** The physical and chemical properties of elements are the periodic functions of their atomic numbers.
- **Modern periodic table**: It is based on arranging different elements in the increasing order of their atomic numbers.
- **Long form of periodic table:** It is also known as the extended form of the periodic table or Moseley's periodic table. It was prepared by Niels Bohr and developed by Range, Werner and Bury on the basis of the electronic configuration of elements.
- The long form of the periodic table has seven horizontal rows known as periods. The first period is the shortest period having only two elements (H, He). The sixth period is the longest period having 32 elements.
- There are 18 vertical columns and these are known as **groups**. Elements having similar outer electronic configurations in their atoms are arranged in columns, referred to as **groups** or **families**.
- The elements having ns^1 and ns^2 electronic configurations in their outermost shell are called **s-block elements.**
- The elements whose atoms have incomplete *p*-orbitals in their outermost shell or in which the last electron enters any *p*-orbital are known as ***p*-block elements.**
- The ***d*-block** is in between the *s* and the *p*-blocks. These are the elements which have **incomplete *d* orbitals.** The *d*-block elements are called transition elements as their properties lie between those of the *s*- and *p*-block elements.
- The elements placed in two separate rows at the bottom of the periodic table are the ***f*-block elements.** They have incomplete *f* orbitals in their electronic configurations.
- The elements from cerium to lutetium having incomplete 4*f* orbitals are called **lanthanones** or **lanthanoids.**
- The elements from thorium to lawrencium having incomplete 5*f* orbitals are called **actinoids.**

Exercise 1

Section A: Multiple Choice Questions (1 Mark)

1. Identify the group which is not a Dobereiner triad.
   (a) Li, Na, K (b) Be, Mg, Cr
   (c) Ca, Sr, Ba (d) Cl, Br, I

2. Newlands relation is called
   (a) musical law
   (b) law of octaves
   (c) periodic law
   (d) atomic mass law

3. Up to which element was the law of octaves found applicable?
   (a) oxygen (b) calcium
   (c) cobalt (d) potassium

4. At the time of Mendeleev, the number of elements known was
   (a) 63 (b) 65
   (c) 62 (d) 64

5. Which option describes the achievements of the Mendeleev's periodic table?
   (a) prediction of noble gases

(b) it eliminated the blank spaces left in the table
(c) predicting that the elements can be arranged based on their properties
(d) an element in a trend has an average atomic mass of the elements above and below it

6. In Mendeleev's periodic table, gaps were left for the elements to be discovered later. Which of the following elements found a place in the periodic table later on?
(a) chlorine (b) silicon
(c) oxygen (d) germanium

7. The properties of eka-aluminium predicted by Mendeleev are the same as the properties of the element ________ discovered later.
(a) scandium (b) germanium
(c) gallium (d) aluminium

8. Which of the following forms the basis of the modern periodic table?
(a) atomic mass (b) atomic number
(c) number of nucleons (d) all of these

9. The arrangement of elements in the modern periodic table is based on their
(a) increasing atomic mass in the period
(b) increasing atomic number in the horizontal rows
(c) increasing atomic number in the vertical columns
(d) increasing atomic mass in the group

10. What is the other name for group 18 elements?
(a) noble gases (b) alkali metals
(c) alkali earth metals (d) halogens

11. Which groups of elements are called transition metals?
(a) group number 1 to 2
(b) group number 13 to 18
(c) group number 3 to 12
(d) group number 1 to 8

12. In the modern periodic table, the shortest and longest period contain ________ number of elements respectively.
(a) 8, 32 (b) 2, 32
(c) 2, 18 (d) 8, 18

13. Which option arranges the elements of period 4 in the correct sequence?

(a)

Mn	Cr	V	Ti	Sc	Ca	K
54.94	50.20	50.94	47.90	44.96	40.08	39.102

(b)

K	Ca	Ti	Sc	V	Cr	Mn
39.102	40.08	47.90	44.96	50.94	50.20	54.94

(c)

K	Ca	Sc	Ti	V	Cr	Mn
39.102	40.08	44.96	47.90	50.94	50.20	54.94

(d)

Mn	Cr	V	Sc	Ti	Ca	K
54.94	50.20	50.94	44.96	47.90	40.08	39.102

14. The 14 elements after actinium are called
(a) lanthanides
(b) actinides
(c) *d*-block elements
(d) *p*-block elements

15. Identify the wrong sequence of the elements in a group
(a) Ca, Br, Ba (b) Cu, Au, Ag
(c) N, P, As (d) Cl, Br, I

16. Which of these belong to the same period?

Element	A	B	C
Atomic number	2	10	16

(a) A, B (b) B, C
(c) C, A (d) A, B and C

17. A student studies about Mendeleev's periodic table and lists some statements.
P. No fixed position was given to carbon in the periodic table.
Q. The atomic masses do not increase in a regular manner.
R. Isotopes of an element have different chemical properties but similar atomic masses.
Which option lists the limitations of the Mendeleev's periodic table?
(a) only P (b) only R
(c) both P and Q (d) both Q and R

18. Which of the following statements are correct?
(i) Modern periodic table is able to remove the demerits of Mendeleev's table.

(ii) The position of hydrogen with alkali metals is justified as it has only one valence electron.
(iii) The oxide of hydrogen is acidic like the oxides of halogens.
(iv) The long form has four blocks—*s, p, d, f.*

(a) (i), (iii) (b) (ii), (iii)
(c) (i), (ii), (iv) (d) (ii), (iv)

19. Which of the following statements are correct?
(i) Mendeleev's periodic table cannot explain the anomalous pairs properly.
(ii) The modern periodic table cannot justify the position of isotopes.
(iii) The inert gas group of the 18th column was added without any disturbance.
(iv) In the modern periodic table, elements in the same group show similar properties.

(a) (i), (iii) (b) (ii), (iii)
(c) (i), (ii), (iv) (d) (i), (iii), (iv)

20. Match the following.

Column I	Column II
1. Mendeleev	(i) Triad system
2. Moseley	(ii) Law of octaves
3. Newlands	(iii) Periodic properties are a function of the atomic mass
4. Dobereiner	(iv) Periodic properties are a function of the atomic number

(a) (1)—(iv), (2)—(iii), (3)—(i), (4)—(ii)
(b) (1)—(iv), (2)—(iii), (3)—(ii), (4)—(i)
(c) (1)—(iii), (2)—(iv), (3)—(ii), (4)—(i)
(d) (1)—(i), (2)—(ii), (3)—(iii), (4)—(iv)

Assertion Reason Questions

Direction: In the following questions, two statements (Assertion) A and Reason (R) are given. Mark
(a) if A and R are both correct and R is the correct explanation of A;
(b) if A and R are both correct but R is not the correct explanation of A;
(c) A is true but R is false;
(d) A is false but R is true.

Assertion	Reason
1. Mendeleev used atomic number to arrange elements.	**1.** Classification of elements makes their study easier.
2. Co, Ni are an anomalous pair in Mendeleev's periodic table.	**2.** The modern periodic table justifies their position on the basis of atomic number.
3. Periodicity is the resemblance of properties after a definite interval.	**3.** Periodicity in properties is actually due to repetition of similar outer electronic configuration.
4. Elements of atomic number 58 to 71 are lanthanoids.	**4.** Elements of atomic number 90 to 103 are actinoids.
5. B, Si, As, Te are metalloids.	**5.** Their properties lie in between those of non-metals and metals.
6. Isotopes are placed in the same column and same period.	**6.** Isotopes have different atomic numbers.
7. The first period is the shortest period.	**7.** It has only 2 elements.

Section B: Very Short Answer Questions (2 Marks)

1. Define Dobereiner's triad system.
2. Write down three elements that are a Dobereiner's triad.
3. Why was the system of classification of elements into triads not found suitable?
4. If the atomic weights of sulfur (S) and tellurium (Te) are 32 and 127, find the atomic weight of selenium (Se) using the law of triads.
5. Define Newlands' law of octaves.
6. Why did Newlands' law of octaves fail?
7. What was the need for classification of elements?
8. Define Mendeleev's periodic law.
9. Which important property did Mendeleev use to classify the elements in his periodic table?
10. State the modern periodic law of classification of elements.
11. What is the basic difference between Mendeleev's and the modern periodic laws?

12. Why did Mendeleev leave some places empty in his periodic table?

13. Write the number of horizontal rows in the modern periodic table. What are these rows called?

14. Write the number of vertical columns in the modern periodic table. What are these columns called?

15. Write the total number of periods in the modern periodic table. **[Board Term II, 2015]**

16. Why should hydrogen be placed in group I?

17. Name two elements of the first period in the modern periodic table.

18. Give reasons why noble gases are placed in a separate group in the modern periodic table.

19. Which group of elements could be placed later on in Mendeleev's periodic table without disturbing the original order?

20. The three elements predicted by Mendeleev from the gaps in his periodic table were called eka-boron, eka-aluminium and eka-silicon. What names are given to these elements respectively?

21. Give the number of elements in the 2nd and the 5th period in the modern periodic table.

Case Study or Passage Based Questions

The long form of the periodic table is based on the electronic configuration of elements so it is the best possible arrangement of placing elements. It contains horizontal rows and vertical columns. All the elements present in a vertical column resemble in properties. This table is divided into four blocks – *s*, *p*, *d*, *f* – and the elements present in them are known as the *s*, *p*, *d* and *f*-block elements respectively.

22. Which periods are called long periods?

23. In the long form of the periodic table, which period is the longest and how many elements are present in it?

24. Which block elements are called representative elements?

25. Which group elements are known as inert gases?

26. Why are lanthanides called rare earth elements?

27. On which side of the periodic table do you find (i) metals, (ii) non-metals?

Section C: Short Answer Questions (3 Marks)

1. (i) What were the limitations of Dobereiner's triads?
   (ii) Lithium, sodium and potassium form a Dobereiner's triad. The atomic masses of lithium and potassium are 7 and 39 respectively. Predict the atomic mass of sodium.

2. (i) Lithium and beryllium will resemble which elements in properties respectively, according to the law of octaves?
   (ii) Write down two drawbacks of Newlands' law of octaves.

3. Fill in the blanks.
   (i) The concept of grouping elements into triads was given by ________.
   (ii) Mendeleev's basis for the periodic table is ________.
   (iii) The basis for the modern periodic table is ________.
   (iv) Isotopes belong to the same ________ in the periodic table.

4. What is the basic theme of organisation in the periodic table?

5. What is the basic difference in approach between Mendeleev's periodic law and the modern periodic law?

6. (i) Who proposed the modern periodic law?
   (ii) What was the basis of arranging the elements in it?
   (iii) State the modern periodic law.

7. How do the early attempts to classify the chemical elements differ from the modern attempt? Which of the two is better and why? Give examples.

8. State the main aim of classifying elements. Which is the more fundamental property of elements that is used in the development of the modern periodic table ? On which side of the periodic table can one find metals, non-metals and metalloids?

9. Why do elements in the same group have similar physical and chemical properties?

10. Why is the assignment of position to hydrogen in the periodic table considered anomalous?

11. State Mendeleev's periodic law. Write two achievements of Mendeleev's periodic table.

12. Write two limitations of Mendeleev's periodic table.

13. (i) What are metalloids? (ii) Name any four metalloids.

14. Give an example of:
    (i) a metal that is liquid at room temperature,
    (ii) a non-metal that is liquid at room temperature,
    (iii) an inert gas (At. no < 20).
    [Board Term II, 2015]

15. How do the early attempts to classify the chemical elements differ from the modern attempt? Which of the two is better and why? Give examples.

16. Write any three demerits or limitations of Mendeleev's periodic table.

17. Justify the position of hydrogen with alkali metals and halogens separately.

18. Define periodicity. What is the cause of periodicity?

19. How does the modern periodic table justify the anomalous position of some pairs of elements?

20. Write any three demerits of the long form of the periodic table.

21. What is the name given to the elements present from column 3 to column 12 why they are given this name?

22. What are lanthanoids and actinoids? Where are they placed in the long form of the periodic table?

23. What are metalloids? How many electrons can be present in the valence shell of their atoms? In what part of the periodic table are these located? What type of oxides are formed by these elements?

Section D: Long Answer Questions (5 Marks)

1. What are the limitations of Newlands' law of octaves?

2. Which important property did Mendeleev use to classify the elements in his periodic table and did he stick to that?

3. State two main properties of elements on which Mendeleev's periodic classification was based. Why is the position of hydrogen not fixed in his periodic table?

4. (i) How does the electronic configuration of an atom relate to its position (period and group) in the modern periodic table?
   (ii) Would you place the two isotopes of chlorine, Cl-35 and Cl-37, in different slots because of their different atomic masses or in the same slot because their chemical properties are the same? Justify your answer.

5. How does the electronic configuration of an atom of an element relate to its position in the modern periodic table? Explain with one example.

6. Write down five major differences between Mendeleev's periodic table and the modern periodic table.

7. How could the modern periodic table remove the various anomalies of Mendeleev's periodic table?

8. How does the electronic configuration of an atom relate to its position in the modern periodic table?

Section E: Case Study or Passage Based Questions (4 Marks)

Classification of elements is necessary not only to make their study easier, but also helps in correcting atomic mass and discovering new elements. The modern periodic table or long form of periodic table is the best classification of elements as it is based on the electronic configuration of elements. Most of the elements are placed perfectly in particular columns. Horizontal columns are periods while vertical columns are groups. This table also addresses the demerits of Mendeleev's periodic table.

1. The number of horizontal rows and vertical columns in the long form of the periodic table are respectively,
   (a) 7, 16 (b) 7, 18
   (c) 7, 9 (d) 8, 16

2. The elements present in column 3 to 12 are known as
(a) representative elements
(b) transition metals
(c) lanthanoids
(d) actinoids

3. The maximum number of elements in any horizontal and vertical column are respectively
(a) 18, 18 (b) 18, 32
(c) 32, 32 (d) 16, 32

4. Which block contains soft metals?
(a) *s*-block (b) *p*-block
(c) *d*-block (d) *f*-block

5. Which of the following sets of elements are metalloids?
(a) C, N, O (b) Be, B, C
(c) Li, Be, B (d) B, Si, Ge

High Order Thinking Skills (HOTS) Questions

1. (i) The modern periodic table has been evolved through the early attempts of Dobereiner, Newlands and Mendeleev. List one advantage and one limitation of all three attempts.
(ii) Name the scientist who first of all showed that the atomic number of an element is a more fundamental property than its atomic mass.
(iii) State the modern periodic law.
[CBSE 2018]

2. (i) Why did Mendeleev have gaps in his periodic table?
(ii) State any three limitations of Mendeleev's classification.
(iii) How does the electronic configuration of atoms change in a period with the increase in atomic number?

3. (i) Name the metals among the first five elements of the modern periodic table.
(ii) Write their chemical symbols.
(iii) Write the formula of their oxides.
[CBSE 2016]

4. Answer questions (a)–(d) on the basis of your understanding of the following paragraph and the related studied concepts. Around the year 1800, only 30 elements were known. Dobereiner in 1817 and Newlands in 1866 tried to arrange the then known elements and framed laws which were rejected by the scientists. Even after the rejection of the proposed laws, many scientists continued to search for a pattern that correlated the properties of elements with their atomic masses.

The main credit for classifying elements goes to Mendeleev for his most important contribution to the early development of a periodic table of elements wherein he arranged the elements on the basis of their fundamental property, the atomic mass and also on the similarity of chemical properties. The formulae of their hydrides and oxides were treated as the basic criteria for the classification of the elements. However, Mendeleev's classification also had some limitations as it could not assign a position to isotopes. He also left some gaps in the periodic table.
(a) State Mendeleev's periodic law.
(b) Why did Mendeleev leave some gaps in the periodic table?
(c) If the letter R was used to represent any of the elements in the group, then the hydride and oxide of carbon would respectively be represented as
(i) RH_4, RO (ii) RH_4, RO_2
(iii) RH_2, RO_2 (iv) RH_2, RO
(d) Isotopes are
(i) atoms of an element with similar chemical properties but different atomic masses
(ii) atoms of different elements with similar chemical properties but different atomic masses
(iii) atoms of an element with different chemical properties but same atomic masses
(iv) atoms of different elements with different chemical properties but same atomic masses
[CBSE 2020]

5. Ria and Reena are the students of Class X. Ria is very well organised whereas Reena is an unorganised student and always faces a lot of problems in handling life situations.
(i) How does organisation help in daily life?
(ii) How can you relate the above fact with the chapter 'Classification of Elements'. How does classification of elements help us to study them properly?

6. A part of the periodic table is given below:

Periods \ Group	1	2	3 to 12	13	14	15	16	17	18
2		P					S	T	U
3		A						V	X
4		B		Q	R			W	
5		C							

On the basis of this table answer the following questions.

(i) Find the most reactive non-metal and metal here.

(ii) The family in which A, B, C are present is known as ___________.

(iii) The group in which element E is present is known as ___________.

(iv) Which of the following are also present in the atmosphere?

(v) The elements for which gaps were left by Mendeleev in his table are ___________.

Answers

Section A: Multiple Choice Questions

1. (b)	**2.** (b)	**3.** (b)	**4.** (a)
5. (a)	**6.** (d)	**7.** (c)	**8.** (b)
9. (b)	**10.** (a)	**11.** (c)	**12.** (b)
13. (c)	**14.** (b)	**15.** (b)	**16.** (a)
17. (a)	**18.** (c)	**19.** (d)	**20.** (c)

Assertion Reason Questions

1. (d)	**2.** (a)	**3.** (a)	**4.** (b)
5. (a)	**6.** (c)	**7.** (a)	

Section B: Very Short Answer Questions

1. See text (Section 5.1.1).

2. Lithium, sodium, potassium.

3. It is because all the elements discovered at that time could not be classified into triads.

4. ^{32}S $^{?}Se$ ^{127}Te

Using law of triads

Atomic weight of Se $= \dfrac{32+127}{2} = 79.5$

5. See text (Section 5.1.1).

6. It failed as, during classification of elements, only the atomic masses are considered and not their properties.

7. To make the study of these elements easy, they have been divided into a few groups in such a way that elements in the same group have similar properties.

8. See text (Section 5.1.1).

9. His table is based on the chemical properties of elements.

10. It states that, properties of elements are the periodic functios of their atomic number.

11. Mendeleev's periodic law is based on atomic weight while the modern periodic law is based on atomic number.

12. Mendeleev left some places empty in his periodic table so that newly discovered elements can be given places.

13. There are seven horizontal rows in the modern periodic table. These rows are called periods.

14. There are 18 vertical columns in the modern periodic table and these are known as groups.

15. The total number of periods in the modern periodic table is seven.

16. Hydrogen should be placed in group 1, since it has only one electron in its outermost shell.

17. Hydrogen and helium.

18. It is because they resemble each other but do not resemble other group elements.

19. Zero group.

20. Scandium (Sc), gallium (Ga) and germanium (Ge) respectively.

21. The 2nd period has 8 elements and the 5th period has 18 elements.

Case Study or Passage Based Questions

22. IV and V periods are called long periods.

23. In the long form of the periodic table, VI period is the longest and it has 32 elements.

24. *s*- and *p*-block elements are called representative elements.

25. Zero group or 18 column elements are called inert or noble gases.

26. As lanthanides are present in nature in low abundance, they are called rare earth elements.

27. Metals are placed on the left side while non-metals are placed on the right side of the periodic table.

Section C: Short Answer Questions

1. (i) Dobereiner's system of classification failed to arrange all the elements in the form of triads having similar chemical properties. Dobereiner could identify only three triads from the elements known at that time.

(ii) Atomic mass of Na $= \frac{7+39}{2} = \frac{46}{2} = 23$

2. (i) Lithium resembles sodium, and beryllium resembles magnesium.

(ii) (a) The law was applicable to elements up to calcium (Ca) only.

(b) Properties of newly discovered elements did not fit into the law of octaves.

3. (i) Dobereiner
(ii) atomic mass
(iii) atomic number
(iv) position (group)

4. The basic theme of organisation of elements in the periodic table is to simplify and systematise the study of the properties of all the elements and millions of their compounds. On the basis of similarities in chemical properties, the various elements have now been divided into different groups. This has made the study simple because the properties of elements are now studied in the form of groups rather than individually.

5. Mendeleev's periodic law states that the physical and chemical properties of the elements are a periodic function of their atomic weight while the modern periodic law states that physical and chemical properties of elements are a periodic function of their atomic numbers. Thus, the basic difference in approach between Mendeleev's periodic law and the modern periodic law is the change in the basis of classification of elements from atomic weight to atomic number.

6. (i) Henry Moseley
(ii) Atomic number
(iii) See text (Section 5.2).

7. The early attempts to classify elements were made on the basis of atomic masses of the elements. The modern basis is the atomic number. This is better because now it has been established that the atomic number of an element is the fundamental property. Further there are certain anomalies in Mendeleev's periodic table such as the position of isotopes and misplacement of certain elements; for example, argon is placed (Atomic mass = 40, atomic number = 18) before potassium in Mendeleev's periodic table (Atomic mass = 39, At. no = 19).

8. (i) **Aim of classification:** Systematic study of the known elements.
(ii) **Basic property:** Atomic number.
(iii) Metals on the left.
(iv) Non-metals on the right.
(v) Metalloids at the border of metals and beginning of non-metals.

9. Elements in the same group have similar electronic configuration and hence have similar physical and chemical properties.

10. Electronic configuration of hydrogen resembles that of the alkali metals. Like the alkali metals, hydrogen combines with halogens, oxygen and sulfur to form compounds having similar formulae. On the other hand, just like halogens, hydrogen also exists as diatomic molecules and it combines with metals and non-metals to form covalent compounds. So, no fixed position can be given to hydrogen in the periodic table.

11. **Mendeleev's periodic law:** Properties of elements are the periodic function of their atomic masses.

Achievements: (i) It could classify all the elements discovered at that time.
(ii) It helped in the discovery of new elements.
(iii) It helped in the correction of atomic mass of some of the elements. **(Any two)**

12. Two limitations of Mendeleev's periodic table are: (i) Position of isotopes could not be explained.
    (ii) No fixed position for hydrogen.

13. (i) Borderline elements which are intermediate in properties between metals and non-metals are called metalloids.
    (ii) Boron, silicon, germanium and arsenic.

14. (i) Mercury (ii) Bromine
    (iii) Helium or neon or argon **(Any one)**

15. The early attempts to classify elements were made on the basis of atomic masses of the elements. The modern basis is the atomic number. This is better because, it is now established that the atomic number of an element is the fundamental property. Further there are certain anomalies in Mendeleev's periodic table such as the position of isotopes and misplacement of certain elements. For example, argon is placed (Atomic mass = 40, atomic number = 18) before potassium in Mendeleev's periodic table (Atomic mass = 39, At. no = 19).

16. See text (Section 5.2).

17. See text (Section 5.2).

18. See text (Section 5.2).

19. See text (Section 5.2).

20. See text (Section 5.2).

21. See text (Section 5.2).

22. See text (Section 5.2).

23. Metalloids are elements which resemble both metals and non-metals. The valency shell of metalloids contain 3, 4, 5, 6 elements starting from periods 2 to 5 respectively. They are found on the right side of the table in a zig-zag manner; for example, boron, silicon, germanium, arsenic, and so on. They form amphoteric oxides.

Section D: Long Answer Questions

1. The limitations of Newlands law of octaves are:
   (i) The law was applicable to elements up to calcium (Ca) only.
   (ii) It contained only 56 elements. Further it was assumed by Newlands that only 56 elements existed in nature and no more elements would be discovered in the future.
   (iii) In order to fit elements into the table, Newlands adjusted two elements in the same column as fluorine, chlorine and bromine which have very different properties than these elements. Iron, which resembles cobalt and nickel in properties, has been placed differently away from these elements.

2. Mendeleev used atomic weight as the basis of classification of elements in the periodic table. He arranged the known elements in order of increasing atomic weight grouping together elements with similar properties. He sincerely stuck to this basis leaving blank spaces where elements were not known at that time. For example, the element gallium after aluminium, and germanium after silicon, were not known at the time Mendeleev prepared his periodic table. He called these elements as eka-aluminium and eka-silicon. Later, these elements were discovered and their properties were found to be similar to those predicted by Mendeleev.

3. (i) The physical properties of elements are the periodic function of their atomic masses.
   (ii) Similar formulae of oxides and hydrides formed by the elements of the same group was selected as the chemical property used as the basis of classification.
   The atomic number of hydrogen is 1 and hence it should occupy the first position in group 1 of the periodic table. But because of its chemical similarity with elements of group 1 and 17, the position of hydrogen in the periodic is not fixed.

4. (i) In the periodic table, elements are placed according to their electronic configuration. If an element has only one shell in its electronic configuration, it is placed in the first period. If the element has two shells, then it is placed in the second period, and so on. Vertical columns in the periodic table are called groups.

There are eighteen groups and in a group, all the elements have the same number of valence electrons.

(ii) They will be placed at the same slot as their atomic number is the same, the valence electrons and valency are also same and they have the same chemical properties.

5. The position of elements depends upon number of valence electrons which depend upon the electronic configuration. Those elements which have the same number of valence electrons, occupy the same group. For example, those elements which have one valence electron belong to group 1 and elements which have two valence electrons belong to group 2.

The period number is equal to the number of shells. For example, the atomic number of sodium (Na) is 11, so the electronic configuration will be 2, 8, 1. Sodium has one valence electron in the valence shell so it belongs to group 1. As sodium has three shells, it belongs to the 3rd period.

6.

S. No.	Mendeleev's periodic table	Modern periodic table
(i)	Elements have been arranged in the increasing order of atomic masses.	Element have been arranged in the increasing order of atomic number.
(ii)	It consists of 8 groups.	It consists of 18 groups.
(iii)	No sub-groups.	All the groups from I to VIII are divided into sub-groups.
(iv)	He left some gaps in the filled table for undiscovered elements.	All elements are filled in a regular pattern and there are no gaps.
(v)	Isotopes were placed in different groups.	Isotopes were placed in the same group.

7. **Modification in Mendeleev's periodic table:** With the adoption of the atomic number as the basis for classification of elements, some of the anomalies in the Mendeleev's periodic table have now disappeared. These are given below.

(i) **Position of hydrogen:** The atomic number of hydrogen is 1 and hence, it should occupy the first position in group 1 of the periodic table. But because of its chemical similarity with elements of group 1 and group 17, the position of hydrogen in the periodic table is still not confirmed.

(ii) **Position of isotopes:** As the isotopes of an element have the same atomic number, they should occupy the same place in the periodic table.

(iii) **Position of anomalous pairs:** The position of anomalous pairs of elements, that is, pairs of elements in which the element with the higher atomic mass preceeds the element with lower atomic mass, is also decided. When these elements are arranged according to their atomic number, they occupy their right places in the periodic table.

8. (i) The group number of elements having up to two valence electrons is equal to the number of valence electrons.

(ii) The group number of elements having more than two valence electrons is equal to the number of valence electrons plus 10.

For example:

Number of valence electrons	Group number in periodic table
1	1 } Group number = number of electron
2	2 } Group number = number of electron
3	3+10 = 13 } Group number = number of electrons +10
4	4+10 = 14 } Group number = number of electrons +10

(iii) For those elements, which do not have the same number of valence electrons but have the same number of shells, the number of shells is equal to the period number to which they belong in the periodic table. For example,

2nd period:

Li	Be	B	C	N	O	F	Ne
2,1	2,2	2,3	2,4	2,5	2,6	2,7	2,8

Section E: Case Study or Passage Based Questions

1. (b) 2. (b) 3. (c) 4. (a)
5. (d)

High Order Thinking Skills (HOTS) Questions

1. (i) **Dobereiner triads**
Advantage: The three elements of a triad were found to possess similar properties.
Limitation: Some elements which are dissimilar were being grouped into a triad.
Newland's octaves
Advantage: Newlands was the first one to come up with the idea of periodicity while classifying the elements.
Limitation: It was applicable only for lighter elements up to calcium.
Mendeleev's attempt:
Advantage: Although noble gases were discovered very late, they can be placed easily in Mendeleev's periodic table without disturbing the order of the rest of the elements.
Limitation: (i) The position of hydrogen was not explained. (ii) Henry Moseley showed that atomic number of an element is a more fundamental property than its atomic mass. (iii) It states that the properties of the elements are the periodic function of their atomic number.

2. (i) Gaps were left for undiscovered elements in the periodic table.
(ii) (a) Position of hydrogen was not justified.
(b) Increasing order of atomic mass could not be maintained.
(c) Isotopes have similar chemical properties but different atomic masses, they cannot be given separate places.
(iii) The number of shells remains the same, the number of valence electrons goes on increasing from left to right in a period till the octet is complete as shown below.

Li	Be	B	C	N	O	F	Ne
2,1	2,2	2,3	2,4	2,5	2,6	2,7	2,8

3. (i) Lithium and beryllium.
(ii) Li and Be.
(iii) Li_2O and BeO.

4. (a) According to Mendeleev's periodic law, the physical and chemical properties of elements are the periodic function of their atomic masses.
(b) Mendeleev predicted the existence and the properties of some elements, which were not known at that time. Therefore, he left some gaps in the periodic table.
(c) (ii) RH_4, RO_2
(d) (i) Atoms of an element with similar chemical properties but different atomic masses are isotopes.

5. (i) Organisation is all about keeping things in proper order. It plays a significant role in helping us in achieving our goals. It gives us a sense of control and allows for increased productivity.
(ii) Classification of elements is based on organisation of the elements in a proper manner that helps to make the study of all the elements easier. To make the study of all the elements in the periodic table easy, these elements have been divided into a few groups in such a way that elements in the same group have similar properties.

6. (i) The most reactive non-metal is U, that is, fluorine and the most reactive metal is C, which is rubidium.
(ii) Alkali metals.
(iii) Alkaline earth metals group.
(iv) S (nitrogen), T (oxygen) and X (argon) are present in the atmosphere.
(v) Q (gallium) and R (germanium).

Hints

Multiple Choice Questions

18. The oxide of hydrogen (H_2O) is neutral.

19. It justifies the position of isotopes on the basis of atomic number.

Case Study or Passage Based Questions

3. The sixth and seventh period (horizontal row) and third column (vertical row) both contain 32 elements each.

5.4 Position of an Element in the Periodic Table

In order to decide the position of any element in the periodic table, the electronic configuration of the element plays a vital role. By writing the electronic configuration of any element, we can tell its period number, its group number, number of valence electrons and valency. We have to first write the electronic configuration of an element using the rules to write the electronic configuration and then to find period number and group number we have to see the orbit number in which the last electron is filled and the total number of valence electrons respectively.

Period number = Largest value of '*n*' (last orbit number)

Group number = Sum of valence electrons

Note: In the case of elements with atomic number between 21 and 30, the group number is found as follows.

Group number = number of electrons in last two orbits – 8

For example, in the case of iron (Fe = 26)

Fe = 2, 8, 14, 2 Period = IV

Group number = (14 + 2) – 8 = VIII

TEST YOUR KNOWLEDGE

5.5 If the atomic number of Ca is 20, predict its period and group number.

Solution:

$$Ca \rightarrow 2, 8, 8, 2$$

Period = Largest '*n*' value = 4 = IV

Group number = Number of valence electrons = 2 = II

5.6 Find the period and group number of zinc (atomic number = 30).

Solution:

Zinc is present in IV period and in VIII group.

$$Zn \rightarrow 2, 8, 18, 2$$

Period = Largest '*n*' value = 4 = IV

Group number = Number of valence electrons = 2 = II

5.5 Trends in Periodic Properties

The physico-chemical properties of an element mainly depend upon its outer electronic configuration. As the outer electronic configuration changes on moving from left to right across a period, within the same period the elements show a variation both in their physical and chemical properties (Fig. 5.13). On the other hand, if we talk about a group, the elements have the same outer electronic configuration which means that most of their chemical properties are similar. However, their physical properties show a regular variation on moving down the group because of a corresponding increase in the number of filled inner orbits.

The properties of the elements which vary periodically with their atomic numbers when we move from left to right across the period or top to bottom in any group are known as periodic or atomic properties. Some common periodic properties are valency, atomic size, metallic nature, non-metallic nature, reactivity, *electronegativity, *ionisation energy and *electron-affinity. (* marked properties are not in your syllabus.)

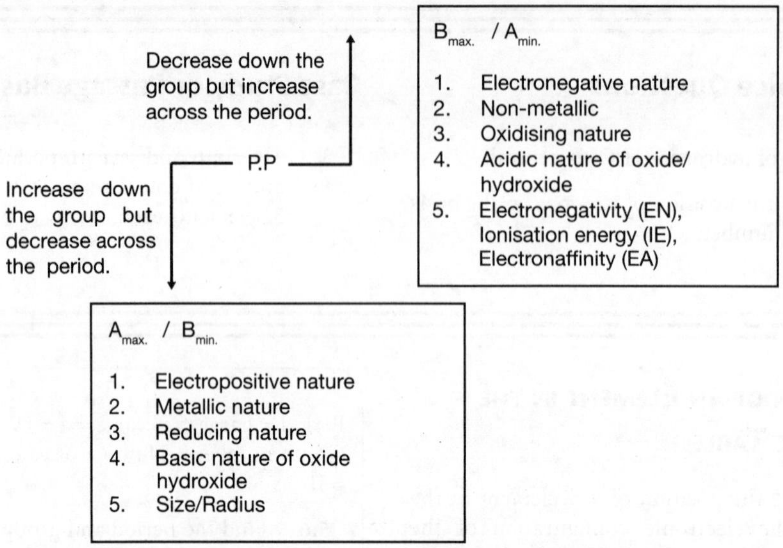

Fig. 5.13 Variation of periodic properties

5.5.1 Valence Electrons and Valency

The number of electrons present in the outermost orbit of an element represents the valence electrons. The number of valence electrons represents the group number. For example, if the number of valence electrons is two, the group number will be II. On moving left to right across any period the number of valence electrons increases from 1 to 8 while it remains same on moving down the group (Table 5.5).

Table 5.5 Valence electrons and valency

	Group	I	II	III	IV	V	VI	VII	VIII
Second period	Element	Lithium	Beryllium	Boron	Carbon	Nitrogen	Oxygen	Fluorine	Neon
	Symbol	Li	Be	B	C	N	O	F	Ne
	Atomic number	3	4	5	6	7	8	9	10
	Electronic configuration	K, L 2, **1**	K, L 2, **2**	K, L 2, **3**	K, L 2, **4**	K, L 2, **5**	K, L 2, **6**	K, L 2, **7**	K, L 2, **8**
Third period	Element	Sodium	Magnesium	Aluminum	Silicon	Phosphorus	Sulfur	Chlorine	Argon
	Symbol	Na	Mg	Al	Si	P	S	Cl	Ar
	Atomic number	11	12	13	14	15	16	17	18
	Electronic configuration	K, L, M 2, 8, **1**	K, L, M 2, 8, **2**	K, L, M 2, 8, **3**	K, L, M 2, 8, **4**	K, L, M 2, 8, **5**	K, L, M 2, 8, **6**	K, L, M 2, 8, **7**	K, L, M 2, 8, **8**
Fourth period	Element	Potassium	Calcium	Gallium	Germanium	Arsenic	Selenium	Bromine	Krypton
	Symbol	K	Ca	Ga	Ge	As	Se	Br	Kr
	Atomic number	19	20	31	32	33	34	35	36
	Electronic configuration	K,L,M,N 2,8,8,**1**	K,L,M,N 2,8,8,**2**	K,L,M,N 2,8,18,**3**	K,L,M,N 2,8,18,**4**	K,L,M,N 2,8,18,**5**	K,L,M,N 2,8,18,**6**	K,L,M,N 2,8,18,**7**	K,L,M,N 2,8,18,**8**

From this table it is clear that both Li and Na have one valence electron as they are in the same group while on moving from Na to Ar the number of valence electrons increases from 1 to 8.

Valency: The combining capacity of an element is known as its *valency.* It is measured in terms of the number of valence electrons (present in the outer orbit), or the number of H atoms or O atoms or Cl atoms linked with one atom of an element.

Valency = number of valence electron = number of H-atoms or O-atoms or Cl-atoms linked with one atom of an element.

For example, (i) the valency of lithium and sodium is 1 as both elements have only one valence electron. (ii) The valency of beryllium and magnesium is 2 as both have two valence electrons. (iii) The valency of boron and aluminium is three as both have three valence electrons. (iv) The valency of carbon and silicon is four as both have four valence electrons. (v) In CH_4, the valency of carbon = 4 as the carbon atom has 4 valence electrons or is linked with 4 H atoms. (vi) In NH_3, the valency of nitrogen = 3.

Variation in Valency in a Period

In the case of representative elements (*s*- and *p*-block) it increases as the number of valence electrons increases from 1 to 8 across the period (Table 5.6).

Table 5.6 Variation in valency across a period

Group	I	II	XIII	XIV	XV	XVI	XVII	XVIII
Number of electrons	1	2	3	4	5	6	7	8
Valence on hydrogen scale	1	2	3	4	3	2	1	0
Valence on oxygen scale	1	2	3	4	5	6	7	0

On the hydrogen scale, the valency increases from 1 to 4 and then decreases from 4 to 0, while on the oxygen scale, it increases from 1 to 7.

	1	2	13	14	15	16	17
3rd period	NaH	MgH_2	AlH_3	SiH_4	PH_3	H_2S	HCl
	1	2	3	4	3	2	1
	Na_2O	MgO	Al_2O_3	SiO_2	P_2O_5	SO_3	Cl_2O_7
	1	2	3	4	5	6	7

* Due to zero valency, inert gases are chemically unreactive and found to be in the isolated state.

Valency in a Group

In a particular group, all the elements have the same valency which is equal to the group number as all of them have the same number of valence electrons. That is, the valency does not change in a specific group.

Tendency to lose or gain electrons: Every element or atom tries to attain the octet state (8 valence electrons) or the zero valency state or the nearest inert gas electronic configuration for stability by losing or gaining electrons. Elements with one or two or three valence electrons can easily lose electrons to attain the octet state while elements with five or six or seven electrons can easily gain electrons to attain the nearest inert gas configuration. Elements having four electrons undergo sharing of electrons to complete their octet. Losing or gaining of electrons is decided on the basis of energy considerations (lower energy value). For example,

Cations (formed by losing electrons)	Anions (formed by gaining electrons)
$\underset{2,8,1}{Na} \longrightarrow \underset{2,8}{Na^+} + e^-$ Valency = 1	$F + e^- \longrightarrow F^-$ ($\underset{2,7}{F}$, $\underset{2,8}{F^-}$) Valency = 1
$\underset{2,8,2}{Mg} \longrightarrow \underset{2,8}{Mg^{2+}} + 2e^-$ Valency = 2	$\underset{2,6}{O} + 2e^- \longrightarrow \underset{2,8}{O^{2-}}$ Valency = 2
$\underset{2,8,3}{Al} \longrightarrow \underset{2,8}{Al^{3+}} + 3e^-$ Valency = 3	$\underset{2,5}{N} + 3e^- \longrightarrow \underset{2,8}{N^{3-}}$ Valency = 3

Silicon (Si) has four valence electrons, which means, it can get the stable octet state configuration either by losing four electrons or by gaining four electrons but considering energy requirements it is not favourable as high energy is required. In such a case, sharing of electrons takes place. For example,

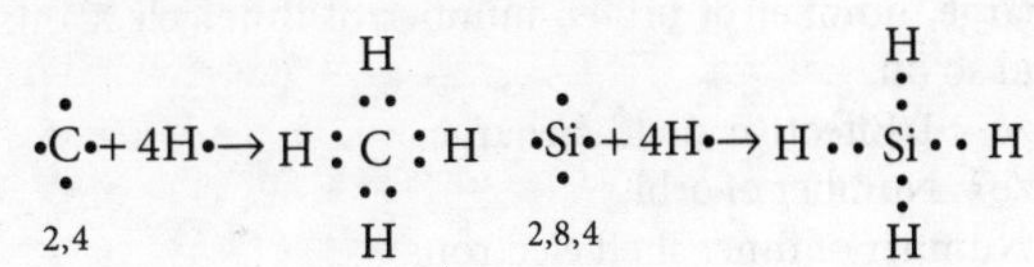

Competition Edge

If the number of valence electrons is more than 4, valency = valence electrons – 8

For example, in H_2S

Valency of sulfur = 6 – 8 = –2 = 2

Valency is simply a number with no significance of + and – signs.

5.5.2 Atomic Size (Radius)

If an atom is considered to be a sphere, the atomic size can be given by the radius of the sphere and is commonly known as its *atomic radius.* It can be defined as *the distance between the centre of the nucleus of an atom to its outer most shell of electrons, that is, penultimate shell electron.*

It is usually expressed in meter (m), centimeter (cm), angstroms (Å) and picometers (pm). 1 meter = 100 cm, 1 Å = 10^{-8} cm = 10^{-10} m and 1 pm = 10^{-10} cm = 10^{-12} m.

The absolute value of the atomic radius cannot be determined because it is not possible to locate the exact position of electrons in an atom as an orbital has no sharp boundaries. It is not possible to isolate an individual atom for its size determination due to its small size. As the absolute value of atomic size cannot be determined, it is expressed in terms of the operational definitions such as *ionic radius, *covalent radius, *van der Waals radius and *metallic radius (* marked: you will study these in class XI).

Competition Edge

Effective nuclear charge (Z_{eff}) and screening effect: The force of attraction present between the nucleus and the valence electrons is reduced by the force of repulsion exerted by the inner shell electrons on these valence electrons. It is called *screening* or *shielding effect* and the net force is called effective nuclear charge.

Factors Affecting Size

The atomic radius depends upon the effective nuclear charge, number of orbits, number of inner electrons, and so on.

Size ∝ 1/Effective nuclear charge

Size ∝ Number of orbits

∝ Number of inner shell electrons

∝ Electronic repulsion

$$\propto \frac{1}{\text{Number of valence electrons}}$$

Variation in the Value of Atomic Radius

In a period: On moving from left to right across any period, the atomic size decreases as the number of atomic orbits remain the same, while the number of valence electrons increases. As a result, the effective nuclear charge increases and the size decreases. In general, the size decreases as follows:

IA (group 1) > IIA (group 2) > IIIA (group 13) > IVA (group 14) > VA (group 15) > VIA (group 16) > VIIA (group 17)

For example, (i) in the second period, the atomic radius decreases as follows:

Li > Be > B > C > N > O > F

Lithium is the largest while fluorine is the smallest.

(ii) In the third period, the atomic radius decreases as follows:

Na > Mg > Al > Si > P > S > Cl

Sodium is the largest while chlorine is the smallest.

In a group: On moving down a group, the atomic radius increases from top to bottom as the number of shells or orbits and screening effect (inner electron repulsion) increases (Fig. 5.14). Which means, the effective nuclear charge decreases so the valence electrons are away from the nucleus.

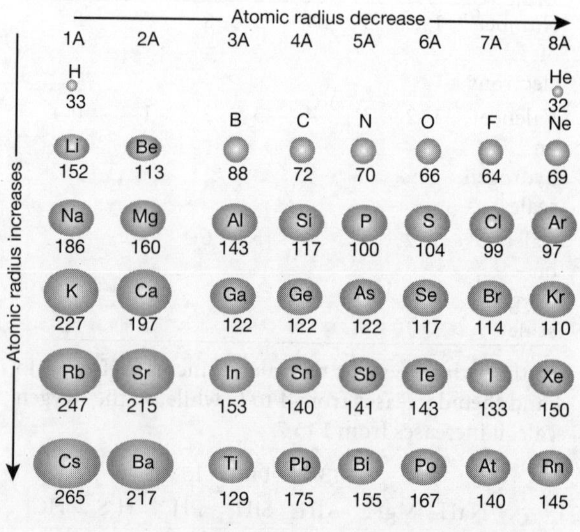

Fig. 5.14 Variation of atomic radius in periods and groups

For example, (i) in the first group, the atomic radius increases from lithium to cesium as follows:

$$Li < Na < K < Rb < Cs$$

(ii) In the seventh group, the atomic radius increases from fluorine to iodine as follows:

$$F < Cl < Br < I$$

Size of cation is smaller than its atom: During cation formation, the outermost orbit becomes empty and the number of valence electrons decreases, so, the effective nuclear charge increases and the size decreases. For example,

Na	>	Na^+
2,8,1		2,8
Orbits 3		2

The sodium atom is larger than the sodium ion since the atom has more number of orbits and less effective nuclear charge.

Size of anion is greater than its atom: During anion formation, electrons are taken up; so, the effective nuclear charge decreases and the size increases. For example,

Cl^-	>	Cl
2,8,8		2,8, 7
Orbits 3		3

The chloride ion is larger than the chlorine atom as electron repulsion increases the size.

Competition Edge

Size of cation α 1/magnitude of positive charge $M^{+3} < M^{+2} < M^+ < M$ Example, $Fe^{+3} < Fe^{+2} < Fe$	Size of anion α magnitude of negative charge $M^{-4} > M^{-3} > M^{-2} > M^- > M$ Example, $O^{-2} > O^- > O$	In general anion > atom > cation Example, $X^- > X > X^+$

TEST YOUR KNOWLEDGE

5.7 Explain why the potassium atom is larger than the potassium cation.

Solution:

The potassium atom (K) is larger than a potassium cation (K^+) since it has more number of orbits and less effective nuclear charge.

$$\underset{2,8,8,1}{K} \xrightarrow{-e^-} \underset{2,8,8}{K^+}$$

	K	K^+
Orbits	4	3
Z_{eff}	less	more

5.8 Write in decreasing order of size.

(a) F^-, N^{3-}, O^{2-}

(b) Na^+, Al^{3+}, Mg^{2+}

Solution:

(a) $N^{3-} > O^{2-} > F^-$ (More the –ve charge, larger the size)

(b) $Na^+ > Mg^{2+} > Al^{3+}$ (More the +ve charge, smaller the size)

Competition Edge

Electronegativity: According to Pauling (1931), it is the power or tendency of an atom in a molecule to attract the shared pair of electrons towards itself.

Ionisation enthalpy: It is defined as 'the energy required to remove the most loosely bound electron from an isolated atom in the gaseous state resulting in the formation of a positive ion'.

$$M - 1e^- \rightarrow M^+ - I_1$$

Electron gain enthalpy or electron affinity (EA): It is the amount of energy released when a neutral isolated gaseous atom accepts an extra electron to form a gaseous anion.

$$M + 1e^- \rightarrow M^- + E_1$$

In general, all these three periodic properties increase from left to right across the period and decrease down the group (just opposite to atomic radius).

5.5.3 Metallic and Non-metallic Nature

In the modern or long form of the periodic table, most of the elements are metals followed by non-metals and a few metalloids. Except hydrogen (non-metal), all the elements of group I and II are soft metals, cuttable with a knife. Generally those elements which have 1, 2 or 3 valence electrons are metals and are electropositive in nature as they can easily lose these valence electrons to form cations. Metals are electropositive by nature and form cations.

Generally, non-metals are elements having 4 to 8 valence electrons and they are electronegative in nature as they accept electrons and form anions. Most of the elements placed on the right hand side of the table, that is, *p*-block elements are non-metals.

Variation in a period: On moving from left to right across a period, the metallic nature decreases and the non-metallic nature increases. That is, the electropositive nature decreases and the electronegative nature increases. On moving left to right in a period, the atomic size decreases and force of attraction increases; so, the removal of electrons becomes tough and tendency to accept electrons increases. Thus, the metallic and electropositive nature decreases and non-metallic and electronegative nature increases. For example,

Elements of second period

Li Be (Metals) B (Metalloid) C N O F (Non-metals)

Nature of elements

Metallic and electropositive character decreases →
Non-metallic and electronegative character increases

Elements of third period

Na Mg Al (Metals) Si (Metalloid) P S Cl (Non-metals)

Nature of elements

Metallic and electropositive character decreases →
Non-metallic and electronegative character increases

Variation in a group: On moving down a group, the metallic nature increases (non-metallic nature decreases); which means the electropositive nature increases down the group while the electronegative nature decreases. This is because, on moving down a group, the atomic size increases and the force of attraction decreases, so the removal of electrons become easier. Metallic and electropositive nature increases and non-metallic and electronegative nature decreases. For example,

Elements of first group Li, Na, K, Rb, Cs, Fr

Nature of elements

Metallic and electropositive character increases →

Elements of seventh group F, Cl, Br, I, At

Nature of elements

Non–metallic and electronegative character decreases →

* Among all the elements Fr is the most electropositive while F is the most electronegative.

5.5.4 Chemical Reactivity

All the elements in the periodic table are not chemically equally reactive; which means, some of them are more reactive than the others and some may even be unreactive (inert gases). The chemical reactivity of an element in fact depends upon its position in the periodic table which in turn depends upon its outer electronic configuration.

Variation in a period: On moving from left to right in a period, the chemical reactivity of elements first decreases and then increases. For example, in the case of the third period, sodium is more reactive than magnesium and magnesium is more reactive than aluminium. Similarly, aluminium is more reactive than silicon. From phosphorus onwards, the reactivity increases upto chlorine. That is, chlorine is more reactive than sulfur and sulfur is more reactive than phosphorus. However, argon is unreactive (inert) due to its stable octet configuration.

This can be explained with the help of the outer electronic configuration of these elements. Sodium has only one valence electron, magnesium has two valence electrons, aluminium has three valence electrons. This means, the removal of the electron is easier in sodium with respect to magnesium, and in magnesium it is easier with respect to aluminium (Fig. 5.15).

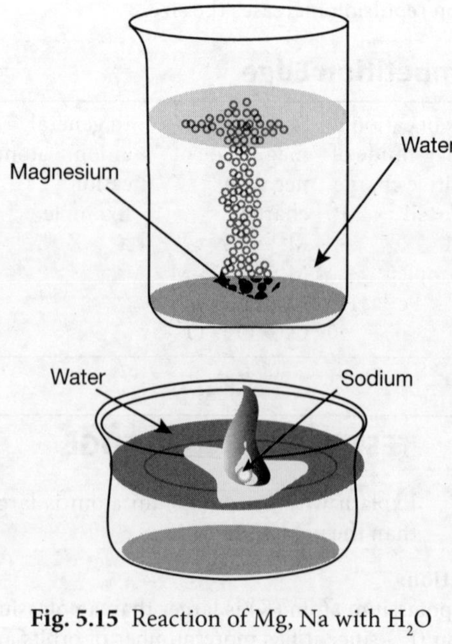

Fig. 5.15 Reaction of Mg, Na with H_2O

So, sodium is the most reactive among these three and aluminium is the least. In the case of silicon there are four valence electrons, so both addition and removal of electrons is not easier due to energy considerations. Hence, sharing of electrons is possible and its reactivity is less than that of aluminium. This fact is supported by the fact that sodium reacts with water so vigorously that the evolved hydrogen catches fire, magnesium can react with boiled water, aluminium reacts with water very slowly and silicon does not react with water. From phosphorus, the chemical reactivity again increases. It can be explained as follows.

As phosphorus, sulfur and chlorine have 5, 6, 7 valence electrons respectively, in order to attain the octet, these elements need 3, 2, 1 electrons respectively. As phosphorus needs 3 electrons, it is tough which means less reactivity; sulfur needs 2 electrons so it is less tough than phosphorus which means it is more reactive than phosphorus; in the case of chlorine only 1 electron is needed so it is most reactive (Table 5.7).

Variation in a group: All the elements in a particular group have a similar outer electronic configuration. That is, they have the same number of valence electrons so they show similar chemical properties. It is important to note that the chemical reactivity of elements varies gradually in a group.

Reactivity of metals in a group: In a group, the chemical reactivity of metals increases from top to bottom as the chemical reactivity of metals depends upon their tendency to lose electrons. On moving down the group, the tendency to lose electrons increases as the size or radius increases and the effective nuclear charge decreases which means removal of electrons becomes easier.

Fig. 5.16 Reaction of Li, Na, K with H_2O

First group:	Li	Na	K	Rb	Cs	Fr
Second group:	Be	Mg	Ca	Sr	Ba	Ra

Chemical reactivity increases →

*As the reactivity increases down the group, when alkali metals react with water, the reaction becomes more and more vigorous and explosive too (Fig. 5.16).

Reactivity of non-metals in a group: In a group, the chemical reactivity of non-metals decreases from top to bottom since it depends upon their tendency to accept electrons and on moving down the group the tendency to accept electrons keeps on decreasing as size increases and effective nuclear charge decreases, which means accepting of electrons gets tough.

Sixth group:	O	S	Se	Te	Po
Seventh group:	F	Cl	Br	I	At

Chemical reactivity decreases →

Table 5.7 Variation in chemical reactivity of the elements of the second, third and fourth period

No. of valence electrons	1	2	3	4	5	6	7	8
Elements of II period	Li	Be	B	C	N	O	F	Ne
Elements of III period	Na	Mg	Al	Si	P	S	Cl	Ar
Elements of IV period	K	Ca	Ga	Ge	As	Se	Br	Kr
Chemical reactivity	Very reactive	Less reactive	Still less reactive	Least reactive	Quite reactive	More reactive	Very reactive	Unreactive

Reactivity decreases → Reactivity increases →

5.5.5 Nature of Oxides

Variation in a period: We know that the oxides of metals are basic and those of non-metals are acidic as on moving from left to right across a period, the metallic nature decreases and the non-metallic nature increases. On moving left to right in a period, the basic nature decreases while the acidic nature increases. The same trend is observed in the case of hydroxides also. For example,

Na_2O,	MgO,	Al_2O_3,	SiO_2,	P_2O_5,	SO_3,	Cl_2O_7
Strongly basic	Basic	Amphoteric	Weakly acidic	Moderately acidic	Strongly acidic	Very strongly acidic

$NaOH > Mg(OH)_2 > Al(OH)_3 > Si(OH)_4$

Most basic Most acidic

Basic nature decreases and acidic nature increases →

Variation in a group: On moving down a group, the basic nature increases while the acidic nature decreases as the metallic nature and electropositive nature increases while non-metallic and electronegative nature decreases. For example,

Li_2O	BeO	B_2O_3	CO_2	N_2O_5
Weakly basic	Amphoteric	Feebly acidic	Weakly acidic	Moderately acidic

*In group IA (1), Cs_2O and $CsOH$ will be most basic.

Periodic table and chemical bonding: We know that elements combine together to form molecules by chemical bonding—either ionic or covalent bonding. By using the periodic table, we can find out which type of bond is formed between two elements and what type of compound (ionic or covalent) is formed. An ionic bond is formed when a metal transfers its valence electrons to a non-metal. Metals are placed on the left hand side of the periodic table and non-metals are present at the right hand side of the periodic table. When a left hand side element (IA, IIA and IIIA) combines with a right hand side (column 15, 16, 17) element an ionic bond is formed and the compound formed is ionic in nature. Metals can lose 1, 2 and 3 electrons to form a cation and to attain the stable nearest inert gas configuration while non-metals can take these electrons and form anions to attain their octet state. For example,

$Na \rightarrow Na^+ + e^-$

2,8,1 2,8

(Sodium cation)

$Cl + e^- \rightarrow Cl^-$

2,8,7 2,8,8

(Chloride anion)

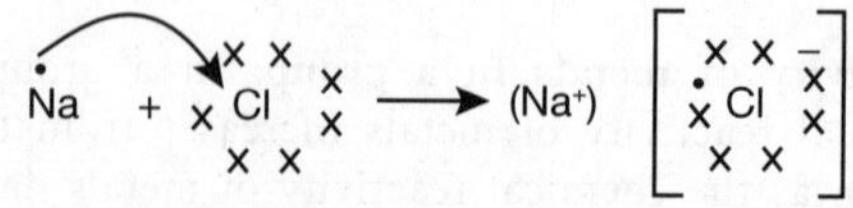

Competition Edge

An ionic compound is formed by the combination of a metal cation (M^{x+}) and non-metal anion (E^{y-}).

Symbol M E

Charge +x −y

Valency x y Formula = M_yE_x

Compound:	Sodium chloride			Magnesium chloride			Calcium oxide	
Symbol	Na	Cl	Symbol	Mg	Cl	Symbol	Ca	O
Charge	+1	−1	Charge	+2	−1	Charge	+2	−2
Valency	1	1	Valency	2	1	Valency	2	2
	Formula = NaCl			Formula = $MgCl_2$			Formula = CaO	

TEST YOUR KNOWLEDGE

5.8 Write the decreasing order of basic nature. Na_2O, MgO, K_2O, CaO

Solution:
The basic nature decreases left to right across the period and increases down the group.

So, $K_2O > Na_2O > CaO > MgO$.

5.9 An element is placed in the 2nd group and 3rd period of the periodic table, burns in the presence of oxygen to form a basic oxide.
(a) Identify the element.
(b) Write the electronic configuration.
(c) Write the balanced equation when it burns in the presence of air.
(d) Write a balanced equation when this oxide is dissolved in water.
(e) Draw the electron dot structure for the formation of this oxide.

Solution: (a) Magnesium
(b) 2, 8, 2
(c) $2Mg + O_2 \rightarrow 2MgO$
(d) $MgO(s) + H_2O(e) \rightarrow Mg(OH)_2(aq)$
(e)

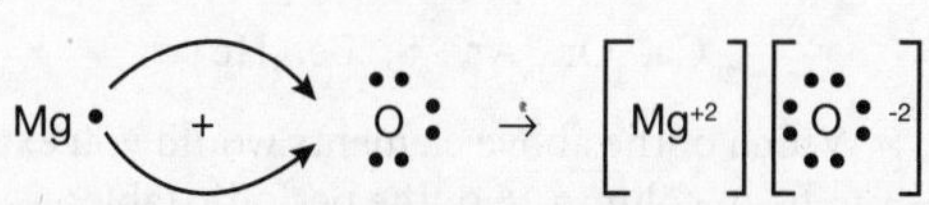

5.10 Based on the group valency of elements, state the formula for the following giving justification for each.
(i) Oxides of 1st group elements
(ii) Halides of the elements of group 13
(iii) Compounds formed when an element of group 2 combines with an element of group 16.

Solution: (i) Valency of group 1 elements = 1; Valency of oxygen = 2
Oxides of group 1 elements: Formula of the oxides of group 1 is M_2O, where M is the group 1 element and O is oxygen.
(ii) Valency of group 13 elements = 3; Valency of halogens = 1
Halides of group 13 elements: Formula of the halides of group 13 is MX_3, where M is the group 13 element and X is the halogen.
(iii) Valency of group 2 elements = 2; Valency of group 16 elements = 2
Compounds of group 2 and group 16 elements: Formula of the compounds of group 2 and 16 is MN, where M is the group 2 element and N is the group 16 element.

5.11 Four elements P, Q, R and S have atomic numbers 12, 13, 14 and 15 respectively. Answer the following questions giving reasons:
(i) What is the valency of Q?
(ii) Classify these elements as metals and non-metals.
(iii) Which of these elements will form the most basic oxide?

Solution: (i) The valency of Q is 3 as its valence shell has three electrons.
(ii) Elements P and Q are metals as they have 2 electrons in their valence shell and they are positively charged ions whereas elements R and S are non-metals as they gain electrons to complete their octet.
(iii) P and Q will form basic oxides as they are metals.

Quick Review

- **Valency** = number of valence electrons = number of H atoms or O atoms or Cl atoms linked with one atom of an element.
- **Atomic size or radius:** It is the distance between the centre of the nucleus of an atom to its outermost shell of electrons, that is, penultimate shell electron.
- **Size of cation is smaller than its atom**: During cation formation, the outermost orbit is destroyed and the number of valence electrons decreases; so, Z_{eff} increases and size decreases.

- **Size of anion is greater than its atom:** During anion formation, electrons are taken up, so Z_{eff} decreases and the size increases.
- **Ionisation energy** or ionisation potential is defined as the energy required to remove the most loosely bound electron from an isolated atom in the gaseous state resulting in the formation of a positive ion.
- **Electronegativity** is the power or tendency of an atom in a molecule to attract the shared pair of electrons towards itself.
- **Electron gain enthalpy** is the amount of energy released when a neutral isolated gaseous atom accepts an extra electron to form a gaseous anion.
- On moving left to right across a period, the atomic size, metallic nature, reactivity of metal, basic nature of oxide and hydroxide decrease while the non-metallic nature, reactivity of non-metal and acidic nature of oxide increases.
- On moving down the group, the atomic size, metallic nature, reactivity of metal, basic nature of oxide and hydroxide increase while the non-metallic nature, reactivity of non-metal and acidic nature of oxide decreases.

Exercise 2

Section A: Multiple Choice Questions (1 Mark)

1. What is the atomic number of the elements of period 3 and group 17 of the periodic table?
(a) 10 (b) 4
(c) 17 (d) 21

2. Name the neutral atom in the periodic table which has the same number of electrons as K^+ and Cl^-.
(a) helium (b) argon
(c) neon (d) krypton

3. An atom of an element has the electronic configuration 2,8,2. To which group does it belong?
(a) 4th group (b) 6th group
(c) 3rd group (d) 2nd group

4. Where would you locate the element with electronic configuration 2, 8 in the modern periodic table?
(a) group 8 (b) group 2
(c) group 18 (d) group 10

5. Element X forms a chloride with the formula XCl_2, which is a solid with high melting point. X would most likely be in the same group of the periodic table as:
(a) Si (b) Mg
(c) Al (d) Na

6. Carbon belongs to the second period and group 14. Silicon belongs to the third period and group 14. If the atomic number of carbon is 6, the atomic number of silicon is
(a) 7 (b) 14
(c) 24 (d) 16

7. An element X has a mass number 40 and contains 21 neutrons in its atom. To which group of the periodic table does it belong?
(a) group 1 (b) group 4
(c) group 2 (d) group 3

8. Consider the following elements

$$_{20}Ca,\ _{8}Or\ _{18}Ar,\ _{16}S,\ _{4}Be,\ _{2}He$$

Which of the above elements would you expect to be in column 16 of the periodic table?
(a) $_{20}Ca$ and $_{16}S$ (b) $_{20}Ca$ and $_{8}O$
(c) $_{18}Ar$ and $_{16}S$ (d) $_{8}O$ and $_{16}S$

9. The elements A, B and C belong to groups 1, 14 and 17 respectively of the periodic table. Which two elements will form ionic compounds?
(a) A and B
(b) A and C
(c) B and C
(d) both option (a) and (b)

10. An element X from group 2 of the periodic table reacts with Y from group 17 to form a compound. Give the formula of the compound.
(a) XY_2 (b) XY
(c) X_2Y (d) $(XY)_2$

11. A metal M is in the first group of the periodic table. What will be the formula of its oxide?
(a) MO (b) M_2O
(c) M_2O_3 (d) MO_2

12. An element X combines with oxygen to form an oxide XO. This oxide is electrically conducting.

Write the formula of the compound formed when X reacts with chlorine.

(a) XCl_3 (b) XCl

(c) XCl_2 (d) XCl_5

13. An element A belongs to the third period and group 16 of the periodic table. Find out the valency of A.

(a) valency = 6 (b) valency = 2

(c) valency = 1 (d) valency = 3

14. Which of the following sets of elements is written in order of their increasing metallic character?

(a) Na, Li, K (b) C, O, N

(c) Mg, Al, Si (d) Be, Mg, Ca

15. Pick out the chemically most reactive elements from the given triads.
Li, Na, K and F, Cl, Br.

(a) Li and F (b) Li and Br

(c) K and F (d) K and Br

16. The atom of an element has the electronic configuration 2, 8, 7. To which of the following elements would it be chemically similar?

(a) N (7) (b) P (15)

(c) Na (11) (d) F (9)

17. Which one of the following statements is correct about the trends in the properties of the elements of a period on going from left to right?

(i) The oxides become more acidic.
(ii) The elements become less metallic.
(iii) There is an increase in the number of valence electrons.
(iv) The atoms lose their electrons more easily.

(a) (i), (ii) (b) (ii), (iii)

(c) (i), (ii), (iii) (d) (i), (ii), (iii), (iv)

18. Which one of the following statements is correct about the trends in the properties of the elements of a group on going down in a group?

(i) The valence electrons increase.
(ii) The metallic character of the elements increases.
(iii) The size of the atom increases.
(iv) The chemical reactivity of the metals increases.

(a) (i), (ii) (b) (ii), (iii)

(c) (i), (ii), (iii) (d) (ii), (iii), (iv)

19. Match the following.

Column I	Column II
1. Fluorine	(i) Metalloid
2. Cesium	(ii) Inert gas
3. Neon	(iii) Most electropositive metal
4. Boron	(iv) Most reactive non-metal

(a) (1)—(iv), (2)—(iii), (3)—(i), (4)—(ii)
(b) (1)—(iv), (2)—(iii), (3)—(ii), (4)—(i)
(c) (1)—(iii), (2)—(iv), (3)—(ii), (4)—(i)
(d) (1)—(i), (2)—(ii), (3)—(iii), (4)—(iv)

20. Match the following.

Column I (Decreasing order)	Column II (Elements)
1. Electronegativity	(i) $K_2O > CaO > Al_2O_3$
2. Size	(ii) $SO_3 > CO_2 > Al_2O_3$
3. Acidic nature	(iii) F > O > N
4. Basic nature	(iv) Li > Be > B

(a) (1)—(iv), (2)—(iii), (3)—(i), (4)—(ii)
(b) (1)—(iv), (2)—(iii), (3)—(ii), (4)—(i)
(c) (1)—(iii), (2)—(iv), (3)—(ii), (4)—(i)
(d) (1)—(i), (2)—(ii), (3)—(iii), (4)—(iv)

Assertion Reason Questions

Direction: In the following question, two statements (Assertion) A and Reason (R) are given. Mark

(a) if A and R are both correct and R is the correct explanation of A;
(b) if A and R are both correct but R is not the correct explanation of A;
(c) A is true but R is false;
(d) A is false but R is true.

Assertion	Reason
1. The valency of the element is equal to the number of valence electrons.	**1.** The valency of elements can change on moving down the group.

2. CsF is the most ionic compound.	2. Cesium is the most electropositive and fluorine is most electronegative element.
3. Inert gases have zero electronegativity.	3. Halogens have the highest electronegativity in their periods.
4. On moving down the group, the metallic nature decreases.	4. On moving across the period, the non-metallic nature increases.
5. Cu, Ag, Au are called coinage metals.	5. Cu, Ag, Au are transition metals.
6. Al_2O_3 is an amphoteric oxide.	6. It can react with both acids and bases.
7. Fluorine is the most reactive non-metal.	7. It is the most electronegative element.
8. CO, H_2O are neutral oxides.	8. Cs_2O is the most basic oxide.

Section B: Very Short Answer Questions (2 Marks)

1. What is the group number and period for calcium atom?
2. What do you mean by valency?
3. Name the element having an electronic configuration of 2, 8, 3. What is its valency?
4. The atomic number of an element is 17. To which group and period does this element belong? Determine its valency.
5. State the similarity in the electronic configuration of all the elements present in group 1 of the periodic table.
6. An element has an atomic number 3. To which period of the periodic table does it belong?
7. State one reason for placing Mg and Ca in the same group of the periodic table. **[Board Term II, 2015]**
8. The atomic numbers of three elements A, B and C are 12, 18 and 20 respectively. State giving reasons, which two elements will show similar properties.
9. Name any two elements that have two electrons in their valence shell.
10. What is the electronic configuration of an element with atomic number 10? What will be its valency?
11. There are 7 electrons in the outermost *L* shell of an element. Predict the period and group in the periodic table this element belongs to.
12. The electronic configuration of two elements X and Y are 2,8,7 and 2,8,8,3 respectively. Write the atomic numbers of X and Y.
13. Write any one difference between the electronic configurations of group 1 and group 2 elements.
14. Write the atomic numbers of two elements X and Y having electronic configurations 2, 8, 2 and 2, 8, 6 respectively.
15. The atomic number of three elements X, Y and Z are 3, 11 and 17 respectively. State giving reasons, which two elements will show similar chemical properties.
16. Why is lead a metal and silicon a metalloid?
17. Write two reasons responsible for the late discovery of noble gases.
18. How does the metallic character of a metal depend on the size of the atom?
19. How does the electronegativity change as we move from left to right across a period?
20. Write the element with the smallest size and highest electronegativity.
21. Out of Li and K, which has strong metallic character and why? **[CBSE 2016]**
22. Atomic radius of hydrogen is 37 pm. Express it in metres.

Case Study or Passage Based Questions

I. On the basis of the electronic configuration of an element, we can decide its position in a period, in a group and its valency (equal to the number of valence electrons present in it). Valence is the combining capacity of an element. It varies on moving from left to right across a period but remains the same within a group. When two elements combine together to form a compound, the formula of the compound is obtained by interchanging the valencies of the two elements.

23. How does the valency of the elements vary while going from left to right along a period in the modern periodic table?

24. An element has an electronic configuration 2, 8, 6. Explain its position in the periodic table.

25. (i) The atomic number of Mg and Al are 12 and 13, respectively. Write down their electronic configuration.
(ii) Mention the period of the modern periodic table to which the above two elements belong. Give reasons for your answer.

26. Element Y with atomic number 3 combines with element A with atomic number 17. What would be the formula of the compound?

27. Name any two elements which can show similar chemical properties to Mg also give the reason of these similarities.

28. Why are He, Ne and Ar called inert gases?

II. Those properties which show a regular variation on moving from left to right across any period or moving down in any group are known as periodic properties. These include the metallic, non-metallic nature, electropositive nature, electronegative nature, atomic size, electronegativity, reactivity, acidic and basic nature, and so on. A few properties such as the atomic radius and metallic nature decrease across the period while they increase down the group. A few properties such as the non-metallic nature and electronegativity increase across the period but decrease down the group.

29. The size of the sodium atom is bigger than that of the hydrogen atom. Why?

30. Arrange the following elements in the descending order of atomic size and give a reason for your answer. Mg, Cl, P, Ar. (Atomic numbers of the above elements are 12, 17, 15, 18 respectively.)

31. How would the tendency to lose electrons change as you go:
(i) from left to right across a period?
(ii) down a group?

32. The atomic radii of three elements A, B and C of the periodic table are 186 pm, 104 pm and 143 pm respectively. Giving a reason, arrange these elements in the increasing order of atomic numbers in the period.

33. Out of the elements X and Y, which has a bigger atomic radius? Give a reason to justify your answer.
(i) X has an atomic number 18 and atomic mass 40.
(ii) Y has an atomic number 20 and atomic mass 40.

34. Write the decreasing order of electronegativity of Ne, Na, F, O, N.

Section C: Short Answer Questions (3 Marks)

1. Write the electronic configuration of two elements P (atomic number 17) and Q (atomic number 19) and determine their group numbers and period numbers in the modern periodic table.

2. The atomic numbers of three elements X, Y and Z are 3, 11 and 17, respectively. State giving reasons which two elements will show similar chemical properties.

3. Elements in the periodic table show periodicity of properties. List any four such properties.

4. Four elements P, Q, R and S have atomic numbers 12, 13, 14 and 15 respectively. Answer the following.
(i) What is the valency of Q?
(ii) Classify these elements as metals and non-metals.
(iii) Which of these elements will form the most basic oxide?

5. An element X with electronic configuration (2, 8, 2) combines separately with two radicals, $(NO_3)^-$ and $(SO_4)^{2-}$.
(i) Is X a metal or a non-metal? Write the nature of its oxide.
(ii) Write the formula of the compounds of X formed by the combination of these radicals. Are these compounds covalent or electrovalent?

6. Consider the following elements (atomic numbers are given in parenthesis): Ca (20); K (19); F (9); Be (4).
(i) Select:
(a) the element having one electron in the outermost shell.
(b) two elements of the same group. Write the number of this group.
(ii) Write the formula of the compound formed by the union of Ca (20) and the element X (2, 8, 7).

7. (i) Predict which of the following elements will form cations and which will form anions.
(a) Na (b) Al (c) Cl (d) O
(ii) Name two elements that are inert in nature. [Board Term II, 2015]

8. From the following elements:

$_4Be$, $_9E$, $_{19}K$, $_{20}Ca$

(i) Select the element having one electron in the outermost shell.
(ii) Two elements of the same group.

Write the formula and mention the nature of the compound formed by the union of $_{19}K$ and element X (2, 8, 7). [CBSE, 2015]

9. An element M has an atomic number 11.
(i) Write its electronic configuration.
(ii) State the group to which M belongs.
(iii) Is M a metal or a non-metal?
(iv) Write the formula of its chloride.

10. The elements $_4Be$, $_{12}Mg$ and $_{20}Ca$, each having two valence electrons in their valence shells, are in periods 2, 3 and 4 respectively, of the modern periodic table. Answer the following questions associated with these elements, giving reasons in each case.
(i) In which group should they be?
(ii) Which one of them is least reactive?
(iii) Which one of them has the largest atomic size? [CBSE, 2015]

11. How many groups and periods are there in the modern periodic table? How do the atomic size and metallic character of elements vary as we move:
(i) down a group and
(ii) from left to right in a period? [CBSE 2015, 17]

12. List any two properties of the elements belonging to the first group of the modern periodic table.

13. In the modern periodic table how does the metallic trend of the elements change as we move
(i) from left to right in a period, and
(ii) down a group?

14. Explain why cations are smaller and anions are larger in radii than their parent atoms?

15. Na, Mg and Al are the elements of the same period of modern periodic table having one, two and three valence electrons respectively. Which of these elements (i) has the largest atomic radius, (ii) is the most reactive? Justify you answer stating a reason for each case. [CBSE, 2015]

16. The position of three elements X, Y and Z in the periodic table are shown below.
(i) Which type of ion, cation or anion, will be formed by the element X?
(ii) Which element, Y or Z, has an atom of comparatively larger size?

Give reasons in each case.

Group 1	Group 2
-	-
X	-
-	-
Y	Z

17. (i) How does the tendency to gain electrons change as we go down the group? Give reasons.
(ii) Given below is a part of the periodic table. How does the valency vary, as we move vertically downward from Li to Fr. Give reasons.

Li	Be
Na	
K	
Rb	
Cs	
Fr	Ra

18. Given below are atomic radii of some elements of the second period.

Element	B	O	N	C
Atomic radii (pm)	88	66	74	77

Arrange these elements in the increasing order of their atomic number. Give reasons for your answer.

19. 'Electronegativity is the case of gaining electrons.'
(i) Which type of elements will have high electronegativity?

(ii) How does electronegativity change as we move (a) across a period, (b) down a group? **[Board Term II, 2015]**

20. An element X (Atomic number = 20) burns in the presence of oxygen to form a basic oxide.
(i) Identify the element and write its electronic configuration.
(ii) State its group number and period number in the modern periodic table.
(iii) Write a balanced chemical equation for the reaction when this oxide is dissolved in water. **[Board Term II, 2016]**

21. An element X belongs to the third period and the second group of the modern periodic table.
(i) Write its electronic configuration.
(ii) Is it a metal or non-metal? Why?
(iii) Write the formula of the compound formed when X reacts with an element,
(a) Y of electronic configuration 2, 6 and (b) Z with electronic configuration 2, 8, 7. **[Board Term II, 2016]**

22. The atomic number of an element X is 19. (i) Write its electronic configuration.
(ii) To which period of the modern periodic table does it belong and what is its valency?
(iii) If X burns in oxygen to form its oxide, what will be its nature—acidic, basic or neutral?
(iv) Write a balanced chemical equation for the reaction when this oxide is dissolved in water. **[Board Term II, 2016]**

23. Four elements A, B, C and D have atomic numbers 12, 13, 14 and 15 respectively. Answer the following questions giving reasons.
(i) What is the number of valence electrons and the valency of D?
(ii) Which of them will have the largest atomic radius?
(iii) Which of these elements will form the most basic oxide? **[Board Term II, 2016]**

24. The elements 4Be, $_{12}$Mg and $_{20}$Ca each having two valence electrons in their valence shells are in periods 2, 3 and 4 respectively of the modern periodic table. Answer the following questions associated with these elements, giving reasons in each case.
(i) In which group should they be?
(ii) Which one of them is the least reactive?
(iii) Which one of them has the largest atomic size? **[Board Term II, 2015]**

25. The elements Li, Na and K, each having one valence electron, are in periods 2, 3 and 4 respectively of the modern periodic table.
(i) In which group of the periodic table should they be?
(ii) Which one of them is least reactive?
(iii) Which one of them has the largest atomic radius? Give reasons to justify your answer in each case.

26. What is electronegativity? How it is different from electron affinity?

Section A: Long Answer Questions (5 Marks)

1. What is meant by 'group' in the modern periodic table? How do the following change on moving from top to bottom in a group?
(i) Number of valence electrons
(ii) Number of occupied shells
(iii) Size of atoms
(iv) Metallic character of element
(v) Effective nuclear charge experienced by valence electrons

2. The position of eight elements in the modern periodic table is given below where the atomic numbers of elements are given in the parenthesis.

Period No.		
2	Li(3)	Be(4)
3	Na(11)	Mg(12)
4	K(19)	Ca(20)
5	Rb(37)	Sr(38)

(i) Write the electronic configuration of Ca.
(ii) Predict the number of valence electrons in Rb.
(iii) What is the number of shells in Sr.
(iv) Predict whether K is a metal or a non-metal.
(v) Which one of these elements has the largest atoms in size?

(vi) Arrange Be, Ca, Mg and Rb in the increasing order of the size of their respective atoms. **[Board Term II, 2016, CBSE 2016]**

3. An element X has a mass number 35 and the number of neutrons is 18. Write the atomic number and electronic configuration of X. Also write group number, period number and valency of X. **[CBSE 2016]**

4. An element X belongs to the 3rd period and group 16 of the modern periodic table.
(i) Determine the number of valence electrons and the valency of X.
(ii) Write the molecular formula of the compound when X reacts with hydrogen and write its electron dot structure.
(iii) Name the element X and state whether it is metallic or non-metallic. **[CBSE 2016]**

5. An element X (atomic number = 20) burns in the presence of oxygen to form a basic oxide.
(i) Identify the element and write its electronic configuration.
(ii) State its group number and period number in the modern periodic table.
(iii) Write the formula of its oxide. **[CBSE 2016]**
(iv) Write a balanced chemical equation for the reaction when this oxide is dissolved in water. **[CBSE 2016]**
(v) Will its atomic radius be larger or smaller than that of potassium with atomic number 19?

6. Two elements A and B belong to the 3rd period of the modern periodic table and are in group 2 and 13 respectively. Compare their following characteristics in tabular form. **[CBSE 2016]**
(i) Number of valence electrons in their atoms.
(ii) Size of their atoms.
(iii) Their tendencies to lose electrons.
(iv) The formula of their oxides.
(v) The formula of their chlorides.

7. Three elements X, Y and Z have atomic numbers 7, 8 and 9 respectively.
(i) State their positions (group number and period number both) in the modern periodic table.
(ii) Arrange these elements in the decreasing order of their atomic radii.
(iii) Write the formula of the compound formed when X combines with Z. **[CBSE 2016]**

8. Predict the formula of the stable binary compounds that would be formed by the combination of the following pairs of elements.
(i) lithium and oxygen
(ii) magnesium and nitrogen
(iii) aluminium and iodine
(iv) silicon and oxygen
(v) phosphorus and fluorine
(vi) element 71 and fluorine

9. Answer the following questions based on the elements with atomic number 3 to 9.
(i) Name the element with smallest atomic radius.
(ii) Name the element which shows maximum valency.
(iii) Name the element which is a metalloid.
(iv) Name the element which is the most electropositive.
(v) Write the chemical formula of the compound formed when the elements of atomic number 6 and 8 react together. **[CBSE 2019]**

10. The atomic number of elements A, B, C, D and E are given below

Element	A	B	C	D	E
Atomic no.	7	10	12	4	19

From the above table, answer the following questions.
(i) Which two elements are chemically similar? (ii) Which is an inert gas?
(iii) Which element belongs to the 3rd period of the periodic table?
(iv) Which element among these is a non-metal?

11. An element X belongs to the 3rd period and group 16 of the modern periodic table.
(i) Determine the number of valence electrons and the valency of X.
(ii) Give the molecular formula of the compound when X reacts with hydrogen and write its electron dot structure.
(iii) Name the element X and state whether it is metallic or non-metallic. **[Board Term II, 2016]**

12. In the following table, the positions of six elements A, B, C, D, E and F are given as they are in the modern periodic table.

Group→ Period↓	1	2	3-12	13	14	15	16	17	18
2	A			B		C			D
3					E				F

On the basis of the above table, answer the following questions.

(i) Name the element which forms only covalent compounds.

(ii) Name the element which is a metal with a valency of 3.

(iii) Name the element which is a non-metal with a valency of 3.

(iv) Out of B and C, which one has a bigger atomic radius and why?

(v) Write the common name for the family to which the elements D and F belong.

[Board Term II, 2015]

13. The electrons in the atoms of four elements A, B, C and D are distributed in three shells having 1, 3, 5 and 7 electrons in the outermost shell respectively. State the period in which these elements can be placed in the modern periodic table. Write the electronic configuration of the atoms of A and D and the molecular formula of the compound formed when A and D combine.

14. Study the following table in which positions of six elements A, B, C, D, E and F are shown as they are in the modern periodic table.

Group→ Period↓	1	2	3-12	13	14	15	15	17	18
1.	A					B			C
2.				D	E				F

On the basis of the above table, answer the following questions.

(i) Name the element which forms only covalent compounds.

(ii) Name the element which is a metal with a valency of 3.

(iii) Name the element which is a non-metal with a valency of 3.

(iv) Out of D and E, which is bigger is size and why?

(v) Write the common name for the family to which the elements C and F belong.

15. Atoms of eight elements A, B, C, D, E, F, G and H have the same number of electronic shells but are different in their outermost shells. It was found that elements A and G combine to form an ionic compound which can also be extracted from sea water. The oxides of the elements A and B are basic in nature while those of E and F are acidic. The oxide of elements D is almost neutral.

Answer the following questions based on the information given.

(i) To which group or period of the periodic table do the listed elements belong?

(ii) Which one of the eight elements is likely to be a noble gas?

(iii) Which one of the eight elements would have the largest atomic radius?

(iv) Which two elements amongst these are likely to be non-metals?

(v) Which one of these eight elements is likely to be a semi-metal or metalloid?

Section E: Case Study or Passage Based Questions (4 Marks)

There are a few properties such as metallic nature, electropositive nature, reactivity of metals and size of elements which decrease across the periods but increase down the group. Some other properties like non-metallic nature, electronegative nature, reactivity of non-metals, electronegativity, ionisation energy increase on moving across the period from left to right but decrease on moving down the group. The valency first increases across the period and then decreases, while it remains the same in a group.

1. Which of the following groups will have the highest electronegativity?
(a) alkali metals
(b) inert gases
(c) halogens
(d) alkaline earth metals

2. Which is the most electronegative and smallest in size here?
(a) oxygen (b) nitrogen
(c) fluorine (d) chlorine

3. Which is the correct decreasing order of size?
(a) $F > O > N > C$ (b) $C > N > O > F$
(c) $F > N > O > C$ (d) $O > F > N > C$

4. Which of the following oxides is the most basic?
(a) SO_2 (b) CO_2
(c) K_2O (d) CaO

5. Which of the following pairs of elements are chemically inert?
(a) Li, Be (b) Na, Ne
(c) F, O (d) Ne, Ar

6. Which of the following elements can be present in air?
(a) Na, O, F (b) N, O, Ar
(c) Al, Mg, O (d) F, O, N

High Order Thinking Skills (HOTS) Answer Questions

1. An element M with an electronic configuration (2, 8, 2) combines separately with $(NO_3)^-$, $(SO_4)^{2-}$ and $(PO_4)^{3-}$ radicals. Write the formula of the three compounds so formed. To which group and period of the modern periodic table does the element M belong? Will M form covalent or ionic compounds? Give reasons to justify your answer. **[HOTS; CBSE 2016]**

2. (i) On moving down the group a sharp increase in the atomic radius is found from Li to K but later on, a gradual change in atomic radius value is observed. Why?
(ii) Consider the following species. N^{3-}, O^{2-}, F^-, Na^+, Mg^{2+} and Al^{3+}. (a) What is common among them? (b) Arrange them in the order of increasing ionic radii.

3. Four elements A, B, C and D along with their electronic configurations are given below.

Elements	A	B	C	D
Electronic configuration	2,1	2,8	2,8,1	2,8,8

Now answer the following questions.
(i) Which two elements belong to the same period?
(ii) Which two elements belong to the same group?
(iii) Which element out of A and C is more reactive and why?

4. Two elements X and Y belong to group 1 and 2 respectively in the same period of the periodic table. Compare these elements with respect to:
(i) number of electrons in their outermost orbit (ii) their valences (iii) metallic character (iv) their atomic size (v) formula of their chlorides (vi) formula of their sulfates

5. From the following part of the periodic table, answer the following questions.

1 Lithium	2	13	14 Carbon	15	16 Oxygen	17 Fluorine
X			P			Q
Y						R
Z						T

(i) Which is the most reactive metal?
(ii) Name the family of fluorine Q, R, T.
(iii) Name one element each of group 2 and 15.
(iv) Give the name of the element P placed below carbon.
(v) Compare X and P with respect to the size of atoms.

6. The positions of three elements A, B and C in the periodic table are shown below.

Group→ Periods↓	1	2	3
I			A
II			B
III		C	
IV			
V			
VI			
VII			
VIII			

Give reasons to explain the following.
(i) Element A is a metal.
(ii) Element B has larger size than C.
(iii) Element C has a valency –3.

7. (i) What is the basic difference between a group and a period in terms of electronic configuration?
(ii) An element X is liquid, non-metallic. Can you predict its position in the periodic table?

(iii) Given below are the melting points and atomic radii of three elements X, Y and Z of the periodic table having '*n*' electrons in the outermost shells of their atoms.

	X	Y	Z
Melting point (°C)	100·5	97·5	63·5
Atomic radii (Å)	1·33	1·54	1·96

State four inferences that can be drawn about the elements on the basis of the above data.

8. The table given below shows a part of the periodic table.

H							He
Li	Be	B	C	N	O	F	Ne
Na	Mg	Al	Si	P	S	Cl	Ar

Using this table explain why,

(i) Li and Na are considered as active metals.

(ii) the atomic size of Mg is less than that of Na.

(iii) fluorine is more reactive than chlorine.

9. *Atomic size refers to the radius of an atom. Khushi, when she read this statement understood that it is the distance between the centre of the nucleus and the outermost shell (valence shell) of an isolated atom. She also came to know that atomic size generally increases from top to bottom in a group, due to addition of a new shell and decreases along a period from left to right, due to increase in nuclear charge which pulls the electrons towards it.

(i) Khushi has to arrange the following elements in the increasing order in a group in accordance of their atomic size Li, Rb, K, Cs, Na.

(ii) Atomic number is considered to be a more appropriate parameter than atomic mass for classification of elements in the periodic table. Give any two reasons to justify the statement.

10. *Some metals react with water and produce metal oxides or hydroxides and liberate hydrogen gas. Metals like sodium and potassium react violently with cold water. However, calcium reacts less violently with cold water because it is less reactive as compared to Na and K. Magnesium does not react with cold water. It reacts with hot water to form magnesium hydroxide and hydrogen gas. Metals like aluminium, iron and zinc do not react with cold as well as hot water but they react with steam to form metal oxide and hydrogen gas.

(i) Which gas is produced when a reactive metal reacts with water?

(ii) How can we extinguish a fire?

(iii) Why is sodium kept under kerosene?

(iv) Which metal did not react with water even in the form of steam?

(v) How can we prevent iron from rusting?

Answers

Section A: Multiple Choice Questions

1. (c)	2. (b)	3. (d)	4. (a)
5. (b)	6. (b)	7. (a)	8. (d)
9. (b)	10. (a)	11. (b)	12. (c)
13. (b)	14. (d)	15. (c)	16. (d)
17. (c)	18. (d)	19. (b)	20. (c)

Assertion Reason Questions

1. (c)	2. (a)	3. (b)	4. (d)
5. (b)	6. (a)	7. (a)	8. (b)

Section B: Very Short Answer Questions

1. It is present in group 2 and the IVth period.

2. The combining capacity of an element is known as its valency.

3. Name of the element is aluminium (Al) and its valency is 3.

4. It belongs to group 17 and the 3rd period, and its valency is 1.

5. Elements in a group have similar electronic configuration; so all of them have 1 valence electron.

6. Second period.

7. Due to the presence of 2 electrons in the valence shell and similar chemical properties they are placed together.

8. A and C will show similar properties because they have the same number of valence electrons (2).

9. (i) Magnesium: 2, 8, 2 (ii) Calcium: 2, 8, 8, 2

10. Electronic configuration of the element with atomic number 10 = 2, 8
Valency: Zero.

11. As electronic configuration of element = 2, 7, the period is 2 and group is 17.

12. Atomic number of X = 2, 8, 7 = 17
Atomic number of Y = 2, 8, 8, 3 = 21

13. Group 1 elements have 1 valence electron while group 2 elements have 2 valence electrons.

14. Atomic number of X = 2 + 8 + 2 = 12
Atomic number of Y = 2 + 8 + 6 = 16

15. X and Y will show similar chemical properties as these have the same number of valence electrons.
X = 2, (1) and Y = 2, 8, (1)

16. Lead is a metal and silicon is a metalloid (on moving down the group, the metallic nature increases).

17. (i) They are unreactive.
(ii) They have zero valency.

18. The metallic character of a metal increases with an increase in the size of its atom.

19. Electronegativity increases on moving from left to right across a period.

20. Fluorine has the smallest size and highest electronegativity.

21. K has strong metallic bonds because it is larger in size and can lose electrons very easily.

22. (i) 1 pm = 10–12 m
Atomic radius = 37 × 10–12 m

Case Study or Passage Based Questions

23. On moving from left to right in the periodic table, the valency increases up to 4 and then decreases.

24. Third period and group 16.

25. (i) Mg: 2, 8, 2 Al: 2, 8, 3
(ii) 3rd period; because, electrons in these atoms are filled in the K, L, M shells.

26. Electronic configuration of the element Y with atomic number 3 = 2, 1.
Element A with atomic number 17 = 2, 8, 7.
Formula of the compound: YA.

27. Be and Ca can show similar chemical properties to Mg as all these elements have the same number of valence electrons (2) and are present in the same group (IIA).

28. He, Ne and Ar are called inert gases because their outermost shell is complete and they do not lose or gain electrons in their outermost shell, hence they have zero valency.

29. Size increases down the group as new shells are added.

30. Mg, P, Cl, Ar. An increase in the atomic number means an increase in nuclear charge which tends to pull the electrons closer to the nucleus and thus reduces the size of the atom.

31. **Tendency to lose electrons:**
(i) Decreases with an increase in atomic number in a period from left to right.
(ii) Will increase down a group.

32. Order of atomic number of element A < C < B. Because atomic size generally decreases along a period, B has the highest atomic number followed by C and A.

33. Radius of Y is bigger than that of X.
In X, the number of shells is three while in Y it is four.

34. F, O, N, Na, Ne.

Section C: Short Answer Questions

1. Electronic configuration of P: 2, 8, 7.
Group number: 17; Period number: 3

Electronic configuration of Q: 2, 8, 8, 1.
Group number: 1; Period number: 4

2. The two elements X and Y will show the same chemical properties because they have the same number of valence electrons in the group. They form positively charged ions by losing one electron.

3. (i) Atomic size
(ii) Valency or combining capacity
(iii) Metallic property
(iv) Non-metallic property

4. (i) The valency of Q is 3.
(ii) Elements P and Q are metals. Elements R and S are non-metals.
(iii) P and Q elements will form basic oxides.

5. (i) X is a metal. Nature of its oxide is basic.
(ii) $X(NO_3)_2$, XSO_4
These compounds are ionic/electrovalent.

6. (i) (a) K (potassium: 2, 8, 8, 1)
(b) Be and Ca are in the same group because both have the same number of valence electrons in their outermost shell.
(ii) Ca X

Valency 2 1

$Ca_1X_2 = CaX_2$

7. (i) Cations are formed by Na and Al (sodium and aluminium) while anions are formed by Cl and O (chlorine and oxygen).
(ii) Inert elements are He and Ne (helium and neon).

8. (i) $_{19}K$ has the electronic configuration 2, 8, 8, 1. It has one valence electron.
(ii) $_4Be$ and $_{20}Ca$ belong to the same group, that is, 2.
$_{19}K = 2, 8, 8, 1$

$$\underset{(2,8,8,1)}{K} \rightarrow \underset{(2,8)}{K^+} + e^-$$

$$\underset{(2,8,7)}{X} + e^- \rightarrow \underset{(2,8,8)}{X^-}$$

$(K^+)(X^-)$, that is, KX is the formula of the compound and it is ionic in nature.

9. (i) 2, 8, 1 (ii) Group 1
(iii) Metal (iv) MCl

10. (i) They should be in group 2, since they have 2 valence electrons.
(ii) Be is the least reactive because of more effective nuclear charge due to smaller size and least tendency to lose electrons.
(iii) Ca has the largest atomic size due to the presence of 4 shells.

11. There are 18 groups and 7 periods in the modern periodic table.
(i) Atomic size and metallic character both increase on moving down the group.
(ii) Atomic size and metallic character both decrease along a period while going from left to right.

12. First group elements are also known as alkali metals. They possess following the properties.
(i) These elements exhibit +1 valency.
(ii) These are very reactive and thus are not found freely in nature.

13. (i) On moving across a period, the nuclear pull increases due to the increase in atomic number; thus, the atomic size decreases. Hence, elements cannot lose electrons easily. The metallic nature decreases across a period, on moving from left to right.
(ii) On moving down a group, the atomic size increases and the nuclear charge also increases. The effect of an increased atomic size is greater compared to the increased nuclear charge. Therefore, the metallic nature increases as one moves down a group.

14. The ionic radius of a cation is always smaller than the parent atom because the loss of one or more electrons increases the effective nuclear charge. As a result, the force of attraction of the nucleus towards the electrons increases and hence the ionic radii decrease. In contrast, the ionic radius of an anion is always larger than its parent atom because the addition of one or more electrons decreases the effective nuclear charge. As a result, the force of attraction of the nucleus towards the electrons decreases and hence the ionic radii increase.

15. (i) Na has the largest atomic radius because it has 11 protons and 11 electrons, and the least effective nuclear charge among these elements.

(ii) Al is the least reactive element because it is smallest in size, therefore, has the most effective nuclear charge and least tendency to lose electrons.

16. (i) The valency of X is 1. It has a tendency to lose one electron. Therefore, X will form a cation.
(ii) Y, because the atomic radius decreases across a period due to increase in the electrostatic force between the electrons and the nucleus.

17. (i) The tendency to gain electrons decreases on moving down the group because with an increase in size, the effective '+ve' nuclear charge decreases.
(ii) The valency remains constant as we move downwards. Elements of the same group have the same number of valence electrons.

18. B, C, N, O.
In a period, as the atomic number increases, the size of the atom decreases. This is because an increase in the nuclear charge tends to pull the electrons closer to the nucleus and reduces the size of the atom.

19. (i) Non-metals (ii) (a) Increases (b) Decreases

20. (i) X (20): 2, 8, 8, 2. It is calcium.
(ii) Second group, fourth period.
(iii) $CaO + H_2O \rightarrow Ca(OH)_2$

21. (i) Electronic configuration: 2,8,2.
(ii) Metal, as it can easily lose electrons (from the outermost orbit).
(iii) X Y XZ
2 2 2 1
Compound: XY XZ2

22. (i) Electronic configuration of X (19): 2,8, 8, 1.
(ii) Fourth period, valency 1.
(iii) Basic oxide (X_2O).
(iv) $X_2O + H_2O \rightarrow 2\,XOH$

23. (i) Valence electrons in D = 5 and valency of D = 3.
(ii) A will have the largest atomic radius because the atomic radius decreases across a period from left to right.
(iii) A will form the most basic oxide as it is the most metallic.

24. (i) They all belong to group 2 because all three have 2 electrons in their outermost shell.
(ii) Be is the least reactive because it has 2 shells and due to a higher nuclear charge, it is not easy to remove electrons from it.
(iii) Ca is the element having the largest atomic radius because it has 4 shells.

25. (i) They are in group 1 because all the elements in group have 1 valence electron. They form positively charged ions by losing one electron.
(ii) Lithium is the least reactive because reactivity of an element depends upon the ability of its atoms to donate or accept electrons.
(iii) Potassium will have the largest atomic radius due to the addition of a new shell. Atomic size generally increases from the top to bottom in a group.

26. Though both electronegativity and electron affinity refer to the attraction of the electron, they are used in different contexts. Electronegativity refers to the attraction of the bonding electron or shared pair of electrons whereas electron affinity refers to the attraction towards the electron by an isolated gaseous atom.

Section D: Long Answer Questions

1. Vertical columns of the periodic table are known as groups.
(i) The number of valence electrons remains constant when we move down the group.
(ii) The number of occupied shells increases down the group.
(iii) The size of the atoms increases down the group.
(iv) The metallic character of elements increases down the group.
(v) The effective nuclear charge decreases down the group.

2. (i) 2, 8, 8, 2
(ii) It has one valence electron.
(iii) 2, 8, 18, 8, 2. It has five shells (orbits).
(iv) 'K' is a metal.
(v) Rubidium (Rb) has the largest atom in size.
(vi) Be > Mg > Ca > Rb is the increasing order of atomic size.

3. X has a mass number = 35
Number of neutrons = 18
Number of protons = Mass number – Number of neutrons = 35 – 18 = 17
Atomic number = Number of protons = 17
Electronic configuration is 2, 8, 7.
Hence, it belongs to group 17 and the 3rd period.
Its valency is equal to 1.

4. (i) X has electronic configuration of 2, 8, 6, therefore it has 6 valence electrons and its valency is 2.
(ii) When X reacts with hydrogen, it forms a compound H_2X.

$H^1 \quad X^{2-}$

$H_2 \quad X$

Its electron dot structure is

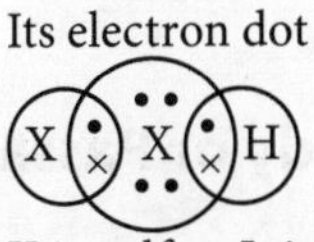

(iii) X is sulfur. It is non-metallic.

5. (i) The element is calcium. Its electronic configuration is 2, 8, 8, 2.
(ii) Group number is 2 and period number is 4.
(iii) The formula of its oxide is CaO.
(iv) $CaO + H_2O \rightarrow Ca(OH)_2$
(v) It will be smaller due to greater effective nuclear charge.

6.

S. No.	Characteristics	Element A	Element B
1.	No. of valence electrons	2	3
2.	Size	Bigger than B	Smaller than A
3.	Tendency to lose electrons	More	Less
4.	Formula of oxides	AO	B_2O_3
5.	Formula of chlorides	$AlCl_2$	BCl_3

7. (i) X, Y and Z belong to group 15, 16 and 17 respectively. They belong to the same period (3rd).
(ii) X > Y > Z is the decreasing order of their atomic radii.
(iii)

$X^3 \quad Z^1$

$X \quad Z_3$

So, XZ_3 is the formula of the compound formed.

8. (i) Lithium is an alkali metal (group 1). It has only one electron in the valence shell, therefore, its valency is 1. Oxygen is a group 16 element with a valency of 2. Therefore, the formula of the compound formed would be Li_2O (lithium oxide).
(ii) Magnesium is an alkaline earth metal (group 2) and hence has a valency of 2. Nitrogen is a group 15 element with a valency of 8 – 5 = 3. Thus, the formula of the compound formed would be Mg_3N_2 (magnesium nitride).
(iii) Aluminium is a group 13 element with a valency of 3 while iodine is a halogen (group 17) with a valency of 1. Therefore, the formula of the compound formed would be AlI_3 (aluminium iodide).
(iv) Silicon is a group 14 element with a valency of 4 while oxygen is a group 16 element with a valency of 2. Hence the formula of the compound formed is SiO_2 (silicon dioxide).
(v) Phosphorus is a group 15 element with a valency of 3 or 5 while fluorine is a group 17 element with a valency of 1. Hence, the formula of the compound formed would be PF_3 or PF_5.
(vi) The element with atomic number 71 is a lanthanoid called lutetium (Lu). Its common valency is 3. Fluorine is a group 17 (halogen) element with a valency of 1. Therefore, the formula of the compound formed would be LuF_3 (lutetium fluoride).

9. (i) Element with smallest atomic radius: Fluorine /F
(ii) Element with maximum valency: Carbon /C
(iii) Element which is metalloid: Boron /B
(iv) Element which is most electropositive: Lithium /Li
(v) CO and CO_2

10. (i) C and D (ii) B (iii) C (iv) A

11. (i) Electronic configuration of X = 2,8,6.
Valence electrons = 6.
Valency = 8 – 6 = 2

(ii) Formula with hydrogen: H_2X or H_2S

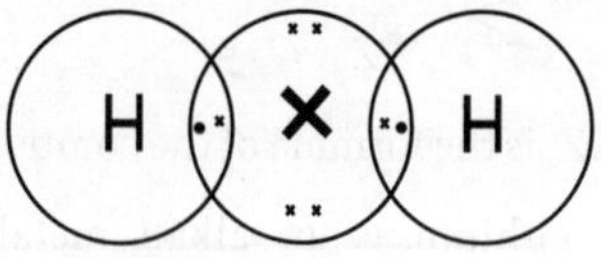

(iii) Sulfur; non-metal.

12. (i) E (ii) B (iii) C

(iv) B, because the atomic radius decreases from left to right due to an increase in the nuclear charge.

(v) Noble gases.

13.

Elements	Valence electrons	Period
A	1	3
B	3	3
C	5	3
D	7	3

Electronic configuration of A: $1s^2\ 2s^2\ 2p^6\ 3s^1$

Electronic configuration of D: $1s^2\ 2s^2\ 2p^6\ 3s^2\ 3p^5$

Molecular formula of the compound formed with A and D

Atomic number of A = 11; Electronic configuration A = 2, 8, 1

Number of valence electrons of A = 1; Valency of A = 1

Atomic number of D = 17; Electronic configuration D = 2, 8, 7

Number of valence electrons of D = 7; Valency of D = 8 – 7 = 1

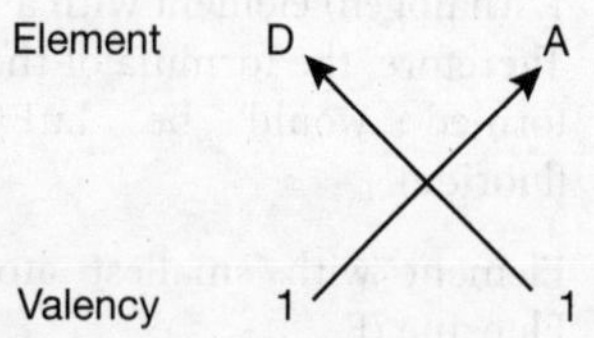

So, the formula of the compound formed when elements A and D combine is AD.

14. (i) Element E will form covalent compounds by sharing its four valence electrons.

(ii) Element D is a metal with a valency of 3.

(iii) Element B is a non-metal with a valency of 3.

(iv) We know that the size of the elements decreases on moving left to right in a period. Therefore, element D is greater than element E.

(v) The name of the family to which elements C and F belong is the noble gas family (group 18).

15. (i) A and B belong to group 1 and 2 because they form basic oxides. C belongs to group 13 as it has 3 valence electrons. D belongs to group 14 as it forms an almost neutral oxide. E and F belong to group 15 and 16 as they form acidic oxides. G belongs to group 17 as it has 7 valence electrons, and H to group 18. They belong to the 3rd period of the periodic table.

(ii) H belongs to the noble gases

(iii) A has the largest atomic radius

(iv) E and F are likely to be non-metals

(v) D is likely to be a metalloid or a semi-metal

Section E: Case Study or Passage Based Questions

1. (c) **2.** (c) **3.** (b) **4.** (c)
5. (d) **6.** (b)

High Order Thinking Skills (HOTS) Questions

1. The three compounds formed are: $M(NO_3)_2, MSO_4, M_3(PO_4)_2$

M has 2 valence electrons, therefore, its valency is equal to 2. M is Mg and the compounds so formed are $Mg(NO_3)_2$, $MgSO_4$, $Mg_3(PO_4)_2$.

Here, M belongs to group 2 and the 3rd period of the modern periodic table.

M will form ionic compounds because it can lose 2 electrons easily to become stable.

2. (i) On moving from Li to K, a sharp increase in atomic radius is observed as the number of orbits increases; from K to Cs this change is gradual as screening by the inner orbit electrons is not effective which means Z_{eff} is more.

(ii) (a) Each one of these ions contains 10 electrons and hence all are isoelectronic ions.

(b) The ionic radii of isoelectronic ions decreases with an increase in the magnitude of the nuclear charge.

For example, consider the isoelectronic ions N^{3-}, O^{2-}, F^-, Na^+, Mg^{2+} and Al^{3+}. All these ions have 10 electrons but their nuclear charges increase in the order: N^{3-} (+7), O^{2-} (+8), F^- (+9), Na^+ (+11), Mg^{2+} (+12) and Al^{3+} (+13). Therefore, their ionic radii decrease in the order $N^{3-} > O^{2-} > F^- > Na^+ > Mg^{2+} > Al^{3+}$.

3. (i) A and B, C and D are in the same period.
(ii) A and C have one valence electron, so they belong to the same group.
(iii) C is more reactive because C is placed below A in the periodic table and the reactivity increases down the group.

4. (i) **Number of electrons in the outermost orbit of** X = 1 and Y = 2.
(ii) Valency of X = 1 and Y =2.
(iii) Metal X is more metallic than Y.
(iv) Atomic size of X is bigger than that of Y.
(v) Chloride XCl; YCl_2
(vi) Sulfate X_2SO_4; YSO_4

5. (i) Z
(ii) Halogens
(iii) Magnesium and nitrogen
(iv) Silicon
(v) X has a bigger size than P because X has less effective nuclear charge.

6. (i) Element A is a metal because it is present in group 1 (1 valence electron) and can lose the electron easily.
(ii) Element B has a larger size than C because it has more number of shells than C. Also, B lies in the third period and has three shells whereas C lies in second period and has two shells.
(iii) C being an element of group 3, has three valence electrons; therefore, its valency is 3.

7. (i) In a group, though the number of shells increases as we move downward, the number of electrons in the outermost shell remains the same.

In a period, as we move from left to right, the outermost shell or valence shell gets progressively filled from group 1 to group 18 and the shell does not change.

(ii) As the element X is non-metallic, it may be belonging to group 17. In group 17, the first two elements are gases. The metallic character of the elements (non-metals) increases on moving down the group. Therefore, the element X may be located in group 17 and the 3rd or 4th period of the periodic table.

(iii) (a) Since each element contains the same number of valence electrons, *n*, in the outermost shells of their atoms, X, Y and Z, belong to the same group.
(b) The atomic radii increase from X to Z, so the metallic character increases from X towards Z.
(c) X, Y and Z have melting points above 25°C (room temperature). Therefore X, Y and Z are solid at room temperature.
(d) The ionisation potential decreases from X to Z because they belong to the same group in the descending order.

8. (i) They can lose electrons easier. Due to their bigger size, the energy required to remove electrons is less.
(ii) It is because of greater effective nuclear charge in Mg.
(iii) Fluorine has a very small size and just needs one electron to get itself into a stable noble gas configuration. Hence, they react spontaneously with elements to achieve that stable state, contributing to its high reactivity.

9. (i) Elements arranged in the increasing order in a group are Li, Na, K, Rb, Cs.
(ii) (a) Although most of the elements were placed in the order of increasing atomic masses, increasing order could not be maintained in all cases.
(b) There was no place for isotopes in Mendeleev's periodic table, although they differ in atomic mass.

10. (i) Hydrogen gas.
(ii) By carbon dioxide gas.
(iii) Because sodium reacts violently with water and air.
(iv) Lead, copper or gold.
(v) By painting iron articles. **(Any three)**

Hints

Section A: Multiple Choice Questions

1. This element is chlorine and its atomic number is 17.

2. Both K^+ ($Z = 19$), Cl^- ($Z = 17$) have 18 electrons and argon also has 18 electrons.

5. As element X forms XCl_2, it is bivalent. Here magnesium is bivalent.

7. Atomic number = mass number – number of neutrons
40 – 21 = 19
Configuration of the element X = 2, 8, 8, 1
As it has only 1 valence electron, it belongs to group 1.

8. As both contain 6 valence electrons, they belong to group VIA (group 16) or column 16.

12. As element X is bivalent and chlorine is monovalent, they will form XCl_2.

15. Among metals, the reactivity increases down the group. So, K is most reactive and in the case of non-metals, the reactivity decreases down the group so fluorine is most reactive.

18. The number of valence electrons remains the same in a group.

Section E: Case Study or Passage Based Questions

1. Electronegativity increases from left to right across the period and the electronegativity of inert gases is zero.
Halogens > Alkaline earth metals > Alkali metals > Inert gases

4. Basic nature decreases from left to right across the period so $K_2O > CaO > CO_2 > SO_2$.

6. Nitrogen (N_2), oxygen (O_2) and argon (Ar) are present in air.

Activities with Discussion and Conclusion

Activity 5.1

- Looking at its resemblance to alkali metals and the halogen family, try to assign hydrogen a correct position in Mendeleev's periodic table.

Now answer:

- To which group and period should hydrogen be assigned?

Discussion: Hydrogen has the lowest atomic number ($Z = 1$) and lowest atomic mass (1.008 u). Obviously, it is placed in group 1 and period 1. As the properties of hydrogen resemble that of the alkali metals and the halogens, it can be placed either with the alkali metals of group IA or the halogens of group VIIA of Mendeleev's periodic table.

Conclusion: No correct position can be assigned to hydrogen in the periodic table since the position of hydrogen in the periodic table is anomalous or controversial. Usually it is placed in group 1 and period because it has the lowest atomic number (1) and has the lowest atomic mass (1.008 u).

Activity 5.2

- Consider the isotopes of chlorine, Cl-35 and Cl-37.

Now answer:

- Would you place them in different slots because their atomic masses are different?
- Or would you place them in the same position because their chemical properties are the same?

Discussion: In Mendeleev's periodic table, elements are arranged in order of their increasing atomic masses. Therefore, in principle, the two isotopes of chlorine, Cl-35 and Cl-37, should be placed at different positions because their atomic masses (35 u and 37 u) are different but there is a no vacant position available for Cl-37 between Cl-35 of group VII A and K (39.1 u) of group IA.

However, in the long form of the periodic table, elements are arranged in the order of their increasing atomic number. Hence, the two isotopes of chlorine, Cl-35 and Cl-37, have the same atomic number ($Z = 17$) and hence have similar chemical properties. Thus, they can be placed at the same position in group 17 of the long form of the periodic table.

Conclusion: As isotopes have the same atomic number, they are placed at the same position in a particular group of the long form of the periodic table.

Activity 5.3

- How were the positions of cobalt and nickel resolved in the modern periodic table?
- How were the positions of isotopes of various elements decided in the modern periodic table?
- Is it possible to have an element with atomic number 1.5 placed between hydrogen and helium?
- Where do you think should hydrogen be placed in the modern periodic table?

Discussion: In the modern periodic table, elements are placed in the increasing order of their atomic number. As the atomic number of Co is 27 and that of Ni is 28, cobalt (group 9) is placed before nickel (group 10) thus their positions are justified in the modern periodic table.

- In the modern periodic table, elements are arranged in the increasing order of their atomic numbers. As the various isotopes of an element have the same atomic number, they are assigned the same position in the modern periodic table.
- An element with atomic number 1.5 is not possible as the atomic numbers of elements are always in whole numbers (natural number).
- Refer to answer to activity 5.1.

Activity 5.4

- Look at group 1 of the modern periodic table.

Now answer:

- Name the elements present in it.
- Write down the electronic configuration of the first three elements.
- What similarity do you find in their electronic configuration?
- How many valence electrons are present in these three elements?

Discussion: The names of the elements present in group 1 are hydrogen (H), lithium (Li), sodium (Na), potassium (K), rubidium (Rb), cesium (Cs) and francium (Fr).

- Electronic configurations of elements are:

Elements:	H	Li	Na	K
No. of shells:	K	KL	KLM	KLMN
Electronic configuration	1	21	281	2881

- All these elements have one electron in their respective valence shells.

Activity 5.5

- If you look at the long form of the periodic table, you will find that the elements Li, Be, B, C, N, O, F and Ne are present in the second period.

Now answer:

- Write down their electronic configurations.
- Do these elements also contain the same number of valence electrons?
- Do they contain the same number of shells?

Discussion: The electronic configuration of the elements of second period are:

Elements:	Li	Be	B	C	N	O	F	Ne
Electronic	KL	KL	KL	KL	KL	KL	KL	KL
configuration:	2, 1	2, 2	2, 3	2, 4	2, 5	2, 6	2, 7	2, 8

- These elements do not contain the same number of valence electrons. The number of valence electrons actually increases from 1 to 8.
- All these elements contain the same number of shells or orbits, that is, two—K and L.

Activity 5.6

- How do you calculate the valency of an element from its electronic configuration?
- What is the valency of magnesium with atomic number 12 and sulfur with atomic number 16?
- Similarly find out the valencies of the first twenty elements.
- How does the valency vary in a period on going from left to right?
- How does the valency vary in going down a group?

Discussion: For elements of group 1, 2, 13 and 14, the valency is equal to the number of valence electrons and for elements of group 15–18, the valency is equal to 8 – number of valence electrons.

- Electronic configuration of Mg with atomic number 12 is

K	L	M
2	8	2

  Hence the number of electrons in the valence shell = 2.

  As the element belongs to group 2, its valency is equal to the number of electrons in the valence shell = 2.

  The electronic configuration of sulfur with atomic number 16 is

K	L	M
2	8	6

  Hence, the number of electrons in the valence shell = 6.

  As this element belongs to group 16, its valency = 8 – 6 = 2.
- Please note that, in the case of helium, the valency is equal to two minus the number of valence electrons, 2 – 2 = 0.

  The valency of the first 20 elements has been calculated as shown below.

Element	Atomic number	Group number	Electronic configuration K L M N	Number of valence electrons	Valency of the element
H	1	1	1	1	1
He	2	18	2	2	2 – 2 = 0
Li	3	1	2 1	1	1
Be	4	2	2 2	2	2
B	5	13	2 3	3	3
C	6	14	2 4	4	4
N	7	15	2 5	5	8 – 5 = 3
O	8	16	2 6	6	8 – 5 = 2
F	9	17	2 7	7	8 – 7 = 1
Ne	10	18	2 8	8	8 – 8 = 0
Na	11	1	2 8 1	1	1
Mg	12	2	2 8 2	2	2
Al	13	13	2 8 3	3	3
Si	14	14	2 8 4	4	4
P	15	15	2 8 5	5	8 – 5 = 3
S	16	16	2 8 6	6	8 – 6 = 2
Cl	17	17	2 8 7	7	8 – 7 = 1
Ar	18	18	2 8 8	8	8 – 8 = 0
K	19	1	2 8 8 1	1	1
Ca	20	2	2 8 8 2	2	2

- In a period, the valency first increases from 1 to 4 and then decreases from 4 to zero.
- In a group, the number of valence electrons remains the same, so all the elements in a group have the same valency. For example, the valency of all the elements of group 1 is 1; that of group 2 is 2; that of group 13 is 3; and so on.

Activity 5.7

- Atomic radii of the elements of the second period are given below:

Period 2 elements:	B	Be	O	N	Li	C
Atomic radius (pm):	88	111	66	74	152	77

Now answer:

- Arrange them in decreasing order of their atomic radii.
- Are the elements now arranged in the pattern of a period in the periodic table?
- Which elements have the largest and the smallest atoms?
- How does the atomic radius change as you go from the left to the right in a period?

Discussion: The decreasing order of atomic radii of the elements of second period is

Li	Be	B	C	N	O
152	111	88	77	74	66

- Yes, the above elements are arranged in the pattern of a period in the periodic table.
- Li has the largest (152 pm) and O has the smallest (66 pm) atoms.
- The atomic radii decrease as we move from the left to the right in a period. This is due to an increase in the nuclear charge which tends to pull the electrons closer to the nucleus and so the atomic size decreases.

Activity 5.8

- Study the variation in the atomic radii of the first group elements given below and arrange them in the increasing order.

Group 1 element:	Na	Li	Rb	Cs	K
Atomic radius (pm):	186	152	244	262	231

Now answer:

- Name the elements which have the smallest and the largest atoms.
- How does the atomic size vary as you go down a group?

Discussion: The atomic radii (in pm) of the element of group 1 increase in the order:

Li (152), Na (186), K (231), Rb (244) and Cs (262).

- Li has the smallest (152 pm) atom while Cs has the largest (262 pm) atom.
- As we go from top to bottom in a group, the atomic size increases as the number of inner filled shells increase.

Activity 5.9

- Examine the elements of the third period and classify them as metals and non-metals.

Now answer:

- On which side of the periodic table do you find the metals?
- On which side of the periodic table do you find the non-metals?

Discussion: The elements of the third period from left to right are:

$$\underbrace{Na \quad Mg \quad Al}_{Metals} \quad \underbrace{Si}_{Metalloid} \quad \underbrace{P \quad S \quad Cl}_{Non-metals} \quad Ar$$

- Metals lie on the left side of the periodic table.
- Non-metals lie on the left side of the period table.

Conclusion: Metals lie on the left side and non-metals lie on the right side of the periodic table.

Activity 5.10

- How do you think the tendency to lose electrons will change in a group?
- How will this tendency change in a period?

Discussion:

- As we move from top to bottom in a group, the atomic size increases due to a corresponding increase in the number of inner filled shells. As a result, the force of attraction of the nucleus towards the electrons decreases and hence the tendency to lose electrons increases.
- As we move from left to right in a period, the nuclear charge increases regularly by one unit and hence the force of attraction of the nucleus towards the electrons increases. As a result, the tendency to lose electrons decreases.

Conclusion: The tendency to lose electrons increases down a group but decreases across a period.

Activity 5.11

- How would the tendency to gain electrons change as we go from left to right across a period?
- How would the tendency to gain electrons change as we go down a group?

Discussion:

- As we go from left to right across a period, the nuclear charge increases regularly by one unit, so the force of attraction for the incoming electron increases. Consequently, the tendency to gain electrons increases across a period.
- As we go down a group, the atomic size progressively increases. As a result, the force of attraction of the nucleus towards the incoming electron decreases. Consequently, the tendency to gain electrons decreases.

Conclusion: The tendency to gain electrons increases across a period and decreases down a group.

Exemplar Problems

1. The three elements A, B and C with similar properties have atomic masses X, Y and Z respectively. The mass of Y is approximately equal to the average mass of X and Z. What is such an arrangement of elements called? Give one example of such a set of elements.

2. Elements have been arranged in the following sequence on the basis of their increasing atomic masses. F, Na, Mg, Al, Si, P, S, Cl, Ar, K
   (a) Pick two sets of elements which have similar properties.
   (b) The given sequence represents which law of classification of elements?

3. Can the following groups of elements be classified as Dobereiner's triads?
   (a) Na, Si, Cl
   (b) Be, Mg, Ca
   Atomic mass of Be 9; Na 23; Mg 24; Si 28; Cl

35; Ca 40
Explain by giving reasons.

4. In Mendeleev's periodic table, the elements were arranged in the increasing order of their atomic masses. However, cobalt with an atomic mass of 58.93 amu was placed before nickel having an atomic mass of 58.71 amu. Give reasons for this.

5. 'Hydrogen occupies a unique position in the modern periodic table'. Justify the statement.

6. Write the formulae of the chlorides of eka-silicon and eka-aluminium, the elements predicted by Mendeleev.

7. Three elements A, B and C have 3, 4 and 2 electrons respectively in their outermost shells. Give the group number to which they belong in the modern periodic table. Also, give their valencies.

8. If an element X is placed in group 14, what will be the formula and the nature of bonding of its chloride?

9. Compare the radii of two species X and Y. Give reasons for your answer.
   (a) X has 12 protons and 12 electrons.
   (b) Y has 12 protons and 10 electrons.

10. Arrange the following elements in the increasing order of their atomic radii.
   (a) Li, Be, F, N
   (b) Cl, At, Br I

11. Identify and name the metals out of the following elements whose electronic configurations are given below.
   (a) 2, 8, 2 (b) 2, 8, 1
   (c) 2, 8, 7 (d) 2, 1

12. Write the formula of the product formed when the element A (atomic number 19) combines with the element B (atomic number 17). Draw its electronic dot structure. What is the nature of the bond formed?

13. Arrange the following elements in the increasing order of their metallic character Mg, Ca, K, Ge, Ga.

14. Identify the elements with the following property and arrange them in increasing order of their reactivity.
   (a) An element which is a soft and reactive metal.
   (b) The metal which is an important constituent of limestone.
   (c) The metal which exists in the liquid state at room temperature.

15. Properties of the elements are given below. Where would you locate the following elements in the periodic table?
   (a) A soft metal stored under kerosene.
   (b) An element with variable (more than one) valency stored under water.
   (c) An element which is tetravalent and forms the basis of organic chemistry.
   (d) An element which is an inert gas with atomic number 2.
   (e) An element whose thin oxide layer is used to make other elements corrosion resistant by the process of 'anodising'.

Long Answer Questions

1. An element is placed in the 2nd group and 3rd period of the periodic table, burns in presence of oxygen to form a basic oxide.
   (a) Identify the element.
   (b) Write its electronic configuration.
   (c) Write the balanced equation when it burns in the presence of air.
   (d) Write a balanced equation when this oxide is dissolved in water.
   (e) Draw the electron dot structure for the formation of this oxide.

2. An element X (atomic number 17) reacts with an element Y (atomic number 20) to form a divalent halide.
   (a) Where in the periodic table are elements X and Y placed?
   (b) Classify X and Y as metal(s), non-metal(s) or metalloid(s).
   (c) What will be the nature of the oxide of element Y? Identify the nature of bonding in the compound formed.
   (d) Draw the electron dot structure of the divalent halide.

3. The atomic numbers of a few elements are 10, 20, 7, 14.
   (a) Identify the elements.

(b) Identify the group number of these elements in the periodic table.

(c) Identify the periods of these elements in the periodic table.

(d) What would be the electronic configuration for each of these elements?

(e) Determine the valency of these elements.

4. Complete the following crossword puzzle.

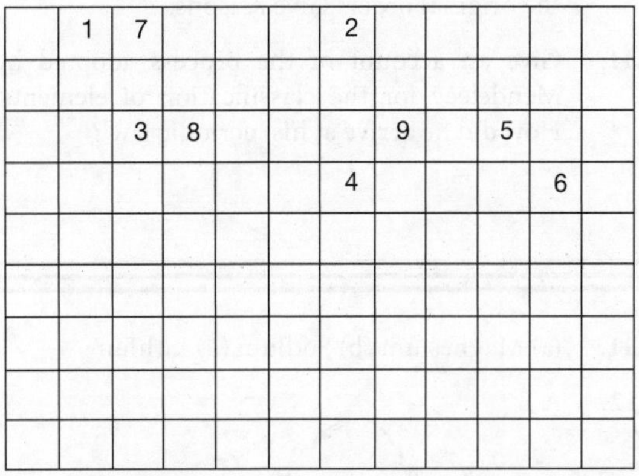

Across:

(1) An element with atomic number 12.

(3) Metal used in making cans and member of group 14.

(4) A lustrous non-metal which has 7 electrons in its outermost shell.

Down:

(2) Highly reactive and soft metal which imparts a yellow colour to the flame and is stored in kerosene.

(5) The first element of the second period.

(6) An element which is used in making fluorescent bulbs and is the second member of group 18 in the modern periodic table.

(7) A radioactive element which is the last member of the halogen family.

(8) A metal which is an important constituent of steel and forms rust when exposed to moist air.

(9) The first metalloid in the modern periodic table whose fibres are used in making bullet-proof vests.

5. (a) In this ladder, symbols of elements are jumbled up. Rearrange these symbols in the increasing order of their atomic number in the periodic table.

(b) Arrange them in the order of their group also.

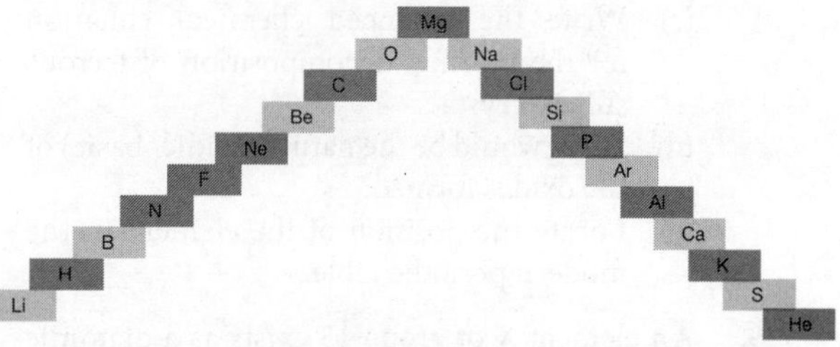

6. Mendeleev predicted the existence of certain elements not known at that time and named two of them as eka-silicon and eka-aluminium.

(a) Name the elements which have taken the place of these elements.

(b) Mention the groups and the periods of these elements in the modern periodic table.

(c) Classify these elements as metals, non-metals or metalloids.

(d) How many valence electrons are present in each one of them?

7.
- The electropositive nature of the element(s) increases down the group and decreases across the period.
- The electronegativity of the elements decreases down the group and increases across the period.
- Atomic size increases down the group and decreases across a period (left to right).
- Metallic character increases down the group and decreases across a period.

On the basis of the above trends of the periodic table, answer the following about the elements with atomic numbers 3 to 9.

(a) Name the most electropositive element among them.

(b) Name the most electronegative element.

(c) Name the element with the smallest atomic size.

(d) Name the element which is a metalloid.

(e) Name the element which shows maximum valency.

8. An element X which is a yellow solid at room temperature shows catenation and allotropy. X forms two oxides which are also formed during the thermal decomposition of ferrous sulfate crystals and are the major air pollutants.

(a) Identify the element X.

(b) Write the electronic configuration of X.

(c) Write the balanced chemical equation for the thermal decomposition of ferrous sulfate crystals.
(d) What would be the nature (acidic/ basic) of the oxides formed?
(e) Locate the position of the element in the modern periodic table.

9. An element X of group 15 exists as a diatomic molecule and combines with hydrogen at 773 K in the presence of a catalyst to form a compound, ammonia, which has a characteristic pungent smell.
(a) Identify the element X. How many valence electrons does it have?
(b) Draw the electron dot structure of the diatomic molecule of X. What type of bond is formed in it?
(c) Draw the electron dot structure for ammonia. What type of bond is formed in it?

10. Which group of elements could be placed in Mendeleev's periodic table without disturbing the original order? Give reasons.

11. Give an account of the process adopted by Mendeleev for the classification of elements. How did he arrive at his 'periodic law'?

Answers

1. The arrangement of these elements represents a Dobereiner's triad. For example, lithium, sodium and potassium.

2. (a) (i) F and Cl (ii) Na and K
(b) Newlands' law of octaves

3. (a) No, because all these elements do not have similar properties although the atomic mass of silicon is the average of the atomic masses of sodium (Na) and chlorine (Cl).
(b) Yes, as they have similar properties and the mass of magnesium (Mg) is roughly the average of the atomic masses of Be and Ca.

4. Elements with similar properties can be grouped together.

5. Hydrogen resembles alkali metals as well as halogens.

6. $GeCl_4$, $GaCl_3$

7.

Element	Group no.	Valency
A	Group 13	3
B	Group 14	4
C	Group 2	2

8. XCl_4; Covalent bonding

9. Radius of Y is less than that of X because Y is the cation of X.

10. (a) F < N < Be < Li
(b) Cl < Br < I < At

11. (a) Magnesium (b) Sodium (d) Lithium

12. A B

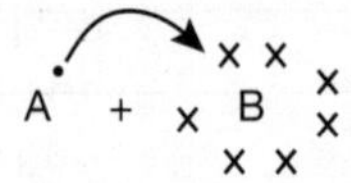

Ionic bond.
A = K (potassium) B = Cl (chlorine)

13. Ge < Ga < Mg < Ca < K

14. (a) Na or K (b) Ca (c) Hg
Hg < Ca < Na < K

15. (a) Sodium (Na), group 1 and period 3; or, potassium (K), group 1 and period 4.
(b) Phosphorus (P), group 15 and period 3.
(c) Carbon (C), group 14 and period 2.
(d) Helium (He), group 18 and period 1.
(e) Aluminium (Al), group 13 and period 3.

Long Answer Questions

1. (a) Magnesium (Mg)
(b) K, L, M
2, 8, 2
(c) $2Mg(s) + O_2(g) \rightarrow 2MgO(s)$
(d) $MgO(s) + H_2O(l) \rightarrow Mg(OH)_2(aq)$
(e)

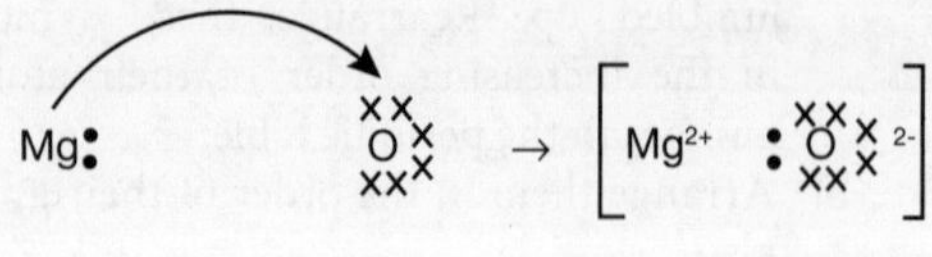

2. (a) X belongs to group 17 and the 3rd period. Y belongs to group 2 and the 4th period.
(b) X is a non-metal and Y is a metal.
(c) Basic oxide; ionic bonding.
(d)

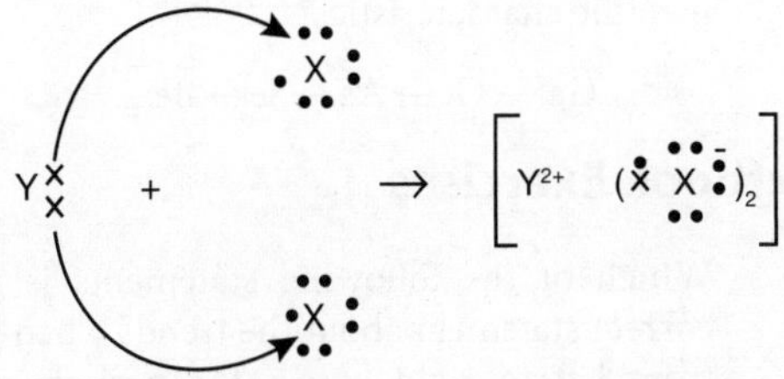

3. (a) Elements: Neon (Ne), calcium (Ca), nitrogen (N), silicon (Si)
(b) Groups: 18, 2, 15, 14
(c) Periods: 2, 4, 2, 3
(d) Electronic configuration: (2, 8); (2, 8, 8, 2); (2, 5); (2, 8, 4)
(e) Valency: 0, 2, 3, 4

4.

	[1]M	[7]A	G	N	E	[2]S	I	U	M		
		S				O					
		[3]T	[8]I	N		D	[9]B		[5]L		
		A	R			[4]I	O	D	I	[6]N	E
		T	O			U	R		T	E	
		I	N			M	O		H	O	
		N					N		I	N	
		E							U		
									M		

5. (a) H, He, Li, Be, B, C, N, O, F, Ne, Na, Mg, Al, Si, P, S, Cl, Ar, K, Ca
(b) Group 1: H, Li, Na, K
Group 2: Be, Mg, Ca
Group 13: B, Al
Group 14: C, Si
Group 15: N, P
Group 16: O, S
Group 17: F, Cl
Group 18: He, Ne, Ar

6. (a) Germanium (Ge) and gallium (Ga).
(b) Group 14, period 4 and group 13, period 4.
(c) Ge: metalloid; Ga: metal.
(d) Ga: 3 Ge: 4.

7. (a) Lithium
(b) Fluorine
(c) Fluorine
(d) Boron
(e) Carbon

8. (a) Element X is sulfur (atomic no. 16)
(b) K, L, M
2, 8, 6
(c) $2FeSO_4(s) \xrightarrow{\text{Heat}} Fe_2O_3(s) + SO_2(g) + SO_3(g)$
(d) Acidic
(e) 3rd period, group 16

9. (a) Nitrogen (atomic no. 7)
2,5; it has 5 valence electrons

(b)

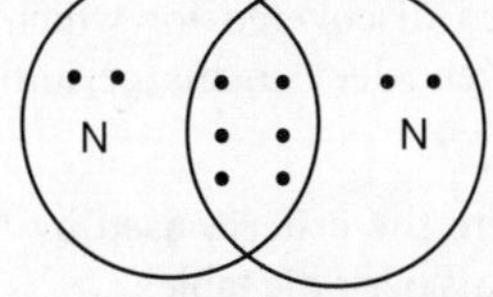

Triple covalent bonds

(c)

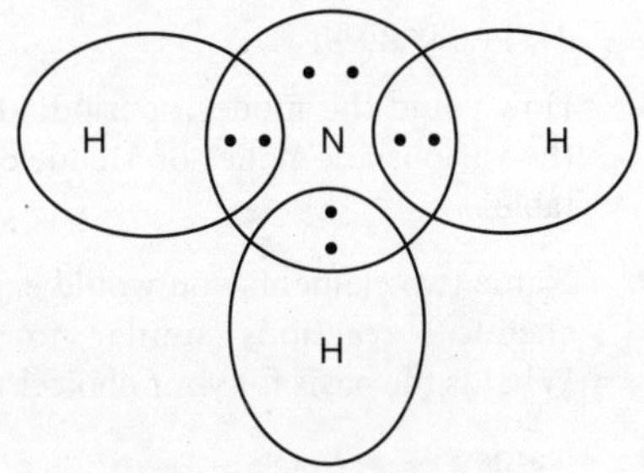

3 single covalent bonds

10. Noble gases: According to Mendeleev's classification, the properties of elements are the periodic function of their atomic masses and there is a periodic recurrence of elements with similar physical and chemical properties. Noble gases being inert, could be placed in a separate group without disturbing the original order.

11. (63 elements were known.) Compounds of these elements with oxygen and hydrogen were studied (formation of oxides and hydrides). Elements with similar properties were arranged in a group. Mendeleev observed that elements were automatically arranged in the order of increasing atomic masses.

NCERT CORNER

In-Text Questions

1. Did Dobereiner's triads also exist in the columns of Newlands' octaves? Compare and find out.

2. What were the limitations of Dobereiner's classification?

3. What were the limitations of Newlands' law of octaves?

4. Use Mendeleev's periodic table to predict the formulae for the oxides of the following elements: K, C, Al, Si, and Ba.

5. Besides gallium, which other elements have since been discovered for which blanks were left by Mendeleev in his periodic table? (any two)

6. What were the criteria used by Mendeleev in creating his periodic table?

7. Why do you think the noble gases are placed in a separate group?

8. How could the modern periodic table remove the various anomalies of Mendeleev's periodic table?

9. Name two elements you would expect to show chemical reactions similar to magnesium. What is the basis for your choice?

10. Name:
    (i) Three elements that have a single electron in their outermost shells.
    (ii) Two elements that have two electrons in their outermost shells.
    (iii) Three elements with filled outermost shells.

11. (i) Lithium, sodium and potassium are all metals that react with water to liberate hydrogen gas. Is there any similarity in the atoms of these elements?
    (ii) Helium is an unreactive gas and neon is a gas of extremely low reactivity. What, if anything, do their atoms have in common?

12. In the modern periodic table, which are the metals among the first ten elements?

13. By considering their position in the periodic table, which one of the following elements would you expect will possess maximum metallic characteristics?

$$Ga \rightarrow Ge \rightarrow As \rightarrow Se \rightarrow Be$$

Textbook Exercises

1. Which of the following statements is not a correct statement about the trends when going from left to right across the periods of the periodic table?
   (a) The elements become less metallic in nature.
   (b) The number of valence electrons increases.
   (c) The atoms lose their electrons more easily.
   (d) The oxides become more acidic.

2. Element X forms a chloride with the formula XCl_2, which is a solid with a high melting point. X would most likely be in the same group of the periodic table as
   (a) Na (b) Mg
   (c) AI (d) Si

3. Which element has
   (i) two shells, both of which are completely filled with electrons?
   (ii) an electronic configuration 2,8,2?
   (iii) a total of three shells, with four electrons in its valence shell?
   (iv) a total of two shells, with three electrons in its valence shell?
   (v) twice as many electrons in its second shell as in its first shell?

4. (i) What property do all elements in the same column of the periodic table as boron, have in common?
   (ii) What property do all elements in the same column of the periodic table as fluorine, have in common?

5. An atom has an electronic configuration 2, 8, 7.
   (i) What is the atomic number of this element?
   (ii) To which of the following elements would it be chemically similar?
   (Atomic numbers are given in parentheses.)
   N (7), F (9), P (15), Ar (18).

6. The position of three elements A, B and C in the periodic table are shown below.

Group 16	Group 17
–	–
–	A
–	–
B	C

(i) State whether A is a metal or a non-metal.
(ii) State whether C is more reactive or less reactive than A.
(iii) Will C be larger or smaller in size than B?
(iv) Which type of ion, cation or anion, will be formed by element A?

7. Nitrogen (atomic number 7) and phosphorus (atomic number 15) belong to group 15 of the periodic table. Write the electronic configuration of these two elements. Which of these will be more electronegative? Why?

8. How does the electronic configuration of an atom relate to its position in the modern periodic table?

9. In the modern periodic table, calcium (atomic number 20) is surrounded by elements with atomic numbers 12, 19, 21 and 38. Which of these have physical and chemical properties resembling calcium?

10. Compare and contrast the arrangement of elements in Mendeleev's periodic table and the modern periodic table.

Answers

In-Text Questions

1. Yes, Dobereiner's triads (example, Li, Na and K) existed in Newlands' octaves. If we consider lithium as the first element, the 8th element from it will be sodium. Again, considering sodium as the 1st element, the 8th element from it is potassium. This means that according to the Newlands' law of octaves, lithium, sodium and potassium will have similar properties.
According to the law of triads, the properties of sodium should be between those of lithium and potassium. Thus, lithium, sodium and potassium form a Dobereiner's triad having similar chemical properties. This proves that Dobereiner's triads also exist in the columns of Newlands' octaves.

2. Dobereiner's classification had a big limitation. All the known elements couldn't be grouped into sets of 3 elements with similar properties. In fact, only 3 triads were formed, and the rest of the elements couldn't be accommodated.

3. The limitations of Newlands' law of octaves were:
   - The new elements which were discovered later could not be fitted into the law of octaves.
   - Newlands' law was applicable only up to calcium. After Ca, every eighth element does not possess properties similar to the first.
   - The wrong order of arrangement of elements was made for a few elements. For example, Co and Ni do not resemble the halogens but were found together in the same slot. Fe was similar to Co and Ni but was separated and kept in a different slot.

4.

Element	Group no.	Formula
K	1	K_2O
C	4	CO_2
Al	3	Al_2O_3
Si	4	SiO_2
Ba	2	BaO

5. Besides gallium, germanium and scandium have also been discovered to fit the gaps that were left by Mendeleev.

6. The criteria used by Mendeleev in creating his periodic table are:
   (i) Elements were arranged in the increasing order of their atomic masses.
   (ii) All elements in a group should have similar properties.

7. Noble gases are inert. They do not resemble other elements as they are highly unreactive and rarely get involved in any chemical

reaction. Since, all the noble gases show the same properties, they are grouped separately.

8. The modern periodic table is based on the atomic number of elements; therefore,
   (i) the problem of positions of isotopes was solved because isotopes have the same atomic number,
   (ii) the wrong order of Ar, K, Co, Ni was removed in the modern periodic table.

9. Calcium and barium are two elements that will show chemical reactions similar to magnesium. The reason being, that both of them belong to same group as magnesium. Ba and Ca have the same number of valence electrons (2) as Mg and hence will show same chemical properties as magnesium.

10. (i) Li (2,1), Na (2,8,1) and K (2,8,8,1)
    (ii) Be (2,2) and Mg (2,8,2)
    (iii) He (2), Ne (2,8) and Ar (2,8,8)

11. (i) All the given metals are highly reactive. So, these metals can react with water to liberate hydrogen gas. They have the same number of valence electrons (1) and can readily lose them to become positive ions.
    (ii) Helium and neon have completely filled outermost shells which makes them exceptionally unreactive and thus, stable. They belong to the noble gas group.

12. Among the first ten elements, lithium and beryllium are metals.

13. The metallic character decreases from the left to the right in a period and increases down the group. As Be belongs to the extreme left of the periodic table, Be is the most metallic among the given elements.

Textbook Exercises

1. (a) Metals lose their electrons easily. On moving from left to right across the period, the tendency to lose electrons decreases, hence, the non-metallic character increases.
   (b) On moving from left to right across the period, the number of valence electrons increases.
   (c) That the atoms can lose their electrons more easily is incorrect because the tendency to lose electrons decreases and hence, it gets more difficult to lose electrons when moving from left to right in a period.
   (d) Metal oxides are basic and non-metal oxides are acidic or neutral in nature. On moving from left to right in any period, the non-metallic character increases so the acidic nature of the oxides increases.

2. The number of valence electrons in Cl is 7; so, one Cl atom needs to accept 1 electron in the ionic bond with X. As there are two Cl atoms, it implies that X will supply two electrons. Thus, X must have 2 electrons in its outermost shell.
   Na has 1 electron in its valence shell.
   Mg has 2 electrons in its valence shell.
   Al has 3 electrons in its valence shell.
   Si has 4 electrons in its valence shell.
   X would most likely be in the same group of the periodic table as magnesium (Mg) and the formula will be $XCl_2 = MgCl_2$.

3. (i) Neon ($_{10}$Ne) has two shells, both of which are completely filled with electrons (2 electrons in the K shell and 8 electrons in the L shell).
   (ii) Magnesium ($_{12}$Mg) has the electronic configuration of 2,8,2.
   (iii) Silicon ($_{14}$Si) has a total of three shells, with four electrons in its valence shell (2 electrons in the K shell, 8 electrons in the L shell and 4 electrons in the M shell).
   (iv) Boron ($_5$B) has a total of two shells, with three electrons in its valence shell (2 electrons in the *K* shell and 3 electrons in the *L* shell).
   (v) Carbon ($_6$C) has twice as many electrons in its second shell as in its first shell (2 electrons in the *K* shell and 4 electrons in the *L* shell).

4. (i) All the elements in the same column as boron have the same number of valence electrons (3).
   B: 2,3
   Al: 2,8,3
   Ga: 2,8,18,3
   In: 2,8,18,18,3
   Tl: 2,8,18,32,18,3
   Hence, they all have a valency equal to 3.
   (ii) All the elements in the same column as fluorine have the same number of valence electrons (7).

F: 2,7
Cl: 2,8,7
Br: 2,8,18,7
I: 2,8,18,18,7
At: 2,8,18,32,18,7
It means, they all have a valency equal to 1.

5. (i) Atomic number is equal to the number of protons, which is equal to number of electrons in a neutral atom. The total number of electrons will be 17 (2 + 8 + 7); so, the atomic number of this element will be 17.

(ii) It would be chemically similar to F (9) with a configuration of 2,7 as both of these have 7 valence electrons.

6. (i) Group 16 and 17 elements are non-metals.

Group 16	Group 17
O	F
S	Cl (A)
Se	Br
Te (B)	I (C)

Hence, A is a non-metal.

(ii) C is less reactive than A, as reactivity decreases down the group for halogens.

(iii) C will be smaller in size than B as moving across a period, the nuclear charge increases so, electrons are pulled closer to the nucleus.

(iv) A will form an anion as it has 7 electrons in its valence shell and thus, needs 1 more electron to complete its octet.

$$A + e^- \rightarrow \underset{\text{Anion}}{A^-}$$

7.

Element	K L M
Nitrogen	2,5
Phosphorus	2,8,5

Nitrogen is more electronegative than phosphorus. On moving down the group, the number of shells increase; so, the valence electrons move away from the nucleus and the effective nuclear charge decreases. This causes a decrease in the tendency to pull an electron in its outermost shell and hence, the electronegativity decreases.

8. In the modern periodic table, atoms with similar electronic configurations are placed in the same column and atoms with the same number of shells are placed in the same period. If an element has an electronic configuration of, say, 2, 8,3, it will be placed in the 3rd group and in the 3rd period as it has 3 valence electrons and a total of 3 shells.

9. The element with atomic number 12 [Mg (2,8,2)] and the element with atomic number 20 [calcium (2,8,8,2)] have the same number of valence electrons (2). Hence, they will have similar chemical properties.

10.

	Mendeleev's periodic table		Modern periodic table
1.	Elements are arranged in the increasing order of their atomic masses.	1.	Elements are arranged in the increasing order of their atomic numbers.
2.	There are a total of 7 groups (columns) and 6 periods (rows).	2.	There are a total of 16 groups (columns) and 7 periods (rows).
3.	Elements having similar properties were placed directly under one another.	3.	Elements having the same number of shells are present in the same period while elements having the same number of valence electrons are present in the same group.
4.	The position of hydrogen could not be explained.	4.	Hydrogen is placed above the alkali metals.
5.	No distinguished positions for metals and non-metals.	5.	Metals are present at the left-hand side of the periodic table whereas non-metals are present at the right-hand side.